Vegetable Science and Technology

Vegetable Science and Technology

Pranab Hazra

Dean, Post Graduate Studies
Professor, Department of Vegetable Science
Faculty of Horticulture
Former Dean, Faculty of Horticulture, Bidhan Chandra Krishi Viswavidyalaya
Mohanpur-741252, West Bengal, India

A Paperback Division of
NEW INDIA PUBLISHING AGENCY
101, Vikas Surya Plaza, CU Block, LSC Market
Pitam Pura, New Delhi 110 034, India
Phone: + 91 (11)27 34 17 17 Fax: + 91(11) 27 34 16 16
Email: info@nipabooks.com
Web: www.nipabooks.com

Feedback at feedbacks@nipabooks.com

ISBN 978-93-87973-24-4

Composed, Designed & Printed in India

Dedicated to my father

Late Nabaghanashyam Hazra

Preface

"Hidden hunger" or micronutrient deficiency is a pernicious problem around the world particularly in the under-developed and developing countries that is caused due to lack of vitamins and minerals in the human diet which affects the health of about three billion people worldwide. Vegetables are considered essential for well-balanced diets since they supply vitamins, minerals, dietary fibre and phyto-chemicals. In the daily diet, vegetables have been strongly associated with improvement of gastrointestinal health, good vision, reduced risk of heart disease, stroke, chronic diseases such as diabetes and some forms of cancer. Vegetable farming not only stipulates higher per capita income but also attract investment, entrepreneurship and business based on vegetable crops. India envisaged spectacular increase in the vegetable production from 15 million tonnes during 1950 to 169 million tonnes at present. However, the most critical challenge is the attainment of enhanced productivity in the farmers' fields through input use-efficient and environment friendly manner.

This book has been designed to cater the needs of undergraduates and postgraduates of State and Central Agricultural Universities studying vegetable science and horticultural science. This book has been framed to provide the principles for environmental and growth factors, seedling and graft production, nutrient and water management, organic and protected farming, crop protection, post-harvest management and marketing of vegetable crops. Every production aspect of 42 major and minor vegetable crops grown in the tropical and subtropical regions along with information regarding origin and taxonomy, importance and uses, botany, nutritional and medicinal values, plant protection measures and post-harvest management have been provided. Long experience acquired through teaching and research on different aspects of vegetable science in the State Agricultural University was put to structure this book however, the contents have been enriched with the information from the number of books and research articles. I gratefully acknowledge the contribution of all the authors of those publications mentioned in the last part of the book.

I am confident that its users (students, vegetable specialists, horticulturists, agricultural research scientists and extension personnel) would find this book very informative as well as practical.

This work would not have been possible without the blessings of my mother (Sandhya) and sustained support of my wife (Antimoni) and son (Soham).

I thank Messers New India Publishing Agency for bringing out this publication.

Kalyani, 2018

Pranab Hazra

Contents

Chapter 1

Vegetable Production Scenario in India

Historical Account on Vegetable Cultivation in India

The vegetables have chequered history in India which dates back to pre-historic period before the arrival of Aryans in India (c. 1500-1000 BC). During the pre-historic times before the arrival of Aryans in India, the Proto-Australoids (Nisada) used brinjal, bottle gourd and watermelon as vegetables. Pea existed in India much before the Aryans came to the country and it was already present in Western Asia prior to its cultivation. The pea, most likely the field pea (*Pisum sativum* subsp. *sativum* var. *arvense* L.) was eaten during the Harappan civilization as evidenced from the archaeological excavations of Mohenjo-daro and Harappa (2500-1750 B.C.) in which carbonized pea seeds were discovered. Pea seeds were also found from Neolithic site (last stage of the Stone age) of Chirand in Bihar (situated 10 Km south-east of Chapra), Chacolithic site (period or Copper Age) of Maheshwar Nevdatoli (c. 1200 BC.) in Madhya Pradesh and part of Osmanabad district in Maharashtra. The presence of seeds of melon along with wheat was also recorded from the Harappan excavation. Plant remains of *Chenopodium album*, *Portulaca oleracea* and a few species of Solanaceae, Polygonaceae and Labiatae family during Mesolithic times (Middle Stone Age) were found at Mesolithic site of Damdama, Uttar Pradesh (c. 7000-6000 BC). Wood charcoals of plant remains of *Lablab purpureus* were found from the early phase of Indus-Saraswati (Sarasvati) or Indus Ghaggar-Hakra civilization (c. 3100-2800 BC). Kundru (*Coccinia indica*), cowpea and field pea were recorded from Hulas in Saharanpur district, Uttar Pradesh, late Indus valley civilization archeological site. Carbonized seeds of hyacinth bean (dolichos bean), peas, *Phaseolus, Vigna* and melon from Rojdi in Gujarat and *Chenopodium* and amaranth from Surkotada, Gujarat (c. 2300-1800 BC), both being the archaeological sites belonging to the Indus valley civilization also suggested the existence of vegetables in India in the pre-historic times. Among the vegetables eaten during the pre-historic period, like brinjal, bottle gourd, watermelon, peas and melon as mentioned earlier, only brinjal has its origin in India while the other were brought from other countries much before the arrival of the Aryans. It is difficult to understand how a few important

Indian native vegetables, like cucumber, *Luffa* (smooth and ridge gourd), pointed gourd and Dolichos bean *(Lablab purpureus)* were not known at that time. Similarly okra which has a secondary centre of origin in India has not been mentioned in any ancient literature. However, *Abelmoschus moschatus* called "Lata kasturika" in Sanskrit as a source of perfume has been recorded in Carak Samhita. The use of bottle gourd, which is not a native vegetable, is known from pre-historic times. Its dried fruit-shell is used as a hand jug (utensils) by saints and monks and in musical string instruments, like ektara, tamboora, tanpura, sitar and veena. Even mythological paintings have always delineated Naradmuni with tamboora in his hand and the goddess Saraswati with Veena which suggest the antiquity of bottle gourd in India. However, its date of introduction and the place from where it came remain unrecorded. It is also not mentioned in any iconography of the Medieval period (9th and 10th Century AD).

With the advent of the Aryan culture and Sanskrit literature in the ancient period much information became available on different kinds of vegetables consumed by the Aryans. The vegetables were mainly collected from the forests for eating or medicinal purpose. In the beginning, like in other countries, in India also most of the vegetable crop species were known as source of medicine rather than for human consumption. The two most important treatises on medicinal uses of plant species including vegetables are the Carak Samhita written around 600 BC. and Susruta Samhita in the 3rd to 4th century AD. Paleo-ethnobotanical and archaeo-botanical information available have indicated the utilization of the Indian plant wealth by the ancient humankind for about 8000 years and its movement to neighboring countries and inflow and exchange of plants from there. The Rig Veda (c. 3700-2000 B.C) mentions several vegetables like, green leafy vegetables, melons, pumpkins, gourds, lotus stem, cucumber, bottle-gourd, water chestnut, bitter gourd (karavella), radish, brinjal, some aquatic plants (avaka, andika), fruits such as mangoes, oranges and grapes along with cereals (rice) and pulses like, masha (black gram, urad), mudga (green gram, mung) and masura (lentil). Spices such as coriander, turmeric, pepper, cumin, asafoetida, cloves, sesame and mustard were well known at that time. Bottle gourd and cucumber was mentioned in Yajurveda and wax gourd and onion in Athravaveda, both written before 800 BC. As we move further down to the period of the Ramayana and Mahabharata (probably around 1400 BC, to around 400 BC), we find a far richer mentioning of food items. Lords Rama, Lakshmana and Devi Sita ate a vast menu that contained fruits, leafy vegetables, rice and meat. The Epic Ramayana (c. 1400 BC) had records of brinjal, dolichos bean, radish and bottle gourd while *Coccinia* had a reference in the Mahabharata. The vegetables referred to in other important ancient literature are onion in Apastamba Dharma Sutra–I (c. 860 BC-300 BC), wax gourd, pointed gourd, pumpkin, bottle gourd and brinjal in early Buddhist canonical works, Sutras and Jatakas (c. 500 BC) and pea, coriander, snap melon and *Luffa* in Kautilya's Arthshastra (c. 400 BC). Later, Kasyapa Samhita (c 2nd century BC-400 AD.) mentioned bitter gourd and garlic, Patanjali (c. 200 BC.) mentioned onion, early Jain canonical works before 5th century AD. mentioned bitter gourd, palak, brinjal and bathua and Dhanvantari, an Avatar of Vishnu from the Hindu tradition prior to 800 AD mentioned fenugreek (methi). The famous Kalidas plays (c. 500 AD) also described lotus and bottle gourd. During the Gupta period (c. 350-650 AD) brinjal and bottle gourd have been delineated in the Ajanta paintings. There are references of medicinal uses of several vegetables such as, brinjal, okra, ivy gourd lotus stalk, snake gourd, snap melon, coriander, cucumber, carrot, ivy gourd, pea, bitter gourd, spine gourd (*Momordica cochinchinensis)*, wax gourd, garlic, kulfa (Indian purslane), radish, palak, onion, pointed gourd and amaranthus. Indian spinach (*Basella)* and *Chenopodium* (bathua) was mentioned in Carak Samhita (c. 600 B.C.). Similarly Susruta Samhita (c. 3rd – 4th century AD.) mentions *Coccinia*, pea, watermelon, bitter gourd, melon, *Luffa*, fenugreek, onion and cucumber. The Chinese travelers Yuan Chwang (c. 606-648 AD.) and Itsing, a Tang Chinese Buddhist monk (c. 671-695 AD.) who came to India during the Gupta period have also mentioned the use of onion, garlic, lotus fibre and melons as vegetables. Much later, after 1300 AD, the Vrikshayurveda (the Science of Plant Life) written by Surapala has references of the crop husbandry of different vegetable crops like, pumpkin, brinjal, brihati (*Solanum indicum* L.), the probable progenitor species of cultivated brinjal, coriander, ridge gourd, bitter gourd, wax gourd, radish, pointed gourd, pea, garlic and cucumber.

The famous memoir, Ain-I-Akbari (Constitution of Akbar), written by his vizier, Abu'l-Fazl ibn Mubarak during the period of the Mughal emperor Akbar (1556-1605 AD.) has references of vegetables like, onion, garlic, chilli, carrot, turnip, radish, palak, amaranth, pointed gourd, snake gourd, ash gourd, bitter gourd and melon. The 325 year old text, "Nuskha Dar Fanni-Falahat" (The Art of Agriculture) written in Persian (c. 1692 AD.) by Mughal Prince Dara-Shikoh, son of Shahjahan, mentioned the growing of melon, cucumber, pumpkin, brinjal, cabbage, beet root, carrot, onion, garlic, celery, coriander, lettuce and parsley.

Portuguese, French and British traders and missionaries introduced some important vegetables, like potato, chilli, cauliflower, cabbage, tomato, French bean, sweet pepper, pumpkin, summer squash, sweet potato etc. from Europe mainly during the 16th century to 19th century AD. Chilli was brought to the south western coast of India by the Portuguese, Vasco-da-Gama, in late 15th century. Pumpkin, summer squash, sweet potato and potato were also introduced by the Portuguese. Potato was first mentioned in Terry's account of a banquet at Ajmer hosted by Asaph Khan, the Mughal king, to Sir Thomas Roe in 1615 AD. in which potato was served as a cooked vegetable. Cauliflower was introduced from South Africa and Britain during 1822, cabbage and tomato around 1828. During the British period, there was a beginning of regular introduction of germplasm of different vegetable crops into India from abroad. The East India Company and the Royal Agri-Horticultural Soceity, Calcutta during the British period used to bring in seeds of the European vegetables (cabbage, cauliflower, leek, asparagus, lettuce, European carrot, European radish, garden beet, tomato, etc.) from England and other parts of Europe for growing in India.

Present Status of Vegetable Cultivation in India

India's varied climate ensures successful cultivation of almost all kinds of vegetables in different parts of the country. Total production of horticultural crops in the country during 2015-16 is estimated to be around 286 million tonnes which is 2.0% higher than the previous year. India is the second largest producer of vegetables (169.0 million tonnes) next to China in the world. In India, vegetables are grown in a area of 9.575 million hectares with average productivity of 17.65 t/ha which contributes 14% of the total production of vegetables in the world. At present, the top five countries that produce most vegetables in the world are China, India, USA, Turkey and Iran with the production share of 46.1, 14.0%, 4.7, 3.1 and 2.08 per cent, respectively.

Total vegetable production in India before independence was 15 million tonnes and since independence for decades the growth rate was stabilized near 0.5%. The impetus on vegetable research and policy intervention to promote vegetable crops witnessed a sudden shot up of growth rate of 2.5%, a hike of five times in the last decade giving total annual production of over 169.0 million tonnes of vegetables during 2015-16 which is about 1.5% higher than the previous year. The increase in total production mainly came through increase in area under vegetable cultivation as well as enhancement in productivity. During the last decade between 2004-05 to 2014-15, area under the crop increased by 64% and productivity by 17%. This production scenario amply suggests that small and marginal farmers of India, in particular, opted for vegetable farming in a big way for their livelihood security. Rapid urbanization and enhancement in per capita income have also given a fillip for increased demand of vegetables.

Table 1. All India area, production and productivity of vegetables

Year	Area (in '000 ha)	Production (in '000 t)	Productivity (t/ha)
1991-92	5593	58532	10.5
2001-02	6156	88622	14.4
2002-03	6092	84815	13.9
2003-04	6082	88334	14.5
2004-05	6744	101246	15.0
2005-06	7213	111399	15.4
2006-07	7581	114993	15.2
2007-08	7848	128449	16.4
2008-09	7981	129077	16.2
2009-10	7985	133738	16.7
2010-11	8495	146555	17.3
2011-12	8989	156325	17.4
2012-13	9205	162187	17.6
2013-14	9396	162897	17.3
2014-15	9471	166566	17.6
2015-16	9575	169000	17.65

Source: Horticulture Statistics at a glance, 2015, Dept. of Agriculture and Co-operation, Govt. of India

Among the different states in India, West Bengal ranks top contributing 16% of the total production followed by Uttar Pradesh (14%), Bihar (8.75%), Madhya Pradesh (8.60%), Gujarat (7.0%), Odisha (6.0%) and Karnataka (5.0%). In India, 40 different kinds of vegetables are cultivated in different parts of the country however, the major grown vegetables are potato (27%), onion (11%), tomato (10%), brinjal (7%), cabbage (5%), cauliflower (5%), peas (3%) and other vegetables including wide array of cucurbitaceous vegetables (32%). Top potato growing states are Uttar Pradesh (29%) and West Bengal contributing 29 and 26 percent production, respectively. In tomato production, Andhra Pradesh is the leader (18% production) followed by Madhya Pradesh (11% production) and Karnataka (11% production. In brinjal, cabbage and cauliflower, West Bengal is the leader accounting about 25% of the total production. In onion, Maharashtra is the leading producer (29% production) followed by Karnataka (16% production) and Madhya Pradesh (16% production).

Table 2. Ranking of first five countries in production of vegetables

Vegetable	First	Second	Third	Fourth	Fifth
Potato	China	India	Russia	Ukraine	United States
Tomato	China	India	United States	Turkey	Egypt
Eggplant	China	India	Iran	Egypt	Turkey
Onion	China	India	United States	Egypt	Iran
Cabbage and other Brassicas	China	India	Russia	Japan	South Korea
Cauliflower and Broccoli	China	India	Italy	Mexico	France
Cucumber	China	Turkey	Iran	Russia	United States
Pumpkin, squash and gourd	China	India	Russia	Iran	United States
Green bean	China	Indonesia	India	Turkey	Thailand
Dry Bean	Myanmar	India	Brazil	China	Mexico
Carrot	China	Russia	United States	Uzbekistan	Ukraine
Turnip	China	Uzbekistan	Russia	United States	Ukraine
Lettuce	China	United States	India	Spain	Italy
Spinach	China	United States	Japan	Turkey	Indonesia
Okra	India	Nigeria	Iraq	Ivory Coast	Pakistan
Soybean	United States	Brazil	Argentina	China	India
Ginger	India	China	Nepal	Nigeria	Thailand
Cassava	Nigeria	Thailand	Vietnam	Indonesia	Costa Rica
Yam	Nigeria	Ghana	Ivory Coast	Benin	Togo
Sweet potato	China	Uganda	Nigeria	Indonesia	Tanzania

Source: FAOSTAT, Food and Agriculture Organization of the United Nations, 2013

India ranks first in okra production and second largest producer of brinjal, potato, onion, cauliflower, cabbage and tomato. Low productivity is the typical characteristic of vegetable production scenario in India. Low productivity is caused mainly due to severe degradation of ecosystem services *viz.,* agricultural production, supply of water, control of climate and nutrient cycles coupled with widespread subsistence production with low levels of technology, unsustainable use of natural production factors such as soil, biological diversity and water; insufficient preparatedness to cope with unpredictability and adaptation to climate change and poor post-harvest and marketing facilities.

Table 3. Average and highest productivity of major vegetables in the world

Vegetables	India (t/ha)	World (t/ha)	Potential productivity (t/ha)	Maximum productivity (t/ha)
Tomato	19.50	32.80	60-80	70.45 (USA)
Brinjal	16.08	17.48	40-50	34.7 (Japan)
Chilli	9.18	14.4	30-40	44.5 (Spain)
Okra	11.60	6.90	15-20	17.78 (Jordan)
Pea	9.50	8.35	18-20	20 (Lithuania)
Melons	20.48	20.95	30-40	45.83 (Cyprus)
Gourds	9.72	12.97	25-30	41.33 (Israel)
Cucumber	6.67	16.98	40-50	67.67 (Korea)
Watermelon	12.75	27.13	30-40	40.96 (Spain)
Cabbage	21.50	27.70	30-40	42.59 (Japan)
Cauliflower	18.30	16.90	35-40	45.25 (New Zealand)
Onion	14.20	19.10	40-50	60.33 (Korea)
Garlic	4.17	12.37	15-20	23.23 (Egypt)

(*Source:* NHB, Indian Horticulture Database-2012)

In present situation, it will be very difficult to increase the area under vegetable crops further due to rapid shrinkage in cultivable land due to use of land for rapid urbanization, development of raods and other infrastructure for the industry. Hence, the focus should be on the enhancement of productivity through the use of improved varieties and hybrids and judicious management of natural resources (soil, water), plant nutrition, irrigation water and plant protection measures.

There is a vast gap between actual and potential productivity which indicates ample scope for development of technologies to harness the potential to meet domestic and export demand. Although commendable progress was made on research front and a number of technologies have been developed, several target oriented research strategies are needed to get the answer of some of the unsolved questions.

Table 4. Priorities for developing multiple resistant variety/ hybrid with premium attributes

Crop	Targeted biotic stress	Priortised trait (s) to be incorporated
Tomato	TLCV + early blight + bacterial wilt + RKN	Tolerant to high temperature, high TSS and lycopene
Brinjal	Phomopsis + bacterial wilt + fruit- and shoot-borer	Early maturily
Chilli	Leaf curl virus + thrips + mites + anthracnose	High oleoresin and less capsaicin
Sweet pupper	*Phytophthora* + thrips + mites	Adapted to tropical regions of India
Okra	YVMV + ELCV + fruit-borer	Dark green pods with five ridges
Onion	Stemphyllium + purple blotch + thrips	Brown bulb, white bulb with high TSS
Cucumber	DM + mosaic	Pickling type
Muskmelon	PM + DM + anthracnose + fusarium	High TSS
Watermelon	PM + DM + anthracnose	High TSS
Cabbage	Black rot + diamond-back moth	Tolerant to high temperature
Cauliflower	Black rot + diamond-back moth	Tolerant to high temperature

Note: TLCV, Tomato leaf curl virus; RKN, root-knot nematode; YVMV, yellow-vein mosaic virus; ELCV= Enation leaf curl virus; DM, downy mildew; PM, powdery mildew

Poor adoption of technologies has always been a major handicap in increasing productivity in India. Therefore, promotion of developed technologies through various schemes is the key to harness available technologies ensuring increased productivity and profitability to the farmers. Likewise, pro-active government policies for development of infrastructure and law enforcement to support and execution of different promotional activities are equally pertinent to achieve higher productivity.

Export and Industrial Importance

The vast production base offers India tremendous opportunities for export. India exports both fresh and processed vegetables. Amongst the fresh vegetable items identified as having good export potential are onion, potato, okra, bitter gourd, chilli, other seasonal and non-traditional vegetables, organically grown vegetables and hybrid seeds. During 2015-16, total vegetable exported from India worth INR 4866.91 crore and the major importing countries were United

Arab Emirates, Bangladesh, Malaysia, Netherland, Sri Lanka, Nepal, United Kingdom, Saudi Arabia, Pakistan and Qatar accounting 55% of total export from India.

Perishable nature of vegetables demands comprehensive planning for movement, storage, processing and distribution of products. The growth of vegetable industry as a commercial proposition largely depends on allied enterprises like storage, processing, marketing and maintenance and service enterprises to encourage vegetable growing. Basic reasons for growth in demand of processed food in India are: growth in urban population and increase in employment which egg on to live a fast-paced life; increased number of women joining the work force which lead them to compromise on household chores such as cooking; growth in economy which results in the growth of income and increase in organized retail chain.

India accounts for only about 1.5% of the total international food trade, despite its rich heritage in agriculture and agricultural products. It has been estimated that the total annual consumption of processed fruits and vegetable products in the country comes only 50,000 tonnes of which defence sector and star hotels account for 15,000 tonnes and the remaining 35,000 tonnes are available to the public accounting per capita consumption of only 40 g/year. Indian food processing industry is primarily export oriented. India's geographical situation gives it the unique advantage of connectivity to Europe, the Middle East, Japan, Singapore, Thiland, Malaysia and Korea. However, India has extremely low levels of food processing as compared to other countries. In India about 4000 processing plants are at present functioning but capacity most of these plants are low. For this reason, only about 2.2% of the total fruits and vegetables are processed in India contrast to 70 to 80% in the different developed and developing countries which signals tremendous potential for exporters and investors in this trade.

Central sector organizations for research in vegetable crops

- Indian Agricultural Research Institute (IARI), New Delhi
- Indian Institute of Horticultural Research (IIHR), Bangalore
- Indian Institute of Vegetable Research (IIVR), Varanasi
- Central Institute of Temperate Horticulture (CITH), Srinagar
- Central Potato Research Institute (CPRI), Kufri, Shimla
- Central Tuber Crops Research Institute (CTCRI), Thiruvananthapuram, Kerala

- Directorate of Onion and Garlic Research, Pune, Maharastra
- National Horticultural Research and Development Foundation, Nasik Maharashtra
- ICAR–Research Complex for Eastern Region, Patna, Bihar
- ICAR–Research Complex for North-Eastern Hill Region, Barapani Meghalaya
- Vivekananda Parvatiya Krishi Anusandhan Sansthan, Almora, Uttarakhand
- Central Agricultural Research Institute, Port Blair
- ICAR–Central Coastal Agricultural Research Institute, Goa.

Chapter 2

Nutritional and Medicinal Values of Vegetables

Food, at the fundamental level, is viewed as a source of nutrition to meet daily requirements at a minimum in order to survive but with an ever greater focus on the desire to thrive. Nutritional security denotes the consumption and physiological use of adequate quantities of safe and nutritious food by every member of the family and encompasses the process of equitable distribution among members of household and communities. This would entail the need to ensure a varied food intake, comprising all the essential macro- and micronutrients (protein, carbohydrates, minerals and vitamins), plant pigments (lycopene, β-carotene, anthocianin, lutein, capsanthin, zeaxanthin, etc.) and other functional phytochemicals (secondary metabolites, like flavonoids, isothiocyanates, glucosinolates, etc.) through a diversified diet. The growing awareness in recent years of the health promoting food has directed increased attention to vegetables as vital components of daily diets.

Main Nutrient Components of Vegetables

Nutrient components of our food can be grouped into five main divisions, namely, (i) Carbohydrate, (ii) Protein, (iii) Fat, (iv) Vitamins and (v) Minerals. Balanced diet is the food containing all these nutrient components in balanced form to run the physiological process in good tune. So, only eating to one's heart's content may not be the criterion of a balanced diet. Our dietary constitution which is mostly cereal based is really alarming. Cereals, which supply chiefly carbohydrate, constitute only a part of the diet. Moreover, the present average carbohydrate supply through cereals is also not adequate. Chief deficiencies in our diet are calories, protein, vitamin A and riboflavin. According to the studies of Indian Council of Medical Research, New Delhi and National Institute of Nutrition, Hyderabad, meager intake of vegetables, the low cost protective foods, is largely responsible for malnutrition among the majority of our population. The dieticians advocate intake of 125g leafy vegetables, 75g other vegetables and 100g root and tuber vegetables (total 300g) everyday to make our diet balanced. In households below the poverty line, per capita vegetable consumption is very low, apprehended to be even lower than 40g per day.

Vegetables provide all the nutrient components, like carbohydrate, protein, fat, vitamins, minerals and water along with roughages which are the essential constituents of a balance diet. Calorific value of the vegetables is not much in comparison to the cereals and animal products. However, calorie requirement can well be supplemented by carbohydrate-rich vegetables like, potato, cassava, taro, yam, sweet potato, etc. Vegetables abounding in vitamins and minerals are, rightly called as protective foods. Raw vegetables are particularly useful because many vitamins, such as, vitamin C and vitamin B complex are not stable at cooking temperatures. Mineral complexes of the products of animal origin create an excess of acid inside the body which disrupts protective mechanisms and metabolic processes. Vegetables neutralize these substances and provide alkaline reactions for normal metabolism. Cellulose, pectin and other constituents present in the fibers of vegetables help to clear the bowels, reduce constipation and also promote digestion. Consumption of sufficient quantities of vegetables very much reduces the possibility of cancer in the intestine and colon. Cancer specialists advocate consumption of sufficient vegetables, particularly, cabbage, cauliflower, carrot, hyacinth bean, sweet potato, green papaya and bottle gourd. The vegetables also contain significant levels of biologically active components that impart health benefits beyond basic nutrition through their antioxidant properties. Pointed gourd, carrot and bean contain active chemical substances having cancer protection properties. Green leafy vegetables constitute a major part of any balanced diet and are good sources of minerals and vitamins. The ethno-botanical reports offer information on the medicinal properties of green leafy vegetables which include details on their antidiabetic, antihistaminic, anticarcinogenic and antibacterial activities. Diets rich in fruits and vegetables contain high amounts of antioxidant compounds like vitamin C, vitamin E, carotenoids, flavonoids, tannins and other phenolic constituents which boost the immune mechanism.

Carbohydrates

Carbohydrates serve as the chief source of energy in the food. It is also called protein sparing food. The carbohydrates may be classified into groups according to their complexity. The monosaccharides (e.g. glucose, fructose, mannose, etc., $C_6H_{12}O_6$) may be defined as single straight chain molecules that cannot be hydrolyzed further into simpler substances. The union of two or more monosaccharide units results in the formation of oligosaccharides (e.g. sucrose, lactose, maltose, $C_{12}H_{22}O_{11}$ etc.) and polysaccharides (e.g. starch, cellulose, insulin etc. $(C_6H_{10}O_5)_x$.

Carbohydrates in the form of starch and cellulose on complete hydrolysis with acids, yield only glucose and hence are called homopolysaccharides. Simple sugars like glucose and complex carbohydrates like starch are completely digested and are the important source of energy. Carbohydrates are absorbed through the intestinal wall into the blood mainly in the form of glucose. Glucose is the favourite body fuel and upon oxidation (burning) during respiration, energy is liberated along with the formation of CO_2 and H_2O. Upon burning 1g glucose 4.0 calories of energy is liberated. Non-digestible carbohydrates like, cellulose, hemicelluloses, gums, pectin and lignin provides roughage and dietary fibre.

Daily requirement of carbohydrates is 400-500g. Simple carbohydrates or simple sugars i.e., glucose, fructose and maltose are naturally found in many vegetables like, beet, pea, sweet corn, ripe pumpkin, winter squash, corn starch, etc. Polysaccharides consist of up to hundreds or even thousands molecules of simple sugars. Some tuber and corm producing vegetables (e.g., potato, sweet potato, cassava, yam, taro, Jerusalem artichoke, Chinese potato, etc.), dry beans, green banana, sweet corn, etc. are good sources of starch. Potato is the important starchy food crop in both subtropical and temperate regions. Similarly, sweet potato is the important carbohydrate supplement crop throughout the tropical and subtropical countries. Infact, sweet potato is the cheapest source of calorie.

Protein

Protein is extremely complex nitrogen-containing organic compounds. They constitute a major part of the protoplasm. It is the important constituent of cells, tissues and vital body fluids like blood. Function of proteins in diet is not primarily to supply energy on oxidation but to supply certain essential components of the living tissue of the organism itself. The capacity of the living organism for storing proteins is relatively small as compared to that of carbohydrates and fats. Hydrolysis of large, colloidal, non-diffusible protein molecules leads to formation of simple, crystalloidal and diffusible final products – the amino acids. Amino acids, the building unit of protein molecules, have an amino (NH_2) group, attached to the same carbon atom that holds the carboxyl (COOH) group.

$$\begin{array}{c} NH_2 \\ | \\ R - C - COOH \\ | \\ H \end{array}$$

Human can adequately produce most amino acids, except for nine types that are required from foods and these (leucine, isoleucine, valine, lysine, methionine, threonine, tryptophan, phenylalanine and histidine) are called essential amino acids. Proteins in the diet completely broken down in the gastrointestinal tract to individual amino acids are absorbed into the blood and are passed to various parts of the body. Protein in the form of enzymes and hormones regulate a wide range of vital metabolic processes. Proteins as antibodies help the body to defend against infection. Amino acids are utilized by the body for various functions like tissue building and replacement of depleted protein. Proteins, though not considered to be primarily a body fuel, are also utilized for the production of energy. Metabolism of 1g protein liberates 4.0 calories of energy.

Daily requirement of protein is 60-70g. Vegetables contain less protein compared to the product of animal origin. However, protein quality (composition of amino acids) is quite good although sulfo-amino acids (methionine, cystine) are most of the time limited in vegetable proteins. Good sources of protein are pea, cowpea, broad bean, lima bean, leaves of fenugreek, mustard, pumpkin, pointed gourd, drumstick, agathi, celery, garlic, Brussels sprouts, leafy crucifers, etc.

Fat

Fats and oils are lipids (esters of fatty acids or substances capable of forming such esters) in which excess nutrients and heat energy are stored in human body. Apart from the ingested lipid foods, fats also arise from the metabolism of carbohydrate and protein. Lipids with carbohydrate and protein form an essential part of the colloidal complex of cytoplasm. The fats and oils present in vegetables and other foods are largely mixtures of palmitin, stearin and olein. Fats are hydrolysed with the liberation of glycerol ($C_3H_5(OH_2)_{14}$.COOH) and stearic acid ($CH_2.(CH_2)_{16}$.COOH) and unsaturated, e.g., oleic acid ($C_{18}H_{34}O_2$). For healthy living, we need to obtain 20-30% of our total daily energy intake from fat. Humans cannot synthesize certain types of fatty acids namely, linoleic acid and alpha-linolenic acid but these are needed for bodily functions and are known as "essential fatty acids". After absorption, fat passes from the blood to the tissues where it is either burned or stored for further use. Fats have a high fuel value and upon oxidation, 1g of fat liberates 9 calories of energy. Apart from supplying more than twice the energy liberated by either protein or carbohydrate, it helps in the absorption of the soluble vitamins like vitamin A, vitamin E ($\alpha-$ tocopherol) and vitamin K. Different essential fatty acids hydrolysed from fat perform vitamin like functions in the body.

Daily requirement of fats and oils is 50g and most of this requirement is met through cooking oil and animal fats. Vegetables contain very low fat which mostly ranges between 0.1 and 0.2%. Mature seeds of some legumes like, French bean, hyacinth bean and cucurbits are good sources of fat. Other important sources of fat are small bitter gourd, chilli, sweet corn, cluster bean, globe artichoke, palak, spinach, fenugreek leaves, mustard leaves, coriander leaves, Bengal gram leaves, radish top and mint.

Vitamin contents of vegetables

Vitamins are biologically active compounds, occurring in varying and minute proportions in foods. Only 13 vitamins are essential for health and must be consumed as part of a healthy diet in order to maintain normal physiological process. Vitamins functions as essential building block of certain coenzymes which are indispensible for normal vital activities. Vitamins are also involved in the utilization of major nutrients like proteins, fats and carbohydrates. For this reason, specific vitamin deficiency produces characteristic symptoms. Unlike amino acids, vitamins do not enter into the tissue structure hence, function like a catalyst in the physiological processes. Different compounds show vitamin activity. Based on their solubility, vitamins are classified into two groups.

Water-soluble vitamins

Water-soluble vitamins are vitamin B complex, including Thiamine (vitamin B_1), Riboflavin (vitamin B_2), niacin, B_6 (pyridoxine), folate, B_{12} (cobalamin), biotin, and pantothenic acid and vitamin C. These vitamins have no, or very little, storage capacity in the body, since they are mainly excreted with urine. Consequently, they are required on a daily basis, otherwise symptoms associated with their deficiency will arise and related functions can be affected. Each vitamin may work independently or in combination with other vitamins to serve specific functions. For example, the vitamin B group that acts as a coenzyme in energy metabolism includes vitamins B_1, B_2, niacin, biotin, and pantothenic acid. Another vitamin B group includes vitamins B_6, B_{12}, and folate which work together to produce and maintain red blood cells, in addition to their own specific activities. Vegetables are good sources of water-soluble vitamins, except for vitamin B_{12} which is found only in animal sources and is produced by microorganisms.

Fat-soluble vitamins

Fat-soluble vitamins dissolve in fat, oils or organic solvents and can be stored in bodily organs that accumulate fat, such as the liver. Four types of fat-soluble vitamins exist viz., A, D, E, and K. Vegetables are called protective foods because of high vitamin and mineral contents in it.

Vitamin A

This fat soluble vitamin is obtained from food in two forms: as retinol or pre-formed vitamin A and as β–carotene or provitamin A, which gets converted into vitamin A in the liver and intestine. Retinol is a light sensitive pigment, also called visual purple. Pre-formed vitamin A is found only in foods of animal origin and the richest source is liver, fish liver oil and dairy foods. β – Carotene, the best known form of pro-vitamin A is available in yellow and dark green vegetables and fruits.

Vitamin A is essential for clear vision in dim light and at night. It maintains integrity of epithelial tissues, maintains healthy mucous membranes of eyes, mouth, gastro intestinal, respiratory and genitor-urinary systems. Daily requirement of vitamin A is 5000 IU (0.6 µg β – carotene = 1 IU vitamin A). Abundant vitamin A (more than 2500 IU per 100g) is present in carrot, amaranth, palak, spinach, Indian spinach, water spinach, fenugreek leaves, mustard leaves, pumpkin leaves, drumstick leaves, portulaca, cowpea leaves, beet top, coriander leaves, broccoli, kale, teasle gourd, muskmelon, winter squash, pumpkin, etc.

Deficiency symptoms

i) Night blindness (Nyctalopia) as early sign of deficiency.

ii) Xerophthalmia in infants and young children causing keratinization in epithelial cells of eyes.

iii) Dryness in skin.

Thiamine

Principal synonyms are vitamin B_1, Aneurin. Oryzanin. Daily requirement is 1.2 mg. This water soluble vitamin is essential for the proper utilization of carbohydrates in the body. Most of the vegetables contain 0.05 to 0.06 mg thiamine per 100g fresh. Important sources of thiamine are palak, pea, ripe tomato, chilli, muskmelon, garlic, leek, pea, hyacinth bean, giant taro, tannia, asparagus, globe artichoke, palak and agathi.

Deficiency symptoms

i) Thiamine deficiency results in beri-beri disease. Muscular weakness and loss of weight, neuritis, fluid in tissues and body cavities (oedema) and impaired cardiac function are the symptoms of beri-beri.

ii) Loss of appetite

iii) Dilation of heart

Vitamin B_2 Complex

a) **Riboflavin :** This vitamin, synonymous to vitamin B_2, is essential as a part of coenzymes for several oxidation processes inside the cell and is concerned with energy and protein metabolism. Daily requirement is 1.7 mg. Range of riboflavin content in most of the vegetables is 0.01 to 0.08mg per 100g of edible part. Important sources of riboflavin are palak, chilli, sweet pepper, cauliflower, knolkhol, broccoli, Brussels sprouts, teasle gourd, garlic, lettuce, celery, okra, winged bean, asparagus, portulaca, pigweed, radish top, beet top, mint, cowpea leaves, chicory, watercress, endive and parsley.

Deficiency symptoms

i) Dark red inflamed tongue similar to niacin deficiency
ii) Dermatitis (inflammation of skin)
iii) Loss of hair and dry sclay skin
iv) Diarrhoea (sometimes occur)
v) Cracks in corners of mouth (angular stomatitis)
vi) Ulcers in oral cavity
vii) Cracked lips

b) **Niacin (Nicotinic acid) :** This vitamin is involved as component of a coenzyme in oxidative reactions and is concerned with the metabolism of carbohydrate, fat and proteins. This vitamin can be synthesized in the body from tryptophan. Daily requirement is 19mg. Most of the vegetables contain niacin to the tune of 0.1 to 0.7 mg per 100g fresh. Important sources of niacin are palak, amaranth, bitter gourd, teasle gourd, pointed gourd, bottle gourd, pumpkin, chilli, radish, lettuce, carrot, pea, cowpea, hyacinth bean, okra, sweet potato, spinach and fenugreek leaves.

Deficiency symptoms

i) Pellagra showing inflammation and changing of skin in hand, feet, leg and neck, particularly in light exposure
ii) Nervous breakdown
iii) Stomach and intestinal disorder

c) **Pyridoxine (Vitamin B_6) :** The co-enzyme form of this vitamin, pyridoxal phosphate is required for the metabolism of amino acids and conversion of tryptophan to nicotinic acid. It is also associated with the metabolism of essential fatty acids. It is widely distributed in vegetable source.

Deficiency symptoms

i) Cracking at the corners of lips
ii) Retarded growth and anaemia
iii) A skin disease called acrodynia

d) **Folic acid (Folicin) :** It is associated with reduced risk of cardiovascular disease and colon cancer. Folic acid lowers homo-cysteine that has an adverse effect on the lining of the arteries. It is also required for the multiplication and maturation of red cells and is particularly important for women. Requirement of folic acid is 50-100mg depending on age. Vegetables like spinach, palak, lettuce, cabbage, cowpea, French bean are good sources of folic acid.

Deficiency Symptoms

i) Different forms of anaemia (megaloblastic) which is more acute in pregnant women

ii) Impaired growth and nervous breakdown

Vitamin E (α-Tocopherol)

It posses anti-oxidant property and prevents oxidation of β carotene and vitamin A in the intestine. It also prevents oxidation of polyunsaturated fatty acids (PUFA) in cells and thereby maintains integrity of cell membranes. Daily requirement is 5.0 mg. Green leafy vegetables and sweet corn are good sources of vitamin E.

Deficiency symptoms

i) Degeneration of kidney

ii) Necrosis of the liver

iii) Reduction in the capability of reproduction

Vitamin K

Principal synonyms are coagulation vitamin and anti-hemorrhagic vitamin. It is required for the formation of pro-thrombin (blood clot) and for the normal function of liver. Daily requirement is 0.115mg. Green leafy vegetables contain vitamin K. Moreover, intestinal bacteria produce this vitamin.

Deficiency Symptoms

i) Delayed and faulty coagulation of blood in cut wounds

ii) Hindrance in normal secretion of bile from liver

Vitamin C (Ascorbic acid)

Ascorbic acid or vitamin C can be synthesised by many animals but not human beings. It acts as a coenzyme in collagen synthesis, which facilitates wound repair and stimulates the immune system. Vitamin C is also known as a potent antioxidant and good enhancer for non-heme iron absorption. It maintains connective tissues in blood capillaries thus hastens the process of healing wound and prevents hemorrhages of small blood vessels. About 50mg of vitamin C is required daily. Vegetables particularly chilli, sweet pepper, cabbage, broccoli, kale, leafy greens, drumstick, radish top, coriander leaves, bitter gourd , amaranth and Indian spinach are good source of vitamin C.

Deficiency symptoms

i) Scurvy characterized by weakness, bleeding gums, dropsy and defective bone growth

ii) Delayed healing of wounds and susceptibility to cough and cold

iii) Reduced resistance to diseases

Some other vitamins which are not generally associated with the vegetables are pantothenic acid, biotin (vitamin H), vitamin B_{12} (cobalamin), cholin (sinkalin), inositol and vitamin D. These vitamins may be present in the product of animal origin (e.g., vitamin D) and may be synthesized by intestinal bacteria (e.g. pantothenic acid, biotin).

Mineral contents of vegetables

Minerals are inorganic, single atoms that the body needs for adequate functioning. Many minerals are co-factors for different enzymes and required for energy utilization, fertility, mental stability, and the immune system. From nutritional point of view, calcium, iron, iodine and zinc are important since deficiencies of these elements occur in humans. Minerals are classified into two types: macro elements and trace elements based on the amount that the body requires for maintaining its normal functions.

Macro-elements

Macro elements are minerals that the body needs in amounts greater than 100 mg per day and these are calcium, magnesium, phosphorus, potassium, sodium and chloride. Similar to vitamins, minerals can work individually or in combination to maintain bodily functions. Vegetables are good sources of many macro-elements however, absorption of minerals in the body is sometimes interfered due to the presence of certain phyto-chemicals like oxalate, phytate, etc.

Trace elements

Trace elements are minerals that the body requires in amounts less than 100 mg per day and these elements include copper, iodine, iron, zinc, fluoride, selenium, manganese, and chromium. In most developing countries, iron, zinc and iodine are often problem nutrients due to inadequate intake. Phytate is the most common inhibitor that is found widely in vegetables which affect absorption of non-heme iron however, vitamin C enhances iron absorption. Zinc is found in some vegetables but its bioavailability is quite low due to the inhibitive effect of phytate. Minerals are the components of various vital body constituents as documented through several research works.

a) Calcium, phosphorus and magnesium are essential components for building strong bones and teeth

b) Sodium and potassium are needed for muscle and nerve activities

c) Chloride is a component of hydrochloric acid that is needed for digestion of foods in the stomach

d) Different minerals like Na^+, K^+, Ca^{++}, Zn^{++}, Cu^+, Mg^{++}, Cl^-, HCO^-_3, HPO^-_4, $H_2PO^-_4$, etc. are important constituents of fluids present outside and within the cell. Proper concentration of these electrolytes inside and outside the cell is essential to maintain osmotic balance and keep cells in proper shape. They also provide proper nerve stimulation and permeability of cell membranes

e) Iron is the important component of hemoglobin. Iron content of crystalline human hemoglobin is 0.34 per cent

f) Iron is needed for the synthesis of neuro-transmitters - serotonin, dopamine, norepinephrine and DNA and is needed for immune function

g) Phosphorus along with C, H, N and O are the components of the deoxyribonucleic acid (DNA), the basis of life. Phosphorus is also the constituent of phospholipids

h) Iodine is the vital constituent of thyroid hormones, thyroxin and triiodothyrone and its deficiency interferes with the function of thyroid gland resulting in swelling of this gland (goiter)

i) Magnesium is required for cellular metabolism and is implicated to have role in cardiovascular diseases

j) Zinc is a cofactor of many enzymes and involved in the production of proteins that are partly responsible for growth and maintaining the immune system

Calcium

Daily requirement is 500-600 mg. Different *Brassica* vegetables (e.g. kale, Chinese cabbage, cabbage, cauliflower, broccoli, etc.), ivy gourd and fresh winged bean are good sources of bio-available calcium. These vegetables contain low or no oxalate that normally interferes with calcium absorption. Other good sources of calcium are curry leaves, hyacinth bean, amaranth, palak, Indian spinach, fenugreek leaves, Bengal gram leaves, pumpkin leaves, pointed gourd leaves, radish top, agathi, carrot top, mint, cowpea leaves, drumstick leaves, coriander leaves, pigweed, portulaca, beet top, Ceylon spinach, Swiss Chard, young green onion, chow-chow and parsley. However, some leafy greens like, spinach, palak, etc. may not provide adequate levels of bio-available calcium due to presence of oxalate which interferes with calcium absorption.

Iron

Daily requirement is 10 mg. Hemoglobin in blood is decreased with its deficiency. Most of the vegetables other than leafy greens contain 0.4 to 1.7 mg/100g. Important sources of iron are amaranth, Bengal gram leaves palak, spinach, fenugreek leaves, mustard leaves, coriander leaves, portulaca, radish top, mint, cowpea leaves, lettuce, green banana, hyacinth bean, pigweed, agathi, drumstick leaves, giant spine gourd, watermelon, sorrel and water spinach.

Phosphorus

Daily requirement is 1000 mg. Normal development is retarded due to its deficiency. Good sources of phosphorus are garlic, pea, lima bean, taro, drumstick, pumpkin leaves, carrot top, chilli, cauliflower, broccoli, Brussels sprouts, bitter gourd, parsnip, cowpea, hyacinth bean, winged bean, globe artichoke, drumstick leaves and mushrooms.

Iodine

Daily requirement is 0.1 to 0.2 mg. it is required for the synthesis of thyroxin hormone from the thyroid gland. Iodine content of vegetables is generally low. Okra, tomato, sweet pepper, carrot, onion, garlic, beet, agathi leaves, etc. contain appreciable quantity of iodine.

Sodium

Daily requirement is 4000-6000 mg. Its deficiency is manifested by weight loss and nervous breakdown. Generally, sodium requirement is met through intake of common salt. Some vegetables like, celery, green onion, Chinese cabbage, radish, etc. are good sources of sodium.

Phytochemical or Nutraceutical content

Phytochemicals are bioactive compounds which are either plant pigments (lycopene, β-carotene, anthocyanin, lutein, capsanthin, zeaxanthin, etc.) or secondary metabolites (flavonoids, isothiocyanates, glucosinolates, etc.) found in most of the vegetables which provide colour, flavour and texture and protect the plants against insects, microbes, and the oxidation stress created by sunlight and oxygen. Several research studies have noted that these compounds significantly benefit human health by reducing risks associated with non-communicable chronic or degenerative diseases such as hypertension, diabetes mellitus, coronary heart diseases and cancer. The beneficial effects of phytochemicals on health appear to result primarily from the complex interaction among them in foods. Different group and possible actions of different phytochemicals that are beneficial for human health are as follows.

Antioxidants

Most phytochemicals have antioxidant activity and protect cells against oxidative damage and reduce the risk of developing certain types of cancer. Some phytochemicals which shows antioxidant activity are allyl propyl sulfides, carotenoids (β–carotene, lycopene, zeaxanthin, cryptoxanthin, lutein, etc.), flavonoids and polyphenols.

Flavonoids

Flavonoids are a large family of low molecular weight polyphenolic compounds ubiquitous in plant occurring as glycosides which include flavones, flavonols, flavonones, isoflavones, flavan-3-ols, anthocyanins, anthocynidins, catechins, biflavans and isoflavonoids. Potential flavonols present in vegetables are quercetin, kaempferol, myricetin and luteolin. The level of kaempferol is high in kale, broccoli and endive. Chilli is also rich in polyphenols, particularly the flavonoids, quercetin and luteolin. Garlic, broccoli, potato, mushroom, white cabbage and cauliflower, kidney and pinto beans, beans, beet and corn have been reported to have high antioxidant activity. Other vegetables such as kale, spinach, Brussels sprouts, sprouting broccoli, beets, red bell-pepper, onion, corn, eggplant, cauliflower and cucumber are also rich source of flavonoids hence, have potent antioxidant activity. High levels of quercetin have been found in onion, kale, tomato and certain varieties of lettuce. In carrot, β chlorogenic acid, caffiec acid and its derivatives are reported to have significant anti oxidant activity.

Quercetin and other flavonoids frequently exist as their glycoside forms in vegetables such as in onion. Glucoside-bound quercetin is absorbed and hydrolyzed in small intestine through glucose-transport system or passive transport after lactase phlorizin hydrolase (LPH)-dependent deglucosidation. Quercetin glycosides other than glucoside forms are likely to be absorbed in large intestine after hydrolysis with entero-bacteria. Polyphenols, the secondary plant metabolites and high molecular weight flavonoids are generally considered as antineutrinos but recognition of antioxidative properties of these phenolics has evoked a rethinking towards the health benefits of these secondary metabolites.

Plant pigments

Plant pigments are edible colours found in tissue of plants which include anthocyanins, betalains, carotenoids and chlorophylls. These pigments play important ecological and metabolic functions in the plants and are more frequently exploited as the source of nutraceuticals to address a number of human ailments.

Lycopene

The compositional fruit quality of tomato has received increasing interest because of the nutritional importance of lycopene, flavonoids and chlorogenic acid in the human diet. There is considerable interest in the dietary role of lycopene in inhibition of heart disease and reducing the risk of certain cancers, including prostate cancer and breast cancer.

Betalains

Betalains have been widely used as natural colourants for many centuries but their attractiveness for use as colourants of foods (or drugs and cosmetics) has increased recently due to their reportedly high anti-oxidative free radical scavenging activities.

Chlorophylls

Chlorophyll is the most important plant pigment and a 'real life force' that nature uses to explode plants into greenery. The anti-mutagenic properties of chlorophylls have been demonstrated in various assays and clearly, intake of chlorophyll has potential to act as a chemo-preventive compound in humans.

Anthocyanin

Anthocyanins are natural pigments belonging to the flavonoid family. They are responsible for the blue, purple, red and orange colour of many fruits and vegetables. Anthocyanins are capable of acting on different cells involved in the development of atherosclerosis, one of the leading causes to cardiovascular dysfunction. It has been found that both the anthocyanins and cyanidin aglycone from tart cherries reduced cell growth of human colon cancer cell lines. Black carrots are a good source of anthocyanin pigments containing to the tune of 1750 mg kg^{-1} fresh weight. Anthocyanin fruit mutant (*Aft*) of tomato containing anthocyanin in the fruit has been utilized in developing purple fruited variety or hybrids of tomato.

Table 1. Different colour rich vegetables

Colour	Pigments	Vegetables
Red	Lycopene	Tomato and watermelon
	Betacyanins	Beet root
Orange	β-carotene	Carrot, cantaloupe, pumpkin, sweet potato
Blue/purple	Anthocyanins	Brinjal, purple tomato
Yellow	Lutein	Yellow corn
Green	Chlorophyll	Leafy greens, broccoli, kale, spinach, cabbage, asparagus, etc.
Black	Anthocyanins	Black carrot

Compounds showing cyto-protective activity

- Onion and garlic contain several sulphur compounds such as allicin, allistalin, garlicin, diallyl disulphide, diallyl trisulphide and allyl propyl disulphide which are effective in reducing harmful blood cholesterol, thus preventing coronary thrombosis, heart attack and stroke. They are also effective against some bacteria. Allicin, produced when onion bulb or garlic cloves are crushed, spontaneously decomposes to form sulphur-containing compounds with chemo-preventive activity.
- Mushrooms, white cabbage, cauliflower and garlic had been shown to have strong protective activity against a number of diseases.
- Spinach is regarded as the *brain food* and is needed to avoid memory loss and Alzheimer's disease. It is believed that the phytochemicals present in these extracts may have properties that increase cell membrane fluidity, allowing important nutrients and chemical signals to pass in and out of the cell and thereby, reducing inflammatory processes in the tissues.
- Bitter gourd contains substances with antidiabetic properties such as charantin, vicine, and polypeptide-p, as well as other unspecific bioactive components however, the mechanism of action, whether it is *via* regulation of insulin release or altered glucose metabolism and its insulin-like

effect, is still under debate. White cultivars of brinjal are high in alpha-glycosides inhibitor activity making it a potential food crop for the management of type -2 diabetes. Diphenyl amine found in onion is also effective against diabetes.

- Leguminous vegetables reduce blood cholesterol concentration thus preventing heart attack and stroke. Celery contains 3-n-butyl pthalide which is effective against hypertension. The decholestrolizing action of brinjal is attributed to the presence of poly-unsaturated fatty acid which includes linoleic and lenolenic acid.

Hormones

Some phytochemicals can act as the female hormone, estrogen and are known as phytoestrogen which may help to reduce menopausal symptoms and decrease the risk of osteoporosis, heart disease, and some cancers. Nuts and beans, especially soybeans, contain these phytochemicals.

Anticarcinogenic phytochemicals

Since last decades, special attention has been given to the vegetables that are rich in plant secondary metabolites responsible for the induction of detoxifying enzymes, e.g., glutathione-S-transferase, quinone reductase and epoxide hydrolase, which inactivate reactive carcinogens by destroying their reactive centres or by conjugating them with endogenous ligands, thereby triggering their elimination from the body. In cole crops like, broccoli, cabbage, cauliflower, etc., this inducer activity of detoxifying enzymes is principally the highly reactive thiocyanates. Glucosinolates are not directly bioactive but are very stable precursors of isothiocyanates and their hydrolysis by myrosinase is a prerequisite for observed biological activity. Sulforaphane, the hydrolysis product of the glucosinolate, glucoraphanin is highly potent at up regulating detoxification enzymes in cell culture. However, the extent to which vegetable *Brassicas* protect against cancer probably depends on the genotype of the consumer, in particular the allele present at the *GSTM1* locus. This gene codes for the enzyme glutathione transferase which catalyses the conjugation of glutathione with isothiocyanates. Approximately 50% of humans carry a deletion of the *GSTM1* gene which reduces their ability to conjugate process and excrete isothiocyanates. Brinjal is also rich in bioflavonoids which can prevent stomach cancer. Flavonoids are also associated with anticarcinogenic activity in animal and cell systems.

Table 2. Source and potential health benefits of important phytochemicals

Group	Sources	Potential health benefits
Carotenoids: β – carotene	Dark green leafy vegetables, cantaloupe, carrot, muskmelon, broccoli, red pepper, spinach, sweet potato, pumpkin, summer squash	Visual function, antioxidant, anticarcinogenic, gene expression, immune functioning, prevention of certain type of cancers, cardiovascular disease, aged-related macular degeneration.
Carotenoids: α-carotene	Sweet potato, pumpkin, canta-loupe, green beans, lima beans, broccoli, Brussels sprouts, cabbage, kale, lettuce, peas, spinach, summer squash, carrots	It prevents tumor growth
Carotenoids: Lycopene	Tomato, watermelon	Antioxidant, anticarcinogenic, prevention of certain type of cancers particularly prostate cancer, cardiovascular disease, aged-related macular degeneration.
Flavonoids: Quercetin, kaempferol, Catechin	Kale, spinach, broccoli, Brussels sprouts, beet, red bell pepper, onion, corn, brinjal, cauliflower cucumber, garlic, beans, mushroom, lettuce, endive, chilli	Antioxidant, oxidation of LDLs, prevention of arteriosclerosis and heart disease, inhibition of cell proliferation in the develop-ment of cancer
Phytoestrogens: Isoflavonones (Genistein, daidzein), coumestans, lignans	Soy bean, pumpkin seeds, peas, other legumes	Aantioxidant, prevention of osteoporosis and improvement of bone health, easing symptoms of menopause, prevention of heart disease (lowering plasma lipids and lipoproteins).
Glucosinolates and Isothiocyanates (ITC): sulforaph-anes, phenethyl (PEITC), benzyl (BITC)	Broccoli, Brussels sprouts, cabbage, cauliflower, kale, mustard green, radish, turnip and watercress	Cancer prevention (protection against carcinogenesis by induction of cytoprotective cressenzymes), inhibition of inflammatory processes, modulation of signalling pathways in prostate tissue
Monoterpenes: d-Limonene, perillic acid	Dill, garlic, celery, maize, rosemary, ginger, basil	Cancer prevention, treating gallstones

Organosulfur compounds: Allicin, diallyl sulphide, diallyl disulfide, allyl mercaptan	Chive, garlic, onions and other edible *Alliums*	Cancer prevention, lower LDLs, promote the immune system
Saponins	Beans, herbs	Hypercholesterolemia, hyperglycemia, antioxidant, cancer prevention, anti-inflammatory
Capsaicinoids	Chili peppers	Topical pain relief, cancer prevention, cancer cell apoptosis

Dietary Fibre Content of Vegetables

Additionally, vegetables are packed with soluble as well as insoluble dietary fibre known as non-starch polysaccharides (NSP) such as cellulose, mucilage, hemi-cellulose, gums, pectin, xyloglucan, xylan, mannan, free arabinan, etc. These substances absorb excess water in the colon, retain a good amount of moisture in the fecal matter, and help its smooth passage out of the body. Thus, sufficient fibre offers protection from conditions like chronic constipation, hemorrhoids, colon cancer, irritable bowel syndrome, and rectal fissures. Dietary fibre has also attracted global attention due to their role in protection against certain types of cancer, regulation of transit, lowering of blood cholesterol, etc. It binds to bile salts and thus prevents their re-absorption which ultimately leads to reduction in cholesterol level in blood resulting reduced incidence of coronary heart disease. The gums and pectins reportedly reduce the post-parandial levels of glucose in blood. Diets naturally high in fibre can be considered to bring about several main physiological consequences mentioned below :

- ❒ It helps to prevent constipation.
- ❒ It reduces the risk of colon cancer.
- ❒ It improves in gastrointestinal health.
- ❒ It improves the glucose tolerance and the insulin response.
- ❒ It reduces hyperlipidemia, hypertension, and other coronary heart disease risk factors.
- ❒ It reduces the risk of developing some cancers.
- ❒ It increase satiety and hence manages weight to some degree.

Current recommendations from the United States National Academy of Sciences, Institute of Medicine, suggest that adults should consume 20–35 grams of dietary fibre per day and The British Nutrition Foundation has recommended a minimum fibre intake of 18 g/day for healthy adults. Vegetables are good source of dietary fibre ranging from 0.2g/100g edible portion in watermelon to 3.0 g/100g in pointed gourd.

Anti-nutrient factors

Plants produce many defence strategies to protect themselves from predators and many of these such as, resveratrol and glucosinolate which are primarily pathogen-protective chemicals also have demonstrated beneficial effects for human health. Many, however, have the opposite effect. For example, phytate, a plant phosphate storage compound, is an antinutrient as it strongly chelates iron, calcium, zinc and other divalent mineral ions making them unavailable for uptake. Different anti-nutrient compounds (phytates, oxalates, trypsin inhibitors, lectins, etc.), food allergens (albumins, globulins, etc.) and toxins (glycoalkaloids, cyanogenic glucosides, phyto-hemagglutinins) in crop plants need to be reduced to enhance nutrient potential of the vegetables.

Factors affecting nutrient content in vegetables

Several factors can directly or indirectly affect the nutritional quality of crops. Among these are soil factors, such as pH, available nutrients, fertilizer applications and cultural practices, texture, organic matter content and soil-water relationships; weather and climatic factors, including temperature, rainfall and light intensity; the crop and cultivar; maturity at harvest, postharvest handling and storage, anti-nutritive components, and residues of chemical fertilizers and pesticides.

Safety of vegetables

Chemical hazard

To achieve nutrition security, foods must be safe for consumption. Indiscriminate use chemical pesticides to control the biological enemies may ensure higher production but make the product unsafe. Injudicious application of chemical fertilizers without taking care of the replenishment of organic matter in the soil causes deterioration of soil quality and thereby reduce nutrient load in the produce. In most developed countries, agricultural chemicals are applied under good agricultural practices or GAP system which controls the safety of vegetables at the production sites or farms. Vegetables that are grown under GAP are normally safe from chemical hazards.

Biological hazards

Unclean vegetables can be contaminated with pathogens and parasites from soil and water as well as manures. Vegetables grown under GAP may be free from chemical hazard but they may not be free of biological hazard, especially from manure and soil. Vegetables should be cleaned with a surfactant solution in order to remove the contaminated soil. Treatment with potassium permanganate solution, chlorinated or ozone treated water is the best option for disinfection.

Chapter 3

Classification of Vegetable Crops

Classification of Vegetable Crops

Vegetables are defined as herbaceous plant origin product with no calorific value and providing micronutrients, vitamins, antioxidants, fibre and alkaline reaction inside human body. Vegetable crops are not only botanically different but also their cultural and climatic adaptation also shows wide variability. It is, therefore, necessary to classify the vegetable crops for their easy recognition. Vegetable crops can conveniently be classified "Botanically" for the purpose of vegetable breeding and according to "Method of culture" from the point of crop husbandry.

Botanical classification

Botanical classification involves grouping of plants into kingdom, division, sub-division, phylum, sub-phylum, class, sub-class, order, family, genera, species, sub-species and variety. The broadest group in which vegetables are discussed is family. The genus and species constitute the scientific name. Scientific names are accepted worldwide and there cannot be any confusion as per their nomenclature. All vegetable crops belong to the division Angiospermae. The division Angiospermae has two classes: Monocotyledoneae and Dicotyledoneae.

The crops used for vegetable purpose in the world belong to 1200 species under 78 families and of them, more than 860 species under 59 families belong to dicotyledoneae and about 340 species under 19 families belong to monocotyledoneae. In the tropical and subtropical parts of the world, about 90 species of vegetable crops are cultivated, but hardly 25 of them are commercially important in terms of market demand, development of variety/hybrid and seed production. In the present botanical classification, over 140 different kinds of vegetable crops under 7 monocotyledoneae and 20 dicotyledoneae families have been included.

Table 1. Botanical classification of vegetable crops

Name	Botanical name	Chromosome number (2n)
	MONOCOTYLEDONEAE	
	1. Amarylidaceae (Alliaceae)	
Onion	*Allium cepa* L.	16
Multiplier onion	*A. cepa* var. *aggregatum* L.	16
Top onion	*A. cepa* var. *viviparum* (Metz.) Alef.	16
Garlic	*A. sativum* L.	16-32 (2x–4x)
Leek	*A. ampeloprasum* L. var. *porrum*	32, 48 (4x, 6x)
Welsh onion	*A. fistulosum* L.	16
Shallot	*A. ascalonicum* L.	16
European chive	*A. schoenoprasum* L.	16
Chinese chive	*A. tuberosum* Rottler ex Spreng	32
Kurrat	*A. kurrat*	32 (4x)
	2. Araceae	
Taro	*Colocasia esculenta* (L.) Scott	-
Eddoe type	*C. esculenta* var. *antiquorum*	28,42 (2x, 3x)
Dasheen type	*C. esculenta* var. *esculenta*	42 (3x)
Giant taro	*Alocasia macrorrhiza* (L.) Scott, *A. indica* (Roxb.) Scott.	26,28
Swamp taro	*Colocasia esculenta var. stolonifera*	26
Giant Swamp taro	*Cyrtosperma chamisonis*	26,28
Tannia	*Xanthosoma sagittifolius* (L.) Scott.	26
Elephant foot yam	*Amorphophallus paeoniifolius* (Dennst.) Nicolson	26,28
	3. Bambusaceae	
Bamboo	*Bambusa vulgaris* Schrader ex Wendland	82
	4. Dioscoreaceae	
Yam	*Dioscorea* spp.	30-80 (3x-8x)
Greater yam	*D. alata* L.	40 (4x)
Lesser yam	*D. esculenta* (Lour.) Burkill	40 (4x)
White yam	*D. rotundata* (L). Poir.	40 (4x)
	5. Gramineae (Poaceae)	
Sweet corn	*Zea mays* L. var. *rugosa*	20
	6. Liliaceae	
Asparagus	*Asparagus officinalis* L.	20

	7. Nelumbonaceae	
Lotus	*Nelumbo nucifera* Gaertn.	16
	DICOTYLEDONEAE	
	1. Aizoaceae	
New Zealand spinach	*Tetragonia expansa* (Murr.)	32
	2. Amaranthaceae	
Leafy amaranth	*Amaranthus tricolor* L., *A. blitum* L., *A. viridis* L., *A. tristis* L., etc.	32 or 34
	A. dubius L.	64 (4x)
	3. Basellaceae	
Indian spinach	*Basella alba* (green type)	48
	B. rubra (red type)	48
	4. Chenopodiaceae	
Garden beet	*Beta vulgaris* L. subsp. *vulgaris*	18
Palak	*B. vulgaris* L. var. *bengalensis* Hort.	18
Chard	*B. vulgaris* L. var. *cicla*	18
Spinach	*Spinacea oleracea* L.	12
French spinach	*Artiplex hortensis* L.	12
Pigweed (Bathua)	*Chenopodium album* L.	36 (4x)
	5. Compositae (Asteraceae)	
Lettuce	*Lactuca sativa* L.	
Head type	*L. sativa* var. *capitata* L.	18
Leaf type	*L. sativa* var. *crispa* L.	18
Cos type	*L. sativa* var. *longifolia* L.	18
Asparagus type	*L. sativa* var. *asparagina* L.	18
Chicory	*Cichorium intybus* L.	18
Endive	*Cichorium endivia* L.	18
Globe artichoke	*Cynara scolymus* L.	34
Jerusalem artichoke	*Helianthus tuberosus*	102 (6x)
Terragon	*Artemisia dracunculus* L.	18
Salsify	*Tragapogon porrifolius* L.	12
Black salsify	*Scorzonera hispanica* L.	22
Spanish salsify	*Scolymus hispanica* L.	14,28
Dandelion	*Taraxacum officinale* Weber ex Wiggers	16
	6. Convolvulaceae	
Sweet potato	*Ipomea batatas* (L.) Lam.	90 (6x)
Water spinach	*Ipomea aquatica* Forsk	30

7. Cruciferae (Brassicaceae)

Cabbage	*Brassica oleracea* L. var. *capitata* L. *f. alba* DC	18
Red cabbage	*B. oleacea* L. var. *capitata f. rubra* (L.) Thell	18
Savoy cabbage	*B. oleacea* var. *sabauda* L.	18
Cauliflower	*B. oleacea* L. var. *botrytis* L.	18
Brussels sprouts	*B. oleracea* L. var. *gemifera* DC.	18
Sprouting broccoli	*B. oleracea* L. var. *italica* Plenck	18
Knolkhol (Kholrabi)	*B. oleracea* var. *gongylodes* L.	18
Kale/collard	*B. oleracea* var. *acephala*	18
Chinese cabbage	*B. rapa* var. *pekinensis*	20
Chinese kale	*B. oleracea* var.*alboglabra* Bailey	18
Mibuna	*B. rapa* var. *nipposinica*	20
Turnip	*B. rapa* L. var. *glabra* Kitamura	20
Rutabaga (swede)	*B. napus* var. *napobrassica* Peterm.	38 (amphidiploid)
Leaf mustard	*B. juncea* (L.) Czern. & Coss. var. *cuneifolia* Roxb.	36 (4x)
Curled mustard	*B. juncea* var. *crispifolia*	36 (4x)
White mustard	*Sinapis alba* L.	24
Radish	*Ranphanus sativus* L.	18
European radish	*R sativus* L. var. *radicula* DC	18
Japanese & Chinese winter radish	*R sativus* L. var. *raphanistroides* Makins	18
Black or Spanish radish	*R. sativus* L var. *niger* (Mill.) Pers.	18
Oilseed radish	*R sativus* L var. *caudatus* (L) Hooker et Anderson	18
Horse radish	*Armoracia rusticana* Gaertn.Mey. & Schred.	32 (4x)
Water cress (brahmi sag)	*Rorippa nasturtium-aquaticum* (Syn. *Nasturtium officinale* R. Br.)	32
Garden cress	*Lepidium sativum* L.	16,32
Upland cress	*Barbarea vulgaris* R. Br.	16
Sea kale	*Crambe maritime*	60
Catran	*Crambe tatarica* Busch	60
Tuberous rooted Chinese mustard	*Sinapis juncea* var. *napiformis* Paill & Bois.	24

8. Cucurbitaceae

Cucumber	*Cucumis sativus* L.	14
Muskmelon	*Cucumis melo* L.	24
Snap melon	*C. melo* L. var. *momordica*	24
Long melon	*C. melo* L. var. *utilissimus* Duth.& Full.	24
Snake melon	*C. melo* L. var. *flexuosus* Naud	24
Netted melon	*C. melo* L. var. *reticutalus*	24
Pickling melon	*C. melo* L. var. *conomon* Mak.	24
Cantaloupe	*C. melo L.* var. *cantalupensis* Naud.	24
Winter melon	*C. melo* L. var. *inodorus*	24
Mango melon	*C. melo* L. var. *dudain*	24
Small gourd	*C. melo* L. var. *agretis* Naud.	24
Gherkin	*Cucumis anguria* L.	24
Watermelon	*Citrullus lanatus* (Thumb.) Matsum & Nakai	22
Round melon	*Praecitrullus fistulosus (*Stocks) Pang.	24
Pumpkin	*Cucurbita moschata* (Duch.) Poir.	40
Summer squash	*C. pepo* L.	40
Winter squash	*C. maxima* Duch.	40
Buffalow gourd	*C. ficifolia* Bouche	40
Bottle gourd	*Lagenaria siceraria* (Molina) Stadl.	22
Bitter gourd	*Momordica charantia*	22
Balsam apple	*M. balsamina* L.	22
Giant Spine gourd	*M. cochinchinensis* (Lour.) Spreng	28
Teasle gourd	*M. subangulata* ssp. *renigera* (G. Don) de Wilde	56
Spine gourd	*M. dioica* Roxb. Ex Willd.	28
Ridge gourd	*Luffa acutangula* (L.) Roxb.	26
Sponge gourd	*L. cylindrica* Roem.	26
Pointed gourd	*Trichosanthes dioica* Roxb.	22
Snake gourd	*T. anguina* L.	22
Wax gourd	*Benincasa hispida* (Thunb.) Cogn.	24
Ivy gourd	*Coccina grandis* (L.) Voigt.	24
Chow Chow	*Sechium edule* (Jacq.) Swartz.	28
Slipper gourd or Kaywa	*Cyclanthera pedata* (L.) Schrader	32

9. Euphorbiaceae

Cassava	*Manihot esculenta* Crantz.	36
Chekurmanis	*Sauropus androgynus* (L.) Merr.	24

	10. Labitae	
Chinese potato	*Solenostemon rotundifolius (Poir) Morton* (Syn. *Coleus parviflorus* Benth.)	64, 84
Chinese artichoke	*Stachys siebeldi* Miq.	16
Hoary basil	*Ocimum americanum* L.	24
	11. Leguminoceae (Fabaceae)	
Garden pea & snow pea	*Pisum sativum* L. sub sp. *hortense*	14
French bean	*Phaseolus vulgaris* L.	22
Lima bean	*P. lunatus* L.	22
Runner bean	*P. coccineus* L.	22
Tepary bean	*P. acutifolius* Grey	22
Thicket bean	*P. polystachyus*	22
Adzuki bean	*P. angularis* (Wild.) Wright	22
Sword bean	*Canavalia gladiata* (Gucav.) DC.	22,24
Gotani bean	*C. plagosperma*	22
Jack bean	*C. ensiformis* (L.) D.C.	22
Valvet bean	*Mucuna deeringiana* (Bor.) Merr.	22
Cluster bean	*Cyamopsis tetragonolobus* (L.) Taub.	14
Winged bean	*Psophocarpus tetragonolobus* (L.) Taub.	18,22
Broad bean	*Vicia faba* L.	12
Hyacinth bean	*Lablab perpureus* (L.) Sweet.	22,24
Vegetable type	*Lablab perpureus* (L.) var. *typicus*	22,24
Pulse type	*Lablab perpureus* (L.) var. *lignosus*	22,24
Horse gram	*Dolichos uniflorus* Lam.	22,24
Cowpea	*Vigna unguiculata*	22
Small podded vegetable and pulse type	*Vigna unguiculata* sub sp. *unguiculata*	22
Long podded vegetable type	*V. unguiculata* sub sp. *sesquipedalis*	22
Soybean	*Glycine max* (L.) Merrill	40
Yam bean	*Pachyrhizus erosus* Urbn.	22
African yam bean	*Sphenotylis stenocarpa*	22
African-locust bean	*Parkia clappertonia*	24
Ground bean	*Voandzeia subterranea* (L.) Thou.	22
Marama bean	*Tylosema esculentum* (Burchell) Schreiber	44

Agathi	*Sesbania grandiflora* Poit.	24
Fenugreek	*Trigonella foenum-graceum* L.	16
	12. Malvaceae	
Okra	*Abelmoschus esculentus* (L. Moench)	126,130,132 (Allopolyploid)
	13. Martynaceae	
Martynia	*Proboscidea jussieeui* (Keller)	30
	14. Moringaceae	
Drumstick	*Moringa oleifera* Lamk.	28
	15. Polygonaceae	
Rhubarb	*Rheum rhaponticum* L., *Rheum rhabarbarum* L.	22, 44 (4x)
Sorrel	*Rumex acetosa* L.	XX+12; XY_1Y_2+12
Patience dock	*R. patientia* L.	80
Bladder dock	*R. vesicarius* L.	18
Buck wheat	*Fagopyrum tatariocum* (L.) Gaertn.	16
	16. Portulaceceae	
Pusley	*Portulaca oleracea* L.	54
Ceylon spinach	*Talinum triagulare* Willd.	24
	17. Rutaceae	
Curry leaf	*Murraya koenigii* (L.) Spreng	18
	18. Solanaceae	
Potato	*Solanum tuberosum* L.	48 (auto 4x)
Brinjal	*Solanum melongena* L.	24
Fruit round or egg-shaped	*S. melongena* var.*esculentum*	24
Fruit long & slender	*S. melongena* var. *serpentinum*	24
Plant dwarf	*S. melongena* var. *depressum*	24
Tomato	*Solanum lycopersicum* L.	24
Currant tomato	*Solanum pimpinellifolium* (Juslen) Mill.	24
Cherry tomato	*Solanum lycopersicum* var. *cerasiformae*	24
Sweet pepper & Chilli	*Capsicum annuum* L.	24
Tabasco pepper	*C. frutescens* L.	24
Husk tomato	*Phesalis pubescens* L.	24
Tomatillo	*P. ixocarpa* Brot.	24
Tree tomato	*Cyphomandra betacea*	24
	19. Tiliaceae	
Jute (jute mallow)	*Corchorus olitorius* L.	14

	20. Umbelliferae (Apiaceae)	
Carrot	*Daucus carota* L.	18
Celery	*Apium graveolens* L. var. *dulce* (Mill.) Pers.	22
Celeriac	*A. graveolens* L. var. *rapaceum* (Mill.) Gaud.	22
Leaf celery	*A. graveolens* L. var. *secalinum*	22
Parsley	*Petroselinum crispum* (Mill.) Nym. Ex. Hill.	22
Turnip-rooted parsley	*P. crispum* var. *tuberosum*	22
Parsnip	*Pastinaca sativa* L.	22
Turnip-rooted chervil	*Chaerophyllum bulbosum* L.	22
Skirret	*Sium sisarum* L.	40
Chervil	*Anthriscus cerefolium*	18
Dill	*Anethum graveolens* L., *A. sowa* L.	22
Coriander	*Coriandrum sativum* L.	22

Classification based on method of culture

Botanical classification is useful to the students of Vegetable Science and academicians but is of little value to the vegetable grower. For example, potato and tomato though belong to the same botanical family but have entirely different cultural requirements. This classification is the best for the vegetable growers because the vegetable crops requiring the same cultural practices (but not same season for cultivation) are grouped together though they maybe divergent botanically. The crop species in the groups like cole crops, bulb crops, cucurbits, peas and beans belong to the same family however, the crops under the groups like salad crops, pot herbs and greens belong to different family.

Table 2. Classification of vegetable crops based on method of culture

Group 1	:	**Potato**
Group 2	:	**Solaneceous fruit crops:** Tomato, brinjal, chilli, sweet pepper.
Group 3	:	**Cole crops:** Cabbage, cauliflower, broccoli, Brussels sprouts, knolkhol, kale, Chinese cabbage.
Group 4	:	**Cucurbits:** Pumpkin, summer squash, winter squash, muskmelon, watermelon, cucumber, bottle gourd, ash gourd, ridge gourd, snake gourd, sponge gourd, bitter gourd.
Group 5	:	**Root crops:** Radish, carrot, beet, turnip, horse radish, rutabaga, salsify, turnip-rooted chervil, Spanish salsify, black salsify.

Group 6	:	**Bulb crops:** Onion, garlic, leek, shallot, welsh onion, chive.
Group 7	:	**Salad crops:** Lettuce, celery, endive, chicory, parsley, sorrel, water cress, garden cress, upland cress.
Group 8	:	**Greens and pot herbs:** Vegetable amaranth, palak, spinach, New Zealand spinach, coriander, Indian spinach, French spinach, dandelion, pigweed, tarragon, parsley, Ceylon spinach, fenugreek.
Group 9	:	**Peas and beans:** Garden pea, French bean, cowpea, hyacinth bean, asparagus bean, cluster bean, winged bean, lima bean, velvet bean.
Group 10	:	**Tuber crops other than potato:** Taro, giant taro, tannia, yam, cassava, elephant-foot yam, Chinese potato.
Group 11	:	**Sweet potato**
Group 12		**Okra**
Group 13	:	**Pointed gourd**
Group 14	:	**Chayote (chow chow)**
Group 15	:	**Tropical perennial vegetable crops:** Drumstick, agathi, chekurmanis, curry leaf
Group 16	:	**Temperate perennial vegetable crops:** Asparagus, rhubarb, seakale, globe artichoke.

Classification based on plant parts used

This classification is important from the consumer and post-harvest handling point of view. For example, leafy vegetables are highly perishable and cannot be stored for longer periods because for their very high respiratory rate. After harvest, these vegetables need to be immediately cooled and stored under refrigerated hold. On the other hand, tuber and bulb vegetables can be stored for a considerable period of time because of their very low respiratory rate. On the basis of edible plant part, vegetables can be classified into the following groups;

Leaves: Cabbage, Chinese cabbage, Brussels sprout, kale, sea kale (shoot), lettuce, celery (crisp petiole), rhubarb (leaf stalks), spinach, French spinach, New Zealand spinach, Indian spinach, Ceylon spinach, water spinach, amaranth, palak, fenugreek, coriander, chekurmanis, parsley, chard, endive, mustard green, collard, water cress, chicory, sorrel, garden cress, dandelion, pigweed (*Chenopodium*), tarragon, pulsey (*Portulaca*), curry leaf, chervil, dill, etc.

Flower/ flower bud: Broccoli (flower buds, stem tissue, and some small leaves), globe artichoke (flower bud), Agathi (bloomed flower)

Stem/modified stem: Knol-khol (swollen above ground stem), asparagus (young shoot called spears), bamboo shoots (young shoot), swamp taro (stolon), cauliflower (proliferated pre-floral apical meristem and flower tissue), etc.

Fruits (unripe/ripe): Watermelon, muskmelon, snap melon, pumpkin, bottle gourd, bitter gourd, ridge gourd, sponge gourd, ivy gourd, wax gourd, cucumber, gherkin, pointed gourd, teasle gourd, tomato, brinjal, chilli, bell pepper, okra, drumstick, cowpea, hyacinth bean, French bean, cluster bean, winged bean, velvet bean, snow pea, lima beans, broad bean, etc.

Underground modified tap root: Radish, turnip, carrot, beetroot, rutabaga, parsnip, Chinese potato, salsify, turnip-rooted chervil, Spanish salsify, black salsify, etc.

Underground modified adventitious root (tuber): Sweet potato, cassava, Jerusalem artichoke (tuberous root).

Underground modified stem (corm /tuber/rhizome): Potato, taro, giant taro, tannia, elephant foot yam, yam, Chinese artichoke (rhizome), ginger, turmeric.

Bulbs: Onion, garlic, leek (swollen leaves with a bit of stem), etc.

Seeds (immature/matured/dried): Pea, French bean, hyacinth bean, cowpea, lima bean, sweet corn, fenugreek, coriander.

Chapter 4

Influence of Environment on Vegetable Production

Development can be defined as the process and timing of differentiation of the component events causing qualitative changes in form and function of the plant and therefore, manifestation of economic plant parts. Sowing is the beginning of the process, after which plants emerge, grow and develop with or without transplanting and may be harvested in vegetative (e.g., cole, bulb, tuber, root, salad crops) or reproductive phase (solanaceous fruits, peas and beans, cucurbits, okra, etc.). Growth and development of plants is dependent on abiotic (physical) and biotic (biological) factors. Abiotic factors include the physical environmental conditions and biotic factors include animals, insects, and diseases. Each plant has certain environmental requirements. To attain the highest potential yields a crop must be grown in an environment that meets these requirements. A crop can be grown with minimal adjustments if it is well matched with its climate or growing condition. Unfavourable environmental conditions can produce a stress on plants resulting in lower yield. In such cases the environment can be artificially modified, as done in protected structures like poly house, shade net house, etc. to meet the crop requirements.

Different abiotic environmental factors *viz*., light, temperature, water and soil greatly influence plant growth and geographic distribution. These factors determine the suitability of a crop for a particular location, cropping pattern, management practices, and levels of inputs needed. A crop performs best and is least costly to produce if it is grown under the most favourable environmental conditions. To maximize the production of any crop, it is important to understand how these environmental factors affect plant growth and development.

Temperature

Temperature affects the growth and development of plants primarily by controlling the manner and speed of biochemical processes. Temperature influences photosynthesis, water and nutrient absorption, transpiration, respiration, and enzyme activity. These factors govern germination, flowering, pollen viability, fruit set, rates of maturation and senescence, yield, quality, harvest duration, and shelf life. Different plants have different temperature requirements. Most plants function in a relatively narrow range of temperatures. Biological activity of plants is limited on the lower side by freezing point of water (0°C) and on the upper side by denaturation of protein (50°C). Plants can only grow and survive within these temperature limits, if during the seasonal cycle sufficient time is available for efficient growth.

Air Temperature

Air temperature is the most commonly measured weather parameter. More specifically, temperature describes the kinetic energy or energy of motion of the gases that make up air. As gas molecules move more quickly, air temperature increases. Air Temperature is a measure of the hotness or coldness of the air. It is also termed as surface temperature in meteorology. Air temperature is the temperature indicated by a thermometer exposed to the air but sheltered from direct sun exposure. Air temperature affects the growth and reproduction of plants, with warmer temperatures promoting biological growth. Air temperature also affects nearly all other weather parameters *viz*.

- The rate of evaporation
- Relative humidity
- Wind speed and direction
- Precipitation patterns and types (rain, snow or sleet).

Dynamics of leaf temperature

Leaf temperature is correlated positively with atmospheric temperature and is expressed as the result of energy distribution, i.e., heat gain and loss. A leaf may absorb 75 to 90 per cent of incident energy in the wavelength of 400-700 nanometre, nm (10^{-9} meter). Of the absorbed heat, 20 per cent becomes effective for photosynthesis and the rest 80 per cent must have to be dissipated. However, light absorption depends on orientation, surface and geometry of leaf. As for example, vertically oriented leaves minimise the heat load effectively. A change of leaf orientation with the response of high temperature and drought as seen in cowpea is termed as 'paraheliotrophy'. Heat budget of the leaves indicate successful interruption of solar radiation to satisfy the photosynthetic requirements and minimising heat load.

Heat gain depends on interception and absorption of light energy along with some adjustments to ambient temperature. Heat loss from leaf depends on the combination of conduction, convection, re-radiation and transpirational cooling. In general, leaf surface temperature is always higher than atmospheric temperature under cool and moderate temperature condition. In shady condition, however, leaf temperature would be close to atmospheric temperature, or slightly below it. At higher atmospheric temperature, the leaf temperature may be lower than that of atmosphere due to transpirational cooling, provided there would be sufficiently high transpiration. As a consequence, fully exposed succulent leaves of vegetable and other crops maintain their temperature below ambient.

At night, particularly clear night, leaf temperature falls below atmospheric temperature due to re-radiation of heat (infrared or heat-ray energy, above 700 nm wavelength) from leaf and simultaneous absorption of radiant energy by night sky. So, the exposed foliage at the top of a plant may actually freeze, even though air temperature does not come to 0°C. Cloud cover suppresses such radiant cooling of leaf. For this reason, poly house-grown plants are deprived of such night time leaf cooling in the late spring and summer months sub-tropical humid climatic condition. On the other hand, poly house prevents the adverse effect of low night temperature during winter months due to this typical green house effect which increase the fruit yield substantially in many warm-loving crops like, sweet pepper, tomato, cucumber, etc. in the winter months.

Influence of air temperature temperature on plant physiology

Temperature is the most important environmental factor for plant growth and development because temperature has unique regulatory influences. Molecules are in continuous motion at any temperature above absolute zero and this motion is accentuated in proportion to the amount of heat supplied. Although biologically important reactions are limited to a small range of temperature, the plants display remarkable thermal resilience. Plants, temperature response curve for growth rises rapidly from 0° to 15°C followed by a steady increase above 15°C to an optimum value between 20° and 30°C. However, optimum temperature range depends on the particular crop. In most of the vegetable crops, growth declines above 35°C and thermal death point for most of the crops including the vegetable crops reaches at temperature nearing 50°C.

Different metabolic responses like enzyme activity, membrane permeability, substrate concentration and cumulative reactions are influenced by temperature. Plants adopt its physiology according to temperature experience. Water uptake by roots is greatly affected by temperature which implies some metabolic involvement and membrane permeability behind this physiology. Transpiration which is the driving force for water uptake within the root system is a function of leaf temperature. The biochemical reactions leading to the metabolic responses do not take place at temperature below the freezing point of water as water changes to solid state and the catalytic action and other properties of water are lost. At 35°C or above, the molecular structure of the enzymes and protein systems are irreversibly altered leading to stunting of the growth processes.

Base temperature

The base temperature, sometimes referred to minimum temperature, is an essential paratmeter for the determination of the beginning and the end of the growing season. According to Sitte *et al.* (1999), it is the lowest temperature where metabolic processes result in a net substance gain in aboveground biomass. Each plant has its own base temperature, e.g., spinach (2°C), pea (4.4°C), lettuce (4.4°C), asparagus (5.5°C), French bean (10°C), pumpkin (13°C), tomato (15°C), etc.

Heat unit and growing degree day

Heat unit systems quantify the thermal environment of these organisms and are commonly used in phenology models—models that relate organism growth and development to local weather/ climate conditions. Growing degree days (GDD), also called growing degree units (GDUs), is a heuristic tool in phenology which is the measure of heat accumulation to predict plant development rates such as date of flowering, harvest maturity of the crop, etc. This system has been used to predict a probable harvest time for several crops particularly peas, muskmelon, etc. However, the system of heat unit works when environmental factors that have a direct influence on plant growth vary in the same direction as temperature.

Optimum temperature range for the crop

The structural and physiological development of the crop plants vary greatly with the temperature of the environment. The minimum and maximum temperatures for growth and development of most of the vegetable crops lie within the range of 7° to 35°C. Each kind of crop grows and develops most rapidly at a favourable range of air temperatures. This is called the optimum air temperature range. For most crops the optimum functional efficiency occurs mostly between 12 and 24°C. Most vegetable crops can be classified according to the temperature requirements of their optimum air temperature range. However they are generally grouped into whether they require low or high air temperatures for growth. Temperature requirements are usually based on average day/night temperature. Those that grow and develop below 18°C are the cool season crops, and those that perform above 18°C are the warm season crops. Crops that originated in temperate climates usually require low temperature, while those that originate in the tropical climates require warm temperature. However, for most crop species, optimum average of the day/night temperatures usually range around 25°C. Optimum temperature for the particular crop is usually based on the average of the day and night temperature requirement. The vegetable crops can be grouped according to the air temperature requirement for optimum growth.

Hot : Night/Day temperature range for growth 18-35°C; optimum range 25-27°C (day/night average). *e.g.*, Okra, watermelon, muskmelon, chilli, sweet potato, yam, cassava, amaranth, winged bean, cluster bean, etc.

Warm : Night/Day temperature range for growth 12-35°C; optimum range 20-25°C (day/night average). *e.g.*, Tomato, sweet pepper, brinjal, lima bean, French bean, hyacinth bean, cowpea, sweet corn, palak, New Zealand spinach, cucumber, pumpkin, pointed gourd, bottle gourd, bitter gourd, ridge gourd, snake gourd, summer squash, drumstick, Indian spinach, portulaca, elephant foot yam, etc.

Cool-hot : Night/Day temperature range for growth 7-30°C; optimum range 20-25°C (day/night average). *e.g.*, Onion, leek, garlic, shallot, chicory, chives, salisfy, pakchoi, globe artichoke, taro, early cauliflower, Asiatic carrot and radish.

Cool-warm : Night/Day temperature range for growth 7-25°C; optimum range 18-25°C (day/night average). *e.g.*, Pea, broad bean, mid and late cauliflower, Chinese cabbage, cabbage, broccoli, Brussel's sprouts, kale, knolkhol, turnip, European or oriental radish and carrot, beet, rutabaga, horse radish, parsnip, potato, parsley, fennel, dil, Swiss chard, spinach, lettuce, celery, celeriac, asparagus, garden cress, mustard, etc.

Soil Temperature

Soil temperature has direct dramatic effects on microbial growth and development, organic matter decay, seed germination, root development, and water and nutrient absorption by roots. In general, the higher the temperature the faster these processes occur. The size, quality, and shape of storage organs are greatly affected by soil temperature. Dark-coloured soils absorb more solar energy than light coloured soils. The capacity of water to move heat from one area to another (conduction) is greater than that of air. Heat is therefore released to the surface faster in clay soils than in dry sandy soils. The lower the air temperature, the more rapid is the loss of heat from soil. Although light-colored sandy soils absorb less solar energy, less heat is also released to the atmosphere because of the low water-holding capacity of the soil.

Chilling Injury

Most crops are injured at temperatures at or slightly below freezing. Tropical or subtropical plants may be killed or damaged at temperatures above freezing but below 10°C and this type of injury is called chilling injury. Susceptibility to cold damage varies with different species and there may be differences among varieties of the same species. The susceptibility to cold damage varies with stage of plant development. Plants tend to be more sensitive to cold temperatures shortly before flowering through a few weeks after anthesis.

Heat Stress

Abiotic stresses either individually or in combination cause morphological, physiological, biochemical and molecular changes that adversely affect plant growth and ultimately yield. Of the major forms of abiotic stress plants are exposed to in nature *viz.,* heat, drought, cold and salinity, heat stress has an independent mode of action on the physiology and metabolism of plant cells. Heat stress is often defined as the rise in temperature beyond a threshold level for a period of time sufficient to cause irreversible damage to plant growth and development. However, heat stress involves the crop, intensity (temperature in degrees), duration, and rate of increase in temperature and the probability and period of high temperatures occurring during the day and/or the night determines its extent. As for example, it has been well documented that heat stress can occur in tomato and sweet pepper at mean daily temperatures of 28-29°C, which are just a few degrees above the optimum temperature range of 21-24°C. Such moderately elevated temperature stress may not disrupt biochemical reactions fundamental for normal cell functioning. Reduced fruit set is the common response to such elevated temperatures mainly due to different reproductive malfunctioning. The susceptibility to high temperatures in plants varies with the stage of plant development however, it depend on the species and genotype, with abundant inter- and intra-specific variations. When air temperatures rise too high to the range of 45-55°C for a persistent period, heat destroys the protoplast which results in cell death. Transpiration from the leaf stomata helps to cool the leaves. It has been calculated that transpiration can reduce heating by about 15 to 25%. Hence, if soil moisture is adequate, injury due to high temperature stress is reduced.

Vernalization

In the biennial and temperate annual vegetable crops like, cabbage, Group III and IV cauliflower, knolkhol, Brussels sprouts, broccoli, kale, Chinese cabbage, onion, leek, beet, turnip, European radish and carrot, parsley, parsnip, celery, Swiss chard, collard, endive, rutabaga, salsify, etc. exposure of the plants to low temperature of 4-7°C (may be more in some crops) for 6-8 weeks is required after they have passed the juvenile phase for the production of flower stalks. Such effect of low temperature exposure on induction of flowering is called vernalization which is common in the species adapted to temperate and Mediterranean climates. This response of temperature may be found in winter annuals, biennials and perennials but it is particularly important in biennials.

A general explanation of vernalization is based on the suppression of vegetative growth but this explanation is not applicable for all the cases. All the plants which require vernalization flower under long-day condition. In fact, the temperate vegetable crops often tend to bolt, i.e., elongate their stems and then flower in response to cool temperatures followed by long days. In broccoli because of the nature of storage organ (loose curd like structure consisting of flower buds), flowering does not involve vegetative stem elongation. However, photoperiodic requirement is not critical as low temperature exposure may replace photoperiodic requirement. In onion, spathe appearance is solely a function of subsequent temperature but floral initiation at the apex is affected by both temperature and photoperiod. Vernalization stimulus is received by the apical meristem and expanding leaves. Some of the vernalization stimulus is related to natural gibberellin substances in the plants. This proposition is based on two main observations: a) gibberellin plays a role in the flowering of many long-day plants, b) vernalization effect could be obtained with gibberellin application in the crops like cabbage, carrot, etc. in some cases. However, it cannot be inferred that application of gibberelic acid can replace vernalization requirement. So, it is suggested that control of flowering by vernalization occurs through multiple interactions between indigenous plant hormones. This vernalization has some crop specific considerations, like (a) juvenile stage for low temperature exposure varies with the crops (in cabbage and cauliflower it is after 10-12 weeks of planting while in Brussels sprouts, it is after 12-16 weeks), (b) sometimes very low temperature shortens the vernalization period, (c) low temperature requirement varies with the crop and cultivars of the crop (cabbage and Snowball group of cauliflower requires 6-7°C for 6-8 weeks, rabi onion varieties of India generally need 10-12°C for 5-6 weeks, beet, turnip, etc., require 8-10°C for 7-8 weeks).

Light

Solar energy, a form of kinetic energy, is produced in the sun due to thermonuclear reactions at the expense of matter (conversion of hydrogen into helium) in accordance with Einstein's famous equation $E = mc^2$. The incident energy from the sun in the form of light exhibits the properties of electromagnetic radiation like wavelength and frequency.

Light impulse for humans, animals and plants

Light to humans and animals is the wavelengths of radiant energy in the electromagnetic spectrum that activates the light receptors in our eyes. When these light receptors are activated, the impulses are interpreted by our brain and we experience vision. Light to plants is the wavelengths of the electromagnetic spectrum including the wavelengths that humans can see (visible light) and some of the wavelengths that humans can't see (such as microwaves and infrared light).

Light in human or animal vision typically acts only as a medium for transferring information about position and movement, shape and colour of material objects. The human and animal's interest in light perception is mainly centred on food, enemies, seeking other members of the same species for reproduction, etc. However, light for the plant is not only used as an informational medium, but also for producing food through the process of photosynthesis. The characteristics of direction and spectral composition of light in the plant's environment is transferred to the plant through the interception and activation of pigment systems within the leaf. This information affects the morphological development (size/proportion of root and shoots) of the plant, imparting to the plant some type of ecological or physiological advantage for survival. Plants also use light for sensing and detecting competitors and keeping track of time.

Effective spectrum as light for plants

As the rays travel to the surface of the sun, some remain as X-rays while others change to various wavelengths: short wavelength of high frequency and high energy cosmic rays, moderately short to moderately long wavelength and moderately frequent ultraviolet rays (15-390 nm wavelength), moderately long wavelength of visible rays (400-700 nm wavelength), long wavelength and low frequency infrared rays (760-2500 nm wavelength) and very long radio waves. The visible spectrum of light (400-700 nm wave length) is the effective spectrum as light spectrum of this wavelength range is used for vision by human beings and animals and for biochemical reactions, like photosynthesis by plant community.

Invisible	Visible spectrum								Invisible
Ultraviolet	V	I	B	G	Y	O	R	FR	Infrared
Wavelength (nm)									
15-390	400	460	490	530	585	610	660	730	760+

Ultraviolet and infrared rays are not at all beneficial to crop growth, rather these are damaging to the plants. Energy level per quantum of light (according to quantum concept, light comes from the sun in the form of tiny particles having certain weight called quanta or photons) increases with the decrease in wavelength. Hence, the ultraviolet rays dump huge packages of energy into cellular components resulting in molecular disorganization. Again infrared rays of long wavelength are absorbed by cellular moisture resulting in damage to the living cells. For this reason, growth and yield of many vegetable crops are substantially enhanced when grown under poly house or poly tunnel with ultraviolet stabilized polyethylene sheet as cladding material which has the property of absorbing ultraviolet and infrared rays.

Photosynthesis: Light energy capture by plants

One of the main roles of light in the life of plants is to serve as an energy source through the process of photosynthesis. 'Photosynthesis is a process used by plants and other organisms to convert light energy into chemical energy that can later be released to fuel the organisms' activities (energy transformation). This chemical energy is stored in carbohydrate molecules, such as sugars, which are synthesized from carbon dioxide and water. Cellular respiration converts sugars into ATP, the "fuel" used by all living things. Plants capture the energy in light using a green pigment called chlorophyll. A very precise number of photons at specific wavelengths (near 680 nm) are required to split a water molecule (H_2O) within the green leaf, which releases oxygen (O_2), and provides chemical energy to continue the long biochemical process to produce more complex molecules such as carbohydrates. Carbon dioxide from the air and water from within the leaf combine to produce oxygen and photosynthates.

Photo-morphogenesis: Light regulated plant development

Photo-morphogenesis is defined as the ability of light to regulate plant growth and development, independent of photosynthesis. Plant processes that appear to be photomorphogenic include internode elongation, chlorophyll development, flowering, abscission, lateral bud outgrowth, and root and shoot growth.

Photo-morphogenesis differs from photosynthesis in several major ways. The plant pigment responsible for light-regulated growth responses is phytochrome. Phytochrome is a colourless pigment that is present in plants in very small amounts. Only the red (600 to 660 nm) and far red (700 to 740 nm) wavelengths of the electromagnetic spectrum appear to be important in the light-regulated growth of plants. The wavelengths involved in generating photosynthesis are generally broader (400 to 700 nm) and less specific. Photomorphogenesis is considered a low energy response process meaning that it requires very little light energy to get a growth-regulating response. Plants generally require greater amount of energy for photosynthesis to occur.

Optimum light intensity

Light intensity is a major factor governing the rate of photosynthesis. Dry matter production often increases in direct proportion with increasing amounts of light. The amount of sunlight received by plants in a particular region is affected by the intensity of the incoming light and the day length. The light intensity changes with elevation, latitude and season as well as other factors such as clouds, dust, smoke or fog. The total amount of light received by a crop plant is also affected by cropping systems and crop density. Plants differ in their light requirements.

Optimum light intensity is the intensity range at which rate of gross photosynthesis is high and rate of respiration is normal resulting in high net photosynthesis in the particular crop. According to Edmond *et al.* (1964), the plant community can be classified, though not very precisely, into four groups.

Shade plants : These plants thrive in 30-50% of full sun but weaken in full sun. These plants require low light intensity of 500 to 1000 foot candles (ft-c), e.g., some ornamental house plants like philodendron, dieffenbachia, begonia, ferns, etc.

Partial shade and sun plants : These plants will produce an edible crop when grown in a shady location however, these plants need at least 50-80% of full sun condition. These plants require moderately high light intensity of 1000--3000 ft-c. e.g., cacao, coffee, tea, vanilla, black pepper and some ornamentals like caladium, sansevieria, peperornia, orchids, etc.

Sun plants : The plants which thrive in full sun but grow poorly in shade. These plants require high light intensity of 3000-8000 ft-c. e.g., most of the vegetable crops like tomato, brinjal, chilli, all cucurbits, peas and beans, sweet potato, all root crops, etc., fruit crops like banana, citrus, coconut, date, fig, olive, papaya, etc. and ornamentals like carnation, chrysanthemum, rose, poinsettia, lily, etc.

Slight shade and direct sun tolerant plants : These plants thrive well over wide range of light intensity of 2000-8000 ft-c. e.g., vegetable crops like cabbage, potato, etc., fruit crops like apple, pear, peach, plum and ornamentals like magnolia, gardenia, forsythia, etc.

Plants within the same group have varying preferences for light intensity. The light saturation point of the plant determines the relative light requirement of plant. The light saturation point is the point above which an increase in light intensity does not result in an increase in photosynthetic rate. Crops such as cucurbits, legumes, potato and sweet potato require a relatively high level of light for proper plant growth while onions, asparagus, carrot, celery, cole crops, lettuce and spinach can grow satisfactorily with comparatively lower light intensity.

Quality of light

Sunlight is often referred to as white light and is composed of all colours of light. A colour of light would be the relative distribution of wavelengths from a radiation or reflective source. Photosynthetic effectiveness varies with the wavelengths of the visible spectrum. As for example, dry matter production in the plants is higher under red or blue light than under white light exposure. Again, more dry weight is added under red light than blue light exposure. On the other hand, blue light is more effective than red in promoting stomata opening. Red light generally stimulates carbohydrate accumulation while blue light stimulates accumulation of protein and other non-carbohydrate substances. Leaves transmit principally green and far-red light whose photosynthetic effectiveness is less than the other wavelength of the visible spectrum. So, less dry matter production is realized when the plants are exposed to green light. However, such exposure of the crops to specific wavelength is impossible in field condition. When illumination is required in glass house or poly house, preferential exposure of red or blue light to the crop plant is possible.

Photoperiodism

Light duration

Due to the tilt of the earth's axis (23° from the vertical) and its travel around the sun, the length of the light period (also called photoperiod or day length) varies according to the season of the year and latitude. It varies from a nearly uniform 12-hour day at the equator ('0' latitude at Entebbe, Uganda) to continuous light or darkness throughout the 24 hours for a part of the year at the poles (70' latitude at North Alaska).

In 21 March and 23 September, the Sun remains directly over the equator resulting in equal length of light (12-hour day) and dark (12-hour night) period all over the world. On 21 December, the Sun remains in the farthest south of the equator, revealing the shortest day and the longest night (10-hour day and

14-hour night) in the regions of northern hemisphere including India, and the reverse situation occurs in southern hemisphere. From 22 December, day length starts increasing with simultaneous shortening of night length in the northern hemisphere. In 21 June, the Sun remains in the farthest north of the equator revealing the longest day and the shortest night (14-hour day and 10-hour night) in India and other regions of northern hemisphere. The opposite will happen in southern hemisphere. From 21 June, day length starts shortening till 21 December when the day length is the minimum in northern hemisphere. However, within the Hemisphere, longest light and shortest dark periods vary with the locations according to their distance from the equator. For this reason, length of the longest day on 21 June in a place of North India will be different from a place in South India. Whatever may be the location-wise variations, pattern of change in the photoperiod remains the same in the particular hemisphere.

Influence of photo-period on growth and development

Photoperiod influences growth and development of the crop in various ways like induction of flowering, carbohydrate production, development of storage organs and sex expression in cucurbits. However, induction of flowering is conspicuously influenced by photo-period.

Induction of flowering

One important plant response to day length in some vegetable crops is flowering while in others, photoperiod has no effect on flowering. Accordingly, the vegetable crops can be classified into three groups: Long-day, short-day and day-neutral plants. All the crops of tropical or subtropical origin are essentially short-day plants (SDPs), whereas the vast majority of the crops of Mediterranean or temperate origin are long-day plants (LDPs). Day-neutral plants (insensitive to photoperiod) are often the product of special selection or the consequence of adaptation to regions outside the latitude of origin. For example, tomato and French bean, the DNPs are essentially SDPs but tend to relatively insensitive to day length as a result of their intensive breeding in temperate latitudes.

Long-day crop

These crops require long light and short night (generally 8-10 hours of continuous dark) for induction of flower buds, e.g., potato, onion, lettuce, cabbage, Chinese cabbage, cauliflower, knolkhol, radish, spinach, palak, beet, turnip, carrot, etc. In general, critical photoperiod for these long-day plants varies from 12-14 hours, which is the minimum length of light period required for differentiation of flower buds in these crops. So the plants will not flower below this critical photoperiod.

Short-day crop

These crops require short light and long dark period (10-14 hours of continuous dark) for the formation of flower buds, e.g. sweet potato, Indian spinach, hyacinth bean, cluster bean, winged bean, etc. In general, critical photoperiod varies from 11-14 hours i.e., it is the maximum length of the light period at which flower buds are developed. So, the plants will not flower when the light period exceeds the critical photoperiod.

Day-night neutral crop

These crops are insensitive to photoperiod for flowering e.g., tomato, brinjal, chilli, sweet pepper, cucurbits, cowpea, okra, French bean, amaranth, etc.

Soil

Soils are produced due to physical and chemical withering of the rocks to different small particles and simultaneous mixing of those inorganic particles with humus, the organic materials and other inorganic constituents. Soils are basically unconsolidated, soft and changeable mass of the earth's crust and capable of supporting plant growth. Soil is a natural medium that provides anchorage for the plant and supplies water and mineral nutrients for normal growth. Soil consists of mineral matter, organic matter, air and water. The proportion of these four constituents and the types of mineral and organic material determines soil properties such as soil type, soil pH and fertility.

Soil type

Soil is made up in part of mineral particles grouped as sand (0.05 to 2 mm), silt (0.002 to 0.05 mm) and clay (<0.002 mm). The ratio of these determines soil types, such as clay, clayey loam, loam, sandy loam and sand. Loam is composed of sand, silt and clay in relatively even amounts, and exerts a greater influence on soil properties than does sand, silt or clay. Soil type determines the soil's capacity to store water and nutrients, aeration, drainage and ease of field operations. Sandy soils are easily tilled, well-drained and aerated but usually have low fertility and water-holding capacity. However, these soils can be improved by incorporating adequate organic matter and fertilizers. Root, tuber and bulb vegetable crops do particularly well in loose textured soils. Clayey soils, on the other hand, are more fertile and have high water retention but are poorly drained and aerated. Different vegetable crops like, brinjal, tomato, chilli, cabbage, etc. generally produce high yield but matures late when grown on clay-loam soils.

Soil pH

Soil pH is a measure of the soil's acidity or alkalinity. Soil pH is determined by the proportion of hydrogen ion to bases absorbed on the clay and organic colloids in the soil. The pH scale has long been used as an index of acidity or alkalinity of soils. A pH value of 7.5 to 8.0 usually indicates the presence of carbonates of calcium and magnesium, and a pH of 8.5 or above usually indicates appreciable exchangeable sodium. Soil pH affects the plant indirectly by influencing the availability of nutrients and the activity of microorganisms. Nutrients are most available at pH levels between 6.5 and 7.5. Nutrients in the soil may be chemically tied up or bound to soil particles and unavailable to plants if the pH is outside this range. Individual plants have pH preferences and grow best if planted in soils that satisfy their pH requirements. Vegetable crops can be classified into three groups according to their tolerance to soil acidity.

- **Less tolerant to soil acidity** (pH 6.8-6.0) **:** Asparagus, Beet, Broccoli, Cabbage, Cauliflower, Celery, Palak, Swiss chard, Cress, Chinese cabbage, Leek, Onion, Lettuce, Muskmelon, New Zealand spinach, Okra, Onion, Orach, Parsnip, Salsify, Spinach, Water cress, etc.
- **Moderately tolerant to soil acidity** (pH 6.8-5.5) **:** French bean, Lima bean, Brussels sprouts, Carrot, Collard, Cucumber, Brinjal, Garlic, Horse radish, Kale, Knolkhol, Parsley, Pea, Chilli, Pumpkin, Radish, Rutabaga, Summer squash, Winter squash, Tomato, Turnip, etc.
- **Very tolerant to soil acidity** (pH 6.8-5.0) **:** Chicory, Dandelion, Endive, Fennel, Potato, Rhubarb, Shallot, Sorrel, Sweet potato, Watermelon, etc.

The adverse effect of acid soils on plant growth is mainly related to nutrient imbalances and slow microbial activity. Low yield of the crops grown on acid soil may be due to i) toxic concentration of aluminium, ii) toxic concentration of manganese, iii) iron toxicity in some soils, iv) deficiency of bases like calcium, magnesium, sodium and potassium, v) molybdenum deficiency, vi) nitrogen, phosphorus and sulphur deficiency, vii) very slow microbial activity and viii) very slow organic matter decomposition.

Optimum availability of all the macro and micro-nutrient elements are realized in weak acidic to neutral soils (pH 6.5-7.0). Availability of phosphorus, calcium, magnesium and molybdenum sharply declines when soil pH falls below 6.5 while that of nitrogen, potassium and sulphur declines sharply below soil pH value 6.0. Availability of micronutrients like boron, copper, zinc and manganese falls sharply when soil pH falls below 5.0.

Sodic or alkali soils have generally high pH values, high exchangeable sodium percentage and are very much deficient in soluble calcium. In non-saline-alkali soil, the exchangeable sodium percentage is greater than 15, and the electrical conductivity of the saturation extract is less than 4 mmhos per cm (at 25°C). The pH values generally range between 8.5 and 10.0. In saline-alkali soil, the exchangeable sodium percentage is greater than 15 and conductivity is greater than than 4 mmhos per cm (at 25°C). The pH value of these soils somewhat exceeds 8.5. Availability of phosphorus, boron, copper, iron, zinc and manganese declines beyond pH 7.0. However, nitrogen availability is reduced only beyond pH 8.0. Most of the vegetable crops can be grown successfully in the weak alkaline soils having pH level up to 7.6. Soils having pH level beyond this limit should be reclaimed before growing vegetable crops.

Soil fertility

Soil fertility is the inherent capacity of soil to provide plant nutrients in adequate amounts and in proper balance for the growth of specific plants. The root zone is the main source of nutrients for plants, which are incorporated by the roots. A fertile soil is usually rich in nitrogen (N), phosphorus (P), potassium (K), calcium (Ca), magnesium (Mg) and sulfur (S) and contains sufficient trace elements like, iron (Fe), copper (Cu), manganese (Mn), zinc (Zn), boron (B), cobalt (Co), molybdenum (Mo), and chlorine (Cl). Humus and other organic matters improve the hydro-physical (soil moisture retention), chemical and biological properties of the soil and thereby increase soil fertility. Soil fertility can be increased by: (a) judicious soil management by incorporating organic matters, lime, etc., (b) tillage, drainage, etc. and (c) supplying nutrients to the soils through inorganic fertilizers and other chemicals. In general, vegetable crops have high nutrient requirements due to their high productivity. Most of the vegetable crops have short growing span (within 4-5 months), while some others have shorter even. Heavy production within such a short period of time also indicates the need for adequate soil fertility for successful vegetable production. Again, vegetative parts are important in some crops like leafy vegetables, cabbage, cauliflower, root crops, bulb crops, etc. while reproductive parts in the others like tomato, brinjal, okra, cucurbits, etc. However, optimum vegetative growth is essential for all the vegetable crops. Excessive vegetative growth sometimes adversely affects yield in some crops like pumpkin, summer squash, tomato, okra, etc. So, soil fertility needs to be handled judiciously according to the need of the crops.

Soil salinity

Soil salinity refers to the presence of excess salts in soil water, which often results from irrigated agriculture. After the plants take up the water, the dissolved salts from irrigated water start to accumulate in the soil. Soil salinity is usually measured as electrical conductivity (EC) of soil solution, and expressed in decisiemens per meter (dS/m). Excess salts generally affect plant growth by increasing osmotic tension in the soil, making it more difficult for the plants to take up water. Excessive uptake of salts from the soil by plants also may have a direct toxic effect on the plants. Soil salinity is most pronounced in arid areas. Not all plants respond to salinity in a similar manner; some crops can produce acceptable yields at much higher soil salinity than others. This is because some crops are better able to make the osmotic adjustments, enabling them to extract more water from a saline soil. For example, turnip and carrot are among the most sensitive vegetables and can tolerate soil salinities of only about 1 dS/m before yield declines. Zucchini, on the other hand, can tolerate soil salinity of up to 4.7 dS/m before yield reduces. The ability of a crop to adjust to salinity is extremely useful. In areas where a build-up of soil salinity cannot be controlled, an alternative crop can be selected that is both more tolerant of the expected soil salinity and able to produce economic yields.Vegetable crops may be categorised into 3 groups according to their salt tolerance.

- **Less tolerant to soil salinity :** Celery, Beans, Radish, Potato, Snake gourd, Sweet pepper, Pea, Brinjal, Artichoke, Sweet potato,Carrot Turnip, etc.
- **Moderately tolerant to soil salinity :** Tomato, Broccoli, Watermelon, Chilli, Cucumber, Amaranthus, Pumpkin, Summer squash, Cauliflower, Cabbage, Muskmelon, Bottle gourd, Spinach, Sweet corn, Onion, etc.
- **Tolerant to soil salinity :** Beet, Kale, Palak, Asparagus, Bitter gourd, Ash gourd, Lettuce, etc.

Water

Plant-water relationship

Water is the most vital component of the living cells and the basis of plant feeding. Water molecules are integral part of living systems being the solvent for metabolites and structural component of proteins and nucleic acids. In fact, biological macro-molecules cannot exist in their complicated secondary folding or spiralled helixes without the support of hydrogen bonding within their water matrix. The water matrix is also critical for metabolic activity because its properties of hydration which permit ready reactivity between molecules in solution as well as between enzymes and their substrates. The biochemical processes are sensitive to tissue hydration. A reduction of only 20-25 per cent below maximum hydration is sufficient to cause metabolic malfunctioning of many enzymes. It is the substrate for photosynthesis and a product of respiration. Water also helps to buffer plant tissue against fluctuations in the external environment. Plant growth including leaf expansion, organ enlargement, reproductive development and stomata function are all clearly associated with moisture status. Hormonal changes (abscisic acid and cytokinins) in plant body occur with brief reduction in water potential. Water is the also major force in shaping the climatic patterns.

The water potential of a solution is quantitatively equivalent to its osmotic potential. However, in plant tissues the differential permeability of cell membranes combined with resilient cell walls lead to some measure of turgor pressure, which makes a positive contribution to water potential. Osmotic influences lower diffusive pressure and in contrast, turgor pressure raises diffusive pressure. So component potentials in plant-water relations may be regarded as the combination of turgid pressure (pressure potential arising from turgidity) and osmotic potential (always less than free water and therefore, negative).

Transpiration lowers water potential at evaporative sites within leaves, and this effect is immediately translated to the root system through creating tension in the plant's vascular system. This demand of water must be satisfied continuously to maintain leaf water potential. The internal water balance of plants depends on the relative rates of water absorption and water loss. Otherwise, development of moisture stress in foliage results different physiological consequences *viz. (a)* decrease in stomatal opening, *(b)* reduction in photosynthesis and transpiration, (c) dehydration of protoplasm, *(d)* reduction in cell division and enlargement, *(e)* increase in respiration initially which later decrease due to less availability of respiratory substrate, (*f*) decrease in total dry matter production and growth, *(g)* hastening in maturity, *(h)* accumulation of sugar particularly during later part of growth, etc. Virtually every facets of the plant's physiology is dislocated to some degree during moisture stress and subsequent recovery. For this reason, actively growing plants as the case of most of the vegetable crops needs to maintain liquid phase continuity from soil water through its vascular system and all the way to evaporative sites in leaves. Vegetables contain large amount of water (more than 85%) and the product quality like, tenderness, succulence, crispness and flavour is very much related to water supply at proper stages. In fact, texture of the vegetables is determined by the combination of tissue structure, cell wall properties and turgor pressure.

Though aerial organs are capable of absorbing from a humid atmosphere or liquid film, virtually all the water and nutrients enter into the plant through root system. Water uptake by roots is greatly affected by temperature which imply metabolic involvement particularly membrane permeability in water uptake by roots. Water potential gradients arising from transpiration are generally assumed to be the driving force behind water absorption by the root system. In other words, xylem sap ascends the stem in a passive response to evaporation demand. As water potential drops, root tip ultimately lose contact with soil particles. Water moves up through adhesion and cohesion forces *via* huge number of individual xylem vessel inside the plants' vascular network, which has high conductivity and the ability to accommodate interruptions in hydraulic continuity. Water gains entry into leaves *via* petioles or leaf sheaths and moves in the liquid phase from vein endings to evaporative sites principally the stomatal cavity. Water evaporates through stomatal openings (transpiration) for the simple reason that relative humidity inside a leaf is higher than atmospheric moisture levels outside. Stomata open when their guard cells achieve a turgor pressure, which is sufficiently greater than the turgidity of the accessory cells. The most convincing theory for the generation of turgidity in the guard cells is based on the formation of solutes in the guard cells followed by passive absorption of water. Sugar, which is converted from starch in the guard cells, potassium ions along with a counter ion like chlorine could be the solutes in the guard cells. Despite stomatal resistance, the steep gradient in vapour pressure from leaf to air favours a rate of transpiration rate well in excess of CO_2 fixation. Transpiration commonly ranges from 500 to 2500 mg H_2O/dm^2-h whereas CO_2 fixation rates generally ranges between 5 to 25 mg CO_2/dm^2-h. This physiological phenomenon amply justifies high water requirement to maintain dry matter productivity. The water requirement is expressed as the units of absorbed water required for the production of one unit of dry matter and the vegetable crops can be classified into four groups.

- **High water requirement :** Palak, Amaranthus, Lettuce, Celery, Other leafy greens, Sweet pepper, Cabbage, Cauliflower, Asparagus, Broccoli, Rhubarb, Brussels sprouts, Horse radish, Radish, Beet, Turnip, Ridge gourd, Rutabaga, Green onion.
- **Moderate water requirement :** Onion (bulb), Cucumber, Chilli, Brinjal, Tomato, Carrot, Potato.
- **Low water requirement :** Pea, French bean, Cowpea, Broad bean, Cluster bean, Winged bean, Hyacinth bean.
- **Very low water requirement :** Watermelon, Muskmelon, Pumpkin, Wax gourd.

Water Balance

Water is essential to photosynthesis, it plays a key role in transpiration, and it regulates the stomata (openings in leaves to liberate leaf moisture to the atmosphere), and thus is crucial to growth and leaf expansion of plants. When water is in balance (the supply is equal to the need), the optimum performance of all components results in steady active plant growth. However, when the balance of water is affected either because there is insufficient available moisture in the soil (or any growing media), or the transpiration of water through the stomata exceeds the plant's capacity to compensate for the internal loss, the plant comes under stress.

Vegetable crops have differing critical growth periods, and if water stress occurs during critical stages of growth, yield is directly affected. When moisture requirements are not met during this critical phase, permanent and irreparable damage usually is the result. The product quality is diminished or yield is reduced.

Table 1. Critical growth stages of some vegetable crops for moisture requirement

Crop	Stages of growth
Tomato	Flower development, fruit set and after each harvest
Brinjal	Flower development, fruit set and after each harvest
Chilli	Tenth leaf to flower upto fruiting and after periodical harvests
Potato	Stolon formation, tuberization and tuber enlargement
Pumpkin, Cucumber, Watermelon, Gourds, Muskmelon, Summer Squash, etc.	Flower bud development and early fruit development
Cabbage	Head formation and enlargement
Cauliflower	Throughout the whole vegetative period
Broccoli	Early stages of development
Onion	Bulb formation and enlargement
Garlic	Bulb formation
Radish, Carrot, Beet, Turnip	Constant water supply during the whole vegetative period
Pea	Beginning of flower bud development and pod filling
French bean, Hyacinth bean	Flowering and pod development
Cowpea	Prior to flowering and after pod set
Lettuce (head type)	Head formation
Celery	Throughout the whole vegetative period
Sweet potato	40-45 days after planting at tuber formation
Okra	After fruiti set
Leafy vegetables	During the whole vegetative period
Elephant-foot yam	Shoot emergence stage

Drought

Drought is generally considered to be a meteorological term and is defined as a period without significant rainfall or supply of moisture to the plants. Drought may lead to plant water stress which severely reduces growth. Periods of even short drought stress can reduce crop growth and yield. The plant may adjust to short-term water stress by closing stomata and thereby reducing water loss through the leaves. However, through manifestation of this mechanism, entry of carbon dioxide through the stomata is reduced or stopped leading to reduction of photosynthesis which slow down the growth. Plant growth and eventually yield is permanently hampered if such conditions are not corrected.

Waterlogging

Under waterlogged conditions, all pores in the soil or soilless mixture are filled with water. As a result, plant roots cannot obtain oxygen for respiration to maintain their activities for nutrient and water uptake. Waterlogging due to lack of oxygen in the soil causes death of root hairs, reduces absorption of nutrients and water, increases formation of compounds toxic to plant growth, and finally retards growth of the plant. Almost all the vegetable crops are very sensitive to waterlogging.

Wind

Light sway of wind is necessary to replenish carbon dioxide (CO_2) near the plant leaf surface. In the poly house CO_2 can be rapidly depleted at the leaf surface, unless there is some ventilation or air infiltration inside the structure. Air movement and distribution within the poly house can be enhanced with horizontal airflow fans. However, in side open poly house horizontal flow of air happens naturally.

Chapter 5

Influence of Plant Growth Regulators on Vegetable Production

The term 'Phyto-hormone' was coined by Thimann in 1948. Phyto-hormones are organic compounds, which are produced naturally in plants, synthesised in one part and usually translocated to the other part where in every small quantity affect the growth and other physiological function of the plants. Auxin was the first hormone to be discovered in plants. At one time, auxin was considered to be the only naturally occurring plant growth hormone however, four other classes of phyto-hormone *viz.*, gibberellins, cytokinin, ethylene and abscisic acid was later discovered. Plant growth regulators include both five specified classes of phyto-hormones (auxin, gibberellins, cytokinin, ethylene and abscisic acid) and other synthetic organic compounds (for example, polyethylene glycol, dinoseb-2-sec butyl 4,6dinitrophenol, proteins and/or amino acids, carboxylic, phenolic acid, humic acids, etc.) having properties of bio-stimulants or bio-inhibitors. Plant growth regulators (PGRs) modify plant physiological processes either by stimulating or inhibiting specific enzymes or enzyme systems to help regulate plant metabolism and thereby enhance yield and product quality. They normally are active at very low concentrations in plants.

Importance of PGRs was first recognized in the 1930s. Since then, wide array of natural and synthetic compounds that alter function, shape and size of crop plants have been discovered. Of the many uses of PGRs, effects on yield are often indirect. A wide range of beneficial effects of PGRs on crop growth and yields are: 1. promotion of germination and/or emergence, 2. stimulation of root growth, 3. promotion, mobilization and translocation of nutrients within plants, 4. modification of flowering, 5. increase in stress tolerance and improvement of water relations in plants, 6. promotion of early maturity, 7. increase in disease resistance, 8. retardation of senescence and 9. improvement of crop yields and/or quality.

Movement of phyto-hormones within plant body

Plants lack glands to produce and store hormones. Animals have two circulatory systems (lymphatic and cardiovascular) powered by a heart that moves fluids around the body. Plants use more passive means to move chemicals around the plant. Phyto-hormones are transported within the plant by utilizing four types of movements. For localized movement, cytoplasmic streaming within cells and slow diffusion of ions and molecules between cells are utilized. Vascular tissues are used to move hormones from one part of the plant to another which include sieve tubes or phloem that move sugars from the leaves to the roots and flowers, and xylem that moves water and mineral solutes from the roots to the foliage.

Classes of phyto-hormone

In general, it is accepted that there are five major classes of phyto-hormones, some of which are made up of many different chemicals that can vary in structure from one plant to the next. The five major classes of phyto-hormones are mentioned below:

1. Auxins

Auxins were the first class of growth regulators discovered. The major naturally occurring auxin is indole acetic acid (IAA). IAA has been implicated in virtually every aspect of plant growth and development, as well as defence responses. This diversity of function is reflected by the extraordinary complexity of IAA biosynthetic, transport and signalling pathways. IAA is synthesized from tryptophan *via* at least two pathways: the tryptamine (TAM) and indole-3-pyruvic acid (IPA) pathways. Tryptamine is a monoamine alkaloid. It contains an indole ring structure, and is structurally similar to the amino acid tryptophan, from which the name derives.

Different synthetic auxins are Indole-3-butyric acid (IBA), 2, 4 dichlorophenoxyacetic acid (2,4-D), 2-(2,4-Dichlorophenoxy) propionic acid (dichlorprop, 2,4-DP), (2,4-Dichlorophenoxy) butyric acid (2,4-DB), 2-Methyl-4-chlorophenoxyacetic acid (MCPA), 2-(2-Methyl - 4 - chlorophenoxy) propionic acids (mecoprop, MCPP), Naphthaleneacetic acid (NAA), 2,4,5-trichlorophenoxyacetic acid (2,4,5-T), 2,4,5-trichlorophenoxpropionic acid (2,4,5-TP), 4-chlorophenoxyacetic acid (4-CPA), β-naphthoxyacetic acid (BNOA), methyl ester of naphthalene acetic acid (MENA), etc.

Functions

Major function include apical dominance, root induction, control in fruit drops, regulation of flowering, parthenocarpy, phototropism, geotropism, herbicides, inhibition of abscission, sex determination, xylem differentiation, nucleic acid activity. Auxins are positively influence cell enlargement, bud formation and root initiation. They also promote the production of other hormones and in conjunction with cytokinins, they control the growth of stems, roots, and fruits and convert stems into flowers. Auxins in seeds regulate specific protein synthesis as they develop within the ovary after pollination, causing the ovary to develop a fruit to contain the developing seeds.

NAA and IBA are also commonly applied to stimulate root growth in cuttings of plants. NAA, 4-CPA, BNOA, 2, 4-D and 2, 4, 5-T are useful for better fruit set. Some synthetic auxins like 2,4-D, 2, 4, 5-T, 2, 4-DP, 2, 4-DB, MCPA and MCPP in higher concentration are toxic to the plants, most toxic to dicots and less so to monocots. 2, 4-D in higher concentration causes twisting, swelling and splitting of soft stems and abnormal growth of leaves often resulting in cupping and inter-veinal chlorosis which eventually lead to death of the plant. Because of this property, 2, 4-D in higher concentration are used as herbicide.

2. Cytokinins

The cytokinins are N6-substituted adenine-based molecules that were originally discovered on the basis of their ability to promote plant cell division which was later called zeatin. Different synthetic cytokinins are 6-benzylamino purine or benzyl adenine (BAP or BA), kinetin, 6-(benzyl-amino)-9-(2-tetrahydropyranyl)-9H-purine (PBA), 1, 3-diphenylurea, thidiazuron (TDZ), etc.

Functions

Cytokinin promotes cell division, cell enlargement and cell differentiation, stimulate bud initiation and root growth, translocation of nutrients, prolong storage life of flowers and vegetables, prevent chlorophyll degradation, morphogenesis, lateral bud development, etc. They have been implicated in a broad range of developmental processes including germination, root and shoot meristem function and leaf senescence. It has an important role in the formation of nitrogen-fixing nodules and other plant-microbe interaction. Cytokinins also help to delay senescence or the aging of tissues, are responsible for mediating auxin transport throughout the plant and affect internodal length and leaf growth. They have a highly synergistic effect in concert with auxins, and the ratios of these two groups of plant hormones affect most major growth periods during a plant's lifetime. Cytokinins counter the apical dominance induced by auxins; they in conjunction with ethylene promote abscission of leaves, flower parts and fruits.

3. Gibberellins

Gibberellins (GAs) are large family of about 136 closely-related tetracyclic, diterpenoid plant growth regulators that stimulate shoot elongation, seed germination, and fruit and flower maturation. Gibberellins are synthesized in the root and stem apical meristems, young leaves, and seed embryos. Gibberellins are synthesized from geranylgeranyl diphosphate in a multi-enzyme pathway that is subject to complex regulation. GA represses expression of several genes whose products are involved in its biosynthesis and promotes expression of genes involved in GA inactivation. GA levels are influenced by other hormones such as auxin and ethylene.

Functions

Like the other phyto-hormones, GAs plays a major role in diverse growth processes including seed development, organ elongation and the control of flowering time. GAs stimulate cell division and elongation, stimulate germination of seeds, stimulates bolting/flowering in response to long days, prevent genetic dwarfism, increase flower and fruit size, dormancy, induces maleness in the plants having dioecious sex form, extension of shelf life of products, etc. Gibberellins are important in seed germination, affecting enzyme production that mobilizes food production used for growth of new cells. They promote flowering, cellular division, and growth after germination. Gibberellins also reverse the inhibition of shoot growth and dormancy induced by ABA.

4. Ethylene

Ethylene is a gaseous plant hormone synthesized from methionine *via* the intermediate 1-aminocyclopropane-1-carboxylic acid (ACC) in a pathway called the Yang cycle. Ethylene is synthesized in essentially all plant organs, although the pathway is regulated by both environmental and developmental factors such as abiotic stress, onset of fruit ripening, etc.

Functions

Ethylene induce uniform ripening in vegetables, promotes abscission and senescence of leaf. It also affects cell growth and cell shape. Synthetic ethylene releasing compound (2-chloroethyl phosphonic acid (Ethrel or Ethephon) is used commercially to regulate fruit ripening and other physiological processes including alteration of sex ratio in monoecious cucurbits.

5. Growth inhibitors and retardants

Abscisic acid (ABA), the only plant hormone in this group is an isoprenoid compound normally produced in the leaves of plants, originating from chloroplast. Different synthetic growth retarding chemicals such as 2-isopropyl-4-dimethylamino-5-methylphenyl-1-piperidinecarboxylate methyl chloride (Amo-1618), 2-chloroethyl trimethyl ammonium chloride (CCC or Cycocel), 2, 4-dichlorobenzyphosphonium chloride (Phosphon D), N-dimethyl-aminosuccinamic acid (B-995 or B-Nine), succinic acid-2, 2-dimethyl hydrazide (SADH), 2, 3, 5-triiodobenzoic acid (TIBA) cause growth inhibition through blocking gibberellic acid biosynthesis and sterol biosynthesis by inhibiting the incorporation of mevalonic acid to sterol (particularly desmethyl sterol). Plant sterols are an essential component of the cell membranes of all eukaryotic organisms. They are either synthesised *de novo* or taken up from the environment. Their function appears to be to control membrane fluidity and permeability. Application of gibberellic acid and sterols completely overcome the effect of the growth retardants. Other growth inhibitors like, isopropyl N-(3-chlorophenyl) carbamate (CIPC) and maleic hydrazide (MH) cause growth retardation (inhibition of sprouting in potato, onion, etc.) by inhibiting mitosis without killing the cells for a period of time.

Functions

Absecisic acid act as plant stress hormone and associated with seed dormancy, drought responses and other growth process. It induces seeds to synthesize storage proteins, seed development and germination, stomata closing.

Application of plant growth regulators in vegetable production

Plant growth regulator affects the physiology of plant growth and influence the natural rhythm of a plant and sometimes it has been found beneficial for crop growth and yield. Some important growth and development that are influenced by plant growth regulators are discussed below.

Seed germination and seedling emergence

Several phyto-hormones have been shown to affect germination of seeds however, the primary event of breaking seed dormancy is stimulated by gibberellins. Seed soaking treatment with gibberellins (5-20 ppm) and different synthetic auxinic chemicals like, IBA (20 ppm) , NAA (20 ppm), BNOA (25-50 ppm), 2, 4- D (0.5 -2 ppm) depending on the crop and variety for 24 hours increase germination percentage, growth and enhance yield in many crops like, tomato, sweet pepper, radish, okra, etc. Seed soaking with succinic acid and fusiococcin (diterpenoid glycosides produced by the fungus *Fusicoccum amygdali*) also increase germination in many crops. Soaking initiates water imbibitions and thereby incorporate the chemical into the seed.

Breaking dormancy

Potato tubers fail to sprout before the termination of rest period. Gibberellins are useful in breaking the dormancy of seed tuber of potato. Soaking the seed tuber in aqueous solution of GA_3 25 ppm for 1-2 hours is the best in breaking the dormancy of seed tuber. It also hastens the sprouting when the dormancy period has nearly been finished. Lettuce is another vegetable in which treatment with GA has been reported to break seed dormancy induced by high temperatures.

Different plant growth regulator combination of GA, ethylene chlorhydrin and thiourea also break the dormancy effectively. Vapour treatment with ethylene chlorhydrin (1 liter per 20 q) followed by dipping in thiourea (1%) for 1hour and finally in GA (1 mg/l) for 2 seconds break dormancy of potato tubers.

Treatment of true potato seeds in aqueous solution of GA_3 1500 ppm for 24 hours before sowing is the best for breaking the dormancy of the true potato seeds.

Mobilization and translocation of nutrients

Plant hormones influence mobilization of inorganic plant nutrients and sugars. Gibberellic acid stimulates phosphate uptake in corn, and sulfate translocation from root to shoot in pea seedlings. Indole acetic acid both depresses and increases potassium uptake in corn seedlings, depending upon concentration.

Induction of flowering

Exogenous application of gibberellins induces flowering in many biennial vegetable crops which require vernalization for induction of flowering and some of the vernalization stimulus is related to natural gibberellin substances in the plants. Treatment of GA_3 would be effective only when the plants have passed the juvenile phase and it would flower normally after vernalization exposure. It has been found that vernalization effect could be obtained with gibberellin application at 100-150 ppm concentration in some crops, like cabbage, European carrot, etc. however, the situation is not one to one in all the cases.

In many short-day and day neutral plants, application of GA_3 usually delays flowering or even stops it. Many long day crops like, lettuce, endive, radish and mustards are induced to flowering under non-inductive conditions by the application of GA_3 at 1000 ppm concentration. MH delays flowering in okra.

Role of auxin in induction of flowering is not clear because it inhibit flowering in some and stimulate in other vegetable crops. Growth inhibitors and retardants at low concentration enhance flowering by retarding vegetative growth however, at higher concentration they generally inhibit flowering.

Alteration of sex ratio in cucurbits

Exogenous application of auxins, gibberellins, growth retardants and ethylene is known to shift the balance of sex expression in cucurbits. Application of these plant growth regulators at 2-4 true leaf stage can alter the sequence of flowering and narrows the ratio of staminate: pistilate / perfect flowers in desired direction in monoecious / andro-monoecious cucurbits. The plant growth regulating chemicals should be applied twice or thrice at an interval of 7 days to get effective result. Effective concentrations of the plant growth regulators for different cucurbits are given below.

- Muskmelon: MH 200 ppm
- Watermelon: TIBA50 ppm, MH 50 ppm
- Pumpkin: Ethrel 250 ppm
- Summer squash: Ethrel 600 ppm
- Ridge gourd: Ethrel 300 ppm
- Bottle gourd: MH 50 ppm, Ethrel 150 ppm, 2-CPA 100-200 ppm
- Bitter gourd: Ethrel 150 ppm, BA 25 ppm, NAA 100 ppm
- Cucumber: Ethrel 100-200 ppm, MH 100 ppm
- Snake gourd: Ethrel 100 ppm

In cucumber, gynoecious lines are used in hybrid seed production. Such gynoecious lines are maintained by spraying GA_3 at 1500 to 2000 ppm, silver nitrate at 300-400 ppm or silver thiosulphate at 300-400 ppm concentration at 2-4 leaf stage of the plant which induces staminate flowers for enabling selfing and subsequent maintenance of the gynoecious line in true to the type condition.

Stimulation of fruit set and development

Pollen and developing seeds contain the plant hormones like, gibberellins, cytokinin and auxin which are involved in seed and fruit growth and development. Pollination and growth of pollen tubes through the styles also cause production of phyto-hormones which stimulate fruit growth. Fertilization and seed production enhance this process and help in the continued growth of the fruits. Auxins, gibberellins and cytokinins generally enhance fruit set and development. Growth retardants and ethylene normally do not help in fruit setting process. As the fruit develop, ethylene increases and induced ethylene production activates auxin production in synergistic manner. Exogenous application of many synthetic auxins like, 4-CPA, BNOA, NAA, 2,4-D and 2,4,5-TP has been found useful in enhancing fruit set in many crops. Poor fruit set is a major problem in many vegetable crops. Effective concentrations of different growth regulators for enhancing fruit set in different vegetable crops are given below :

- Tomato: NAA 20 ppm, PCPA 50-100ppm, BNOA 40 ppm, 4-CPA 25-50 ppm, TIBA 25-100 ppm, 2,4-D 2-5 ppm, GA_3 25 ppm
- Brinjal: 2,4-D 2 ppm
- Cucumber: 4-CPA 20-50 ppm
- French bean: 2,4,5-T 100 ppm

Seed development in carrot, garden beet and cabbage has been found to be better with the application of 100- 1000 ppm GA_3. In onion also, spraying the seed crop with 50 ppm GA_3 or 20 ppm MH significantly increase seed yield, germination and seed vigour.

Fruit ripening

Ethephon, an ethylene releasing compound induces ripening in tomato and sweet pepper. Application of ethephon at 1000 ppm at turning stage of earliest fruits induce early ripening of fruits thus increasing the early fruit yield by 30-35%. Postharvest dip treatment with ethephon at 500-2000 ppm has also been reported to induce ripening in mature green tomatoes.

Induction of parthenocarpy and fruit development

Kinetin, GA_3 and auxins induce parthenocarpy resulting development of fruits without seeds. Kinetin helps in cell division and helps in early fruit development. Gibberellins and auxins promote fruit growth by cell expansion, probably acting in conjugation with one another. Auxins also inhibit abscission and help in retaining the fruits for longer period of time in the plants. PCPA 50-100 ppm induced parthenocarpy in tomato and brinjal, application of 2,4-D at 0.25% in lanolin paste to cut end of styles or foliar sprays to freshly opened flower cluster has been reported to induced parthenocarpy. NAA 200 ppm concentration is effective in producing parthenocarpic fruits in cucumber.

Inhibition of sprouting

Storage loss of onion and potato due to sprouting can be prevented by exogenous application different growth retarding chemicals. Tubers treated with 1% aqueous solution of isopropyl N-(3-chlorophenyl) carbamate (CIPC) is the best for inhibition of sprouting in potato. Application of NAA at 20 ppm as dip at 10°C for 1-3 days and MENA were also found to be useful in inhibition of sprouting in potato.

Most of these compounds excepting MH are not effective in inhibition of sprouting in onion because the growing points remain in protected condition inside the bulb. Exogenous application of MH at 2500 to 3000 ppm concentration 15 days before the harvest is the best for onion. However, application of both pre and post harvest application is banned due to carcinogenic effect of MH.

Extension of shelf-life of vegetables

Deteriorative physio-biochemical processes that go on inside the harvested vegetables are mediated by indigeneous growth hormones, principally auxin, gibberellins and cytokinins. Exogenous application of growth substances can alter different metabolic process, and some of these alterations are beneficial to extending the shelf-life of vegetables. Yellowing due to degradation of chlorophyll is one of the main factors that limit the shelf-life of cabbage, broccoli and Brussels sprout. Senescence which causes yellowing can effectively be retarded by pre or post-harvest application of BA (N6-benzyladenine), a synthetic cytokin compound at lower concentrations (5-15 pm). Cytokinin controls development of chloroplast through stimulating the conversion of proplastids into chloroplasts with grana. Auxinic growth substances like 2, 4-D, 2, 4, 5-T have profound effect on delaying leaf abscission of cauliflower when applied before or after harvest. Ripening of tomato can be delayed and degradation of chlorophyll in leafy greens can be checked by post-harvest application of gibberellins. Growth inhibitors like, CIPC (chloro-iso-propyl carbamate) and MENA (methyl ester of naphthalene acetic acid) are effective in controlling sprouting of tuber and bulb crops. Potatoes treated with 1% aqueous emulsion of CIPC and kept in cold storage do not have any sprouting for 5 months. Synthesis of solanine in potato is inhibited by Alar (N, N-dimethyl aminosuccinamic acid). Cycocel (2-chloroethyl-trimethyl ammonium chloride), another growth retardant can prolong the shelf-life of lettuce, asparagus, broccoli, etc. However, effective concentration and right stage of application is very important for obtaining maximum benefit.

Maleic hydrazide at 1000 ppm when applied 3 times at weekly interval reduces bulb sprouting of onion in storage however, application of MH is banned now due to carcinogenic effect. In broccoli, post-harvest dip in 20 ppm BA for 15 seconds and held at 25°C under polyethylene cover prolongs the storage life by retaining green colour. In Brussels sprouts, spraying with GA 100 ppm extends the shelf-life to 9-10 days at 20°C. In head lettuce, application of 8 ppm kinetin for 30 seconds and packed in perforated polyethylene bags and stored at 6°C and 90% relative humidity retard senescence. In tomato, application of cycocel 100 ppm decreases fruit decay and weight loss. Post-harvest treatment with BA at 5-20 ppm increases the shelf life of radish. In okra, fruit dipping in the solution of 100 pm GA for 10 minutes and packed in polyethylene bags extend the shelf-life up to 9 days. In pointed gourd, fruit dipping in 100 ppm GA solution extends the shelf-life up to 7 days in ambient condition. Application of BA at 50 ppm concentration is effective in reducing decay loss of ridge gourd stored at 0.5°C temperature.

Gametocides to induce temporary male sterility

Gametocides are chemicals, which affect the male parts of the bisexual flower and thereby induce pollen abortion without hampering the female fertility. Different growth regulators show gametocidal properties. Exogenous application of TIBA 50-100 ppm, 2,4-D 10-50 ppm, NAA 50 ppm, MH 500 ppm cause temporary male sterility in tomato. Application of MH 10 ppm has been found effective in onion. Application of MH 600 ppm, 2,4-D 10 ppm and NAA 50 ppm is effective as gametocide in brinjal. GA at 100 ppm can also be used for inducing male sterility in sweet pepper. Application of 2,3- dichloro-isobutyrate (0.2 to 0.8%), MH at 100 to 500 ppm has been found in causing temporary male sterility in okra.

Herbicide

The first widely used herbicide was 2, 4-dichlorophenoxyacetic acid (2, 4-D) which was developed by a British team during World War II with first widespread production and use in the late 1940s. It still remains one of the most commonly used herbicides in the world. Other auxin growth regulator herbicides which kill broad leaved weeds are 2-(2,4-Dichlorophenoxy) propionic acid (dichlorprop, 2,4-DP), (2,4 Dichlorophenoxy) butyric acid (2,4-DB), 2-Methyl-4-chlorophenoxyacetic acid (MCPA) and 2-(2-Methyl-4-chlorophenoxy)propionic acids (mecoprop, MCPP). Isopropyl-N-phenyl carbamate (IPC) shows toxicity to narrow-leaved weeds under the group monocotyledons. The effects associated with auxin herbicides include bending and twisting of leaves and stems which is evident almost immediately after application.

Chapter 6

Seedling Management for Vegetable Crops

It is said that "Success of any production system depends on the kind of seed which are sowing". It is true for the seedlings also beacause healthy seedlings grown in a well managed nursery will decide the yield and consequently the profit.

Definition of Vegetable Nursery

A vegetable nursery is a place or an establishment for raising or handling of young vegetable seedlings until they are ready for more permanent planting.

Why vegetable nursery is needed

Some vegetable crops require special cares during their early growth period. There are some vegetables with very small sized seeds. Seedlings of many seed propagated vegetable crops like, tomato, brinjal, chillies, sweet pepper, cauliflower, cabbage, knolkhol, broccoli, Brussels sprout, kale, endive, chicory, celery, lettuce, parsley, onion, etc. are first raised in the nursery beds before transplanted in the main field. Many seed propagated vegetable crops *viz*., okra, peas, beans, radish, carrot, beet, turnip and all the cucurbits (e.g. bitter gourd, bottle gourd, pumpkin, cucumber, watermelon, muskmelon, snake gourd, sponge gourd, ridge gourd, etc.) are conventionally grown through direct seeding in the main field. However, many cucurbits, particularly in the early season are also grown in India using the seedlings raised in polythene packets.

Classification of vegetable crops in different transplant groups

- ❒ **Plants easy to transplant:** They form new roots rapidly, put forth shoots rather slowly and absorb water readily e.g. beet, broccoli, Brussels sprouts, cabbage, cauliflower, Swiss chard, lettuce and tomato.
- ❒ **Plants transplanted with moderate ease:** This group replace roots and grow shoots a bit slower than the previous group and do not absorb water very efficiently e.g. celery, brinjal, onion, pepper, leek and cauliflower.
- ❒ **Plants transplanted with difficulty, requiring special care:** They are very poor or slow to both root replacement and enhanced shoot growth.

They take long time to cope up root injury and do not survive very ordinary transplanting. They require special care in handling e.g. cucumber, bitter gourd, pumpkin, squash, melons, sweet corn and legumes vegetable crops.

Advantages of nursery management

- It is possible to provide favourable growth conditions i.e. germination as well as growth.
- Better care of younger plants as it is easy to look after nursery in small area against pathogenic infection, pests and weeds.
- Uniform, vigorous and healthy planting materials can be raised for better growth, development and yield.
- Crop grown by nursery raising is quite early and fetch higher price in the market, so economically more profitable.
- There is saving of land and labour as main fields will be occupied by the crops after one month so more intensive crop rotations can be followed.
- More time is available for the preparation of main field because nursery is grown separately.
- As vegetable seeds are very expensive particularly hybrids, seed requirement can be economized by sowing them in the nursery.

Different features of a vegetable nursery

Nursery site

- Area selected should be well drained, and free from water logging.
- The site should receive proper sunlight.
- The nursery should be near the water supply so that irrigation can be easy.
- The area should be well protected from pet and wild animals.

Nursery soil

- Raising of vegetable seedlings requires fertile and healthy soil.
- Preferably, the soil for nursery should be loam to sandy loam, loose and friable, rich in organic matter and well drained.
- The soil p^H should be close to the neutral i.e. about 7.0

Soil preparation

- It needs a deep cultivation of the nursery land either by soil turning plough or by spade and subsequent 2-3 hoeing with cultivator.
- After that, all the clots, stones and weeds from the field should be removed and land should be leveled.
- Two kg well rotten and fine farm yard manure or compost or leaf compost or 500 g vermi compost per square meter need to be mixed in the soil.
- Well rotten and powdered FYM or compost is also used as organic mulch to improve the water holding capacity of soil.
- If the soil is heavy, 2-3 kg sand per square meter need to be mixed so that the seed emergence may not be hampered.

Soil treatment for getting healthy seedlings

For raising the healthy seedlings, soil must be treated to reduce the load of soil-borne pathogen and pest from the root zone of the seedling as far as possible. Different methods adopted for this purpose are mentioned below:

A. Soil solarization

- Suitable time period for soil solarisation is May-June as air temperature rises up to 45°C during this time.
- Finely prepared soil in the seed bed is first saturated with water
- Transparent polythene sheet of 200 gauges is spread on the whole nursery area for about 5-6 weeks.
- The margin of the polythene should be covered by wet soil (compressed mud) to check the entry of air.
- When air temperature raise upto 40-48°C and soil temperature inside the polyethylene reaches upto 48-54°C which is sufficient to kill most of the harmful pathogens, nematodes and weed seeds.
- After 5-6 weeks, polythene sheet is removed from the nursery bed
- Beds are prepared for seed sowing.

B. Formalin solution treatment

- This treatment should be done 15-20 days before seed sowing.
- Formalin solution is prepared by adding 250 ml commercial grade formaldehyde (40%) per 10 litre of water and soil is drenched @ 4.5 litres per square meter so as to saturate top soil surface upto a depth of 15-20 cm.
- The drenched area is covered with 200 gauge polyethylene sheet so that fumes of formalin penetrate inside the soil to kill the pathogens.
- Wet soil is put on the margin of the covered polythene sheet so that it does not allow the polythene film blown away by the wind from the covered area to out side.
- The polyethylene cover is removed after 24 hours and the bed is kept open for two to three weeks prior to seed sowing.
- Beds are prepared for seed sowing.

C. Application of fungicides

- Generally used fungicides for seed bed treatment are Captan and Thiram which kill the soil borne pathogens.
- 5-6 g of any of the fungicides may be used per square meter nursery area as dry treatment.
- Nursery soil may be drenched with or 0.25-0.30% solution of any of the fungicides @ 4-5 litres per square meter.

D. Control of insects

- Presence of certain insect pest and their egg or secondary stage insects present in the soil can infect the seedlings in the later stage.
- To save the seedlings against them, some insecticides are also used as soil treatment.
- Granular insecticides like, Furadon, Phorate 10 G, etc. are applied to the soil @ 4-5 gram per square meter and then thoroughly mixed to a depth of 15 to 20 cm to kill insects, their eggs and nematodes.
- Nursery soil may be drenched with Chlorpyriphos @ 2 ml/ liter of water upto the depth of 15 to 20 cm.
- After 3-4 days of treatment, beds are prepared for seed sowing.

E. Steam treatment

- Hot steam can be used to treat the soil against harmful insect pest.
- Nursery bed is covered with 200 gauge polyethylene sheet to restrict the movement of air in the covered area.
- Hot steam is supplied for at least 4-6 hours continuously.
- Temperature is lethal for most of the harmful pathogens, insect pest, nematodes and weed seeds.

F. Soil treatment with bio-agents

- Sufficient organic matter should be there in the seed beds for rapid multiplication of the bio-agents.
- Different species of *Trichoderma* spp., *Pseudomonas fluorescens* and *Aspergillus niger* are effective biological agents to control majority of the soil borne pathogens.
- These bio-agents are added @ 10-25 gram per square meter and well mixed in the soil.
- Seeds should be sown 2 days after the application of bio-agents.
- 'Success of biological control depends on the activity of the spores, spore counts in the formulation and sufficient organic matter in the beds for rapid multiplication of the bio-agents.

Preparation of nursery beds

- Raised nursery bed should be prepared and soil must be fine and moist enough to ensure adequate seed germination.
- If the seedlings are to be raised in boxes during unfavourable weather condition, the flat earthen pots, polythene bags, potting plugs, wooden treys, etc. may be used. The soil mixture should be prepared in the ratio of 1:1:1 of soil, sand and well rotten FYM or cow dung manure or leaf mould. This growing medium after proper sterilization by formalin treatment or steam exposure for about 1 hour under autoclave is filled in these seedlings raising structure. Arrangement should be made to drain excess water from these structures by making a hole in the bottom of all types of pots.

Raised nursery beds

- Length of the bed may be kept 3 to 5 meter; however, width is restricted to 1 meter only which facilitates intercultural operations.

- The beds are raised 15 to 20 cm high from the ground level and two beds are spaced by 40-50 cm.
- The space between two beds helps in weeding, nursery care against diseases and insect pest and also for draining out the excess rain water from the nursery beds.
- The number of beds depends on the particular crop, season and growing area of crop.
- The beds should be prepared in the east and west direction and line should be made from north to south direction on the beds.

Seed treatment

- Most of the vegetable seeds are treated with Captan or Thiram @ 2.5-3 g/kg seed to disinfect the seed surface.
- Hot water treatment of seeds of cabbage and cauliflower at 50°C for 30 minutes just before sowing prevents seed borne black rot disease caused by the bacteria *Xanthomonas campestris* pv. *campestris*.
- Biological control of the soil borne damping off diseases caused by *Pythium* sp., *Rhizoctonia* sp., *Fusarium* sp., etc. can be done through seed treatment with different antagonistic fungi like, *Trichoderma* spp., *Gliocladum virens*, *Penicillium* spp., and bacteria like, *Bacillus subtilis* and *Pseudomonas fluoroscens*.
- The use of coated or pelleted seeds has now become quite common in vegetable crops especially those having small seeds like, tomato, brinjal, chilli, sweet pepper, lettuce, etc.
- The pelleted seeds are often coated with pesticides for better protection against pest and disease causing pathogens.
- Germination may be enhanced through osmo-conditioning or priming. In this process, seeds are allowed to imbibe water in a temperature controlled dilute aerated solution of an organic or inorganic osmoticum to ensure faster and uniform germination.

Sowing of seeds in the nursery

The treated seeds are sown in the nursery bed either by broadcasting or in lines depending upon the nature and season of crop.

Broadcasting

In broadcasting method, seeds are broadcasted on the well prepared nursery beds and later on the seeds are covered with well rotten, fine sieved and treated FYM or compost. Major disadvantages of this method are:

- Uneven distribution of seeds in the nursery beds.
- Growth and development of seedlings is poor.
- Some times nursery becomes very much dense which create more possibility of damping off disease occurrence.

Line Sowing

- Line sowing is the best method of seed sowing in nursery.
- Lines are made 0.5 to 1.0 cm deep parallel to the width at 5.0 cm spacing and seeds are sown or placed singly at a distance of about 1.0 cm apart in the line.
- Seed cover is prepared with fine mixture of sand, soil and well rotten and sieved FYM or leaf compost (1:1:1) and then the mixture is sterilized by formalin treatment or steam exposure for about 1 hour under autoclave.
- If the seed cover is not sterilized, 3-4 g thiram or captan may be mixed per kg mixture.

Use of mulch

To maintain the soil moisture for seed germination, seed bed should be covered with a thin layer of mulch of paddy straw or sugar cane trash or any organic mulch during hot weather and by plastic mulch (plastic sheet) in cool weather. It has following advantages:

- It maintains the soil moisture and temperature for better seed germination.
- It suppresses the weeds.
- Protects from direct sunlight and raindrops.
- Protects against bird damage.

Removal of mulch

Due attention is given to remove the covered mulch from the seedbed. The seed bed should be observed daily three days after sowing the seeds. As and when the white thread like structure is seen above the ground, the mulch should be removed carefully to avoid any damage to emerging plumules. Always mulch should be removed in the evening hours to avoid harmful effect of bright sun on newly emerging seedlings.

Use of shedding net

During the seedling growth, if temperature reaches above 30°C then the beds should be covered by green or black coloured 50% or 60% shade nets about 1.0 meter above the ground with suitable support to protect the seedlings from direct scorching sunlight. The newly introduced translucent blue shade net gives blue light, which is excellent for plant growth.

Watering

- The nursery beds require light irrigation with the help of rose can from the seed sowing until the seedlings become ready to transplant.
- Young roots of the seedlings absorb water primarily by osmotic forces and anaerobic condition caused by excessive soil moisture retard seed germination.
- Excess rainwater or irrigated water should be drained out from the field as and when it is required otherwise plants may die due to excess of water.

Thinning

- It is an important operation to remove weak, unhealthy, diseased, insect pests damaged and dense plants from the nursery beds keeping distance of about 0.5 to 1.0 cm from plant to plant.
- The thinning facilitates balance light and air to each and every plant and it also helps in watching the diseased and insect pest attacked plants while moving around the nursery.

Weed control

- Timely weeding in nursery is very important to get healthy seedling.
- Weeding should be done manually either by hand or by hand hoe.

Pricking out of seedlings

- It is a method of producing healthy and hardened seedlings by pricking out the young seedlings when first true leaves have come out and transplanting them to another seed bed or seed box or grown individually in small plastic thumb pots of about 5 cm diameter.
- The seedlings should be uprooted carefully with the help of V-notched wooden stick for second transplanting and spaced about 5 cm within a row which are 7-8 cm apart, both in second seed beds or seed boxes.

- The soil mixture generally used to grow the pricked out seedlings is slightly richer than the previous one for first seed bed.
- Seedlings become ready for planting in the main field generally three weeks after first transplanting.

Plant protection

- Adoption of plant protection measures in the nursery against the incidence of insect pest and diseases is very important task to get the healthy seedlings. The pathogens cansing damping off leaf curl, leaf blight diseases and leaf miner and borer insect pests attack the seedling in the nursery.
- Different vegetable crops like, tomato, chilli, sweet pepper, etc are very susceptible to virus particularly leaf curl virus which are usually deposited inside the plant at the seedling stage by the insect vector primarily white fly (*Bemisia tabaci*).
- Seedling of these vegetable crops should be raised inside the fine(50 mesh) mosquito net coupled with spraying of insecticides outside the mosquito net to keep away the infestation of the insect vectors to produce virus free seedlings.
- It is better to spray the combination of fungicides (0.2% chlorothalonil or 0.25% mancozeb + 0.1% carbendazim), insecticides (metasystox 1.5 ml/ litre or imidachloprid 0.2 ml/litre) and sticker (triton or sandovit @ 1 ml/ litre of solution) to control the fungal pathogen and sucking insect pests like, white fly, jassids, mite, thrips, etc.
- Last spray of insecticides should be given 2 days before transplanting.
- Red pumpkin beetle which is the most harmful insect pest of seedling of cucurbits can effectively be controlled by spraying carbaryl 50 WP @ 2 g/ litre at 2-3 leaf stage.

Hardening of the seedlings in the nursery

- The term hardening includes "Any treatment that makes the tissues firm to withstand unfavourable environment like low temperature, high temperature and hot dry wind."
- In this process seedlings are given some artificial shocks at least 7-10 days before uprooting and transplanting.

Effect of hardening

- Hardening is physiological process in which plants accumulate more carbohydrates reserves and produce additional quiticle on the leaves.
- Hardening improves the quality and modifies the nature of colloids in the plant cell enabling them to resist the loss of water.

- Hardening increases the dry matter and decrease the percentage of freezable water and transpiration per unit area of leaf.
- It decreases the rate of growth.
- Hardened plants can withstand better against unfavourable weather conditions like hot day winds or low temperature.
- Hardening of the plants increases the waxy covering on the leaves of cabbage.

Techniques of hardening

- Lowering the available soil moisture to 20% by withholding watering to the plant 5-7 days before transplanting.
- Lowering the temperature which retards the growth.
- By application of 4000 ppm NaCl with irrigation water.
- By spraying of 2000 ppm of cycocel + 0.25% $ZnSO_4$
- By spraying of 0.2% muriate of potash.

Duration and degrees of hardening

- The plants should be hardened according to their kind so that there is an assurance of high percentage of survival and slow growth under the condition to be expected at the time of transplanting.
- Hardening should be gradual to prevent or check the growth.
- Warm season crops like tomato, brinjal and chillies do not favour severe hardening.
- In Indian condition lowering the available soil moisture to 20% by withholding watering to the plant 5-7 days before transplanting is the best.

Selection of seedlings for transplanting

After attaining proper growth, seedlings are transplanted in main field and at the time of transplanting, seedling should be:

- Stocky and sturdy.
- Should have good root system.
- Should be free from any insect pests and diseases.

Production of grafted seedlings

Grafting is a simple method of propagation in which desired rootstocks are utilized to induce vigour, precocity, enhanced yield and quality and better survival under biotic and abiotic stress conditions. Grafts of cucumber, watermelon, tomato, eggplant and pepper has gained much popularity in different countries like, Japan, Korea, Philippines, Greece, United States, etc.

Benefits of grafted seedlings in vegetable crops

Basic objective of using grafted seedlings is to reduce the dependency upon chemicals required to treat the soil to avoid the soil borne diseases which has opened up new vista in organic farming of vegetables. Development of mechanized and robotic grafting technology has given a fillip to this novel eco-friendly approach. Different benefits imparted by the grafted seedlings are mentioned below.

Avoidance of disease and pest incidence : The main objective of graftingis is to avoid soil borne diseases such as *Fusarium* wilt in different cucurbits particularly cucumber, muskmelon, watermelon and bitter gourd and bacterial wilt in tomato and brinjal through the use of resistant rootstocks. Use of resistant rootstocks in grafting in combination with integrated pest management practice is the most viable option to reduce the need for soil fumigation with methyl bromide to kill these soil borne pathogens which also prove to be a boon in organic farming of vegetables.

Avoidance of nematode infestation : Another important objective of grafting is to avoid the infestation of plant parasitic nematodes particularly root knot nematodes (*Meloidogyne* spp.) which harbour different vegetable crops with the use of nematode resistant rootstock through interspecific or intergeneric grafting. Some of such nematode resistant rootstocks are *Sicyos angulatus* and pumpkin for cucumber, Nematex variety of tomato and *Solanum nigrum* for tomato, etc.

Minimizing the autotoxic effect : Recently autotoxic potential of some cucurbits has been pointed out which may pose serious problem of soil sickness in commercial production of cucurbits although there are differences in autotoxic potential among the cucurbits. Watermelon, muskmelon and cucumber exhibits more auto-toxic potential than the others. The autotoxins affect ion uptake, membrane permeability, photosynthesis and phyto-hormone balance. It seems possible to overcome autotoxicity in different cucurbit crops through grafting them on *Cucurbita ficifolia*.

Providing cold hardiness : Grafting is also useful to initiate the flowering and fruit set at low temperature which is useful in saving electricity energy to maintain requisite day/night temperature regime in the polyhouse during winter particularly in the temperate countries. Early yield of cucumber could be obtained by grafting it on *Cucurbita ficifolia* because the grafted plants can survive at very low temperature of 20°C/12°C day/night.

Survival of grafted plants under excessive moisture condition : Flooding or excessive soil moisture condition generally reduces photosynthetic rate, stomatal conductance, transpiration, soluble protein and activity of ribulose-1, 5-bisphosphate carboxylase/oxygenase (Rubisco). Intergeneric grafting imparts the attributes of flood tolerance particularly in different cucurbits by upkeeping these physiological manifestations through sustained enzyme activity like Rubisco activity *via* Rubisco activase enzyme.

Improving quality traits : Improved quality in the grafted seedlings has not yet been achieved prominently. However, more number of commercially acceptable and shiny fruits of cucumber could be obtained by grafting it on some hybrids of summer squash. Enhanced sugar content in Galia and Haon varieties of melon has been found by grafting them on the melon cultivar Suiker. In tomato, grafted seedlings generally produce increased number of marketable fruits with better colour and enhanced lycopene content and decreased number of malformed, under developed and gray mold infected fruits.

Manipulating the harvesting period : It has been found that cucumber grafted on to *Cucurbita ficifolia* generally grows faster and gives high percentage of early yield.

Stionic effect on mineral nutrition : Grafting also influences growth and mineral content of the plant tissues.

Influence on sex expression: Sex expression and flowering pattern of the scion of different cucurbits changes with grafting because root may control floral formation by production of different flower inducing factors.

Table 1. Prime objectives of grafting in different vegetable crops

Vegetables	Objectives of grafting
Watermelon	Tolerance to Fusarium wilt (*F. oxysporum*), low temperatures, wilting due to physiological disorders, drought tolerance.
Cucumber	Tolerance to *Fusarium* wilt, *Phytophthora melonis,* low temperatures.
Melon	Tolerance to *Fusarium* wilt (*F. oxysporum),* low temperatures, wilting due to physiological disorders, *Phytophthora* disease.
Tomato	Tolerance to bacterial wilt *(Ralstonia solanacearum), Fusarium oxysporum,* nematodes *(Meloidogyne* spp.), *Verticillium dahliae.*
Brinjal	Tolerance to bacterial wilt *(Ralstonia solanacearum) Verticillium alboatrum,Fusarium oxysporum,* low temperatures, nematodes and induction of greater vigour.
Bitter gourd	Tolerance to *Fusarium* wilt (*Fusarium oxysporum* f.sp. *momordicae.*

Table 2. Promising rootstocks for grafting different vegetable crops

Scion	Rootstocks
Cucumber	*Cucurbita moschata, Cucurbita ficifolia, Cucurbita maxima, Sicyos angulatus, Lagenaria siceraria*
Muskmelon (for open field)	*Cucurbita* sp., *C. moschata x C. maxima, Cucumis melo*
Muskmelon (for poly house)	*Cucumis melo, Benincasa hispida,Cucurbita* spp. *C. moschata x C.maxima*
Watermelon	*Citrulus lanatus, Cucurbita maxima, C. moschata C. moschata x C. maxima, Lagenaria siceraria*
Bitter gourd	*Cucurbita moschata, Lagenaria siceraria,Luffa aegyptica*
Tomato	*Solanum pimpinellifolium Solanum lycopersicum, Solanum nigrum*
Brinjal	*Solanum torvum, Solanum integrifolium, Solanum melongena, Solanum nigrum*

Infrastructure facilities for grafting in a commercial scale

Grafting should be carried out in a shady place sheltered from the wind to avoid wilting of the grafted plants. At the time of grafting, it is important to increase the chances for vascular bundles of the scion and rootstock to come in close contact by maximizing the area of the cut surface and taking care that the cut surface does not dry out. After grafting it is necessary to keep the grafted plants at about 30°C and with more than 95% humidity for at least three days. Gradually, the relative humidity should be lowered with the increase in light intensity. Successful commercial grafting programme requires two types of facility *viz* screen house and grafting chamber.

***Screen house*:** A screen house is used for growing seedlings prior to grafting and for hardening of grafted plants under controlled light condition prior to disposing the seedlings or transplanting in the main field and to protect the seedlings from rain. Screen house is also used to check the infestation of virus-transmitting insect vectors such as aphids and whiteflies.

***Grafting chamber*:** Grafting chamber is basically a poly house with mist generation facility through micro-sprinkler where the grafted seedlings are kept for about one week under high humidity and reduced light intensity condition. Immediately after grafting, high humidity is required to minimize drying. Shade is required to minimize shock and induce fast healing of the wounds in both stock and scion.

Methods of grafting

Several methods of grafting are practiced, each involving the joining of the leaf bearing part (scion) of one plant with the rootstocks of another. Different grafting methods include cleft grafting, tube grafting, whip and tongue grafting, splice grafting, flat grafting, bud grafting, hole insertion grafting, tongue approach grafting etc and some of the widely used methods of grafting are briefly described.

Cleft grafting: Tomato plants are mainly grafted by this method in which seeds of the rootstocks are sown 5-7 days earlier than those of the scion. The stem of the scion (at four leaf stage) and the rootstock (at the 4-5 leaf stage) are cut at right angles, each with 2-3 leaves remaining on the stem. The stem of the scion is cut in a wedge and the tapered end fitted into a cleft cut in the end of the rootstock. The graft is held firm with a plastic clip.

Tube grafting: This method of grafting requiring tubes with a smaller inside diameter makes it possible to graft small plants grown in plug trays two or three times faster than the conventional method and is quite popular among Japanese seedling producers. The optimum growth stage for grafting varies according to the kind of plug used. First, the rootstock is cut by a slant cut and then scion is also cut in the same way. Plastic tubes with side slit are placed onto the cut end of the rootstock. The cut ends of the scions are inserted into the tube, splicing the cut surfaces of the scions and rootstocks together. While practicing the tube grafting in eggplant, the seeds of *Solanum torvum* must be sown a few days earlier than those of the other rootstock species.

Tongue approach grafting: Melons and other cucurbitaceous vegetable crops are generally grafted by this method. It gives higher survival ratio because the root of the scion remains until the formation of the graft union. In this method, seeds of cucumber are sown 10-13 days before grafting and pumpkin seeds 7-10 days before grafting to ensure uniformity in the diameter of the hypocotyl of the scion and rootstock. The shoot apex of the rootstock is removed so that the shoot cannot grow. The hypocotyl of the scion and rootstock are cut in such a way that they tongue into each other, and the graft is secured with a plastic clip. The hypocotyl of the scion is left to heal for 3-4 days and then crushed between the fingers and later the hypocotyl is cut off with a razor blade three or four days after being crushed.

Mechanized grafting

Grafting is arduous task and efforts are being made to reduce the labour required. Attempts have been made to mechanize grafting since 1987. There are several basic factors which govern the success of grafting by machine or robot such as seedling shape, location of cut, seedling gripping, cutting method, fixing materials and tools, etc. Grafting robots for plug have been developed by combining the adhesive and grafting plates. This robot makes it possible for eight plugs of tomato, eggplant or pepper to be grafted simultaneously. Recently a fully automatic grafting system for cucurbitaceous vegetables has been designed through which seedling quality is estimated before being used for grafting.

Chapter 7

Plant Nutrient Management for Vegetable Crops

Plant nutrition is the study of the chemical elements and compounds necessary for plant growth, plant metabolism and their external supply. In 1972, Emanuel Epstein defined two criteria for an element to be essential for plant growth:

- In its absence the plant is unable to complete a normal life cycle.
- The element is part of some essential plant constituent or metabolite.

This is in accordance with Justus von Liebig's law of the minimum. The essential plant nutrients include carbon, oxygen and hydrogen which are absorbed from the air, whereas other nutrients including nitrogen are typically obtained from the soil.

Essential plant nutrients

At least 17 elements are known to be essential nutrients for plants. In relatively large amounts, the soil supplies nitrogen, phosphorus, potassium, calcium, magnesium, and sulfur; these are often called the macronutrients. In relatively small amounts, the soil supplies iron, manganese, boron, molybdenum, copper, zinc, chlorine, and cobalt, the so-called micronutrients. Sodium (Na) and Silicon (Si) have been found to be essential in some higher plants and hence, cannot be regarded as essential for all the higher plants. Nutrients must be available not only in sufficient amounts but also in appropriate ratios. Growth and yield get restricted, if any of these elements falls short. Balanced nutrition demands supply of all the essential elements. Apart from essential elements, there are some, like Aluminium (A1), Nickel (Ni), Selenium (Se) and Fluorine (Fl) which are toxic to the plants.

Classification of plant nutrients based on the physiological point of view

Group 1: This group contains C, H, O, N and S. These are the major constituents of the organic materials of the plant body. Carbon is taken up in the form of CO_2 from the atmosphere at the time of photosynthesis, and possibly in the form of HCO_3^- ion from the soil solution. Hydrogen is taken up in the form of water which is reduced to H^+ ion at the time of photosynthesis (photolysis). Nitrogen is taken up in the form of NO_3^- or NH_4^{++} ions from the soil solution. The leguminous plants may get elemental N_2 gas from the soil air through the symbiotic nitrogen fixation by *Rhizobium* bacteria. Certain free living bacteria (*Azotobactor, Clostridium butyricum),* blue-green algae and fungi can fix elemental N from soil air into their body tissues which in turn enrich the soil with nitrogen. It is called non-symbiotic nitrogen fixation. Sulphur is not only taken up in the form of $SO_4^=$ or SO ion from the soil solution but can also be absorbed as SO_2 gas from the atmosphere.

Group 2: This group contains P, B and Si. Phosphorus is taken up in the form of $H_2PO_4^-$, $HPO_4^=$ or $PO_4^{\equiv}$ ions, boron in the form of boric acid (H_3BO_3) or borate ($B_4O_4^=$) ions, and silicon in the form of silicate ions from the soil solution.

Group 3: This group contains K, Na, Mg, Ca, Mn and Cl. These nutrient elements are taken up in the form of their ions from the soil solution.

Group 4: This group contains Fe, Cu, Zn and Mo. These elements, with the exception of Mo are taken up as chelate complexes from the soil solution in addition to their ionic forms. A chelate is an organic compound which holds tightly certain cation attached to it. Molybdenum is taken up in the form of molybdate (MoO_4^-) ions from the soil solution.

Movement of nutrient from soil to root

The processes involved in the absorption of inorganic nutrients by the root hairs are exceedingly complex. However, there are three basic methods through which nutrients make contact with the root surface for plant uptake. They are root interception, mass flow and diffusion.

- **Root interception**: Root interception occurs when a nutrient comes into physical contact with the root surface. As a general rule, the occurrence of root interception increases as the root surface area and mass increases, thus enabling the plant to explore a greater amount of soil. Root interception may be enhanced by mycorrhizal fungi, which colonize in the roots and increases root exploration into the soil. Root interception is responsible for an appreciable amount of calcium uptake and some amounts of magnesium, zinc and manganese.

- **Mass flow**: Mass flow occurs when nutrients are transported to the surface of roots by the movement of water in the soil (i.e. percolation, transpiration, or evaporation). The rate of water flow governs the amount of nutrients that are transported to the root surface. Therefore, mass flow decreases when soil water decreases. Most of the nitrogen, calcium, magnesium, sulfur, copper, boron, manganese and molybdenum move into the root by mass flow.
- **Diffusion**: Diffusion is the movement of a particular nutrient along a concentration gradient. When there is a difference in concentration of a particular nutrient within the soil solution, the nutrient will move from an area of higher concentration to an area of lower concentration. The phenomenon of diffusion is observed when sugar is added to water. As the sugar dissolves, it moves through parts of the water with lower sugar concentration until it is evenly distributed or uniformly concentrated. Diffusion delivers appreciable amounts of phosphorus, potassium, zinc, and iron to the root surface. Diffusion is a relatively slow process compared to the mass flow of nutrients with water movement towards the root.

In general, any dissolved minerals to which the root hair membrane is permeable may enter, even though it may be injurious to the plant. In order for a mineral to be absorbed, it must exist in higher concentration in the soil solution than in cell sap of the root hair otherwise root hair would lose this mineral to the soil solution because according to one of the laws of diffusion, a dissolved substance will always move rapidly from a region of high concentration of that substance to one of lower concentration. The minerals are free to diffuse inward to the adjacent cortical cells of the root. As soon as the concentration of a particular mineral becomes greater in the root hair cell than it is in the adjacent cortical cell, this mineral will diffuse into the adjacent cortical cell and ultimately reach the xylem. There is thus established a decreasing concentration gradient of each mineral from the root hair to the xylem, permitting a continuing absorption of mineral salts from the soil solution.

However, minerals are also absorbed against the concentration gradient and this phenomenon is explained by the influence of metabolic activity on the absorption of inorganic substances. Some of the inorganic salts occurring in the soil solution are in ionised condition. Thus sodium nitrate, $NaNO_3$ may ionize as Na^+ and NO_3^- ions. The two ions of a salt are not necessarily absorbed in equal proportions. Any large absorption of one ion in excess of another of opposite charge in equilibrium with it depends upon the exchange of ions between the soil solution and the root hair. If, for example, an excess of K^+ ions is absorbed from potassium sulfate (K_2SO_4), other cations like Na^+, Ca^+ or Mg^+ may be displaced from the root hair, Similarly, NO_3^- or Cl^- ions may be absorbed with the exchange of HCO_3^- ions form the plant, which is produced through reaction of CO_2 with soil water. Hence evolution of CO_2 by roots due to respiration may therefore, play an important role in mineral absorption by the plants.

It is also possible for a particular ion to retard or accelerate the absorption of another ion. A relatively high concentration of Na^+ may depress the absorption of K^+ or Ca^{++} and the presence of Ca^{++} may influence the entrance of Mg^{++}, K^+ or Na^+ ions. The acidity and alkalinity of the soil solution and many other factors also affect the entrance of inorganic solutes into the plant.

General functions of nutrients

Macronutrients

The nutrient elements C, H, O, N, S are the constituents of the organic materials of the plant body. The reactions which result in the incorporation of these elements into the organic molecules are the fundamental physiological processes of the metabolism. So, these elements are involved in the fundamental physio-biochemical processes like photosynthesis, respiration, protein synthesis, oxidation-reduction processes, etc. These nutrients are also involved in the enzymic processes.

Carbon: About 95 to 99.5 per cent of fresh plant tissue is made up of carbon, hydrogen and oxygen. Only about 0.5 to 5.0 per cent remains as ash after fresh plant tissue is dried and burnt. Carbon and oxygen are mainly involved as components of the carboxylic group (–COOH). Hydrogen and oxygen are mainly involved in the oxidation-reduction processes like reduction of NADP to NADPH in photosynthesis. Plant growth is not seriously retarded by lack of C, H, O as long as H_2O is available and CO_2 is abundant in the air. Thus, plant nutrition emphasises the supply of other nutrient elements to the plants.

Nitrogen: Nitrogen is essential for plants because it is present in all living matter as constituent of protoplasm. In combination with hydrogen it constitutes an integral part of amino acids and nucleic acids. It is also a part of the chlorophyll and several alkaloids. Most of the leaf nitrogen is associated with the functioning of the chloroplast. Nitrogen is mainly involved in the form of amines (NH_2^-, $NH^=$). Nitrogen compounds constitute 40-50 per cent of the dry matter of protoplasm. For this reason, it is required in large quantities. It stimulates plant to synthesize proteins and develop fresh vegetative tissues. There is an intimate relation between plant nitrogen status and the carbon metabolism. Hence

N deficiencies in plants induce changes in carbohydrate synthesis and degradation pathways. Plants optimise their carbon gains in relation to N available for photosynthesis. So, supply of nitrogen is related to carbohydrate utilization. All enzymes and coenzymes also contain nitrogen. Excess nitrogen, on the other hand, decreases root growth, delays maturity, renders fruit quality poor and accumulates potentially hazardous concentration of nitrates. Soils have little capacity to retain oxidised forms of nitrogen (NO_2, NO) and most nitrogen is associated with inert organic matter and microorganisms in the soil. So, depletion of organic reserves in the soil limits the availability of nitrogen to the plants.

Phosphorus: Phosphorus is required by all living organisms. It has a vital role in the breakdown of carbohydrates for the release of energy. Phosphorus is an essential ingredient of the nucleo-proteins and, therefore, cell division is not possible without phosphorus. It is also involved in the transfer of inherited characters as phosphorus is the integral part of DNA molecules. Carbon metabolism is also correlated with phosphorus status of the plants. Phosphorus deficiency progressively stops aerial plant growth, decreases P concentration in dry biomass and diminishes photosynthetic activity. It also participates in the numerous energy capture, transfer and recovery of reactions which are vital for plant growth because phosphorus is the constituent of phospholipids, coenzymes, NADP and ATP. It stimulates early root growth and development, encourages fruiting and seed production and hastens maturity of plant. Phosphorus sediments are widely distributed in the lithosphere and in the earth's crust. It occurs most commonly as calcium phosphate.

Potassium: Potassium is an essential element for plant growth. It does not enter into permanent organic combinations but exists as soluble inorganic and organic salts. Therefore, potassium can be called as the 'catalyst' for various reactions. It activates or stimulates enzyme activities that are involved in protein and carbohydrate metabolism. It helps in carbohydrate transportation and pH control. It helps to regulate salt-water balance and stomata opening. It improves water holding capacity of plant tissue and quality of the product. It is essential for cell organization and structure of cell walls and stems. For this reason, it enhances plant's ability to resist diseases, cold and other adverse conditions. It helps in the reduction of nitrates and subsequent synthesis of proteins. It is generally recognised that directly or indirectly, potassium is a factor in the assimilation of CO_2 by the plants which demonstrates a close relationship between carbohydrates and potassium level. For this reason, root and tuber crops require more potassium.

Calcium: Calcium is the constituent of cell wall in the form of calcium pectate. Thus, it helps in normal cell division. It has a beneficial effect on the permeability of cytoplasmic membranes. Root membranes, in particular, break down without calcium. It promotes early root formation and growth, improves general plant vigour, increases stiffness of the stem, influences intake of other nutrients and encourages seed formation. It neutralizes some poisonous organic acids in plant body. It also helps in translocation of carbohydrates. Calcium increases the storability of different fruits and vegetables.

Magnesium: Magnesium is the key element of chlorophyll molecule. In fact, it is the only metallic constituent in chlorophyll. It plays a part in phosphate nutrition and acts as a carrier for phosphorus, particularly into the seeds. It also appears to be a specific activator of a number of enzymes including transphosphorylases, dehydrogenases and carboxylases. So, magnesium is evolved in carbohydrate metabolism, synthesis of nucleic acid, etc. Magnesium also plays a major role as a cofactor in bridging some coenzymes with enzymes. In fact, magnesium is incorporated in the prosthetic group of these enzymes (conjoint proteins like nucleoprotein, lipoprotein, phosphoprotein, metalloprotein on hydrolysis yields amino acids and other products called 'prosthetic group'. Hence, prosthetic group of nucleoprotein is nucleic acid, of lipoprotein is lipid, of phosphoprotein is phosphoric acid or its derivatives and of metalloprotein are metals like Mg, Mn, Fe, Cu, Zn). Magnesium helps in the translocation of sugars within plants. Magnesium with phosphorus appears to participate in respiratory mechanism.

Sulphur: Sulphur plays an important role in protein synthesis, fat metabolism, functioning of several enzyme system and metabolic activities of vitamins like, biotin, thiamine and coenzyme A. Sulphur is mainly involved in the form of sulfhydral (SH^-) group. Like nitrogen, sulphur is a constituent of several amino acids, including methionine, cystine and tryptophan. About 90 per cent of sulphur in plant is found in these three amino acids. It also helps in the synthesis of chlorophyll and in promoting nodulation in the roots of leguminous crops. The volatile compound allyl propyl disulphides and diallyl sulphide responsible for the characteristic pungency of onion and garlic, respectively and isothiocyanates and thiocyanates for sharp odour and pungency of cole crops, turnip and radish contain sulphur. For this reason, cabbage, cauliflower, onion, garlic, turnip and radish are regarded as high sulphur requiring crops. About 70 to 90 per cent of the total soil sulphur is found in organic matter and the remainder is present as slightly soluble sulphites (SO_3^-) and soluble sulphates ($SO_4^=$). Sulphur acts more like nitrogen and nitrogen to sulphur ratio of about 18: 1 in plant material gives a rough estimate of the supply of sulphur in terms of need for plant nutrition.

Micro-mutrients

Boron : Boron is essential for cell division and maintaining cell wall structure. It is involved in protein synthesis, nitrogen and carbohydrate metabolism, root system development, fruit and seed formation and water relations. It increases the permeability in membrane and thereby facilitates transportation of carbohydrate, nitrogen and sulphur. Apart from facilitating sugar translocation it also helps in synthesis of nucleic acids and plant hormones. It also changes the activities of certain enzymes. It is involved in lignin synthesis and other reactions. It affects cell division and is essential for protein synthesis. It is associated with the uptake and utilization of calcium by plant. It also regulates the ratio of potassium and calcium in plants. Relative response of beet, cauliflower, Brussels sprouts, celery and turnip to boron is high, that of broccoli, cabbage, carrot, radish, sweet potato, spinach, lettuce and tomato is medium while that of asparagus, beans, cucumber, onion, pea, potato and sweet corn is low.

Manganese: Manganese acts as a catalyst in several important enzymatic and physiological reactions in plants. It activates decarboxylase, dehydrogenase and oxidase enzymes. It is incorporated in the prosthetic group of some enzymes, like pyruvate carboxylase. It is involved in the water splitting reaction during photosynthesis, nitrogen metabolism and CO_2 assimilation. It controls redox potential in plant cells during the phases of light and darkness. It is involved in the breakdown of carbohydrate and thereby involved in respiratory process. It is important in photosynthesis, nitrogen metabolism, and nitrogen assimilation. It activates the enzymes that are concerned for the synthesis of chlorophyll. Manganese helps in the formation of carotene, riboflavin and ascorbic acid. The crops showing high response to manganese are beans, beet, lettuce, onion, pea, potato, radish and spinach; those showing medium response are broccoli, cabbage, cauliflower, carrot, celery, cucumber, sweet corn, tomato, turnip, mint *(Mentha* sp.) and that showing low response is asparagus.

Iron: Iron very often occurs in plant body in the form of chelates (chelates refers to a ring structure produced when a metal ion combines with two or more electron donor groups to form a single molecule, and metals bound in the chelate rings lose their ability to act as ion). Iron is present in serveral peroxidase, catalase, and cytochrome oxidase enzymes. It is incorporated in the prosthetic group of certain enzymes like catalase, peroxidase, cytochrorne oxidase, etc. It is found in ferredoxin and involved in $NO3^-$, $SO_4^=$ reduction and N fixation. Iron is indispensable for chlorophyll biosynthesis. It is a direct electron donor or acceptor and has role in photosynthesis and nitrogen metabolism. It is involved in oxidation-reduction reactions in respiration. It is the constituent of certain enzymes and proteins. It also plays an essential role in nucleic acid metabolism. The crops showing high response to iron are beans, beet, broccoli, spinach, tomato and cauliflower and those showing medium response are asparagus, cabbage and sweet corn.

Copper: The chief function of copper in plants may be that of an oxidising agent. It is associated with metal protein (prosthetic group) and thus constituent of many enzymes like ascorbic acid oxidase, phenolase, lactase, cytochrome oxidase and other oxidative enzymes. It is important in photosynthesis, protein and carbohydrate metabolism, and probably N-fixation. It also promotes formation of provitamin A in plants. Crops like lettuce, onion, spinach and dill *(Anethum graveolens)* are more responsive to copper and broccoli, cabbage, carrot, cauliflower, celery, cucumber, sweet corn, tomato, turnip and radish show medium response to copper and crops showing low response to copper are asparagus, beans, pea and potato.

Zinc: Zinc is essential for the synthesis of tryptophan which is the precursor of indole acetic acid (IAA), the auxin. For this reason, zinc deficiency causes shortened internode due to non availability of IAA. It is the metal component of some important enzymes like carbonic anhydrase and alcohol dehydrogenase. It plays an important role in the synthesis of nucleic acid, starch and protein. It helps in seed maturation and production. It also helps in the utilization of phosphorus and nitrogen in plants. The crops like beans, onion and sweet corn show high response; beet, potato and tomato show medium response and asparagus, pea and carrot show low response to zinc.

Molybdenum: Molybdenum is the constituent of nitrate reductase and nitrogenase enzymes hence, is involved in the reduction of nitrates for protein synthesis in all plants. It is essential for nitrogen fixation in the root nodules of leguminous plant by *Rhizobium* bacteria. It is also required for the synthesis of vitamin C. Beet, broccoli, cauliflower, lettuce, onion and spinach show high response to molybdenum. Medium response to molybdenum is shown by beans, cabbage, pea, radish, tomato and turnip, while asparagus, carrot, celery, potato and sweet corn show low response to molybdenum.

Chlorine: Chlorine in the form of chloride is found in plants in concentrations higher than that of either phosphorus or sulphur. It activates the oxygen producing enzymes during photosynthesis hence, essential for photosynthesis and enzyme activation. It plays a role in regulating water uptake on salt-affected soils and maintains charge balance during nutrient uptake. It simulates water holding capacity of plant tissue. It is the constituent of chlorine containing auxin, chloroindole-3- acetic acid. Beet, lettuce and tomato are most susceptible to chlorine deficiency.

Silicon: Silicon occurs in plant cells as such or is bound largely by hydroxyl groups of sugars forming silicate esters. Silicon is mostly used to build inactive tissues or incrusting substances.

Nickel: It is the component of some plant enzymes; involved in N metabolism (most notably urease) and biological N fixation. Urease metabolizes urea nitrogen into useable ammonia within the plant. Without nickel, toxic levels of urea can accumulate within the tissue forming necrotic lesions on the leaf tips. In this case, nickel deficiency causes urea toxicity. There is evidence that nickel helps with disease tolerance in plants, although it is still unclear how this happens.

Cobalt: The essentiality of cobalt is unusual in that the requirement is for a cobalt complex known as cyanocobalamin or vitamin B_{12}. It is essential for N-fixation. Bacteria on root nodules of legumes require cobalt to synthesize vitamin B_{12} and fix nitrogen atmospheric nitrogen.

Evaluation of soil fertility

Soil productivity is the capacity of the soil to sustain plant growth. It is determined in terms of yield of a given crop which reflects the combined influence of all the factors that affect plant growth viz., variety/ hybrid, climate, organisms in soil, water, topography, soil characteristics (physical and chemical) and soil fertility.

Soil fertility refers to the ability of a soil to supply the nutrient elements in amounts, forms and proportions required for successful plant growth and development. It is determined in terms of the amount of the available forms of the essential nutrient elements in soil.

Once a plant nutrient element is released from soil minerals by withering, by decomposition of organic matter or is added to the soil as fertilizer, there are several ways of getting it lost from the soil *viz.,* removal by crop, leaching, erosion, volatilization, denitrification, fixation or combination of two or more of these processes which alter soil fertility. Nutrient management of a crop depends on the nutrient requirements of the crop and the nutrient supplying capacity of the soil on which it is grown.

Evaluation of soil fertility based on nutrient deficiency symptoms

The underlying principle of this technique is to study visual growth characteristics of the plant to identify certain symptoms that are associated with deficiency of a particular nutrient element. The effect of a nutrient deficiency can vary from a subtle depression of growth rate to obvious stunting, deformity, discoloration, distress and even death. Visual symptoms distinctive enough to be useful in identifying a deficiency are rare. Most deficiencies are multiple and moderate. Chlorosis of foliage is not always due to mineral nutrient deficiency. Solarization can produce superficially similar effects, though mineral deficiency tends to cause premature defoliation, whereas solarization does not, nor does solarization depress nitrogen concentration. Visual nutrient deficiency symptoms are generally manifested in the leaves. Advantages of such soil fertility evaluation are:

- It is a quick technique and does not require instruments
- Deficiency symptoms within a specific family are generally quite similar in some cases e.g., boron deficiency symptoms in cole and root crops
- The crop may be saved by spraying the nutrients on foliage or adding completely water-soluble nutrients to the soil.

At the same time, there are certain disadvantages like :

- This method of diagnosis may give information too late for a corrective nutrient management
- Certain deficiency symptoms are not very distinctive and may be visible only when the deficiencies are acute
- The diagnosis becomes complicated when two or more nutrients are deficient
- Certain deficiency symptoms, particularly of micronutrients resemble disease or insect damage.

Symptoms of nutrient deficiency

Nitrogen: The deficiency symptoms are mostly generalized over the whole plant; the deficiency results in decrease in carbohydrate synthesis and loss in chlorophyll; slow and stunted growth; stem thin, erect and hard; leaves smaller than the normal; premature yellowing of lower leaves and new leaves are light green in colour; in acute deficiency flowering is greatly reduced. Some specific nitrogen deficiency symptoms are:

- Yellow lower leaves with purple veins and light green young leaves in tomato
- Yellow lower leaves, light green young leaves and slender upright stem in potato
- Older leaves orange to purple pigmented and young leaves pale green in cole crops
- Leaves pale green and turning reddish to purple in beet
- Leaves light green followed by yellowing in carrot and radish
- Stem hard and leaves yellowish green to yellow in cucurbits
- Stiff, upright, short and light green leaves in onion
- Small, sickly and pale green leaves in okra

- Yellowish green leaves with reddish tinged areas in sweet potato
- Yellow older leaves and pale green young leaves in celery and lettuce

Phosphorus: The deficiency symptoms are mostly generalized over the whole plant; the deficiency causes delay in starch development, accumulation of sugars and consequent development of anthocyanin pigment; early growth gets restricted with poor root development; stem thin and shortened; mature leaves show characteristic dark to blue-green colouration and often purple colouration is developed; the purple colouration first appears on underside of leaves and later throughout; lower leaves sometimes turn yellow or greenish-brown in colour; delayed flowering and maturity. Some specific phosphorus deficiency symptoms are:

- Olive-green leaves with purple colouration in the under surface in tomato
- Forward rolling of leaves with marginal scorch in potato
- Dull green leaves with purple colouration in the under surface in cole crops
- Small dull purple leaves in beet
- Dull green leaves with purple colouration in lower surface in carrot and radish
- Dark green leaves with dropping of older leaves in pea and bean
- Tip die back with wilting of older leaves in onion
- Dull green leaves with purple tinge in the lower surface in lettuce

Potassium: Symptoms of potassium deficiency progressively appear on younger leaves; slow and stunted growth with weak stem; in older leaves interveinal chlorosis near the margins occur followed by the development of grey or tan areas near the leaf margin and eventually a scorch appears around the entire leaf margin; leaf edges may become twisted or curled; chlorotic areas may develop throughout the leaf; necrotic areas may develop at the tip of the plant, potassium deficiency interferes the absorption of water and nutrient from soil, so the plant looks withered. Some specific potassium deficiency symptoms are:

- Bronze or greyish green lower leaves with tip necrosis in tomato
- Brownish, cup-shaped leaflets with necrosis starting from tip in potato
- Yellowing of older leaves with browning at margin in cabbage and cauliflower
- Leaves bluish green with surface crinkling in beet

- Leaves bluish green with pale yellow to brown margin in radish
- Dull green leaves followed by browning in carrot
- Leaves bluish green near veins followed by browning and necrosis at margins in cucurbits
- Chlorotic leaflet with necrotic brown areas at margins in beans
- Mottled leaflet followed by necrosis in cowpea
- Young leaves dark green and yellowing or firing of lower leaves at margins in pea
- Young leaves brown-yellow followed by scorching in okra
- Yellowing and drying of older leaves from tip in onion
- Chlorosis followed by necrosis at margin in older leaves in sweet potato
- Dark green leaves followed by curling and browning in celery.

Calcium: The young leaves are affected first which often become small and get distorted and twisted; tips of the innermost unfolding leaves get twisted, gelatinised and finally die back at tips and margin; hooking appearance of the dry leaves of terminal bud is the characteristic symptom; leaf blades may be whitish green and sometimes brown to light brown spots appear on the leaves; stem becomes weak and brittle; root tip dies and root growth is restricted. Some specific deficiency disorders are:

- Blossom end rot of tomato, sweet pepper and watermelon
- Young leaves turn yellow, brown or purple, roll upwards at margins and become necrotic in tomato
- Cavity in root phloem immediately below the epidermis in carrot and parsnip
- Black heart of celery
- Brown heart of endive
- Tip burn of lettuce and internal breakdown of cabbage
- Small, pale green leaves which roll upwards in potato
- Leaves rolled up at margins and become discoloured in cabbage and cauliflower
- Leaves turn pale green around margins which curl upward in beet
- Leaves chlorotic and scorched in carrot

- Narrow white band at leaf margin in radish
- Chlorotic older leaves and curled young leaves in pea
- Young leaves light green and older leaves show some reddish areas in sweet potato

Magnesium: The symptom begins with a loss of green colour at the tips of the lower leaves between the veins; as the deficiency develops, larger portions of the lower leaves are affected, and the leaves further up in the plant become bleached out and turn whitish in colour; leaf becomes lifeless in texture, small in size and tints of orange, red and purple colour may develop unevenly; leaves may curve upwards at margin; older leaves may fall; stem becomes thin and brittle; maturity gets delayed. Some specific magnesium deficiency symptoms are:

- Interveinal chlorosis and necrosis in the older leaves in tomato, brinjal and chilli
- Interveinal chlorosis and puckering of older leaves in cabbage
- Chlorotic leaves with reddish tint between veins in beet
- Pale green leaves with light yellow or brown spots at tips in carrot
- Interveinal chlorosis in the older leaves in radish
- Interveinal yellow spots followed by chlorosis in the leaves in okra
- Leaves die back at tips in onion and browning of leaf tips in pea.

Sulphur: Characteristic uniform yellowish green appearance of the younger leaves that resembles nitrogen starvation; in severe cases, the older leaves may turn pale green; the stem and petiole become brittle and collapse; shoot growth gets restricted; stem becomes stiff, woody and smaller in diameter; the terminal bud commonly remains alive and necrotic spots may or may not be present over the leaf.

Iron: Typical symptoms are distinctive green vein colour and yellow interveinal areas on the young leaves; points and margins of the leaves keep their green colour for long time; in severe cases, the entire leaf, veins and interveinal areas turn yellow which sometimes become bleached; profuse root hairs.

Copper: Yellowing of young leaves; rolled/dead leaf tips; stunted growth; leaves may become elongated and tips may curl; leaf edges may become ragged; plant top may wither; in onion, bulbs become soft with thin pale yellow scales.

Zinc: Deficiency symptoms mostly appear on the 2nd or 3rd fully mature leaves from the top of the plants; Yellowing of leaves; rosetting and dwarfing of leaves due to shortened intetnodes. Some characteristic symptoms are:

- Small reddish brown spots on cotyledonary leaves of bean
- Green and yellow broad striping at the base of the leaves of corn
- Interveinal yellowing with marginal burning in beet
- "Fern leaf" disorder in potato
- Leaves become smaller, chlorotic and inwardly curved in tomato
- Leaves turn yellow in onion
- Whole plant turns yellow in garlic

Manganese: Interveinal yellowing of young leaves; no lateral roots; chlorotic and necrotic spots in the interveinal areas; greyish areas may appear near the base of the younger leaves; leaves become smaller; Some typical symptoms are:

- Densely red foliage in beet
- Narrow, yellow striping in the leaves of onion and corn
- Chlorosis followed by necrosis and yellowing in the leaves of bean
- Small, yellowish leaves with interveinal yellow mottling in cabbage
- Interveinal chlorosis in pea
- Chlorotic and forwardly rolled leaves in tomato

Boron: Death of growing points (roots/shoots); reddish young leaves; malformed buds; necrosis of internal tissue, stubby root system; rosette appearance of the plant due to shortening of terminal growth; young leaves of terminal buds are light green, thick, curled and brittle; root growth gets restricted; flowers become barren. Some typical symptoms are:

- Browning of cauliflower
- Hollow stem in cole crops, particularly in cabbage
- Black heart of beet
- Internal browning of turnip due to beakdown of internal tissues
- Cracked stem knob in knoll khol
- Cracked root of radish
- Cracked stem of celery
- Stem rosetting and open fruit locule in tomato

Molybdenum: The deficiency mimics N-deficiency in legumes; yellowing of young leaves; leaf tip death; impairment of growth; deformation of shoots; older leaves show interveinal yellowing with necrotic margins; leaves may crinkle as found in cole crops. Some typical symptoms are:

- Whiptail leaf and small, open, loose curd in cauliflower
- Bleached, scorched and mottled older leaves in cabbage
- Pale, upwardly rolled leaves with diffused marginal and interveinal yellow mottling in potato.
- Young leaves turn purple in tomato
- Decomposition and granulation in veins of chilli
- Leaf chlorosis with prominent red vein in beet
- Marginal yellowing followed by inward red coloration in carrot
- Leaf cupping, chlorosis, mottling and whiptail in turnip
- Discoloured leaf and death of growing tip in radish
- Yellow or white veins in leaves of pea
- Pale green leaves with interveinal mottling and scorching in bean
- Deep blue leaves with conspicuous yellow and green mottling in onion

Nickel: Chlorosis of new leaves, small leaves, death of growing tissue

Chlorine: Yellow, small leaves, bronzing and necrosis

Evaluation of soil fertility based on soil test values

A soil test value for all nutrients is an index of the availability of that nutrient to plants in the soil being tested. Soil tests and their interpretations are based on the soil samples. It is, therefore, important that the soil samples should be properly collected and should represent the area under soil test. After scrapping the surface litter, uniform core of a thin slice of soil from the surface to plough depth (15-22 cm) from 15 to 20 spots of each sampled field should be taken. Percentage or ppm of nutrient element can be converted to kg of nutrient per hectare. As for example, the soil containing 0.1 per cent potassium means 1 g potassium per kg of soil. One hectare furrow slice, about 15 cm deep of upland mineral soil weights about 2,240,000 kg. So, the conversion of nutrient is 2,240 kg of potassium per hectare.

Table 1. Range in nutrient content commonly found in soils

Nutrient	Normal range	
	Per cent	ppm
Nitrogen	0.02-0.50	200-5000
Phosphorus	0.01-0.20	100-2000
Potassium	0.17-3.30	1700-33000
Calcium	0.07-3.60	700-36000
Magnesium	0.12-1.50	1200-15000
Sulphur	0.01-0.20	100-2000
Iron	0.50-5.00	5000-50000
Manganese	0.02-1.00	200-10000
Zinc	0.001-0.025	10-250
Boron	0.0005-0.015	5-150
Copper	0.0005-0.015	5-150
Chlorine	0.001-0.10	10-1000
Cobalt	0.0001-0.005	1-50
Molybdenum	0.00002-0.0005	0.2-5

Based on the soil tests, soils can be classified into different fertility level categories viz. very high, high, medium, medium low, low and very low. Very high category soils contain maximum available nutrients in the soil and conversely minimum nutrients are available in very low category soils. The higher the soil test level, the lower is the probability of response to applied fertilizer.

Table 2. Soil fertility level on the basis of nutrient content in the soil

Soil fertility level	Nutrient content in soil			
	Organic C(%)	Available N (Kg/ha)	Available P_2O_5(kg/ha)	Available K_2O(kg/ha)
High	>0.75	>450	>90	>340
Medium	0.50-0.75	280-450	45-90	150-340
Low	<0.50	<280	<45	<150

Evaluation of soil fertility based on plant tissue nutrient analysis

Soil tests, although useful in predicting fertilizer and soil amendment needed, are not the final measure of what nutrients a plant will absorb because temperature, moisture regimes, soil acidity and other soil conditions may modify the uptake of different nutrients by plants. It is sometimes necessary to determine nutrient content in the plants to evaluate the actual soil nutrient availability status. Green tissue analysis is the most important diagnostic techniques for determining deficient, sufficient or excessive amounts of essential elements in plant tissue. The basic principle behind this technique is that the nutrient concentration of plants is related to the amount of nutrient element available in soil. Different plant parts of the same plant contain different concentrations of the same nutrient. Nutrient concentration again varies with the stage of the crop. So, leaf samples for analyses should be selected on the basis of physiological age i.e., developmental stage. It is important that the samples must be free from diseases, insect damage and physical or chemical injury. Leaf near the fruit should not be sampled as the nutrients, it might have contained, are often translocated to the fruits.

Table 3. Guidelines for sampling plant tissue for nutrient analysis

Crop	Plant parts to be sampled	Stage of growth
Potato	Fourth to sixth leaf from growing tip	Early growth (35-40 days after planting)
Brinjal	Leaf blades with midribs minus petioles from most recent fully developed leaf	Flower bud to small fruit stage
Tomato	Leaves adjacent to inflorescence	Mid bloom stage
Chilli and Sweet pepper	Young mature leaves	Early fruit set
Cauliflower	Most recent fully matured leaf	Button stage of curd
Cabbage	Wrapper leaf	Prior to heading
Broccoli	Young mature leaves	First bud formation stage
Brussels sprouts	Young mature leaves	Midgrowth
Watermelon, Muskmelon, Cucumber, Pumpkin, etc.	5th leaf from tip on main stem	Early growth Prior to fruit set
Beans (French bean, Cowpea, Lima bean, hyacinth bean, etc.)	Two to three fully developed leaves at top of the plant	Initial flowering
Pea	Leaves from third node down from top of the plant	Initial flowering
Root crops (Carrot, Radish, Beet, Turnip)	Most recent fully matured leaf	Prior to root enlargement

Bulb crops (Onion, Garlic etc.)	Most recent fully matured leaf from centre	Prior to bulbing
Celery	Petiole of youngest fully elongated leaf	Mid-growth (30-35 cm tall)
Lettuce (head type)	Midrib of wrapper leaf	Heading
Lettuce (leaf type)	Youngest mature leaf	Mid-growth
Leafy greens (Palak Spinach, etc.)	Youngest mature leaf	Mid-growth
Sweet potato	Fourth to sixth leaf from the growing tip	Prior to root enlargement
Asparagus	Top 10 cm of the new fern branch	Mid-growth of the fern
Sweet corn	Entire leaf at the ear node	Tasseling

Table 4. General level of nutrients in the leaf tissue of vegetable crops

Nutrient	Per cent or ppm of dry weight of leaf	
	Sufficient level	Critical level*
Nitrogen (%)	1..50-4.00	1.00-1.50
Phosphorus (%)	0.25-0.80	0.20-0.25
Potassium (%)	2.00-9.00	1.50-2.00
Calcium (%)	0.35-2.00	0.35-0.80
Magnesium (%)	0.25-1.00	0.20-0.25
Sulphur (%)	0.16-0.50	0.10-0.16
Manganese (ppm)	30-200	30-50
Iron (ppm)	50-250	50-80
Zinc (ppm)	30-100	20-30
Boron (ppm)	30-60	20-30
Copper (ppm)	8-20	4-8
Molybdenum (ppm)	0.5-5.0	0.2-0.5

* Below critical level of nutrients in leaf tissue, deficiency symptoms may occur

Nutrient requirements of vegetable crops

Nutrient requirement is basically a genetic characteristic of the crop plant and this requirement may vary with the cultivars of the particular crop. Nutrient removal by high yielding cultivars and hybrids is much higher than what are removed by the low yielding cultivars of the same crop. Again, nutrient removal by any crop depends on the nutrient availability and their absorption which is influenced by soil pH, moisture and soil temperature. In general, vegetable crops exert tremendous pressure on the soil for nutrients due to their high productive ability. When the crop is harvested it marks the removal of plant nutrients from the soil. There is no effective way to reduce the nutrient demand because nutrient requirement at a given yield level is quite stable. Nutrient removal is perhaps the most critical factor when sustainability of farming system is considered. If the nutrients removed from the field are not replaced, the farming system will not remain sustainable due to decline in soil productivity. Some idea of the optimum fertility levels needed for vegetable production can be obtained from the quantities of nutrients removed by the crops. Vegetable crops may be categorised into four groups based on the quantity of nutrient removal:

- Very high removal: Asparagus, peas, beans, potato, celery, etc.
- High removal: Cabbage, cauliflower, knolkhol, radish, sweet potato, yam, spinach, carrot, etc.
- Moderate removal: Onion, garlic, tomato, brinjal, beet, leek, muskmelon, okra, cassava, elephant foot yam, lettuce, etc.
- Low removal: Cucumber, pumpkin, bottle gourd, etc.

From the pattern of primary nutrient removal by different vegetable crops, it is apparent that removal of phosphorus is comparatively low while that of potassium is luxuriantly high. This does not mean that application of very high potassium will give good yield in all the vegetable crops. Again, plants require more phosphorus than its removal for giving optimum yield. On the other hand, they require as much sulphur as phosphorus. Nutrient requirements depend on the specific demands of roots, shoots, leaves and fruit development of the crop, although all the vegetable crops are very responsive to nitrogen application. In general, the vegetable crops whose above ground vegetative parts are consumed (cabbage, cauliflower, knolkhol, leafy vegetables, salad crops, etc.) require more nitrogen; the bulb, root and tuberous vegetable crops (root crops, bulb crops, potato, sweet potato, elephant-foot yam, yam, taro, etc.) require more potassium and fruit vegetable crops (tomato, brinjal, chilli, gourds, melons, etc.) require more phosphorus. Peas and beans require sufficient nitrogen but most of the required nitrogen is supplied through symbiotic nitrogen fixation by *Rhizobium* bacteria in their root nodules. In these crops, a low dose of nitrogen should be applied as basal to stimulate early growth. Cabbage, cauliflower, radish, onion, turnip require comparatively more sulphur as the compounds which impart characteristic odour to these vegetable crops contain sulphur.

Table 5. Removal of primary nutrients from the soil by some vegetable crops

Crop	Nutrient removal (kg/100 kg yield)		
	Nitrogen	Phosphorus	Potassium
	Very high removal		
Asparagus	2.40	1.20	3.00
Potato	0.74	0.28	1.39
Pea	1.25	0.45	0.90
Beans	0.87	0.27	1.07
Celery	0.67	0.27	1.01
	High removal		
Cabbage	0.53	0.16	0.50
Cauliflower	0.40	0.16	0.50
Knolkhol	0.50	0.42	0.85
Carrot	0.42	0.18	0.67
Radish	0.60	0.30	0.60
Spinach	0.50	0.15	0.30
Yam	0.57	0.14	0.61
Sweet potato	0.48	0.19	0.85
	Moderate removal		
Muskmelon	0.36	0.11	0.64
Tomato	0.33	0.11	0.44
Brinjal	0.29	0.08	0.50
Beet	0.24	0.16	0.45
Onion	0.30	0.13	0.40
Leek	0.34	0.13	0.80
Okra	0.30	0.13	0.45
Lettuce	0.30	0.12	0.53
Cassava	0.38	0.18	0.88
Elephant-foot yam	0.34	0.08	0.49
	Low removal		
Cucumber	0.18	0.13	0.30
Pumpkin	0.18	0.14	0.32
Bottle gourd	0.15	0.16	0.16

Table 6. Removal of sulphur from soil by some vegetable crops

Crop	Sulphur removal (kg/100 kg yield)
Cabbage	0.11
Cauliflower	0.08
Tomato	0.06
Onion	0.09
Radish	0.20
Cassava	0.03

Nutrient management

Nutrient management is the science and art directed to link soil, crop, weather and hydrologic factors with cultural, irrigation and soil and water conservation practices to achieve the goals of optimizing nutrient use efficiency, yield, crop quality and economic returns and reducing environmental polution. Some important factors that need to be considered when managing nutrients include:

- Nutrient requirement for the particular crop
- Achievable optimum yield and crop quality
- Timing and management of nutrients using a budget based on all sources
- Management of soil, water and crop to maximise the nutrient use efficiency.

Plant nutrition is a difficult subject to understand completely because of the variation between different plants and even between different species or individuals of a particular crop. Elements present at low levels may cause deficiency symptoms and toxicity is possible at levels that are too high. Furthermore, deficiency of one element may present as symptoms of toxicity from another element, and *vice versa*. An abundance of one nutrient may cause a deficiency of another nutrient. For example, K^+ uptake can be influenced by the availability of the amount of NH_4^+ ion.

Management of organic matter in soil

Soil organic matter consists of decomposed plant and animal residues and numerous dead microorganisms particularly which are nitrogen fixers (both free living and symbiotic). It is very active and important portion of soil. After active decomposition, the organic matter residues are collectively called humus. Beneficial effect of soil organic matter can be summarised as:

- It is the nitrogen reservoir and source of 90-95 per cent nitrogen in unfertilized soil.
- It supplies large portions of soil phosphorus and sulphur.
- It is also the source of other macro-and micro-nutrients
- It supplies cementing substances for desirable aggregate formation of soil.
- It loosens up the soil to provide better aeration and water movement mainly due to soil aggregation which increases soil pores and water holding capacity of the soil is thereby increased.
- The large available surfaces of humus have many cation exchange sites that absorb nutrients for eventual plant use.

- Organic matter acts as a chelate and make bond to different micro-nutrient elements like Fe, Zn, Cu, and Mn and the soluble chelates help to mobilize these micronutrients in soils and increase their availability to plants.
- It supplies carbon to many microbes that perform other beneficial functions in soils and growth substances produced by the micro-organisms in soil help plant growth in many ways.
- It reduces soil erosion and keeps the soil cooler in very hot weather and warmer in winter.

For maximum benefit, organic matter must be readily decomposable and continuously replenished with fresh residues.Under stabilized enveronmental condition, cultivated soil contains a fairly constant level of organic matter of less than 1.0 percent. Under other condition, organic matter content may be 10.0% or more. The majority of the tropical and subtropical soils like most of the soils in India have low humus content which is the consequence of its quick year round mineralization under the influence of tillage and favourable weather condition. Moreover, vegetable crops are generally cultivated intensively and the crop residues are not sufficient to replace the organic matter lost from the soil. The soil organic matter can be replenished by adding farmyard manure, oilcakes, compost, green manure and other bulky organic residues.

Table 7. Approximate composition of some manures

Manure	Total nitrogen (% N)	Availablephosphorus (% P_2O_5)	Solublepotassium (% K_2O)
Farmyard manure	0.5-1.0	0.4-0.8	0.5-1.0
Green manures (average)	0.5-0.7	0.1-0.2	0.6-0.8
Rape seed cake	5.1-5.2	1.8-1.9	1.1-1.3
Castor cake	5.5-5.8	1.8-1.9	1.0-1.1
Mahua cake	2.5	0.8	1.8
Sesame cake	6.2-6.3	2.0-2.1	1.2-1.3
Cotton seed cake	6.4-6.5	2.8-2.9	2.1-2.2
Niger cake	4.7-4.8	1.8-1.9	1.3-1.4
Groundnut cake	7.0-7.2	1.5-1.6	1.3-1.4
Neem cake	5.2	1.02	1.4
Bonemeal (raw)	3.0-4.0	0.0-2	-
Fish manure	4.0-10.0	5.0	0.3-1.5
Night soil	1.2-1.3	3.0-9.0	0.4-0.5
Cowdung + Urine	0.6	0.8-1.0	0.45
Pressmud	1.3	0.15	1.5

Town compost	1.2-2.0	3.81.0	1.5
Rural compost	0.4-0.8	0.3-0.6	0.7-1.0
Sludge	2.5	1.0	0.4
Cowdung slurry from biogas plant	1.6-1.8	1.1-2.0	0.8-1.2
Vermi Compost	1.0	0.5	1.5

Fresh farmyard manures often cause burning effect due to excess soluble nitrogen in the manures. In the cultivation of vegetable crops, application of about 20-25 tonnes of manures per hectare is generally advocated. Combined application of manures and inorganic fertilizers always proves better with respect to production and quality of vegetables. However, care should be taken that some manure cannot safely be mixed with some inorganic fertilizers.

Table 8. Mixing ability of different fertilizers and manures

Fertilizer/ Manure	AS	ASN	CAN	U	AC	SP	AP	MOP	B	C	W	L
Ammonium sulphate (AS)	—	Y	Y	Y	Y	Y	Y	Y	Y	N	N	N
Ammonium sulphate-nitrate (ASN)	Y	—	Y	Y	Y	Y	Y	Y	Y	Y	N	N
Calcium ammonium nitrate (CAN)	Y	Y	—	Y	N	Y	Y	Y	Y	Y	N	N
Urea (U)	Y	Y	Y	—	Y	Y	Y	Y	Y	Y	Y	Y
Ammonium chloride (AC)	Y	Y	Y	Y	—	Y	Y	Y	Y	N	N	N
Super phosphate (SP)	Y	Y	Y	Y	Y	—	Y	Y	Y	Y	N	N
Ammonium phosphate (AP)	Y	Y	Y	Y	Y	Y	—	Y	Y	Y	N	N
Muriate of potash (MOP)	Y	Y	Y	Y	Y	Y	Y	—	Y	Y	Y	Y
Sulphate of potash (SOP)	Y	Y	Y	Y	Y	Y	Y	Y	Y	Y	Y	Y
Bonemeal (B)	Y	Y	Y	Y	Y	Y	Y	Y	—	Y	Y	Y
Compost (C)	N	Y	N	Y	N	Y	Y	Y	Y	—	N	N
Wood ash (W)	N	N	N	Y	N	N	N	Y	Y	N	—	Y
Lime (L)	N	N	Y	Y	N	N	N	Y	Y	N	Y	—
Dicalcium phosphate (DCP)	Y	Y	Y	Y	Y	Y	Y	Y	Y	Y	N	N

Note: Y = can safely be mixed, N = should not be mixed

Management of macronutrients

In natural ecosystems, the minerals absorbed by the crops return to the soil after organic matter decomposition. Soil fertility is more or less maintained through nutrient cycling. In cultivated ecosystems like vegetable cultivation, however, all harvested biomass (product and plant residue) withdrawn from the field contains nutrients that no longer return to the soil. Hence, maintenance of soil fertility and crop yield depend on counter balancing of fertilizer inputs.

Nitrogen: It is most often the limiting element in plant growth and the vegetable crops require more nitrogen than any other nutrients. Nitrogen is highly mobile and easily lost from the soil. In cultivable land, crops can utilize only 50-60 per cent of the available nitrogen. When the nitrogen sources are applied to the soil, they all are eventually converted to NH_4^{++} and NO_3^- forms of nitrogen. Plant roots take up these two inorganic forms of nitrogen almost exclusively. Nitrate, if left unused, is lost through downward leaching as negatively charged NO_3^- is naturally repelled by soil particles and remains free to move with soil water but positively charged NH_4^{++} is held by soil particles and generally, are not subjected to leaching. However, both NO_3^- and NH^{++} forms may be consumed by micro-organisms or converted to gaseous nitrogen forms (N_2 or NH_3) and lost to atmosphere. Nitrogen use efficiency in vegetable crops can be increased by i) split application of fertilizers ii) applying the nitrogen fertilizer 5-10 cm deep in the soil, iii) keeping optimum soil moisture in the field and iv) judicious application of phosphorus and potassium fertilizers.

Table 9. Approximate composition of some common fertilizer materials

Fertilizer material	Total nitrogen (% N)	Available phosphorus (% P_2O_5)	Soluble potassium (% K_2O)	Available sulphur (% S)
Ammonium sulphate, $(NH_4)_2SO_4$	21.0	-	-	24.0
Ammonium chloride, NH_4Cl	25.0	-	-	-
Calcium ammonium nitrate, $(CaCO_3, NH_4NO_3)$	25.0	-	-	-
Ammonium phosphate Sulphate, $(NH_4)HPO_4$ $(NH_4)_2SO_4$	16.0-20.0	20.0	-	15.0
Ammonium sulphate nitrate, $(NH_4)_2SO_4.NH_4NO_3$	26.0	-	-	15.0

Ammonium nitrate, NH_4NO_3	33.5			
Urea, $CO(NH)_2$	46.0	-	-	-
Diammonium phosphate, $(NH_4)_2HPO_4$	16.0-18.0	45.0	-	-
Superphosphate, $Ca(H_2PO_4)_2)$				
Single	-	16.0	-	12.0
Triple	-	42.5	-	-
Rock phosphate, $(2Ca_3(PO_2)_2)$	-	20.0-40.0	-	-
Basic slag, $(CaO)_5P_2O_5SiO_2$	-	15.0-25.0	-	-
Potassium chloride, (KCl) (Muriate of potash)	-	-	60.0	-
Potassium sulphate, (K_2SO_4)	-	-	48.0	18.0

Phosphorus: Young plants assimilate phosphorus very rapidly and thereby need water soluble forms that migrate to the meagre root system. Plants generally accumulate as much as 75 per cent of their phosphorus need in the early stages of growth. This emphasises the importance of ample supply of water soluble phosphorus in the soil zone of the germinating seeds. Most soluble phosphorus ($HPO_4^{=}$, $H_2PO_4^{-}$) become fixed (precipitate and form insoluble compounds) before plant can absorb them, and hardly 10-15 per cent of the applied phosphorus is utilized by the plants. So, leaching loss of phosphorus is practically nil. It is lost primarily by erosion of surface soil. In general, phosphorus is more available at a pH of 6.0 to 6.5 than at higher or lower pH values. In acidic soils, low availability of phosphorus is common due to formation of complex compounds of phosphorus with aluminium(Al) and iron (Fe). Phosphorus use efficiency can be increased by i) Maintenance of soil pH at 6.5 to 7.0, ii) application of phosphatic fertilizers as basal near the active root zone and iii) maintenance of adequate organic matter and moisture in the soil.

Potassium: In vegetable crops, luxury consumption of potassium is a very common feature. Rapid uptake of this nutrient occurs in the early stages of growth and it usually occurs faster than either N or P. So, ample potassium should be made available to the seedlings. Potassium fertilizers are less mobile than nitrogen fertilizers but more mobile than phosphatic fertilizers. Potassium fixation embraces (a) immobilization into microbial bodies and (b) becoming entrapped in the clays. It is also leached substantially through the downward movement of water. For this reason, it is often deficient in high rainfall areas. Potassium use efficiency can be increased by i) split application of potassium fertilizers, particularly in loose soils of high rainfall areas and ii) maintenance of soil pH at 6.0-7.0 by liming which reduces leaching.

Sulphur: Crops require as much sulphur as they do phosphorus and, therefore, sulphur is called the fourth major nutrient. Some vegetable crops like cabbage, onion, turnip, etc. have high sulphur requirement. Sulphur content in the soil is being decreased gradually because (a) modern high analysis fertilizers are free of sulphur, (b) modern plant protection chemicals are sulphur-free compounds. Organic sources of sulphur are about 70-90 per cent. Rainfall also dissolve atmospheric SO_2 released during burning of fuels and act as a source of sulphur. Sulphur can be managed in the soil through i) maintenance of soil pH to 6.0-7.0 by liming, ii) application of organic matter to the soil, iii) application of sulphur –containing fertilizers, like ammonium sulphate, super phosphate, potassium sulphate, etc. periodically on acid and sandy soil low in humus and iv) in acute deficiency, elemental sulphur may be applied before sowing / transplanting.

Table 10. Approximate composition of some secondary and micronutrient containing chemicals

Nutrient	Material	Available nutrient (%)
Calcium	Burnt lime, (CaO)	70.00% Ca
	Salked lime, $Ca(OH)_2$	50.0% Ca
	Lime stone, ($CaCO_3$)	36.0% Ca
	Calcium choloride, ($CaCI_2$)	36.1% Ca
	Calcium nitrate, ($Ca(NO_3)_2 2H_2O$)	20.0% Ca
Magnesium	Magnesium sulphate, ($MgSO_4$)	20.0% Ca
	Dolomite, ($CaCO_3.MgCO_3$)	12.0% Mg + 17.0 % Ca
Sulphur	Elemental sulphur, (S)	85.0-100.0% S
Boron	Borax, ($Na_2B_4O_7.10H_2O$)	11.0% B
	Boric acid, (H_3BO_3)	17.0% B
	Solubor ($Na_2B_4O_7.5H_2O+Na_2B_{10}O_{16}.10H_2O$)	20.5% B
Copper	Copper sulphate, ($CuSO_4.5H_2O$)	25% Cu+13%S
	Copper oxide, (CuO)	79.6% Cu
Iron	Ferrous sulphate, ($FeSO_4.7H_2O$)	20.0% Fe+19.0%S
Manganese	Manganese sulphate, ($MnSO_4.4H_2O$)	25% Ma+17%S
Zinc	Zinc sulphate, ($ZnSO_4.7H_2O$)	23.0% Zn
Molybdenum	Sodium molybdate, ($Na_2MoO_4.2H_2O$)	39.7% Mo
	Ammonium molybdate, ($(NH_4)_6Mo_6O_{24}.4H_2O$)	54.0% Mo

Calcium: It is seldom deficient except in sandy and strongly acidic soils. However, calcium is lost in a number of ways like, crop removal, leaching and soil erosion. Leaching of bases like calcium by percolating water coupled with nitrogen and sulphur fertilization contributes to natural soil acidity. Calcium can be managed in the soil by liming. The amount of lime to be added is determined by the pH, texture and organic matter content of the soil.

Magnesium: In light-textured soil, magnesium is leached through the percolating water. Magnesium can be managed in the soil by application of dolomite ($CaCO_3$. $MgCO_3$) which contains 12.0 per cent magnesium or magnesium sulphate containing 20.2 per cent magnesium.

Methods of fertilizer application

In general, of the total quantity of required N, P, K fertilizers, full dose of P and K fertilizers and half-dose of N fertilizer are applied as basal at the time of final land preparation, and the rest half dose of N fertilizer is top dressed in 2-3 splits. In light-textured soils of high rainfall areas, K fertilizers may also be applied in 2 splits. Sulphur, lime or dolomite are also applied by broadcasting as basal before sowing or transplanting.

Broad casting is the common method of application of N, P and K fertilizers where large quantities of high analysis fertilizers are applied without the injury of salt damage to the plants. However, broadcasting is not an effective method of fertilizer application as far as nutrient availability to the plants is concerned. Some of the effective methods of fertilizer application in vegetable crops are (i) placement of fertilizers at the bottom of the plough furrow, (ii) band placement 5-8 cm below the soil surface, (iii) combination of broadcasting or plough furrow placement with band placement at the side of the row at sowing/ planting time and (iv) drilling of fertilizers below the soil surface before sowing/planting. Banding of P fertilizer beneath the soil is particularly important as P is lost primarily by erosion of soil surface. After application of fertilizers, the land must be moistened to make the nutrients available to the plants. Low soil moisture regime increases the concentration of soil solution, and as a result of which osmotic movements of water and nutrients in the plant body get restricted.

Macronutrient fertilizers can also be applied by foliar spray. Most of the essential plant nutrient elements can be taken up by the plant through foliage, but only in small quantities at a time. Nutrients mostly enter the leaf *via* stomata. At the same time, ectodermata, the microscopic openings in the cell walls and cuticle also absorb nutrient solution. The cuticle itself swells on absorbing water and also becomes permeable to dissolved nutrient elements to some extent. Since most of the stomata are present on the under surface of the leaf, foliar spray should be applied to both under and upper surfaces of the leaf as evenly as possible for rapid and complete absorption of the nutrient solution.

There are differences in the rate of entry of nutrient solution in the plant body *via* leaves. Major portion of applied nitrogen is taken up within few hours. For this reason, foliar application of nitrogen in the form of urea solution of 1-2% concentration is most common. Compared to nitrogen, potassium uptake is relatively slow (50% absorption within 1-4 days). Thus, foliar application of potassium is relatively less common. Foliar spray of potassium in the form of aqueous solution of KCI (muriate of potash) or K_2SO_4 at 2% concentration is permissible. Of the macronutrients, phosphate is absorbed most slowly (50% absorption within 1-5 days). For this reason, phosphatic fertilizer is not generally applied as foliar spray. However, tolerable phosphate concentration as orthophosphate is 0.5% for many crops. Foliar fertilization should be done in the early hours of the morning and in the evening when the relative humidity of the air is high. Morning and evening sprays are advantageous because (i) the spray solution evaporates more slowly so there is less danger of the leaves being scorched due to rapid drying of the solution by bright sunlight which eventually increases the concentration of the solution, and (ii) the dried spray deposits get re-dissolved due to high humidity and enter into the leaf.

Starter solution

Starter solutions are mixtures of soluble fertilizer and water which can effectively boost the early growth of many vegetable crops. The fertilizer material easily dissolves in water and the nutrients are readily available for plant uptake. Starter solutions are used primarily for different vegetable crops such as tomato, eggplant, pepper, muskmelon, watermelon, cucumber, cabbage, cauliflower, broccoli, etc. Starter solutions are easy to make and use. It reduces fertilizer applications and decrease fertilizer residues in soil. It increase yield when used in combination with other fertilization methods.

There are many formulations of starter fertilizers. For vegetable production, the recommended starters contain water soluble nitrogen, phosphatic and potassic fertilizers in the ratio of 1:2:1. Application of high nitrogen starters may result in excessive vegetative growth. The solution is made by mixing the dry starter fertilizer with water at a rate of approximately 30 gram of fertilizer mixture per 4 litre of water. The starter solutions are applied to young vegetable seedlings at the time of transplanting or after emergence of seedling in place of normal watering so that the nutrients reach the root zone immediately to help better establishment of the transplants. It is important to use enough solution to saturate the entire root ball so that all of the roots will have access to the starter.

Management of micro-nutrients

The main reasons of accelerated exhaustion of available macronutrients in the soil are

a) Cultivation of high-yielding cultivars and hybrids which demands more micronutrients

b) Application of more N, P and K fertilizers as it decreases the availability of micronutrients,

c) Inadequate application of organic matters to the soil,

d) Intensive cropping which creates deficiency,

e) Leaching loss of micronutrients particularly in light-textured soils of high rainfall areas,

f) Application of compound fertilizers because straight fertilizers may add some micronutrients e.g., superphosphate contains 26 ppm Cu, 50 ppm Zn, 65 ppm Mn, 9.5 ppm B and 3.3 ppm Mo while, compound fertilizers do not contain such micronutrient element

g) Maximum application of phosphatic fertilizers decreases the availability of zinc, whereas application of ammonium sulphate or other sulphate compounds increases the availability of zinc,

h) Inadequate cultural practices decrease the availability of micronutrients.

Primary deficiency: It refers to low total content of the micronutrients which depends on the soil type and other agro-climatic conditions. Sandy and calcareous soils contain less micronutrient elements whereas clay loam and loam soils contain comparatively more micronutrient. Loose-textured soil low in organic matter and soils of high rainfall areas are also deficient in micronutrient. Primary deficiency can be corrected by the application of chemical compounds containing nutrient elements. It should be kept in mind that micronutrients should be applied in deficient condition only, otherwise increase in micronutrient levels in the soil may lead to phytotoxicity.

Secondary deficiency: It implies that total micronutrients contents in the soil may be ample but their availability is limited due to other factors like soil pH, elemental interaction, etc. Availability of Zn, Mn, Cu and B is often reduced in soils exhibiting alkaline reaction whereas Mo is unavailable in acidic soils. Secondary deficiencies of micronutrients may be corrected by proper soil management like liming, application of organic matter to the soil, etc.

Methods of micronutrient application

Vegetable crops response very well to the application of micronutrient under Indian condition. Generally, micronutrients are applied in four different ways:

Soil application: Micronutrient containing chemical compounds can well be applied to soil mixed with other fertilizers.

Seedling root dipping: This method of application is not widely practised. Generally, 0.2- 0.3 per cent solution of zinc sulphate is used for root dipping.

Seed treatment: Seeds may be sown after treating with the chemical compounds containing Cu, Fe, Mo, Zn, B, Mn, etc.

Foliar spray: Most of the micronutrients after foliar spray enters the plant body through leaves within few hours to one day. Foliar application of micronutrients is widely used and at the same time suitable because of (i) convenience in application, (ii) requirement of small quantity, (iii) quick correction of deficiency, and (iv) avoidance of fixation of the micronutrients in soil. Lime should be added to neutralise the solution of ferrous sulphate, manganese sulphate, copper sulphate and zinc sulphate, otherwise foliage may get scorched. Foliar sprays may be given 2-4 times at an interval of 7-10 days to correct the deficiency.

Table 11. Recommended dose for application of micronutrients

Micronutrient	Soil application** (kg of element/ hectare)	Foliar application (concentration)
Iron	0.5-10.0	0.4% ferrous sulphate + 0.2% lime*
Manganese	5.0-12.0	0.4%-0.6% manganese sulphate + 0.2%-0.3% lime*
Zinc	0.5-8.0	0.2%-0.6% zinc sulphate + 0. 1 -0.3% lime*
Copper	3.0-8.0	0. 1-0.2% copper sulphate + 0.5% lime*
Boron	0.5-5	0.5%-0.6% borax
Molybdenum	0.05-1.0	0.05% sodium or ammonium molybdate

* Lime is added to neutralise the solution, otherwise leaves may get scorched. Minimum 450 litres of water must be used per hectare.

** Rate of application varies with the type of soil. As for example, higher rates of manganese is required for both organic soils and sandy soil that are alkaline in reaction; high rate of zinc is applied when soil is neutral to alkaline in reaction.

Integrated nutrient management

Integrated Nutrient Management (INM) refers to the maintenance of soil fertility and of plant nutrient supply at an optimum level for sustaining the desired productivity through optimization of the benefits from all possible sources of organic, inorganic and biological components in an integrated manner. INM adopts a holistic view of plant nutrient management by considering the totality of the farm resources that can be used as plant nutrients.

Concepts of INM

- Regulated nutrient supply for optimum crop growth and higher productivity.
- Improvement and maintenance of soil fertility.
- Improvement of efficiency of fertilizer use and minimization of losses of plant nutrients
- Zero adverse impact on agro – ecosystem quality by balanced fertilization of organic manures, inorganic fertilizers and bio- inoculants

Determinants of INM

- Nutrient requirement of cropping system as a whole.
- Soil fertility status and special management to overcome soil problems, if any
- Local availability of nutrients resources (organic, inorganic and biological sources)
- Economic conditions of farmers and profitability of proposed INM option.
- Social acceptability.
- Ecological considerations.
- Impact on the environment

Advantages of INM

- Ensures productive and sustainable agriculture.
- Reduces expenditure on costs of purchased inputs by using farm manure, crop residue, etc.
- Utilizes the potential benefits of green manures, leguminous crops and bio-fertilizers.
- Enhances the availability of applied as well as native soil nutrients by augmenting nutrient cycle thus limiting losses to the environment
- Synchronizes the nutrient demand of the crop with nutrient supply from native and applied sources.
- Provides balanced nutrition to crops and minimizes the antagonistic effects resulting from hidden deficiencies and nutrient imbalance.
- Improves and sustains the physical, chemical and biological functioning of soil.
- Minimizes the deterioration of soil, water and ecosystem by promoting carbon sequestration, reducing nutrient losses to ground and surface water bodies and to atmosphere
- Prevents degradation of the environment.

Components of INM

There are various components of plant nutrients for INM which can be applied in an integrated way. Besides inorganic fertilizers as the major component, others include farmyard Manure (FYM), composts, green manure crops, crop residues and bio fertilizers.

Soil Source

Use of appropriate crop varieties, cultural practices and cropping system to mobilize unavailable nutrients is important component of INM.

Chemical fertilizer

Chemical fertilizers are rich in nutrients hence, are required in less quantity to supply nutrients as compared to organic manures. However, continuous use of chemical fertilizers deteriorates the soil conditions. Some important chemical fertilizer component of INM are super granules, coated urea, locally available rock phosphate in acid soils, single super phosphate, muriate of potash and chemical compounds containing micronutrients.

Organic sources

Organic manures like farm yard manure (FYM), compost, vermicompost, sewage, sludge, industrial waste, etc. improve the bulk density of soil up to a layer of 25 cm. It reduces the resistance to penetration of roots and supplements the nitrogen requirement of the crop. It increases available N and P use efficiency when combined with 100% of the recommended quantity of NPK and biofertilizers.

Biological sources

Nitrogen fixation is a process by which nitrogen in the earth's atmosphere is converted into ammonia (NH_3) or other molecules available to living organisms. Atmospheric nitrogen or molecular dinitrogen (N_2) is relatively inert. It does not easily react with other chemicals to form new compounds. The fixation process frees nitrogen atoms from their triply bonded diatomic form, $N \equiv N$, to be used in other ways. Different micro-organisms which can fix atmospheric nitrogen and increase the availability of other nutrients like phosphorus and potassium are called biofertilizers. Of the different types of biofertilizers available, the symbiotic nitrogen fixing *Rhizobium* bacteria is relatively more effective and widely used with an average N fixation rate of 25 kg N/ha. Two important free living nitrogen-fixing biofertilizers are *Azotobactor* and *Azospirillum*. Besides, some free-living, non-specific bacteria like *Clostridium* spp., *Bacillus polymyxa,* etc. can fix atmospheric nitrogen. Generally, microbial inoculants substitute 15-40 Kg N/ha. Another important role of these biofertilizers is liberation of growth substances, which promote germination and plant growth.

Phosphate solubilizing bacteria (PSB) are beneficial bacteria (example, *Pantoea agglomerans*, *Microbacterium laevaniformans*, *Pseudomonas putida* strains) capable of solubilizing inorganic phosphorus from insoluble compounds. Phosphate solubilization ability of rhizosphere microorganisms is considered to be one of the most important traits associated with plant phosphate nutrition.

Potash mobilizing bacteria increase the availability of potassium in the soil and thereby promote photosynthesis. Microorganisms like *Acidothiobacillus ferrooxidans*, *Bacillus mucilaginosus*, *B. edaphicus*, *Burkholderia* sp., *Frateuria* sp., *Pseudomonas* sp., *Rhizobium* sp., etc. are known for their potential in releasing insoluble native K-source in soil into a plant available nutrient pool.

Endomycorrhizae is a balanced symbiosis by which both the plants and the fungus get benefited. Endomycorrhizae have often been referred to as vesicular arbuscular mycorrhizae (VAM) as some of these fungi form dichotomously branched haustoria within the host cells called arbuscular. The fungi also bear a second type of structures which have a swollen, thick-walled appearance and are called vesicles. These structures serve to absorb and release phosphorus in the process of transfer between the symbionts. Experimental results indicate the prospect of using these fungal inoculants for obtaining greater phosphorus uptake and yield of different vegetable crops.

Green manure

Incorporation of green manure crops viz., *Sesbania aculeate* (dhaincha) into the soil when they are still green may add even 60 kg inorganic nitrogen per hectare. Incorporation of mung bean after picking of pods results in savings of inorganic nitrogen. Green manuring with the leaves of *Leucaena leucocephala* (Subabul), *Gliricidia sepium* and *Acacia mangium* can add 50-60 kg nitrogen per hectare.

Chapter 8

Water Management for Vegetable Crops

Relationship Between Plant and Water

Water is the most vital component of the living cells and the basis of plant feeding. Water is also characterized with extraordinary physical properties: it has high specific heat, biggest latent heat of vaporization, very high thermal conductivity and greatest surface tension. Large amounts of heat are tied up with water because of very high specific heat, which contribute to both climatic stability and to heat budgets of the plants and other organisms. Water molecules are integral part of living systems being the solvent for metabolites and structural component of proteins and nucleic acids. In fact, biological macro-molecules cannot exist in their complicated secondary folding or spiralled helixes without the support of hydrogen bonding within their water matrix. The water matrix is also critical for metabolic activity because its properties of hydration permit ready reactivity between molecules in solution as well as between enzymes and their substrates, and the biochemical processes are sensitive to tissue hydration. A reduction of only 20-25 per cent below maximum hydration is sufficient to cause metabolic malfunctioning of many enzymes. It is the substrate for photosynthesis and a product of respiration. Water also helps to buffer plant tissue against fluctuations in the external environment. Plant growth including leaf expansion, organ enlargement, reproductive development and stomata function are all clearly associated with moisture status. Hormonal changes (abscisic acid and cytokinins) in plant body occur with brief reduction in water potential. Water is the also major force in shaping the climatic patterns.

An adequate water supply is essential to maximise both quality and crop yield for a given date. However, water need to be applied at the correct time, in sufficient quantities but without waste keeping the worth of this most valuable natural resource in mind. Both crop performance and efficient use of the available water can be optimised by:

- ❒ Knowing the water holding capacity of the soil in each field and the water requirements and response of each crop grown.
- ❒ Using effective soil moisture monitoring system and using it to schedule irrigation accurately.

- Choosing the right application equipment for the particular situation and knowing how to get the best out of it in terms of uniform and timely delivery.
- Managing water application for maximum economic benefit with minimum impact on the environment.

Water Quality

The quality of water to be used for irrigation is an important consideration. The two main aspects of water quality are its chemical and microbiological properties. Other factors can occasionally affect quality, especially from river or reservoir sources, such as the presence of suspended peat or silt particles or other specific pollutants (e.g. herbicides). The chemical parameters important for irrigation water include:

- pH
- Alkalinity or bicarbonate content
- Electrical conductivity
- Chloride level
- Sulphate level
- Iron and other trace element levels

pH and Alkalinity

pH is the measure of how acid or alkaline the water is. However, it is the level of bicarbonates (principally calcium, but also magnesium and sodium) in the water that determines whether it is termed either 'hard' or 'soft'. Water may, for example, have a high pH but low calcium bicarbonate content. It may be worth considering the use of acidifying fertilizer mixes when fertigating with hard water in drip irrigation systems used on high value vegetable crops.

Electrical conductivity and trace elements

The electrical conductivity is a measure of the total salts content of the water which is expressed in micro-siemens per cm. The main contributors to high conductivity levels are nitrate, chloride and sulphate ions. Water with a naturally high E.C. does not usually give rise to a build up of salts in soils used for field vegetable production due to the diluting effect of rainfall.

Rainwater is generally uncontaminated, it has a low conductivity and pH and a very low calcium level. However, borehole water in coastal areas may have a high conductivity due to salt water contamination. A high conductivity due to high chloride levels makes the water unsuitable for irrigation unless treated or mixed with water from a more suitable source. Crops do vary in their susceptibility to chloride levels.

Table 1. Sensitivity of various vegetable crops to chloride levels in irrigation water

Tolerance Group	Crops	Safe levels mg/litre chloride	For limited use mg/litre	Risk of foliar damage mg/ litre
Very sensitive	Peas, French bean	Up to 100	100 - 200	Over 200
Moderately sensitive	Broad bean, Celery, Onion, Lettuce, Radish	Up to 200	200 - 300	Over 300
Slightly sensitive	Carrot, Cauliflower, Cabbage	Up to 400	400 - 500	Over 500
Least sensitive	Kale, Asparagus, Spinach, Garden beet	Up to 500	500 - 600	Over 600

Table 2. Suggested maximum trace element concentrations in water

Trace Element	Concentration (mg/litre)
Arsenic	0.04
Cadmium	0.02
Chromium	2.00
Copper	0.50
Molybdenum	0.03
Nickel	0.15
Selenium	0.02
Zinc	1.00
Lead	2.00
Boron	2.00

Microbial Contamination

The main microbial contaminants of water are fungal spores (e.g. *Sclerotium cepivorum* causing white rot of onion and garlic, *Sclerotinia* rot of *Alliums*). Irrigation can also provide an entry point for bacteria, such as *Escherichia coli*, into the food chain. This is of particular significance where water is applied to crops eaten raw, such as salad crops. A risk assessment of water source and period when contamination is most likely to occur should be undertaken. Irrigation water should be checked to ensure that it has an acceptably low level of microbial contaminants before use. Water obtained from deep boreholes is likely to contain fewer pathogens. Water lifted from rivers and stored in reservoirs prior to application will show a decline in pathogen levels over time. Drip irrigation, because it avoids wetting the aerial parts of the plant, may result in reduced bacterial loadings.

Hand-held equipment is available for monitoring E.C. and pH during the production season, but a full laboratory water analysis is recommended at least every 12 months for each water source used particularly during dry periods.

Consumptive use and Water use Efficiency

Consumptive use

Of the total absorbed water, hardly 0.1 to 0.5 per cent water is actually incorporated into the plant tissues and the remaining water is ward off from the plant body by transpiration. So, consumptive use is the quantity of water lost by evapo-transpiration plus that actually incorporated in the plant tissues.

Water requirement of a crop can be calculated from consumptive use or evapo-transpiration of a crop which is influenced by temperature, irrigation practices, length of growing season, precipitation and other factors. The volume of water transpired by crop plants depends on availability of water, absorption of water by the roots, temperature and humidity of the atmosphere, wind movement, intensity and duration of sunlight, stage of development of plant, type of foliage, nature of the leaves and kind of crop. Variations in consumptive use rate occur from day to day because of changes in climate. It is increased in hot, dry and windy day and is decreased in cool day. Consumptive use also varies with relative growth of the crop. The phenology of the growing plant can be characterized by vegetative, flowering, fruiting and other distinctive characteristics in vegetative stage like, curding in cauliflower, heading in cabbage, lettuce, bulbing in onion and garlic, tuberization in potato, sweet potato, etc. During the vegetative stage, consumptive use continues to increase till the end of this stage of growth. Flowering occurs near and during the peak of consumptive use.The fruiting stage is accompanied by a decrease in consumptive use until the transpiration essentially ceases during the latter part of the development of seeds inside the fruit.

Table 3. Classification of vegetable crops according to their water requirement*

High	Moderate	Low	Very low
Palak	Onion (bulb)	Pea	Watermelon
Amaranthus	Cucumber	French bean	Muskmelon
Lettuce	Chilli	Cowpea	Pumpkin
Celery	Brinjal	Broad bean	Wax gourd
Other leafy greens	Tomato	Cluster bean	
Sweet pepper	Carrot	Winged bean	
Cabbage	Potato	Hyacinth bean	
Cauliflower			
Asparagus			
Broccoli			
Rhubarb			
Brussels sprouts			
Horse radish			
Radish			
Beet			
Turnip			
Ridge gourd			
Rutabaga			
Green onion			

**Note:* The water requirement is expressed as the units of absorbed water required for the production of one unit of dry matter.

The water loss from the mature plants is about 50 to 90 per cent water which is lost from a free water surface on the same area. Evaporation from free water surface is estimated by using "Evaporation pan". Potential evapotranspiration is indirectly estimated from pan evaporation data after necessary adjustment.

Scheduling of irrigation

Frequency of irrigation and amount of water to be applied depend on a number of factors such as, depth of root system, water use efficiency, stages of growth, soil type, prevailing weather condition and actual consumptive use of the crop. It is a rather complex consideration. However, the relationships with different parameters are hereby discussed to frame some consideration.

Judging the critical stages of the crop

When the plants are young although they require and use less water, need highly moist soil because the root system of young plants is weak, sparsely distributed and located in the upper 15 to 20 cm layer of the soil that gets quickly dried. Critical stage is the particular growth stage (s) of the crop when water stress reduces the yield and quality of the product most. However, insufficient soil moisture during flowering and fruiting stages of almost all crops leads to falling off of flowers and poor ripening of fruits. Growth of fruit and other modified storage organs of vegetable crops are largely dependent on the net rate of water accumulation. Not only does water enter fruits and other storage organs in the xylem but it also enters in the phloem along with the assimilates.

Over-irrigation also causes different harms in many vegetable crops like, bursting of heads in cabbage, branching of roots in carrot, fruit cracking in tomato, etc. Abundant irrigation during ripening of fruits causes an increase in the water content of fruits and subsequently reduces their sweetness and storability. That is why, irrigation should be stopped before fruit ripening in watermelon, muskmelon, pumpkin, winter squash, etc. for which sweetness and/or storability of the fruits is the main quality criteria. In potato, onion, elephant-foot yarn, sweet potato, etc. irrigation should be stopped 15 days before the harvest of tuber/bulb/corm which greatly extends their storability.

Table 4. Critical stages of irrigation in some vegetable crops

Crop	Stages of growth
Tomato	Flower development, fruit set and after each harvest
Brinjal	Flower development, fruit set and after each harvest
Chilli	Tenth leaf to flower, to fruiting and after periodical harvests
Potato	Stolon formation, tuberization and tuber enlargement
Pumpkin, Cucumber, Watermelon, Gourds, Muskmelon, Summer Squash, etc.	Flower bud development and early fruit development
Cabbage	Head formation and enlargement
Cauliflower	Throughout the whole vegetative period
Broccoli	Early stages of development
Onion	Bulb formation and enlargement
Garlic	Bulb formation
Radish, Carrot, Beet, Turnip	Constant water supply during the whole vegetative period
Pea	Beginning of flower bud development and pod filling
French bean, Hyacinth bean	Flowering and pod development
Cowpea	Prior to flowering and after pod set

Lettuce (head type)	Head formation
Celery	Throughout the whole vegetative period
Sweet potato	40-45 days after planting at tuber formation
Okra	After fruitiest
Leafy vegetables	During the whole vegetative period
Elephant-foot yam	Shoot emergence stage

Considering the rooting depth of the crop

The depth, extent and configuration of the root system are important in judging water absorptive potential of the particular crop. Roots of the shallow-rooted vegetable crops are confined to the surface soil which remove water much more rapidly and dry the soil hence, require frequent irrigations. On the other hand, crops with extensive root system continue to extract water from greater soil depths. Thus, interval between irrigation for the deep-rooted vegetable crops can be extended. Knowledge of rooting depths is, therefore, important for scheduling the time and quantity of irrigation. Rooting depths vary greatly among the vegetable crops. The vegetable crops can be classified into 5 groups according to their rooting depth, viz., very shallow-rooted, shallow-rooted, moderately deep-rooted, deep-rooted and very deep-rooted.

Table 5. Rooting depth of different vegetable crops

Very shallow rooted (15-30 cm)	Shallow-rooted (30-60 cm)	Moderately deep-rooted (60-90 cm)	Deep-rooted (90-120cm)	Very deep -rooted (120-180 cm)
Onion	Broccoli	French bean	Pea	Artichoke
Lettuce	Brussels sprouts	Beet	Chilli	Asparasgus
Small radish	Cabbage	Carrot	Rutabaga	Lima bean
Chinese cabbage	Cauliflower	Chard	Summer squash	Parsnip
Parsley	Celery	Cucumber	Turnip	Pumpkin
Spinach	Endive	Brinjal		Winter squash
Cowpea	Garlic	Muskmelon		Sweet potato
	Leek			Tomato
	Potato			
	Palak			
	Radish			

The relationship between rooting depth and water requirement of the crop is not always direct. Some shallow-rooted crops like, cabbage, cauliflower, spinach and onion require relatively lesser quantity water as compared to the crops with deep root system like, tomato, muskmelon, etc., which is quite obvious. On the other hand, water requirement of some shallow-rooted crops like, potato, celery, etc., is much higher than some very deep-rooted crops like sweet potato, artichoke, lima bean, etc. Similarly, water requirement of very shallow-rooted crops like, onion and lettuce is same as that of the moderately deep to deep-rooted crops like, pole bean, beet, carrot, cucumber, pea, chilli, etc.

In vegetable crops, a great proportion of the total root weight is frequently found to occur near the upper surface of the soil. Hence, the deep-rooted vegetable crops often extract water mostly from the upper layers of the soil. Water and nutrient absorption also decreases rapidly with the depth of roots. So, moisture extraction depths of shallow- and deep-rooted vegetable crops are sometimes equated. In such situations, water requirement of the shallow- and deep-rooted vegetable crops differs with the parameters other than the rooting depths like, water use efficiency, duration of the crop, consumptive use, etc. Generally, shallow-rooted crops grown in sandy soil require moderate and frequent irrigation while deep-rooted crops in clayey soils need heavy irrigation at longer intervals.

Drip or Trickle irrigation system

Drip (trickle) irrigation offers the potential for precise water management and it is the most water-efficient method of irrigation. It also provides the ideal vehicle to deliver nutrients in a timely and efficient manner. However, intensive management particularly, appropriate irrigation scheduling, both in terms of timing and volume to be applied is required to achieve high water- and nutrient-use efficiency while maximizing crop productivity.

Factors affecting drip irrigation management

Many factors influence appropriate drip irrigation management including system design, soil characteristics, crop and growth stage, environmental conditions, etc. Influences of these factors can be integrated into a practical, efficient scheduling system which determines quantity and timing of drip irrigation. This system combines direct soil moisture measurement with a water budget approach using evapo-transpiration estimates and crop coefficients. However, apart from use on very high value vegetable crops, or where water supplies are very limited, the investment is currently difficult to justify on field-grown vegetable crops.

Basic approaches to scheduling drip irrigation

There are two basic approaches to scheduling drip irrigation: soil-moisture-based scheduling and a water-budget-based approach that estimates crop evapo-transpiration. There are limitations to both methods, but when used together they are the reliable way to determine both quantity and timing of drip irrigation.

Estimating evapo-transpiration : The amount of water evaporated from the soil surface and lost through transpiration of the crop is collectively called evapo-transpiration (ET). With drip irrigation, evaporation from the soil is minimized, particularly under plastic mulch, leaving crop transpiration as the main component of water loss. Environmental variables, primarily solar radiation, air temperature, relative humidity and wind are the driving forces behind evapo-transpiration. The common technique used to estimate relative evaporative demand is pan evaporation (Ep), the daily loss of water from a standardized open, water-filled pan. Neither pan evaporation nor ET calculated by empirical equations accurately reflects actual crop ET in all climates hence, the ET estimates commonly require modification based on local conditions. Generally, Ep averages 20-30% higher than comparable ET estimates.

Utilizing crop coefficients : Ep or ET values provide an estimate of relative evaporative demand. The amount of water actually lost from a particular field is mostly a function of crop growth stage as water loss increases with the expansion of the crop canopy. To account for crop growth stage, crop coefficients (K_C) are used to adjust either Ep or ET to fit current field conditions. Information on K_C values may be compiled from the published research results carried out in a particular agro-climatic condition.

Irrigation Frequency

The water budget scheduling system outline the volume of water required, but it does not suggest with what frequency it should be applied. It is difficult to generalize about drip irrigation frequency because there are a plethora of factors to consider (crop, root depth and distribution, soil water holding characteristics, drip wetting pattern, degree of automation, etc.). However, three basic considerations can simplify the issue:

- ❐ Early in the season when plants are small, it is beneficial to encourage roots to explore as much of the soil profile as possible. This maximizes nutrient uptake and maximize stress tolerance later in the season. The best approach to early season irrigation is to begin with a full soil profile and encourage deep rooting by not watering routinely, but rather waiting until the 20% depletion of available water is reached at the appropriate monitoring depth.

- In full crop growth stage, irrigation should be given when 20-40% of available soil moisture in the most active root zone is depleted. However, for sensitive crops like, celery, lettuce, sweet pepper, etc. depletion to 20% is allowed and for deep-rooted crops which can tolerate some stress like, tomato, muskmelon, etc. greater depletion may be allowed without loss of yield or quality.
- Individual applications may be limited to approximately 0.5 inches or less which limits the degree of root zone saturation after application and minimizes the amount of applied water likely to drain below the active root zone.

Main advantages of drip irrigation

- Potentially it is the most efficient irrigation system
- The uniformity of distribution and retention of water within the rooting zone causing potential water savings
- Appropriate irrigation management and maximising night-time irrigation
- It is very flexible design and layout which fits in almost any shape of field.
- Less labour required during the growing season.
- In extreme cases more saline water can be used because the water is applied below the leaf canopy.

Table 6. Water saving through drip irrigation of vegetables.

Vegetables	Water required (litres) to produce one kg	Water saving using drip irrigation (%)
Tomato	100-140	35
Watermelon	140-150	40
Okra	330-370	22
Brinjal	225-275	53
Bitter gourd	140-165	53
Ridge gourd	110-125	59
Cabbage	135-150	35
Radish	130-140	45
Beet root	145-160	30
Chilli	350-425	35

Main disadvantages of drip irrigation

- High capital cost hence, it need to be considered with crop value.
- Relatively high energy requirements.
- Requirement of skilled person to opreate the system.

Fertigation

In the later part of 1970, the new concept, Fertigation (fertilizers plus irrigation) has been perceived as an effective measure of applying chemicals and fertilizers to the crops using modern irrigation techniques. Fertigation offers an opportunity for precise application of water soluble fertilizers and other nutrients to the soil at appropriate time with desired concentration. The advantage is that the fertilizers are added in small, frequent doses, as plants need them, with less chance of leaching. Fertigation has the potential to increase produce quality and yield at lower cost and using less energy. Usually nitrogen compounds particularly, calcium nitrate, ammonium nitrate, sodium nitrate, urea, diammonium phosphate, potassium sulfate, potassium nitrate and chloride and sulphate salts of iron, zinc, copper and manganese are generally injected using drip irrigation system. Fertigation reduces fertilizer consumption by 30-50 per cent compared to conventional methods.

Specific considerations for fertigation

- The fertilizer type used and amount required must be soluble enough to dissolve completely in the water of the fertilizer tank.
- The drip irrigation system needs to be completely charged or pressurized before fertigation begins.
- The farthest point from the pump should be at full pressure before injection of any fertilizer begins. If the system is not fully charged, air surges could occur and cause uneven application rates.
- The fertilizer should be injected ahead of the filters to ensure that any undissolved particles are filtered out before fertilizer enters the drip system.
- The period of time in which fertilizer is injected into the system must be at least as long as that required to bring the entire drip irrigation system up to full pressure which will allow each dispensing orifice in the drip line to have the same contact time with the fertilizer solution as it passes through the system.
- Any fertigation system should have an anti-siphon device such as a vacuum breaker or backflow preventer to protect the water supply. If backflow prevention is not in place, fertilizer that remains in the system when it is shut down will siphon back through the pump and into the water source.

Selection of a fertilizer

Requirement of fertilizer materials for fertigation depend on the actual nutrient content and solubility of the material in water. Water in which fertilizers are to be dissolved should have pH levels between 5.8 and 7.8.

Table 7. Solubility of some common fertilizers used in drip fertigation

Fertilizer	Nutrient composition			Salt Index	Solubility (kg/100 litre of cold water)
	(% N)	(% P_2O_5)	% K_2O		
Calcium nitrate	15.5	0	0	52	102.11
Potassium nitrate	13.0	0	44.0	73	90.00
Ammonium nitrate	33.5	0	0	100	118.18
Sodium nitrate	16.0	0	0	100	82.02
Urea	46.0	0	0	75	78.19
Diammonium phosphate	16–18	46–48	0	29	42.99

Scheduling the fertilizer application

It can be applied as much fertilizer material with fertigation as with other methods, but better production can be realized with fertigation because of the increased availability of the nutrients to the plant as a result of nutrients being placed in close proximity to the roots. At the same time, less fertilizer can be used in the drip line to get the same production.

Fertilizer requirement: Requirement of fertilizer materials depends on the nutrient content in it. This means that 3 kg of ammonium nitrate or 6.5 kg of calcium nitrate per acre will equal 1 Kg of actual nitrogen per acre of nutrients being placed in close proximity to the roots. Solubility chart tells that 102.11 kg calcium nitrate or 118.18 kg ammonium nitrate can be dissolved in 100 liters of cold water.

Pre-plant application: Using a soil test as a guide, generally the recommended amounts of phosphorous and micronutrients and 20 to 40 percent of the recommended amounts of nitrogen and potassium are applied during final land preparation prior to applying the plastic mulch and drip tape. This "starter" fertilizer dose provides some nutrition to the crop during its early growth.

Injection schedule: The remaining nitrogen and potassium can be applied throughout the season as needed by the crop however, the specific schedule will vary depending on the crop. Weekly leaf tissue analysis is the best way to ensure that a crop is receiving adequate nutrients.

Table 8. General ranges of sufficient nutrients in leaves of some vegetable crops

Nutrient	Musk-melon	Cu-cumber	Brinjal	Sweet pepper	Summer squash	Tomato	Water-melon
				(percentage)			
N	2.00–4.00	5.0–6.0	4.0–6.0	5.00–5.50	4.00–6.00	3.50–5.00	2.50–4.50
P	0.25–0.40	0.3–1.0	0.3–1.0	0.35–0.45	0.25–1.00	0.35–0.45	0.25–0.75
K	1.80–4.00	4.0–5.0	3.5–5.0	4.50–6.00	3.00–5.00	3.50–5.00	2.25–3.50
Ca	1.80–7.00	1.2–3.5	1.0–2.5	1.00–1.50	1.00–2.50	1.00–1.50	1.10–1.50
Mg	0.50–1.50	0.3–1.0	0.3–1.0	0.30–0.80	0.30–1.00	0.30–0.80	0.25–0.80
S	0.20–0.60	0.2–0.8	0.2–0.8	0.15–0.40	0.20–0.75	0.20–0.40	0.20–0.75
				(ppm)			
B	20–60	25–75	25–75	25–95	25–75	30–100	30–75
Fe	40–200	50–200	50–200	40–200	40–200	45–200	40–200
Mn	20–200	25–200	25–200	20–200	25–200	20–200	25–200
Zn	20–60	20–75	20–75	20–60	20–75	20–60	20–75
Cu	4–25	5–35	5–35	5–35	5–35	5–35	4–15

Frequency of fertilizer injection can range from once a week to daily. A general recommendation is to start with small amounts of N and K and increasing the rates as the season progresses and the crop demand increases. However, fertigation is yet to be adopted at a large scale in India mainly due to high installation cost, non-availability of soluble fertilizers and high cost of imported fertilizers.

Chapter 9

Integrated Pest and Disease Management for Vegetable Crops

Most of the insect pest and disease (insects, mites, diseases, nematodes, etc.) management problems arise from relying entirely on pesticides for control. Integrated pest management (IPM) is a strategy that draws on a range of management tools with the goal of using the least ecologically disruptive techniques to manage pests within economically acceptable levels. IPM considers the production system in a holistic manner and looks at all aspects of the farming enterprise as potentially increasing or decreasing pest numbers and, where possible, enhancing the activities that reduce these pest populations.

Strategy for Integrated Pest and Disease Management

The basic IPM strategy involves routine crop monitoring to ensure that pesticides are only applied when needed, as well as to ensure appropriate timing of pesticide applications. The most developed bio-intensive IPM relies primarily on beneficial organisms to manage insect pests. When greater pest control is needed, interventions chosen are complementary to the survival of these beneficial organisms. As other pests are incorporated more and more prevention strategies are adopted, which reduces the need for direct control practices.

Prevention

Where possible, it is preferable to prevent pest problems rather than manage them after they arrive – *prevention is always better than cure*. However, what preventive measures need to be taken will depend on the particular situation and virulence of the pest and disease. The followings are some prevention strategies that can be important for vegetable crops:

- Selection of varieties having resistance or tolerance to important diseases or insect pests in that particular area
- Use of healthy and disease-free (particularly virus) seedlings because both seeds and seedlings can be a source of many insect-pests, nematodes and diseases.

- Controlled irrigation to minimise the period of leaf wetness which will reduce foliar diseases.
- When foliar diseases are present, avoidance of inter-cultural works in the crops while foliage is wet to reduce spread of the disease
- Judicious application of plant nutrients and avoidance of excess nitrogen application which will reduce crop susceptibility to some fungal diseases.
- Removal of weeds from within and around cropping areas because weeds are known hosts of crop diseases or insect pests e.g. sow thistles (*Sonchus oleraceous*) are hosts of lettuce necrotic yellow virus; *Plantago lanceolata* is the alternate host of broad bean wilt virus and *Brassica* weeds are hosts of a range of diseases and insect pests of *Brassica* vegetable crops.
- Rouging of diseased plants and elimination of egg mass and gregarious lepidopteran insect pest at early stage followed by proper destruction by burning and burring under the soil.
- Crop rotation with the inclusion of non-host crops like, wheat, maize, marigold, etc. to reduce the inoculum load of bacterial wilt pathogen (*Ralstonia solanacearum*) and root knot nematodes (*Meloidogyne incognita*, *Meloidogyne javanica*) in the field.
- Use crop records to identify factors or management practices that may be encouraging or discouraging pests because IPM is knowledge-based and relies on local experience.

Crop monitoring

Scouting and monitoring for the incidence of pest and disease is the first and most fundamental step in adopting IPM. Scouting also helps the growers to determine when insect pests become active or arrive and their relative population size in the field. Scouting is also critical for detecting early symptoms of disease both in the seed bed and main field which is important to adjust management tactics. Scouting for diseases includes monitoring the crop for symptoms, like leaf spots and signs like visible fungal colonies. Scouting for insect pests includes scouting fields to detect arrival of critical pest life stages – egg, immature or adult – or signs of their activity, including entry holes and insect excrement. In many cases, monitoring also involves collecting data on insect pest population size or level of damage, and relating these data to economic thresholds to make control decisions.

A basic strategy involves visual monitoring the same number of plants on a regular basis (e.g. 40 seedlings or 20 mature plants). It is recommended that monitoring sites are widely dispersed throughout the planting, including near crop boundaries. Monitoring should be carried out on at least a weekly basis – at critical points in crop development. In pest outbreak situations, crops should be monitored more frequently. Different monitoring tools include vacuum samplers, pheromone traps and sticky traps. Pheromone traps attract male moths, and do not always give a good indication of the populations of female moths that are depositing eggs into the crop.

Sticky traps are also a useful tool for assessing thrips, whitefly and other small flying species. They are usually yellow, although blue traps can be used for monitoring thrips. In areas where tomato spotted wilt virus is common, sticky traps are regularly used to monitor levels of Western flower thrips (WFT), *Frankliniella occidentalis*.

Tools for scouting

- **Hand lens :** It is used for inspecting small insects, mites, insect eggs or feeding damage.
- **Small tally counter :** It is used to keep accurate counts of insect pests.
- **Traps of various forms :** It includes sticky traps, pheromone traps, etc.
- **Camera :** It is used for taking photographs of the insect pests and disease symptoms.
- **Sweep net :** It is used for collecting insects from foliage
- **Shovel or spade and containers (paper bags, cups) :** These tools are used for collecting plant, disease and insect samples.
- **Ice box :** It is used for transporting and preserving samples.
- **Permanent marker :** It is used to label the containers with pest samples.

Methods for scouting

The general recommendations followed for this purpose are:

- Making sections of the whole fields into manageable portions based on location, size and crop or variety for scouting separately
- Walking a path in the field in an X or a W pattern to cover the whole field for scouting
- Weekly scouting of each field.
- Intensive scouting if biological information is available to predict the emergence or arrival of certain pests and disease.

Documentation of the insect pests, signs of insect damage and disease

- Cupped, chlorotic, spotted or malformed foliage.
- Discoloured, damaged, swollen or sunken areas in plant tissues.
- Insect feeding, such as holes in leaves, entrance holes in stems, chewing or rasping damage or their excrement.
- Aggregation of insects.
- Pockets of less vigorous or dying plants.
- Anything out of the ordinary.

Intervention

Decision to take active intervention with pesticides depends on crop monitoring information, past crop records and any pest 'economic threshold' (the point at which the cost of control is equal to yield loss if no control measure is taken). However, economic thresholds are very hard to identify in vegetable crops given the large fluctuations in crop price and quality. Decisions need to be taken largely from experience, keeping records and evaluating the success of particular decisions. The actual control option to be selected should least likely to disrupt beneficial organisms.

Understanding the biology of the pathogen, host-pathogen interactions, and the effect of environmental factors on this dynamic process in time and space (disease epidemiology) is critical for planning and implementing effective and efficient management strategies. The factors most likely to affect disease control decisions are varietal susceptibility, diseases present in the crop, whether the climatic conditions favour spread or development, crop vigour/health/nutritional status, irrigation methods, crop destination, the effectiveness of control options and if the disease is transmitted by an insect vector, the population size and source of that insect.

Evaluation

Intervention by application of pesticides should be evaluated after application because IPM continues to evolve as new pests arrive, new management options become available and new techniques are adopted. Use of a broad-spectrum insecticide may wipe out the target insect pest as well as the beneficial insects present in the crop, which may result in a minor insect pest becoming a serious problem. Over time, an IPM strategy will tend to shift effort from intervention strategies to prevention strategies.

Bio-control agents for pest management

Egg parasitoids : *Trichogramma brasiliensis* is a parasitoid specifically useful to control polyphagus pest fruit borer (*Helicoverpa armigera*) which attacks tomato, peas and beans, etc. With the onset of adult catch in pheromone trap or first flowering, release of the parasitoids as parasitized eggs @50,000/ha with repeated inoculative release at the same rate at weekly interval is very effective. Similarly, *T. chilonis* and *Trichogrammatodea bactrae* are effective to control okra shoot and fruit borer (*Earius* spp.) and diamondback moth (*Plutella xylostella*) on crucifers, respectively. Weekly release of *T. chilonis* @250,000 parasitized eggs /ha has positive effect in control of brinjal shoot and fruit borer.

Larval parasitoids : *Cotesia plutellae* and *Diadegma semiclausum* are dominant larval parasitoids of diamondback moth, among which the later one is susceptible to high temperature and is mostly active in areas having mild climate. Very high natural parasitization by these larval parasitoids indicates the importance of their role in pest suppression if conserved through proper management practices. Use of neem seed kernel extract (4%) helps to control the pest with least impact to these larval parasitoids. *Campoletis chlorideae* is most common larval parasitoid of *Helicoverpa.* Usually second instar larvae are parasitized in the field. This parasitoid is mass multiplied on young larvae of *Helicoverpa armigera* and *Spodoptera litura.*

Predators : Ladybird beetles (*Coccinella septempunctata* and *Chilomenus sexmaculatus*) are natural predators of aphids and other soft-bodied insects. These predators can easily be mass multiplied and used for suppression of aphids on different vegetable crops. Generally, adult beetles released @30 beetles/sq.m are sufficient to control the aphids. Green lace wings (*Chrysoperla carnea*) are commonly found to predate on soft-bodied insects like aphids, thrips, jassids, whiteflies and eggs of some lepidopteran pests. Mass release of early instars of these predators @ 50,000 eggs/ha is very effective in the management of sucking pests on vegetable crops.

Microbial biocontrol agents : Insect pathogenic bacteria, fungi and viruses are mostly used for the management of the pests of vegetable crops. Unlike natural enemies, these agents can be utilized as a prophylactic measure as done in insecticides and have an advantage of field applicability. The bacteria *Bacillus thuringiensis* (*Bt*) is most widely used microbial insecticide which is useful in the management of several leaf feeding lepidopteran pests (caterpillar) of vegetable crops. Several commercial formulations of *Bt* are now available in markets which are generally used as foliar spray @1.0 kg/ha. It is particularly effective against diamondback moth on cabbage however, it is also recommended for the control of fruit borer on okra. Nuclear Polyhedrosis Viruses (NPVs) are highly target specific entomopathogens with least impact on natural enemies. The NPV's have been widely used for the management of fruit borer *(Helicoverpa armigera)* on tomato and leaf eating caterpillar *(Spodoptera litura)* on cauliflower. It is recommended to spray Helicoverpa NPV and *Spodoptera* NPV @250-300 Larval Equivalent (LE) /ha mixed with 1% jaggery and 0.1% Teepol in 250 litres of water on tomato and cauliflower, respectively in the evening hours to get the optimum control of these insect pests.

Examples of IPM way of controlling insect pests in different vegetable crops

Nursery bed management for different vegetable crops

- Soil solarization with transparent polyethylene sheet during summer is ideal to suppress the soil borne pathogens in nursery. Seed treatment with *Trichoderma* spp. @ 5-10 g/kg or soil treatment @ 10-25 g/ sq.in nursery area is effective against damping off diseases.

Brinjal

- Seed treatment with imidacloprid @ 2.5g/kg reduces the little leaf and leafhopper incidence.
- Intercropping of brinjal (2 rows) with coriander (one row) or fennel (1 row) minimizes the incidence of brinjal shoot and fruit borer.
- Integrated pest management of fruit and shoot borer includes raising the nursery under nylon net cover, removal and destruction of infested shoots along with larvae and periodical cleaning of dry leaves and old infested fruits at weekly interval along with the insect larvae starting from 25 days after transplanting and installation of sex pheromone baited plastic funnel traps at 10m distances just above canopy from 20 days after planting and need-based foliar spray of NSKE (4%) during late afternoon.

Tomato

- Seed treatment with imidacloprid @ 2.5g/kg, raising the nursery under nylon net cover, single spray of imidacloprid (Confidor) @ 0.3 ml/litre about 40 days before fruiting and subsequent spraying of 4% neeem seed kernel extract (NSKE) with 0.1% sticker minimizes whitefly population and reduce the incidence of tomato leaf curl virus (TLCV) and leaf miner.
- Planting of two rows of marigold as trap crop with every 14 rows of tomato along with release of *Trichogramma brassiliensis* @50,000 parasitized egg/ha at flowering stage and spraying of *Helicoverpa* NPV @ 350 LE/ha twice after flowering can manage the fruit borer (*Helicoverpa armigera*) satisfactorily.
- Use of synthetic pyrethroids and endosulfan alternatively with NSKE (4%) is also effective against fruit borer (*Helicoverpa armigera*).

Okra

- Seed treatment with imidacloprid @ 2.5 g/kg is effective against leafhopper in okra during early vegetative stage.
- Spraying of NSKE (3%) spray with 0.1% sticker minimizes the incidence of insect pests in the field.
- Spraying of endosulfan and neem formulation at their half doses at flowering stage can reduce borer (*Earius* spp.) infestation.
- Okra sown during 2nd week of June minimizes the incidence of both jassids (*Amrasca biguttula biguttula*) and shoot and fruit borer (*Earius* sp.).

Chilli

- Seed treatment with imidacloprid @ 2.5 g/kg is effective against thrips (*Scirtothrips dorsalis*) and mites (*Polyphagotarsonemus latus*, *Tarsonemus translucens*, etc.) complex during initial stage.
- Alternate spray of dicofol @ 2.75 ml/lit and wettable sulphur @ 2 g/lit is highly effective against mite.

Cucurbits

- Effective fruit fly management include deep summer ploughing to expose the dormant pupae in the field, crop rotation incorporating non-cucurbitaceous crops, food bait consisting of rotten banana or smashed pumpkin pulp (1000g) + yeast (10 g) + malathion (10 ml) + citric acid (3 g) in flat earthen pot to attract and kill the adult flies and bait spray in one plant per 10 m^2 area with 20 ml malathion (50 EC) + 20 liters water +500 g molassess keeping rest of the plants remains free from insecticides to attract adult flies and control the population.
- Spraying of carbaryl 50 WP @ 2 g/lit at cotyledon stage control red pumpkin beetles.

Cabbage

- Diamondback moth (DBM) population in cabbage can be managed by growing paired mustard rows with every 25 cabbage rows. The first rows of mustard should be sown 15 days before and second 25 days after cabbage planting. The trapped DBM larvae on mustard can be killed by spraying 0.1% Dichlorovos or 5% NSKE. Such practices conserve the population of *Cotesia plutellae,* a dominant natural enemy of DBM.

Biological management of diseases in vegetable crops

Many fungi, bacteria and actinomycetes have been found promising against different soil-borne pathogens causing root rot, damping off, collar rot and wilt in different vegetable crops. Important target pathogens are *Pythium* spp.,

Sclerotium rolfsii, Macrophomina phaseolina, Rhizoctonia solani, Fusarium spp., *Phytophthora* spp., *Phoma* sp. *Sclerotinia sclerotiorum, Aspergillus* spp., *Botrytis, Armillaria, Verticillium, Alternaria solani, Colletotrichum capsici and Pleospora beta.* Several species of *Trichoderma* have gained attention for their highly antagonistic property. Some of the antagonists are highly specific to a particular pathogen, but most of the fungal antagonists have broad-spectrum parasitic host range. Some antagonistic fungi also kill some plant parasitic nematodes infecting vegetables.

Potential biocontrol agents

Soil has maximum microbial biodiversity, which can be utilized for plant disease management. Several antagonists have been isolated and many of them are being utilized in management of soil-borne pathogen. The maximum biocontrol agents belong to fungal genera, viz. *Trichoderma* spp., *Aspergillus niger, A. flavus, Pythium nann, Trichothecium, Paecilomyces lilacinus, Penicillium funiculosus, Myriothecium, Corticium, Pythium cligandrum, Peniophora gigantea, Candida oleophila, Sporidesmium sclerotivorum, Coniothyrium minitans, Ampelomyces quisqualis, Chaetomium, Cladosporium, Fusarium semitectum, Tuberculina, Phialophora, Catenaria* and *Verticillium.* Some bacterial antagonists like, *Pseudomonas* spp., *Agrobacterium radiobactor, Bacillus* spp. and Actinomycetes such as *Streptomyces griseus, S. rimosum* have also been utilized for management of diseases. *Trichoderma* spp. has extensively been exploited for the management of soil borne diseases.

Crop	Pathogen	Antagonist
Tomato, brinjal, chilli,bean	*Rhizoctonia solani*	*Trichoderma harzianum, T. viride*
Tomato, chilli, brinjal, bean, pea, cucumber	*Pythium* spp.	*T. virirde, T. harzianum, T. koningii*
Tomato, chilli, brinjal ,bean, pea	*Sclerotium rolfsii*	*T. virirde, T. harzianum, T. koningii*
Pea, radish	*Pythium* spp., *R. solani*	*T. virirde, T. harzianum, T. hamatum*
Tomato, muskmelon	*F. oxysporum, P. aphanidermatum*	*T. virirde, T. harzianum*
Okra	*Macrophomina phaseolina*	*T. virirde, T. harzianum*
Cucumber	*Sclerotinia sclerotiorum*	*T. harzianum*

The best substrate is sorghum or mixture of sorghum with either rice bran or rice husk. Talcum powder is commonly used as carrier in commercial biopesticide product. The USA, Israel, Bulgaria, Australia, Sweden, Canada and India have commercialized the use of biopesticides in management of soil-borne diseases. Successful commercial use of *Trichoderma* spp. and *Pseudomonas fluorescence* has been demonstrated over past ten years in most part of the country. The biopesticides are either mixture of two bioagents or alone are available with different trade names, viz. *T. viride* (Antagon TV, Trichoguard, Niprot, Bioderma Tricontron), *T. harzianum* (Trichodex, Trichodermin, F. stop, Bio-Coat), *T. virens* (Gliogard), *T. viride* + *P. flourescens* (An tag or Combi), *T. harzianum* + *Bacillus subtilis* (Combat), *P. fluorescence* (Biomonas, Dagger G, Conquerer).

Mechanisms of biocontrol

Three antagonistic mechanisms are involved in biological control of soil-borne plant pathogens, viz. competition, antibiosis and mycoparasitism. Competition occurs when there is a demand of two or more microorganisms for the same nutrient resource in limited space and time. This is most potent mechanism to control *Fusarium oxysporum* f. sp. *melonis* by *Trichoderma harzianum*. Antibiosis occurs when production of toxic metabolites or antibiotics by the biocontrol agent has a direct effect on another organism and such toxic metabolites include gliotoxin (sulfur-containing mycotoxin produced by several species of fungi) produced by *T. virens,* viridian (antifungal metabolite) by *T. viride* and trichodermin (A tetracyclic spiroepoxide which acts as an antifungal and protein synthesis inhibitor) by *T. harzianum.* In case of mycoparasitism or hyperparasitism, fungus is parasitized on another by chemotropic growth, recognition, attachment and degradation. Production of lytic enzymes, chitinases and α 1, 3 glucanases are most important for the lyses of pathogen cell wall. *Trichoderma* species compete aggressively, mycelium coils around pathogen hyphae and cause lyses of the cells of *Phytophthora* spp.

Flavobacterium balustinum and *Xanthomonas maltophila* are most effective in inducing suppression of damping off causing fungus *Rhizoctonia* sp. by *induced systemic resistance* (ISR) mechanism. Many of the antagonistic bacteria also act as plant growth promoting rhizobacteria (PGPR) and systemic acquired resistance (SAR) in plants against many disease causing pathogens of vegetable crops.

Application of biocontrol agents

Soil application and seed treatment are only feasible method of using bioagents under the field condition. Seed treatment is preferred in direct sown crops, whereas soil application and seed treatment are feasible in the crops which are transplanted in the main field. Best utilization of any biocontrol agents in soil is only possible when it is supplemented with well-decomposed organic amendments, congenial moisture and protected from extreme weather conditions till it establishes in soil.

Biological seed treatment can be done by priming on seed, soaking seeds, seedling dipping and dry power treatments depending upon the nature of biocontrol agents. Field studies clearly showed that *T. viride* is better than any other biocontrol agents in this respect. Seed treatments with *Trichoderma* @ 6-10g/kg of seed is effective provided spore concentration remain in between 106 and 109/ml. Dose for either seed or soil treatment is influenced by inoculum of pathogen present in soil and nature of seed. Soil application of *Trichoderma* @ 10-25 g along with 100 g neem cake/ sq.m is effective to manage nursery diseases of different vegetable crops.

Incorporation of green manures along with soil application of *Trichoderma* @5 kg/ha give good control of most of the soil borne diseases in *kharif* season grown vegetable crops. The *Trichoderma* must be applied soon after incorporating green manuring crop preferably within three days depending upon moisture status and texture of soil.

Root dipping of vegetable seedlings by *Trichoderma* formulation @ 1% for 10-25 minutes is very effective in checking root rotting after transplanting in *kharif* season. Drenching of *Trichoderma* @ 1% near collar region should be carried out 20 days after transplanting to check root rot, collar rot and wilt incidence. Fast and vigorous seedling growth is recorded in biocontrol treated nursery beds where 20-25 days old seedlings are ready for the transplanting.

Foliar spray of *Trichoderma* on all ground portions in pointed gourd controls *Phytophthora* blight of both vine and fruit. One resident isolate of *Aspergillus niger*-V @ 1% along with sticker and molasses as foliar spray against *Alternaria* blight and anthracnose of chilli gives satisfactory control.

Sensitivity of the bioagents with pesticide

The fungicides, which are completely insensitive to the bioagents, but sensitive to the pathogen, can be used together to increase the efficiency of disease management strategies. The bioagents show tolerable growth after 7 days of incubation with Captan, chlorothalonil and propineb up to 200 ppm concentration.

Copper hydroxide shows synergistic effect at lower concentration, while fungistatic at higher concentration (400 ppm) for five days. Pencycuron is highly effective against *Rhizoctonia* spp. but insensitive to biocontrol agents and gives full protection when integrated both at a time. Different fungicides like, pencycuron, copper hydro oxide, captan, azoxystrobin carboxin, sulfex and antibiotic like streptocycline can safely be used in bio-control strategies with *Trichoderma* for the management of soil-borne diseases in vegetable crops.

Imidacloprid can be integrated with *T. virens* and *T. viride* with slight adjustment in dose. Seed rate should be compensated according to soil type, insect endemic area and viral infection to get desired seedling stand of tomato. Soil application of carbofuran followed by seed treatment by antagonist can be integrated for insect pest and disease management without adverse effect on seed germination. Pesticides like cypermethrin, fenvalerate and herbicide like glyphosate are safe for integration with *Trichoderma* for management of pest and disease.

Chapter 10

Weed Management for Vegetable Crops

Weeds are unwanted and undesirable plants which interfere with the utilization of land and water resources and thus adversely affect crop production.

Classification of Weeds

Weeds can be classified in three different manner based on morphology, life cycle, habitat and origin.

1. Classification based on morphology

- Broad leaf: *Amaranthus spinosus*, *Eclipta alba* , *Portulaca oleracea,* etc.
- Grasses : *Echinochola crusgalli*, *Cynondon dactylon*, etc.
- Sedges: *Cyperus rotundus*, *Cyperus diformis,* etc.

2. Classification based on life cycle

- Annuals: Annual weeds are usually small herbs with shallow roots and weak stem and are propagated by seeds and mature in one season.
  - Rabi season annuals: *Avena fatua*, *Phalaris minor*, *Ageratum conyzoides*, *Chenopodium album*, *Argemone mexicana*, *Cirsium arvensis*, etc.
  - Kharif season annuals: *Dactyloctenium aegypticum*, *Echinochola colonum*, *Amaranthus spinosus*, *Amaranthus viridis*, *Solanum nigrum*, etc.
- Biennials: Biennial weeds require two years to complete their life cycles. In the first year they remain vegetative and in the second year they produce flowers and set seeds. These types of weeds are mainly found in non-cropped areas. Common examples of biennial weeds are wild carrot or Queen Anne's Lace (*Dacus carota*), *Alternanthera echinata*, *Cirsium vulgare* , *Plantago* spp. , *Arctium* spp. etc.

- Perennials : Perennial weeds live more than two years and they reproduce both vegetatively and sexually through producing seeds.
  - Simple perennials : These perennials are propagated only by seeds e.g. *Sonchus arvensis*.
  - Bulbous perennials : These perennials possess a modified stem with scales and reproduce mainly by bulbs and seeds e.g. *Allium* sp.
  - Corm perennials : These perennials possess a modified shoot and fleshy stem and reproduce mainly by corm and seeds e.g.Timothy grass (*Phleum pratense*).
  - Creeping perennials : Creeping perennials are reproduced by seed, rhizome (*Sorghum halpense*), stolon (*Cynodon dactylon*), enlarged root system with numerous buds (*Convolvulus arvensis*), tubers (*Cyperus* sp.), etc.

3. Classfication based on habitat

- Terrestrial weeds : These weeds are mostly found in upland or dry land condition having enough moisture but the land is not water logged e.g. *Chenopodium album*, *Solonum nigrum*, *Celosia argentea*, etc.
- Aquatic weeds : These weeds grow in water and complete at least a part of their life cycle in water which are of several types.
  - Free floating : *Eichhornia crassipes* , *Pistia stratiotes*, etc.
  - Rooted floating : *Trapa bispinosa* . *Ludwigia adscendens* ,etc.
  - Submerged : *Ceratophyllum demersum* , *Hydrilla verticillata* ,etc.
  - Immersed : *Jussieua repens* , *Nelumbium speciosum* ,etc.
  - Amphibious : *Ranunculus aquatilis* , *Scirpus supinus* ,etc.

4. Classfication based on the origin of weeds

- Alien (Foreign origin) : *Parthenium hysterophorus* , *Argemone mexicana*, etc.
- Apophytes (Indigenous) : *Cynodon dactylon*, *Cyperus rotundus*, *Saccharum spontaneum*, *Acalypha indica*, etc.
- Antrophytes (Introduced by man) : *Phalaris minor*, *Corchorus acutangulua*

Table 1. Common weeds found in the field of vegetable crops

Common Name	Scientific Name	Common Name	Scientific Name
Wild oat	*Avena fatua*	Chottis wank	*Echinochola colona*
Gullidanda/Canery grass	*Phalaris minor*	Cow foot grass	*Dactyloetenium aegyptium*
Swank/ Maddal	*Echinochola crusgalli*	Doob grass /Bermudagrass	*Cynodon dactylon*
Motha /Nut grass	*Cyperus rotundus*	Kans	*Saccharum spontancum*
Goose grass	*Eleusine indica*	Bathua	*Chenopodium album*
Chaulai	*Amaranthus viridis*	Tandla	*Digeria arvensis*
Dodhak	*Euphorbia hirta*	Fumatory /Pitpapra	*Fumaria parviflora*
Meadow pea/natri	*Lathyrus aphaea*	Wild onion	*Asphodelus tenuifolius*
Tiger foot morning Glory	*Ipomoea pestigridis*	Con spurry/ Jangli dhania	*Spergula arvensis*
Dendelon/Jangli gobhi	*Launea asplenifolia*	Wild carrot	*Daucas carota*
Milk weed	*Sonchus arvensis*	Jangli palak	*Rumex dentatus*
Thorny amaranthus	*Amaranthus spinosus*	Night shade	*Solanum nigrum*
Johnson grass	*Sorghum halpense*	Oroanche	*Orobanchae aegyptica*

Crop-weed competition

Only a few vegetable crops like, cabbage (*Brassica capitata*), artichoke (*Cynara scolymus*), etc.are good competitors with weed flora because they quickly cover the soil, topping the weed growth. However, most vegetable crops particularly of temperate latitudes such as, onion, leek, garlic, carrot, sweet pepper, etc. grow slowly and they cover the soil very sparsely, suffering strong weed competition not only for water, nutrients and light, but even for space. Thus, if weed control is not carried out timely, there will be severe reduction in production. Most of the vegetable crops are highly sensible to early weed competition which necessitates control weeds at early stages. Weed competition is especially dramatic when vegetable crops are grown by direct seeding. The critical period of weed competition (i.e. the period during which weed control measures must have to be taken) is usually longer in direct-seeded than in transplanted crops. Obviously, weather conditions and weed density have a great influence on the length of critical periods.

Weeds harbour many vegetable diseases, nematodes, mites and insects, especially aphids and thrips that transmit viruses. Growth of weeds in vegetable production systems is enhanced by soil disturbance, irrigation and the application of fertilizers.

Basic Impacts of Weeds on Crop Production and Different farm Management Practices

Increased cost of production

Weed management costs are estimated to be about 11% of total pre-harvest variable production costs. However, cost of weed management varies based on the crop type and other associated costs e.g. labour charge, cost of inputs, etc.

Reduced yield and crop quality

Weeds, when combined with diseases and pests, can reduce yield in vegetable crops by 10 to 70% by competing for water, soil nutrients, light and space that ultimately restricts the development of the plant. However, impact on yield depends on crop type and stage of the crop. For example, if no weed control measures are taken, reduction in yield is approximately 20% in broccoli, 25% in lettuce but up to 90% in carrot. Other factors that influence yield include pest type, soil type, weather conditions, timing of infestation and crop management.

Weeds can host pests and diseases that reduce crop quality and yield, as well as increase the cost of pesticide use. Insect pests spread the majority of viruses within a crop, or from weeds to crops.

Table 2. Examples of some weed host for different insect pests and pathogens

Weed	Insect pest	Crop host
1. Mustards	Cabbage root maggot (*Delia radicum*)	Cabbage, broccoli
2. Chick weed (*Stelaria media*)	Melon aphids (*Aphis gossypi, Myzus persicae*)	Most cucurbits
3. Wild morning glory (*Convolvulus arvensis)*	Aphids (*Aphis gossypi, Myzus persicae*)	Watermelon, pumpkin, ridge gourd, etc.
4. Wild carrot (*Daucus carota)*	Carrot rust fly (*Psila rosae*)	Carrot, celery.
5. Night shade (*Solanum nigrum)*	Onion thrips (*Thrips tabaci*)	Onion
6. Lambs quarter (*Chenopodium album)*	Common stalk borer (*Papaipema nebris*)	Mostly tomato
Weed	**Pathogen**	**Crop host**
1. Pig weed (*Amaranthus viridis)*	Spinach mildew causing fungus (*Peronospora farinosa* f sp. *spinaciae*)	Spinach, Swiss chard
2. Night shade (*Solanum nigrum)*	Verticillium wilt causing fungus (*Verticillium dahliae*) Tomato mosaic virus	Tomato
4. Night shade (*Solanum nigrum)*	Potato mosaic virus	Potato
5. Mustards (*Brassica* sp.)	Clubroot causing fungus (*Plasmodiophora brassicae*)	Cabbage, Cauliflower

Impact on farm management

- Reduced effectiveness of insecticide and fungicide applications due to plant density.
- Increased difficulty in harvesting crop through reduced access of the equipment.
- Limited registered herbicides are available for controlling weeds within the crop rows, particularly for broadleaf weeds after crop emergence which has led to the increase in the use of pre-emergent herbicides, plastic mulch, precision shallow cultivation and hand weeding where feasible.

- Increased vulnerability of the crop before the crop canopy develops within the first few weeks after emergence, for example carrot, cabbage and beet.
- Some crops that don't form a canopy, such as onion, garlic and leek will remain at a competitive disadvantage with weeds and are more challenging to manage.
- Some weeds secrete harmful chemicals that may have adverse effects on crop plants, soil, human being and livestock. In India, *Parthenium hysterophorus* cause contact dermatitis and asthma. *Rhus radicans* (Poison ivy), *Ambrosia sp.* (Ragweed) cause itching, hay fever and other debilitating allergies contributing to chronic human illness and suffering.
- The perennial rhizome producing grasses like, *Imperata cylindrica, Cynodon dactylon, Sorghum halepense, Saccharum spontaneum* cause considerable damage to the land and the land sometimes becomes unsuitable due to heavy infestation of these weeds.
- Presence of spiny and thorny weeds like, *Amaranthus spinosus, Argemone mexicana* etc. restrict movement of farm workers in carrying out farm practices.
- Some prominent aquatic weeds are *Eichhornia crassipes, Typha angustata, Hydrilla verticillata, Ipomoea aquatica,* etc. may block irrigation and drainage channel impeding the flow of water. These weeds may break the canal banks which may cause seepage, flooding, silting, and inadequate delivery of irrigation water to the farms located at a distance from main water source.

Methods of Weed Control

Cultural method

1. Crop rotation

Crop rotation is the programmed succession of crops during a period of time in the same plot or field. It is a key control method to reduce weed infestation in vegetable crops. Crop rotation was considered for a long time to be a basic practice for obtaining healthy crops and good yields. However, over-dependence of agro-chemicals sidelined this cultural concept. At present, crop rotation is gaining interest and is of value in the context of integrated crop management. Classically, crop rotations are adopted with the following principles:

1. Alternating crops with a different type of vegetation: leaf crops (lettuce, spinach, palak, amaranthus, kale, etc.), root and tuber crops (potato, sweet

potato, carrot, radish, beet, turnip, etc.), bulb crops (leeks, onion, garlic), fruit vegetable crops (tomato, brinjal, sweet pepper, watermelon, muskmelon, pumpkin, summer squash, etc.).

2. Avoiding succeeding crops of the same family: Apiaceae (celery, carrot), Solanaceae (brinjal, tomato, sweet pepper, etc.).
3. Alternating poor- (carrot, onion, leek, etc.) and high-weed competitors (maize, cabbage, potato).

Examples of important crop rotations are as follows:

- Melon - beans - tomato
- Lettuce - tomato - cauliflower
- Potato – cole crops - tomato
- Tomato - okra - French bean

2. Land preparation and tillage

Suitable land preparation depends on a good knowledge of the weed species prevalent in the field. When annual weeds are predominant, the objectives are unearthing and fragmentation which must be achieved through shallow cultivation. When perennial weeds are present, adequate tools will depend on the types of rooting. Pivot roots (*Rumex* spp.) or bourgeon roots *(Cirsium* spp.) require fragmentation and this can be achieved by using a rotovator or cultivator fixed with a tractor. Weeds with fragile rhizomes (for example, *Sorghum halepense*) require dragging and exposure at the soil surface for their depletion, but weeds with flexible rhizomes (for example, *Cynodon dactylon*) require dragging and removal from the field. This can be done with a cultivator or harrow fixed with a tractor. Weeds with tubers (for example, *Cyperus rotundus*) or bulbs (for example, *Oxalis* spp.) require cutting when rhizomes are present and need to be dug up for exposure to adverse conditions (frost or drought). This can be done with mouldboard or disk ploughing. Chisel ploughing is useful for draining wet fields and reducing the infestation of deep-rooted hygrophilous perennial weeds (for example, *Phragmites, Equisetum, Juncus*).

3. Mulching

Use of plastic mulching is very popular in many vegetable-growing areas. Non-transparent plastic is used to impede the transmission of photosynthetic radiation through the plastic to the weeds so that the development of weeds is arrested. Other advantages of mulching are better conservation of moisture, soil structure and reduction in nitrogen leaching which increase yield. Inconveniences are mainly the cost of plastic (although it can be reused) as well as management costs. However, some perennial weeds (e.g. *Cyperus* spp., *Convolvulus arvensis*) cannot be controlled by mulching only. It is obligatory to remove the plastic residues from the field in the form of waste (burning is prohibited). Black plastic mulching on the crop rows and inter-row cultivation is a satisfactory option for organic tomato and melon growers in Europe. Other organic materials (bark, straw, plant residues) can be used, especially if there is a cheap source available nearby. Their advantages are similar to plastic, but weeds can easily manage to reach the surface if the layer is not thick enough. Depending on the materials used, there can be specific problems (e.g. danger of fire with the use of straw, and wind or flooding can remove mulching materials). Some materials can increase the population of crop enemies particularly rodents and snails.

Biological method

Biological approaches involves various living agents such as fishes, snails, pathogens, insects, nematodes as well as botanicals i.e. bio-herbicides to control the weeds. Different bio-herbicides like, Devine ® (*Phytophthora palmivora* to control strangler vine (*Morrenia odorata*) in citrus orchards), Biomal ® (*Colletotrichum gloeosporioides* f. sp. *malvae* to control round-leaved mallow (*Malva pusilla*) in wheat, lentils and flax), etc. are commercially available.

Insects belonging to the orders, Lepidoptera, Hemiptera, Coleoptera, Diptera, etc. are the major biological control agents for terrestrial weeds. Control of some perennial weeds has been done successfully *viz*., *Lantana camara* by lace bug (*Teleonemia scrupulosa*); *Cuscuta* by *Melanogromyza cuscutae*, *Imperata cylindrica* by *Orseoliella javanica*; *Cyperus rotundus* by a plant feeder *Bactra vermosana*. However, biological method of weed control has very little practical use in vegetable crops.

Chemical method

Use of herbicides for the control of weeds offers the greatest feasibility for weed control. Chemical method of weed control is less costly than other methods besides its usefulness in minimizing tillage operations.

Herbicide

The term "Herbicide" came from the Latin words *herba* meaning plant and *caldere* meaning to kill. So, herbicide refers to the chemical that disrupt the physiology of a plant over a long enough period to kill it or severely reduce its growth. The first widely used herbicide was 2,4-dichlorophenoxyacetic acid (2,4-D) which was developed by a British team during World War II with first widespread production and use in the late 1940s. It still remains one of the most commonly used herbicides in the world. The 1970s saw the introduction of other herbicide, Atrazine, which has the dubious distinction of being the herbicide of greatest concern for groundwater contamination because it does not break down readily (within a few weeks) after being applied. Instead, it is carried deep into the soil by rainfall causing the ground water contamination. Glyphosate was introduced in the late 1980s for non-selective weed control. It has now become a major herbicide in selective post-emergence weed control in those crop varieties which are are resistant to it. Many modern herbicides for agriculture are specifically formulated to decompose within a short period after application which is desirable as it allows crops which may be affected by the herbicide to be grown on the land in future seasons.

Classification of herbicides

Currently there are more than 250 herbicides in the world market. There are different classes of herbicides with different modes of action for killing weeds, as well as different potentials to have an adverse effect on health and the environment. Herbicides from different classes also differ in their environmental persistence and fate. Almost all the herbicides can cause acute toxicity. Phenoxy herbicides are involved in acute symptomatic illnesses with relative frequency.

Table 3. Different classes of herbicides

Class of Herbicide	Examples
Acetamides and analides	Alachlor, acetochlor, metolochlor, propachlor, propanil
Carbamates and thiocarbamates	Asulam, terbucarb, thiobencarb
Chlorphenoxy compounds	2,4,-D, 2,4-DP, 2,4-DB, 2,4,5-T, MCPA, MCPB, MCPP, Dicamba
Dipyridyls	Paraquat, diquat
Heavy metals	Lead arsenate, arsenicals
Nitrophenolic and dinitrocresolic	Dinitrophenol, dinitrocresol, dinoseb, compounds dinosulfon
Phosphonates	Glyphosate, glyfusinate, fosamine ammonium
Triazines	Atrazine, simazine, cyanazine, propazine
Urea derivatives	Diuron, flumeturon, linuron, rimsulfuron, tebuthiuron

The herbicides are basically classified on the basis of a) method oapplication, b) mode of action and c) mechanism of action

Method of application of herbicides

- **Soil-applied group :** This group of herbicides (Pre-emergent herbicide) are applied before planting and before emergence of weed or crop which are taken up by the roots of the target plant.
- **Foliar-applied group :** This group of herbicides (Post-emergent herbicide) is sprayed on the standing plant part after the crop has emerged.

Mode of action of the herbicides

Mode of action of the herbicides refers to the overall manner in which the herbicide affects a plant at the tissue or cellular level. Herbicides with the same mode of action will have the same translocation (movement) pattern and produce similar injury symptoms. Selectivity on crops and weeds, behaviour in the soil and use patterns are less predictable, but are often similar for herbicides with the same mode of action. The herbicides are classified into foliar and soil applied groups. The foliar applied groups are then divided into three categories according to movement through the plant.

1. Foliar applied herbicides

- Symplastically translocated (source to sink capable of downward movement)
- Apoplastically translocated (capable of only upward movement)
- Those which do not move appreciably (kill very quickly)

2. Soil applied herbicides

Foliar applied herbicides

Symplastically translocated (leaf to growing points)

These herbicides are capable of moving from leaves (sources of sugar production) with sugars to sites of metabolic activity (sinks of sugar utilization) such as underground meristems (root tips), shoot meristems (shoot tips), storage organs and other live tissues. Since movement to sites is essential for continued plant growth, these herbicides have the potential to kill simple perennial and creeping perennial weeds with only one or two foliar applications. Symptoms are evident on new growth first. Pigment loss (yellow or white colouration), stoppage of growth, and distorted (malformed) new growth are typical symptoms.

Most injury appears only after several days or a week after application. Plants die slowly. Herbicides in this group are usually molecular (non- charged) at low pHs found in the cell walls and negatively charged at higher pHs encountered in the cytoplasm of leaf sieve cells of the phloem (the ionization inside the cytoplasm of the phloem accounts for trapping and movement of these herbicides).

1. **Auxin growth regulators :** Synthetic auxins are organic chemicals which inaugurated the era of herbicides. Synthetic auxins mimic this plant hormone. They have several points of action on the cell membrane and are effective in the control of annual, simple perennial and creeping perennial broadleaved (dicot) weeds. 2,4-D, 2,4-DB, MCPP (mecoprop), MCPA, 2,4-DP (dichlorprop), Dicamba (Benzoic acids), Clopyralid and Picloram (Picolinic acids (Pyridines) and relatives), etc. are synthetic auxin herbicides.

Auxins, the class of plant hormones (or plant growth substances) have a cardinal role in coordination of many growth and behavioural processes in the plant's life cycle and are essential for plant body development. Auxin herbicides stimulate a variety of growth and developmental processes when present at low concentrations at the cellular sites of action. However, with increasing concentration and auxin activity in the tissue, growth is disturbed and the plant is lethally damaged. One of the well-known effects of excess amounts of auxin on dicots is to cause them to overproduce the plant hormone ethylene which may result in a number of plant responses, including epinasty and senescence. Action of 2,4-D persists because plants can't break down it. Another effect of excess ethylene production in response to 2,4-D is to stimulate the production of another plant hormone, abscisic acid (ABA). The effects of ABA on the plant may contribute to eventual plant death.

2. **EPSPS inhibitors (Aromatic amino acid inhibitors) :** These herbicides kill the plant through inhibiting the enzyme 5-enolpyruvylshikimate 3-phosphate synthase (EPSPS) which is involved in the biosynthesis of the aromatic amino acids tryptophan, phenylalanine and tyrosine. The enzyme is a target for herbicides as these amino acids are only synthesized in plants and microorganisms. They affect grasses and dicots in the same manner. Glyphosate and sulfosate are the compounds with this mode of action. Glyphosate acts as a competitive inhibitor for phosphoenolpyruvate and is used as a broad-spectrum systemic herbicide. 5-enolpyruvylshikimate-3-phosphate (EPSP) synthase is an enzyme that catalyzes the chemical reaction : phosphoenolpyru- vate + 3-phosphoshikimate $\rightleftharpoons$ phosphate + 5 enolpyruvylshikimate-3-phosphate (EPSP). Thus, the two substrates of this enzyme are phosphoenolpyruvate and 3-phospho-shikimate, whereas its two products are phosphate and 5-enolpyruvylshikimate-3-phosphate.

3. **ALS/ AHAS inhibitors (Branched-chain amino acid inhibitors) :** These herbicides kill the plant through inhibiting the enzyme acetolactate synthase (ALS) (also known as acetohydroxyacid synthase or AHAS) which is involved in the first step in the synthesis of the branched-chain amino acids (valine, leucine, and isoleucine). The ALS/AHAS are the enzymes found in plants and micro-organisms. These herbicides slowly starve affected plants of these amino acids which eventually lead to inhibition of DNA synthesis. They affect grasses and dicots in the same manner. Shoot meristems cease growth; yellow, pink and purple symptoms appear; roots tend to develop poorly; and the secondary roots are shortened and all nearly the same length producing a "bottlebrush" appearance. Complete symptom development is very slow and requires two to three weeks or more.

The ALS inhibitor family includes sulfonylureas (SUs), imidazolinones (IMIs), triazolopyrimidines (TPs), pyrimidinyl oxybenzoates (POBs), and sulfonylamino carbonyl triazolinones (SCTs).

4. **Chlorophyll/Carotenoid pigment inhibitors :** Amitrole is the only compound of this group which moves well in the symplast, however other compounds in the group show initial movement into shoot tips causing new growth to be devoid of green and yellow pigments.

White new growth, sometimes tinged with pink or purple, characterizes the symptoms associated with the pigment inhibitors. New growth initially appears normal except for the conspicuous lack of green and yellow pigments. Uses of this group of herbicide include selective weed control in soybeans and cotton, general vegetation control and aquatic weed control.

5. **ACCase inhibitors (Lipid biosynthesis inhibitors)** : These herbicides generally kill grasses through inhibiting the enzyme Acetyl coenzyme A carboxylase (ACCase) which is involved in the first step of lipid biosynthesis. Acetyl-CoA carboxylase (ACC) is a biotin dependent enzyme that catalyzes the irreversible carboxylation of acetyl-CoA to produce malonyl-CoA through its two catalytic activities, biotin carboxylase (BC) and carboxyl transferase (CT). ACC is a multi-subunit enzyme in most prokaryotes and in the chloroplasts of most plants and algae, whereas it is a large, multi-domain enzyme in the endoplasmic reticulum of most eukaryotes. The most important function of ACC is to provide the malonyl-CoA substrate for the biosynthesis of fatty acids.

Thus, ACCase inhibitors affect cell membrane production in the meristems of the grass plant. The ACCases of grasses are sensitive to these herbicides, whereas the ACCases of dicot plants are not sensitive to these herbicides. The herbicides of this group viz., Aryloxyphenoxypropionates compounds (fenoxaprop, fluazifop-P, quizalofop) and Cyclohexanediones compounds (clethodims, sethoxydim) are more active in post-emergence (foliar) than soil applied.

All the herbicides of this group produce the same symptoms on grass species; namely discoloration and disintegration of meristematic tissue at and above the nodes, including nodes of rhizomes. Leaves turn yellow, redden and sometimes wilt. Seedling grasses tend to lodge by breaking over at the soil. These herbicides have the potential to be used for selective removal of most grass species from any non-grass crop.

Non translocated herbicides (Contact herbicides)

1. **Cell membrane destroyers :** Compounds in this group cause rapid disruption of cell membranes and very rapidly kill the plants. Examples of this group of herbicides are paraquat, diquat (Bipyridyliums), acifluorfen, fomesafen, lactofen , oxyfluorfen (Diphenyl ethers or nitrophenyl ethers).

The bipyridyliums and the diphenyl ethers penetrate into the cytoplasm, cause the formation of peroxides and free electrons (light is required) which destroy the cell membranes almost immediately. Herbicidal compounds dissolve membranes directly. Rapid destruction of cell membranes prevents translocation to other regions of the plant. Severe injury is evident hours after application, first as water-soaked areas which later turn yellow or brown. Maximum kill is attained in a week or less. Partial coverage of a plant with spray results in spotting and/or partial shoot kill. New growth on surviving plants will be normal in appearance.

Upwardly mobile only herbicides (Apoplastically translocated)

These herbicides translocate only apoplastically. Movement is upward with the transpiration stream (water moving through the plant from the soil and evaporating into the atmosphere at the leaf surfaces).

1. **Photosynthetic Inhibitors or Photo system II inhibitors:** These herbicides kill the plant through reducing electron flow from water to $NADPH^{2+}$ at the photochemical step in photosynthesis. Transfer of electrons from Photosystem II to Photosystem I is essential for the production of photosynthetic energy. If applied as foliar spray, these herbicides move through the cuticle into the cell and chloroplast. If soil applied, the herbicide moves into the root and is translocated upward in xylem. It ultimately moves into the cell and chloroplast where it binds to the Qb site on the D2 protein preventing it from accepting and transferring electrons to the plastoquinone (PQ) pool in Photosystem II. Therefore, this group of compounds cause electrons to accumulate on chlorophyll molecules. As a consequence, oxidative reactions occur in excess of those normally tolerated by the cell, and the plant dies.

Examples of this group of herbicides are atrazine, simazine, cyanazine, prometon, metribuzin, hexazinone (Triazines), terbacil, bromacil (Uracils), linuron, tebuthiuron (Phenylureas), etc. Symptoms develop from bottom to top on plant shoots (older leaves show most injury). Chlorosis first appears between leaf veins and along the margins which is later followed by necrosis of the tissue.

Soil applied herbicides

These herbicide groups have little or no foliar activity and are applied mostly pre-plant incorporated and pre-emergence for control of seedling grasses and some annual broadleaved weeds in soybeans, peanuts, dry beans, cole crops, cotton, alfalfa, clovers, lettuce, tobacco, herbaceous ornamentals, established turf, and in woody species (nurseries, orchards, grapes, Christmas trees, etc.).

Cell division inhibitors

Root Inhibitors

These herbicides inhibit the steps in plant cell division responsible for chromosome separation and cell wall formation. Roots are relatively few in number and club shaped. Examples of this group of herbicides are trifluralin, benefin, prodiamine, oryzalin, pendimethalin and ethalfluralin (Dinitrobenzenamines). Except for oryzalin, these compounds have water solubility less than one part per million. They bind to soil colloids and are unlikely to leach. Losses occur through volatilization and photo-degradation on soil surfaces. Incorporation into the soil by mechanical mixing or by overhead irrigation soon after application is routinely suggested. These root inhibitors do not translocate.

Shoot Inhibitors

This group of very volatile herbicides is applied as preplant. Examples of this group of herbicide is butylate, pebulate, cycloate (Carbamothioates). The shoot inhibitors applied in the soil for the control of grasses, some broadleaved weeds and suppression of some perennials from producing tubers and rhizomes. Injury appears as malformed (twisted), dark green shoots and leaves. This herbicide is generally applied in the crops including corn, large seeded legumes, small seeded legumes, beets, spinach, tomato, potato and ornamentals.

Organic herbicides

Organic herbicides are expensive and may not be affordable for general commercial production. They are also much less effective than synthetic herbicides but of course do not inject unnatural chemicals into the environment and can be used in a farming enterprise that has been certified as organic.

- Spice : Different spices are now effectively used as patented herbicides.
- Vinegar : The concentration of 5-20% solutions of acetic acid is effective which mainly destroys surface growth. Hence, repeated sprayings are needed to suppress the new growth.
- Steam : It is applied commercially in organic vegetable farms however, such weed control is uneconomic and also inadequate because it kills only the surface growth without destroying the underground growth.
- Flame : It is considered more effective than steam.

Application of herbicides

Most herbicides are applied as water-based sprays using ground equipment. Ground equipment varies in design, but large areas can be sprayed with the use of self-propelled sprayers equipped with a long boom, of 60 to 80 feet (20 to 25 m) with flat fan nozzles spaced about every 20 inch (500 mm). Towed, handheld, and even animal-drawn sprayers are also used. Inorganic herbicides are generally applied in big farms of the European countries and The USA aerially using helicopters or airplanes. Herbicides can also be applied through irrigation system which is termed as "Chemigation".

Table 4. Effective weed management with herbicides in different vegetable crops

Vegetable crop	Weed control measure	Time of application
Tomato	Metribuzin @ 0.25-0.75kg ai /ha	Post-emergence
	Butachlor or Alachlor@ 2.0 kg ai/ha	Pre-emergence
Brinjal	Pendimethalin 1.0kg ai/ha or Fluchlorin 1.5kg ai/ha orButachlor 2.0kg ai/ha + one hand weeding 30 days after transplanting	Pre-emergence
Chilli	Pendimethalin 1.25 kg ai/ha + one hoeing 40days aftertransplanting	Pre-emergence
	Nitrofen @ 2.5 kg ai/ha or Oxadiazon @ 1.0 kg ai/ha	Pre-emergence
Cabbage and Cauliflower	Pendimethalin @ 0.56kg ai/ha or Oxyfluorfen @ 0.25kg Fluchoralin@ai/ha or 1.2 kg ai/ha or Nitrofen @ 2.0 kg ai/ha + onehand weeding30 days after planting.	One day before transplanting seedlings
Knol khol	Butachlor @ 1.5kg ai/ha or Fluchloralin @ 1.0kg ai/ha or Oxyfluorfen@ 0.25kg a.i./ha	Pre-planting
Onion	Pendimethalin @ 0.75 kg ai /ha or Fluchloralin @ 1.5 kg ai/ha orOxadiazon @ 1.0 kg ai/haor Nitrofen @ 1.0 kg ai/ha + one handweeding 45days after transplanting	Pre-planting
Garlic	Pendimethalin @ 0.75 kg ai/ha or Oxadiazon @ 1.0 kg ai/ha or Fluchloralin @ 0.9 kg ai /ha + one hoeing 45 days after planting cloves	Pre-planting
Carrot and radish	Tenoran @ 2.5 kg ai/ha or Alachlor @ 1.5 kg ai/ha	Pre-emergence, pre-planting
	Dalapon @ 1.0 kg ai/ha or Fluchloralin @ 1.5 kg ai/ha or Nitrofen 2.5 kg ai/ha	One day after transplanting
Beet and turnip	Butachlor @ 1.5 kg ai/ha or Fluchlor alin @ 1.5 kg ai/ha	Pre-emergence
Pea	Alachlor @ 0.75 kg ai/ha or Tribunil @ 1.5 kg ai/ha + one hand weeding 45 days after sowing	Pre-emergence
French bean	Alachlor @ 1.0 kg ai /ha	Pre-emergence
Cowpea	Alachlor @ 2.0 kg ai/ha or Nitrofen @ 2.0 kg ai/ha	Pre-emergence

Okra	Pendimethalin @ 0.56-0.75 kg ai/ha or Alachlor @ 2.0 kg ai/ha or Fluchloralin @ 1.5 kg ai/ha + one hand weeding 45 days after sowing	Pre-emergence
Muskmelon	Alachlor @ 2.0 kg ai/ha or Fluchloralin @ 1.2 kg ai/ha	Pre-sowing
Watermelon	Butachlor @ 2.0 kg ai/ha	Pre-sowing
Bottle gourd	Alachlor @ 2.5 kg ai/ha or Fluchloralin @ 2.0 kg ai/ha or Butachlor @ 2.5 kg ai/ha	Pre-sowing
Bitter gourd	Alachlor @ 2.5 kg ai/ha or Fluchloralin @ 0.75 kg ai/ha or Butachlor @ 2.5 kg ai/ha	Pre-sowing
Pumpkin and summer squash	Butachlor @ 2.0 kg ai/ha	Pre-sowing
Ridge gourd	Alachlor or Butachlor @ 2.0 kg ai/ha	Pre-sowing
Fenugreek	Pendimethalin @ 0.75 kg ai/ha + hoeing	Pre-emergence

Integrated weed management

Integrated Weed Management (IWM) is a sustainable management system that combines all appropriate weed control methods (cultural, biological and chemical) for a particular vegetable crop. The purpose of IWM is to:

- Reduce the possibility of weed control failure.
- Reduce the impact of weed management activities on the environment.
- Increase crop yield and quality, while assisting to manage insect pests, diseases and soil health.

Strategy for integrated weed management

IWM is more than just relying on a few conventional practices, such as herbicides, tillage and hand weeding. The basic IWM strategy for vegetable production is outlined below:

- Application of pre-plant herbicide.
- Use plastic mulch where appropriate.
- Implementation of drip irrigation (subsurface if feasible)
- Control of weeds in the inter-row space before the spread of the crop.
- Hand weeding within the crop beds.

- Use mechanical inter row management.
- Following of proper crop rotation.
- Use cover crops.
- Maintenance of farm hygiene.
- Focus on soil health management.

IWM is also an important component of the integrated crop protection approach, which combines chemical, cultural and biological methods to keep weeds, insect pests and disease pressure low enough to prevent significant economic loss.

Chapter 11

Organic Farming of Vegetable Crops

Green revolution technologies such as greater use of synthetic agro-chemicals like fertilizers and pesticides, adoption of nutrient-responsive, high-yielding varieties of crops, greater exploitation of irrigation potentials etc. has boosted the production output in most cases. However, continuous use of these high energy inputs indiscriminately now leads to decline in production and productivity of various crops as well as deterioration of soil health and environments.

Main challenges for increased productivity in sustainable manner :

- 60 % of ecosystem services viz., Agricultural production, Supply of water, Control of climate and Nutrient cycles are in degraded condition.
- Unsustainable use of natural production factors such as soil, biological diversity and water.
- Intensive agriculture depends on high energy and contribute 1/3 "green house gas" (GHG) emission.
- Agriculture is insufficiently prepared to cope with unpredictability and adaptation to climate change .
- Migration of population to urban slums.

Such degraded ecosystem service condition has given birth to various new concepts of farming such as organic farming, natural farming, biodynamic agriculture, do-nothing agriculture, eco-farming, etc. The essential concept of these practices remains the same, i.e., back to nature, where the philosophy is to feed the soil rather than the crops to maintain soil health and it is a means of giving back to the nature what has been taken from it (Funtilana, 1990).

Definitions of Organic Farming

According to Funtilana (1990), "Organic Farming is giving back to the nature what is taken from it". Hence, it is not mere exclusion of chemical Agriculture; it is a system of farming based on integral relationship. Therefore, one should know the relationship among soil, water, plant and micro-flora and overall relationship between plants animal kingdom. It is the totality of these relationships, which is the backbone of the Organic Farming. Lampkin (1990) has described Organic Farming most comprehensively covering its all essential features. According to Lampkin (1990), Organic Farming is a production system which avoids or largely excludes the use of synthetically compounded fertilizers, growth regulators and live stock feed additives. To the maximum extent, feasible Organic Farming systems rely on crop rotations, crop residues, animal manures, legumes, green manures, off farm organic wastes and aspects of biological pest control to maintain soil productivity and tilth to supply plant nutrients and to control insect pests, diseases and weeds. Thus, the Organic Farming implies recycling of waste and residue to the native soil itself, replenishing the nutrients depleted from the soil during the crop growth, encouraging the growth of micro-organisms which could regulates phased release of stored nutrients in the soil to the crop growth in right proportion, maintaining soil health by balancing the soil moisture and soil aeration and ensuring soil fertility by firmly binding the nutrient elements in the complex organic molecules.

However, success of organic farming depends to a great extent on the efficiency of agronomic management adopted to stimulate and augment the underlying productivity of the soil resources. It is estimated that around 700 million tonnes of agricultural wastes are produced in the country every year which implies a theoretical availability of 5.0 tonnes of organic manure/hectare arable land/year, which is equivalent to about 100 kg NPK/ha/year (Tandon, 1997). However, in reality, only a fraction of these plant nutrients are added back in the field. There are several other alternatives for supply of soil nutrients from organic sources like vermicompost, bio-fertilizers, etc.

Basic concepts of organic farming

- It concentrates on building up the biological fertility of the soil so that the crops take the nutrients they need from the steady turnover within the soil nutrients produced in this way are released in harmony with the needs of the plants.
- Insect-pests, diseases and weeds are largely controlled by developing an ecological balance within the system and by the use of bio-pesticides and botanicals and various cultural techniques such as crop rotation, mixed cropping, and cultivation.
- It relies on recycling of all wastes and manures within a farm.

World status of organic farming

According to the 18th edition of "The World of Organic Agriculture" which was published by FiBL and IFOAM – Organics International in February 2017, Organic agriculture is practiced in 179 countries, and 50.9 million hectares of agricultural land (including conversion areas) are at present managed organically by approximately 2.4 million farmers. Of the 2.4 million producers were reported, more than three quarters are in developing countries. It should be noted that not all certifiers reported the number of producers; their number is probably higher than 2.4 million. At present, 1.1 percent of the agricultural land is organic. In 11 countries 10 percent or more of the farmland is organic. The global sales of organic food and drink reached 81.60 billion US dollars in 2015.

Organic farming in India

In India, vast arable lands are mainly rain-fed where negligible amount of chemical fertilizers are applied for farming. Farmers in these areas often use organic manure as a source of nutrients that are readily available either in their own farm or in their locality. The North Eastern region of India and vast forest fringe areas of the country may be regarded as organic by default. It is estimated that 18 million hectare of such land is available in the North Eastern region which can be exploited for organic production. With the sizable acreage under naturally organic/default organic cultivation, India has tremendous potential to grow crops organically and emerge as a major supplier of organic products in the world's organic market. However, the total area under organic certification is 5.71 million hectares (FIBL and IFOAM year Book, 2015).

Different Private companies have responded to the market demands for the organic products by organizing large scale conversions to organic systems. By going organic, they add more economic value to the crops, which are already cultivated in a manner similar to organic systems. They are actively engaged in promoting organic agriculture for export. Organic products grown in various agro climatic zones are coffee, tea, spices, fruits, vegetables and cereals as well as honey with a price premium of about 20-30% over conventional products. However, domestic organic markets and consumer awareness has not yet been developed in India, although interest for organic products is growing.

Importance of organic vegetable farming

Organic vegetable production has gained importance in many countries because pesticide application in vegetable production system is much more "visible" and closer to the final consumer than cereals or any other agricultural products. This is the reason why organic vegetables are the most important items in the organic food assortment.

Better taste : Organic vegetables have better nutritive, sensory and storage quality attributes than in the conventional crops. Many studies with the panellist have shown that organically grown tomato, potato, carrot, okra, etc. are tastier than the conventionally grown one.

Enhanced nutrient and lower anti-nutrient contents : Human health worldwide is being threatened by excessive nitrate intake through the dietary pathway, *i.e.* vegetables and drinking water, and the former is the primary source. It confirms a general rule that leafy vegetables accumulate more nitrates, followed by root vegetables and potatoes, respectively. It has been found through different studies that different organically grown vegetables like potato, carrot, cabbage, beetroot, celery, leek, parsley, lettuce, etc. contain lower level of nitrates and simultaneously higher content of vitamin C compared to those produced conventionally. The studies clearly showed a higher content of total sugars in organic vegetables and fruit, such as carrot, garden beet, potato, spinach, kale, cherries, red currant and apples. The organic management and fertilization have a positive effect on the accumulation of certain beneficial minerals and phenolic compounds in brinjal. Several studies reported higher average vitamin C content in organic vegetables (especially tomato, lettuce, spinach and cabbage) as well as higher amounts of some nutritionally significant minerals like phosphorous and magnesium in organically grown potato, carrot, lettuce, spinach, and cabbage.

Disease resistance : Sweet peppers grown under organic culture have high levels of phenolic compounds and peroxidase and capsidiol activity that contribute to disease resistance in organic farming.

General aspects of organic vegetable production

Organic vegetable production requires flexibility and the application of new technologies from the producers. The following aspects have to be considered before starting organic vegetable production.

Seed and seedlings

Whether seeds or seedlings are bought or produced by the grower, organic seed/seedlings must be of high quality because poor quality material (infected seed and plants with disease, weak growing plants, *etc.*) in the initial stages of organic production reduces the success of the whole operation. The seed and seedlings must be from organic origins. If there is no seed of organic quality, producers may use conventional seeds, but only if they are not treated with fungicides or insecticides.

Crop rotation

This operation is required under organic certification programme and is considered essential in organic management. The insect pests and disease causing pathogens are often specific to the host (*i.e.,* a particular crop), and will multiply as long as the crop is there. Alternating crops in time (rotations) or space (strip cropping and intercropping) is, therefore, an important tool for controlling pests, and also for maintaining soil fertility. The system includes different crops that are grown at the same time but on different plots (planed sequences of crops for each season). In a rotational cropping system, incorporation of different crops is possible (legumes crops, groundnut, grassland or legumes cover crops like *Crotolaria* sp.) for improving soil fertility, accumulating nitrogen, reducing weeds, pest and diseases.

Plant nutrition

Organic-based plant nutrients are carbon based compounds which help in maintaining the soil structure and increase its nutrient holding capacity. Organic manures are easily bio-degradable and hence do not cause environmental pollution. The efficient use of the farm nutrients sources has an important role in organic vegetable production. The organic-based plant nutrition in the tropics depends on the growing conditions and cropping combinations. The following organic-based plant nutrients are at the organic vegetable grower's disposal:

- **Green manure :** Green manures in organic vegetable production system requires: a) sufficient water supply, b) high seed density, c) good soil structure, d) avoidance of exhausted soils, e) improvement of nutrient content of soil by adding manure or compost and f) usage of native leguminous plants together with local inoculants (*Rhizobium*). The following leguminous crops can be used as green manures in the organic vegetable farming: alfalfa (*Medicago sativa*), desmodium (*Desmodium intortum* (Mill). Urb), indigofera (*Indigofera tinctoria*), soybeans (*Glycinse max* (L.) Merr), broad bean (*Vicia fava* var. major), hairy vetch (*Vicia ssp*), cowpea (*Vigna unguiculata*), Dhaincha (*Sesbania aculeate*), cluster bean (*Cyamopsis tetragonoloba*), etc.
- **Organic mulching :** The crop residues are also an important source of nutrients for the following vegetable crop. One possibility to obtain a faster residue effect is to chop up the residues and incorporate them superficially 15 cm into the soil. About 80% of the bound nitrogen in the organic material

of the crop residues can be mineralized within 6 to 8 weeks after soil incorporation. Crop residues may incorporate 40 to 100 kg nitrogen per hectare, according to the green mass remaining on the field.

- **Animal manure :** The animal manure is an important source of organic matter for the soil and nutrients for the crops. Adequate management of the manure is essential in order to obtain good quality organic fertilizer and to avoid nutrient losses. Cattle dung is relatively low in nitrogen content and high in potassium. Poultry dung has a high content of phosphorus but too low in potassium content. The mineralization of dung is quite slow hence, it should be applied to the field well ahead of sowing or planting. For root vegetables like, radish and carrot, only well decomposed dung manure should be applied to avoid root forking disorder.
- **Liquid manure (slurry) :** Liquid manure can be applied directly to vegetables crops that remain for longer periods in the field. However, for crops such as carrot and onion, it is not suitable. However, in using liquid manures on vegetables, it is important to maintain the minimal hygienic requirements of not contaminating the crop with slurry.
- **Compost:** The compost can be applied as base fertilizer for organic vegetables. About 25-30 tons per ha of compost may improve the soil activity and enhance the mobilization of the soil nutrients from the reserve of the soil.

Weed management

Weed management is an important issue in organic vegetable production. All preventive and direct measures have to be carefully planned and coordinated in order to reduce manual labour to the maximum. Weed management, improvement of soil fertility and plant nutrition are planned together in organic vegetable production. As an alternative to chemical herbicides, organic producers rely on mechanical and thermal methods. Mechanical weed regulation is closely related to soil tillage and soil management. In organic vegetable production, it is important to have a weed free period according to the developmental stages of the crop. This does not mean the complete eradication of weeds during the cropping period, but rather having a period free of weeds during the sensible earlier stages of the crop.

Different strategies may be implemented depending mainly on the vegetable crop, variety, soil type, kind of equipment or equipment combination (expanding harrow, spring harrow, orbital harrow, finger hoe, star hoe, etc.). Mechanical weeding is possible if the adequate machinery already exists on the farm. In the case that no infrastructure is available, manual weeding is an alternative but may increase labour cost considerably. Thermal weed regulation (flame weeding) might be an alternative.

Pest and disease management

Organic management of disease and pests is based on non-chemical sanitation, cultural, physical and biological means as well as applying approved chemicals for organic culture.

❐ **Prevention measures :** Pest and disease management is in many cases a big challenge for organic vegetable producers. Use of resistant varieties is the best prevention strategies against some diseases and pests. If preventive and direct measures are combined in an optimal way, the risk of an infection by pests and diseases can be reduced to a level that does not cause economically harmful yield reductions. Wide and diverse crop rotations reduce the problem of soil-borne diseases and pests (e.g. *Fusarium*, *Sclerotinia*, *Ralstonia*, nematodes, etc.). Excessive nutrient application can affect the health of plants adversely. High contents of nitrogen promote infection of fungi and infestations of different insects. Intercropping methods including special crops in the rotation like *Tagetes* helps to control nematodes efficiently. Almost all foliar diseases (with exception of powdery mildew) need wet leaves for infection hence, the plant must be kept dry. For this reason, drip irrigation and maintenance of wider spacing of plants are the best preventive measures.

❐ **Direct plant protection :** Organic vegetable growers combine a number of direct management methods like, use of biological control methods (release of natural predators, bio-control agents, etc.), use of protective nets and application of natural products as pesticides (sulphur, copper, oils and plant extracts). All these methods are generally less effective compared to chemical pesticides, therefore, a combination of preventive and direct measures is most successful in organic vegetable production.

Organic certification and inspection system in India

Organic certification system is a quality assurance initiative, intended to assure quality, prevent fraud and promote commerce, based on set of standards and ethics. It is a process of certification for producers of organic food and other organic plant products.

Why Certification is required?

❐ Third party assurance from producer to the consumer separated by distance.

❐ For uniform label.

- Assurance to the consumers that its concern for healthy food has been addressed.
- Effective marketing tool for image, credibility, visibility/ transparency.

Why Labelling is required?

- Easy recognition of organic quality and certification system.
- Confirms the fulfilment of the label regulations and of legal rules.
- It helps to achieve a better price for organic products.

Chapter 12

Protected Cultivation of Vegetable Crops

Protected cultivation can be defined as a cropping technique wherein the micro-climate surrounding the plant is controlled partially or fully as per the requirement of the plant species during their growth period. This cropping technique is the most modern approach to grow different horticultural crops mainly, vegetable crops and flowers for enhancement of both productivity and product quality. This technology has spread extensively the world over in the last few decades. This technology involves the cultivation of different crops particularly, horticultural crops in a controlled environment wherein the environmental parameters like the temperature, humidity, light, soil, water, plant nutrition, etc. are manipulated for enhancement of production as well as product quality even during off-season which ensures additional return to the growers.

Scenario of Protected Cultivation of Vegetables

International scenario

Commercial production of food crop, cut flowers and ornamental plants under glasshouse started from 18th century in Europe. The most significant break through in designs came into existence with the development of plastic technology. Polethylene polymer was first developed in the 1930s and came into agricultural use in the early 1950s in the form of plastic films and pipes for water conveyance including sprinkler and trickle irrigation. New kinds of plastic, like polyvinyl chloride, polypropylene and polyesters were developed later which are cheaper, more durable and opaque or transparent as required for greenhouse covers and several other applications. Now-a-days, nearly 90% of the green houses are being constructed by utilizing Ultra Violet (U.V) stabilized polyethylene sheets as the glazing or cladding material.

Vegetables are produced under green house commercially in about 115 countries. Total world green house vegetable area as estimated in 2017 is 489,214 hectares. Nearly half of the world's greenhouse vegetable area located in Asia (224,974 ha) followed by in Europe (173,561 ha), Africa (36,993 ha), South America (12,502 ha), North America (7,288 ha), Oceania (2,036 ha) and

Antarctica (0.02 ha. for research station pilot programme). Top 10 green house vegetable producing countries are China, South Korea, Spain, Japan, Turkey, Mexico, Netherland, France, USA and Poland.

Indian scenario

Indo American Hybrid Seeds Company was the pioneer in India to utilize greenhouse technology for commercial production of seeds, ornamental plants and cut flowers in the year 1965. In the early 1960s, Field Research Laboratory of DRDO at Leh in Ladakh region of Jammu and Kashmir attempted solar greenhouses comprising of polyethylene sheets for vegetable production research. The impetus of this technology in India came in the mid 1980's with the emergence of U.V. stabilized Low Density Poly Ethylene (L.D.P.E.) which was first marketed by IPCL and boosted the greenhouse research and application for raising vegetables by providing UV stabilized film and aluminium poly house structures. The Defence Research and Development Organization (DRDO) provided adequate support and technological help in this project. Research on vegetable production in this region became successful and more than 2000 small and medium sized greenhouses are at present found at Leh area of Ladakh. Importance of plastics in agricultural uses got further fillip with the formation of National Committee on the Use of Plastics (NCPA) in Agriculture in 1981 under the Ministry of Chemicals and Petrochemicals which was later transferred to Ministry of Agriculture under division of Horticulrure in 1993. With a view to promote agricultural productivity with special reference to optimizing the use of available water resources, improving quality of the product and reducing post harvest losses, the Government of India had allocated funds to the tune of 250 crores with provision of subsidy during VIII Five Year Plan period for the horticultural crops and NCPA has been entrusted with the task of implementing the plasticulture promotional programmes like drip irrigation, plastic mulching and greenhouse technology. The NCPA had under its control 22 Plasticulture Development Centres (PDC) which was later renamed as Precision Farming Development Centre (PFDC) located at different State Agricultural Universities/ Institutes. These centres provided research support for use of plastics and introduction of new technologies viz., erection of shade nets, anti hail net, plastic packaging of seeds, etc.

However, protected cultivation technology as a commercial venture came in the being much late during the early nineties. In India, the area under protected cultivation is presently around 25000 ha while the greenhouse vegetable cultivation area is about 2000 ha. This data amply suggested that application of green house technology for commercial purpose is still in its infancy in India. However, cultivation under protected structures (high or medium-cost polyhouse, low-cost poly house cum rain shelter, poly tunnel, shade house, etc.) is gaining lot of importance in the recent years to meet the increasing demand of fresh vegetables throughout the year both for domestic and export market.

Principles behind protected cultivation

Most crops grow best in light whose wavelengths range from 400 to 700 nanometers. The poly house generally reflects back 43% of the net solar radiation incident upon it allowing the transmittance of the photosynthetically active solar radiation in the range of 400-700 nm wave length which increases photosynthetic efficiency of the crops grown inside the poly house. Moderately short to moderately long wave length (15-390 nm) ultraviolet rays damaging to the crops are absorbed by the U.V stabilized polyethylene sheet used as cladding material. For this reason, excellent crop growth and high yield are generally realized in the crops grown under such protected structures. At the same time, sunlight admitted inside the green house is absorbed by the crops, floor and other objects in the frame work of the green house which in turn emit long wave thermal radiation in the infra red region (760-2500nm) and thus temperature inside the greenhouse rise. This phenomenon of trapping the solar energy in the poly house is called the "Green house effect". Such increase in the temperature of the growing environment though unavoidable, is the most important function of a greenhouse for enhancing crop growth of the warm-loving vegetable crops like, tomato, sweet pepper, cucumber, etc. in severe winter condition. However, during summer, inside temperature rise higher than the optimium levels which necessitate the provision of evaporative cooling / ventilation / misting to maintain the temperature inside the structure well below 35°C. The ventilation system can be natural or a forced one. In the forced system, fans are used which draw out hot air from inside to the tune of 7-9m^3 of air / sec / unit of power consumed with necessary charging of air inside. Various methods of cooling systems like roof shading, water film covering and evaporative cooling are generally employed. The most common methods of evaporative cooling system are "Fan and pad cooling system", "High pressure mist system" and "Low pressure mist system".

Advantages of protected cultivation

- The yield of many horticultural crops particularly vegetables crops and flowers may be 10-12 times higher than that of outdoor cultivation depending upon the type of protected structure, type of crop, environmental control facilities.
- Ideally suited for vegetables and flower crops.
- Water economy due to controlled irrigation mainly through drip system.

- Year round production of floricultural crops.
- Off-season production of vegetable and fruit crops.
- Production of quality produce free of blemishes.
- Reliability of crop increases under greenhouse cultivation.
- Commercial production of disease-free and genetically superior transplants.
- Production of grafts of vegetable crops and micro-propagated plant-lets.
- Efficient utilization of chemicals, pesticides to control pest and diseases.
- Hardening of tissue cultured plants.
- Implementation of modern techniques of Hydroponic (Soil less culture), Aeroponics and Nutrient film techniques is possible only under greenhouse cultivation.

Production of Vegetable Crops under Polyhouse

Vegetable crops often need to be protected against a combination of weather conditions. In addition to protection against fluctuating temperatures, protection is also required against solar radiation, heavy rain, hail and strong wind. Selection of suitable location for poly house construction is of utmost importance. Ploy houses should be away from industrial and over populated areas. Levelled ground where the light intensity is at its maximum is the priority. But even if the greenhouse has to be constructed on a slope, care should be taken such that surface runoff is directed away from it. Adequate water supply and a power source should be available nearer to the site of poly house. Different protective structures are utilized for vegetable production including glass house, poly house, insect-proof net house, shade net house, low tunnel poly house, zero energy poly house, etc. Structurally the poly houses are of different kinds, *viz.*, open-ventilated, closed poly house with fan and cooling-pad system, sloped roof, rain shelter, etc. The poly house structures are of different kinds based on shape (Round top / Quonset type, 'A' or Hut type, Lean-to type, Walk in tunnel type, Low tunnel type, Shelter type, Multi-span type, etc.), utility (temperature and humidity controlled), construction (wooden, pipe or truss framed), covering material (glass, fibreglass, plastic-film). Plastic film covering materials are of different types such as acrylic, polycarbonate, fibreglass reinforced polyester, polyethylene film and polyvinyl chloride film. Plastic glazed greenhouses have many advantages over glasshouses, the main one being cost. Plastic is also adapted to various greenhouse designs, usually resistant to breakage, light in weight and fairly easy to use.

Production technologies

Vegetables can be produced in greenhouses under different methods which include planting in the greenhouse soil directly and the latest technique of hydroponic cultivation which involves either planting in different soilless mixtures using containers or using liquid nutrient media.

Selection of crop and variety/ hybrid

By and large, all vegetable crops can be grown in greenhouse condition. However, growing high value crops like cucumber, sweet pepper, tomato, lettuce, savoy cabbage, etc. ensure more return per unit area. The varieties selected for specific crop must be adaptable to particular growing zone and must meet the requirement of local as well as export market demand. Before going to large scale production of a particular crop, suitable variety/ hybrid should be selected through screening. In general, cucumber varieties recommended for growing in poly houses are of gynoecious and parthenocarpic type (plant produce only pistillate flowers and fruit set does not require pollination). Tomatoes generally selected for poly house production are indeterminate types which have long fruiting span. Important other characteristics of tomato suitable for poly house production include high yields, freedom from fruit cracking and green shoulder disorder, disease resistance and uniform ripening character. Desirable characteristics of sweet pepper are uniformity in shape and fruit colour (red, yellow, violet, etc.), tolerant to viral disease, high yielding ability and good fruit quality. Different types and varieties/hybrids of lettuce can be grown under greenhouse conditions which are loose-leaf, butter head and romaine or cos types which are preferred than the crisp head types.

Soil management

The conventional method of growing vegetables in greenhouse involves planting directly into the well drained existing soil. Continuous cropping of the same crop over a period of time might destroy the soil structure in the greenhouse. Moreover, soil-borne pathogens and pests become a major concern if they get in the closed structures. Hence, adequate care should be taken to maintain the soil structure. Soil preparation entails important initial practices like sanitation, solarization, mulching or fumigation. Soil solarization reduces soil-borne pathogens due to soil-heating effects and at the same time, control weed and soil inhabiting insect pests. It can also be applied as a bioremediation tool for pesticide polluted soils. Improving soil structure and replenishing its fertility by incorporation of manure, compost and other organic amendments ultimately increases the soil microbial activities and nitrogen uptake by the plants and reduction of certain soil-borne diseases. However, regular soil testing should be mandatory to determine the soil fertility and the deficiency of nutrients must be supplemented as and when required. The application of different bio-agents viz., *Trichoderma harzianum, Trichoderma asperellum, Pseudomonas fluorescens* and *Aspergillus niger* at 50.0 ml/kg soil either singly or in rational combinations helps in not only managing many soil-borne pathogens including nematodes but also promotes plant growth.

Seedling production

The plug-tray nursery raising technology has been commercialized and is the best to produce healthy and disease-free seedlings with vigorous roots. These seedlings show minimum mortality after transplantation. However, it is important not only to ensure health of the nursery through proper fertigation but also monitoring of growth through hormonal treatment, proper media sterilization and greenhouse management for virus and pest free nursery. Sterilized, soilless growing media mostly comprising coco-peat, perlite and vermiculite in the ratio of 3:1:1 with addition of hydrogel, biological agents and organic amendments not only helps in raising high density seedlings but also evades soil-borne diseases and some insect pests. Seedlings are watered when necessary and liquid fertilizer is applied based on leaf colour and growth analysis. The optimal inclusion of nutritional supplement and correct pH of the growing substrate ensures the seedling sturdiness and uniformity.

Soil-less cultivation

Soil degradation, over-fertilizing and prevalence of soil-borne diseases in greenhouse systems requires frequent replacing of greenhouse soils which ultimately paved the way for soil-less cultivation in which different materials like rock wool (fibre materials that are formed by spinning or drawing molten minerals), peat (partially decayed vegetation or organic matter), perlite (amorphous volcanic glass having high water holding capacity), coconut fibre (natural fibre extracted from the husk of coconut), etc. are used as growing medium. Soil-less cultivation refers to hydroponic cultivation in which either an aerated nutrient solution or the chemically inert growing mediums (peat moss, coir, sand, sawdust, rock wool, perlite, vermiculite, etc.) moistened with nutrient solutions is used for growing the crop. Intense monitoring and reuse of the drainage nutrient solution after proper disinfection in hydroponically grown greenhouse crops is very much essential to prevent environmental pollution. However, the grower must thoroughly be acquainted with the knowledge about plant physiology, growth habit and nutrient requirements of the crop for adopting hydroponics.

Irrigation and fertigation

Efficient water usage and water economy can be achieved through drip irrigation technology under protected cultivation as compared to the crop production in open. Drip irrigation technology also reduces the incidence of many diseases that develop in rather moist conditions. Fertigation is referred to the irrigation combined with fertilizer application. The understanding of plant growth behaviour including nutrient requirements and rooting patterns, soil chemistry (solubility and mobility of the nutrients), fertilizers' chemistry (mixing compatibility, precipitation, clogging and corrosion) and water quality factors (pH, salt and sodium hazards) is required for effective implementation of fertigation. It allows a precise and homogeneous application of nutrients in the area where the active roots are concentrated. Fertigation through the drip irrigation system keeps the foliage dry which avoid leaf burn incidence and delay the development of plant pathogens. Plant response to fertigation is dependent on the crop, fertilizer form, concentration, frequency of application as well as the stage of plant growth.

Fertilizer programmes for soilless-culture systems must supply all nutrients required by the plants. Carbon, hydrogen, and oxygen are provided from water and carbon dioxide in the air while nitrogen, phosphorus, potassium, calcium, magnesium, sulfur, iron, boron, copper, zinc, manganese, molybdenum, and chlorine are supplied through the nutrient solutions prepared from soluble inorganic salts containing various elements. Commonly used chemicals for the macronutrients include potassium nitrate, calcium nitrate, potassium phosphate, and magnesium sulfate. Micronutrients that are typically and essentially incorporated in the hydroponic solutions include iron, manganese, copper, zinc, boron, chlorine, and nickel.

Nutrient requirement for tomato and cucumber is given below as the reference.

Tomato

Fertilizer compound (g per 100 litre)	A Seeding to first fruit (g per 100 litre)	B First fruit to termination
Magnesium sulfate	50	50
Monopotassium phosphate	27	27
Potassium nitrate	20	20
Potassium sulfate[1]	10	10
Calcium nitrate	50	68
Iron (Fe 330)	2.5	2.5
Micronutrient stock	15ml	15ml

Note : Potassium sulfate is optional; 15 ml from the stock solution of 450 ml is used for each 100 litre final nutrient solution

Cucumber

Fertilizer	A Seeding to first fruit (g per 100 litre)	B First fruit to termination (g per 100 litre)
Magnesium sulfate	50	50
Monopotassium phosphate	27	27
Potassium nitrate	20	20
Calcium nitrate	68	136
Chelated iron (Fe 330)	2.5	2.5
Micronutrient stock	15 ml	15 ml

Note: 15 ml from the stock solution of 450 ml is used for each 100 litre final nutrient solution

Table 1. Preparation of micronutrient stock solution for tomato

Chemical compound	Grams per packet
Boric acid (H_3BO_3)	7.50
Manganous chloride ($MnCl_2 \times 4H_2O$)	6.75
Cupric chloride ($CuCl_2 \times 2H_2O$	0.37
Molybdenum trioxide (MoO_3)	0.15
Zinc sulfate ($ZnSO_4 \times 7H_2O$)	1.18

Note : One packet (15.95 g) plus water to make 450 ml stock solution (heat to dissolve). Use 15 ml for each 100 litre final nutrient solution

Note: 1 ppm is equal to 1 milligram/litre, so to convert ppm to millimoles, the ppm concentration is to be divided by the molecular weight of the element that is to be working with. For example:

- 1 ppm of NO_3 = 1 mg/litre.
- 1 mg/ litre of NO_3/62 mg/mole = 0.016 millimoles of NO_3 in one litre.
- 1 ppm of Fe = 1 mg/litre.
- 1 mg/litre of Fe / 56 mg/millimole = 0.018 millimoles of Fe in one litre.
- 1 ppm of magnesium (Mg) = 1 mg/litre.
- 1 mg/litre of Mg / 24 mg/millimole = 0.041 millimoles of Mg in one litre.

As these examples show, a solution containing 1 ppm of various elements or molecules will contain different mole or millimole amounts of these same elements.

- To convert millimoles to ppm:

  ppm = millimoles/litre x molecular weight (mg/millimole)

  Example:

  ppm NO_3 = 0.016 millimoles of NO_3 in one litre x 62 mg/millimole

  = 1 ppm NO_3

Calculating the required amounts of the various fertilizers is dependent on the volume of water to be used. This is determined by the volume of the stock tank (e.g. 200 litres) multiplied by the injection ratio (e.g. 100:1 or 200:1). For example, using a 200 litres fertilizer concentrate stock tank, and a 200:1 injection ratio, the volume of water that will be used to calculate the amount of fertilizer to add will be: 200 litres (stock tank volume) x 200 (injector ratio) = 40,000 litres.

Management of insect-pests and diseases

Greenhouse vegetable crops grown the world over are vulnerable to various diseases and pest attacks as the environment inside is conducive for their rapid multiplication. The losses caused due to pests in greenhouses are tremendous. Crop losses are mainly due to insect-pests like mites, whiteflies, thrips, aphids, nematodes and different diseases caused mainly by virus and soil-borne fungi and bacteria. The extent of losses due to virus disease may vary from 5% to 90% depending on the strain of the virus, variety/ hybrid, age of the plant at infection time and temperature during disease development. The following sanitation and other cultural practices are strongly recommended to manage insect-pests, mites as well as diseases.

- The poly house should be thoroughly cleaned before planting a new or first-time crop by means of burning, burying, or hauling away all leftover roots and other plant parts so that there is no chance that insects in the egg, larval, nymphal, pupal or adult stages could remain inside the structure. Crop residues must be removed immediately after the final harvest.
- Thorough cleaning the walls of the poly house, cracks and crevices and if possible fumigation with appropriate foggers. Broad-spectrum insecticides can also be used to spray cracks, crevices, walls and other potential hiding places. Spraying should be done around doors, windows, and other possible points of entry. In hot weather, closing up the empty greenhouse for a week will help to eradicate insect pests.
- Use of ultra-fine screening (50 mesh nylon net) over all openings to prevent very small harmful insects particularly white flies from entering the structure. Plants should, however, be inspected daily for insect infestations.
- All plant parts after pruning should immediately be disposed. Culls (undesirable) or overripe fruit should be removed from the poly house and surrounding areas because insects are often attracted to and can live for long periods on these plant materials.
- Planting outdoor crops near greenhouses should be avoided which may serve as host plants for insect pests and may be a source of diseases as well. Weeds should not be allowed to grow within the poly house or to become established around the greenhouse. A 10 to 30-foot vegetation-free zone around the greenhouse can be created with a heavy-duty geo-textile weed barrier covered with gravel.
- A clean stock program will aid in keeping insect-pests out of the greenhouse. Plants coming from other greenhouses should be carefully inspected for insects and diseases and temporarily quarantined until it is clear that the plants are free of pests. Poly house workers should avoid wearing yellow clothing because it is highly attractive to insect like whitefly which is the main carrier of viral diseases.
- Addition of an 'air lock' type entrance to each production house so that wind-carried insects, soil, etc. do not enter the production area directly from the outside.
- Prevention of un-sterilized soil from being carried into the production space.
- Restricted access to the visitors in the production and transplant houses.
- Raising transplants at a height of at least 1 foot above the ground to minimize dust or splashing soil contamination of plants.

- Prohibition of cigar, cigarette, snuff, and chewing tobacco use by workers involved in production areas, so as to minimize contamination by viruses present in tobacco.
- Rigid hand-scrubbing rules for personnel involved in pruning, pollinating, tying or harvesting activities.
- Filters on all air intakes to restrict air-blown soil and vector-insect entry.
- Periodic tool, walkway, and bench surface treatment with disinfectants.
- The use of disease resistant varieties when available helps to reduce disease problems.
- Administration of biological control measures in the poly house environment as far as possible and if it is not feasible to use biological controls and if biorational pesticides do not offer sufficient control, then only chemical pesticides may be used with utmost care and precision.

Pollination management

In the poly house, pollination of flowers is mandatorily needed for getting optimal fruit set and production of quality produce. Without pollination, the blossoms drop off before setting fruit. Problem of pollination is acute in closed poly house where there is no breeze inside and bees or insects are completely absent. In the naturally ventilated poly house, mild breeze sometimes is enough to release the pollens from the anther over the stigma in the self-pollinated vegetable crops like, tomato, sweet pepper, brinjal, etc. However, some times, even ventilation is there and fans are on, the breeze inside is typically not strong enough to accomplish pollination. In these self-pollinated crops, shaking or flicking the fruiting stem few times in the morning after anthesis is excellent to accomplish pollination. Battery-operated pollinating tools like, electric vibrators and air blowers are also very much efficient to accomplish pollination. Gynoecious and parthenocarpic cucumber which produce seedless fruits do not need pollination at all to set the fruits. Pistillate flowers of the triploid watermelon need to be pollinated with the pollens from the diploid standard seeded watermelon genotype regularly to get the seedless fruits. The muskmelons and cantaloupe cultivars possessing either monoecious or andro-monoecious sex form require pollination to get fruit set. In the poly house-grown cucurbits like, watermelon, muskmelon, etc. hand pollination is the best however, mechanism of pollination with bumblebees is an effective alternative. Different bumble bee species used in greenhouses are *Bombus terrestris*, *B. impatience* and *B. occidentalis*. Stingless bees (*Scaptotrigona depilis* and *Nannotrigona testaceicornis*) can also be successfully used as pollinators. *Eristalis tenax*, the hoverfly species, in the order Diptera possesses desirable characters for the pollination of *Capsicum annuum* under greenhouse conditions.

Harvesting operations

The primary goal of harvest and post harvest handling operations is to maintain vegetable quality as close as possible to harvest condition through subsequent handling operations. Maintaining high quality is the thread that links each of the components in the harvest/handling system, and carelessness at any stage can quickly change the grade.

Crops must be harvested at the optimal stage of maturity for the intended use. Proper harvesting techniques are also essential for minimizing mechanical injury. Frequent sanitizing of hands/gloves and clippers can assist in reducing damage and disease problem. A reusable picking container must be selected that has smooth inner surfaces, is not too deep and is made of an easily cleanable material, such as plastic Harvest operations should be done during the cooler part of the day and after harvest, the product should be protected from the sun to prevent temperature increase. Harvested material should be pre-cooled as soon as possible. Provision should be made for gently transporting the harvested product to the packing/storage area. Prior to reuse, the containers should be cleansed and sanitized to minimize the spread of decay pathogens.

Harvesting of tomato, pepper and cucumber is a continual process throughout the growing cycle. Tomatoes and peppers are harvested with stems attached and are packed that way to avoid bruising and damage. Care should be taken so that tomatoes are not stored or transported with other vegetables such as peppers, lettuce or cucumbers because ethylene that is evolved during ripening tomatoes can damage other vegetables.

Chapter 13

Production Technology of Vegetable Crops

Potato

Potato, *Solanum tuberosum* L. under the family Solanaceae (Chromosome No. 2n = 4x = 48) is the most important non-grain food crop in the world, ranking 3rd in terms of total production with over 381 million tonnes per year in (FAOSTAT, 2017), after rice and wheat. It is grown in 150 countries spread across both temperate and tropical regions and at elevations from sea level to 4,000 m. It is extensively cultivated in China, Russian Federation, Ukarine, Poland, Ireland, Great Britain, Germany, Netherlands, France, Spain, South America, India and USA. Potato was brought from Peru to Spain by the Spaniards in 1565 when it was widely grown in Spain and Italy. They were probably brought to England in 1586 by Sir Francis Drake. Potato was introduced in India in early 17th century probably by the Portuguese. Now it is one of the principal cash crops of India.

The crop

The subsection Potato is distinguished from all other subsections within the genus *Solanum* by "true potatoes whose tubers are borne on underground stolons, which are true stems, not roots". The series tuberosa is characterised by "imparipinnate or simple leaves, forked peduncle, rotate to pentagonal corolla and round berries".

The species *S.tuberosum* is characterised by "pedicel articulation placed in the middle third, short calyx lobes arranged regularly, leaves often slightly arched, leaflets always ovate to lanceolate, about twice as long as broad, tubers with well marked dormancy period"

Potato is an annual herbaceous plant and growth habit varies between and within species. The plant has a rosette or semi-rosette habit. The stem is erect in the early stage but becomes spreading and prostrate later on. The leaves are compound and alternate, irregularly odd pinnate. The tuber is an enlarged portion of an underground stem or stolon. Tuber eyes are the buds from which next season's growth will emerge. Eyes are concentrated near the apical end of the tuber, with fewer near the stolon or basal end. Eye number and distribution are characteristic of the variety.

Potato is mainly propagated vegetatively by means of tubers and sometimes by botanical seeds, i.e., true potato seeds. Potato has a terminal inflorescence consisting of 1-30 (but usually 7-15) flowers, depending on the cultivar. The five petals give the open flower a star shape. A flower also has a pistil that generally protrudes above a cluster of five large, bright yellow anthers. The corolla colour varies from white to complex range of blue, red and purple. Flowers are self-pollinated but also cross pollination take place by insect. Seeds are produced in the fruit which is botanically called berry. Potato fruits, stems and leaves contain glycoalkaloids, the toxic compounds, of which the most prevalent are solanine and chaconine.

Importance and use

Potato is often used as a substitute for cereals being a major source of carbohydrate. It is one of the major vegetable crops of the world and is grown in almost every country. In India it is considered as a staple vegetable and used either alone or mixed with other vegetables. Potato is used in various forms such as, boiled, mashed, baked, fried and cooked in sambar and different curries of vegetable, fish and meat. Potato is processed into different commercially used dehydrated, canned and fried products like, shredded potato, sliced potato, chips, flakes, French fries, finger chips, granules, disc, cubes, flour, etc. It is used for several industrial purposes such as production of starch, alcohol, dextrin, glucose, etc. Potato starch is a large grained starch and contains 25 % amylose and 73 % amylopectin and high phosphate content.

Nutritional value

Potato can be regarded as a wholesome food. Carbohydrates (20.6%) are the chief dietary constituent of potato. Nutritionally, potatoes are second only to soybean for amount of protein/ha, with the major storage protein being the glycoprotein, patatin, one of the most nutritionally balanced plant proteins known. Patatin constitutes about 20% of the total soluble protein in potato. It exhibit enzymatic activity and is believed to play a role in defense against pests and pathogens. It also may elicit allergic responses in humans.

It also contains essential nutrients, minerals and vitamins. Major minerals present in potato are calcium, phosphorus and iron. It is quite a rich source of vitamins especially vitamin B_1, B_2, B_6 and C. Potato also contain good amount of essential amino acids like, leucine, tryptophan and isoleucine.

Medicinal value

Due to rich carbohydrate contents of potato, it is easy to digest and tasteful as well. This is the reason boiled potato is prescribed for babies, patients or for weak persons who require supply of natural energy from their light diets or for those who cannot take very heavy diet as meal. Potato contains natural fibre so it is helpful for appetite control and easy bowel movement. Potato is good for patients suffering from high blood pressure owing to presence of potassium and small amount of sodium. Potato consumption helps in supply of vitamin C and thus it works effectively against bleeding and spongy gums, frequent attack of cold, etc. Potato is an effective and instant antidote for inflammation. If the extract of potato is smeared on an external swelling it takes instant care for the pain. Similarly, consumption of potato helps in natural curing of digestive and intestinal problem. Consumption of potato for its neutral taste is a safe and comfortable food for the patients suffering from mouth ulcer. The raw pulps of crushed potato with honey make an excellent face and skin mask helping for natural removal of skin spots and pimple infections. Regular application of potato pulp makes the hard area of elbow joint and knees soft and supple. Extract of potato is a good natural healer for minor burn injury. Potato contains high content of magnesium which creates natural prevention against depositions of calcium in kidney.

Origin and taxonomy

The potato is a native of tropical South America region where it grows wild in nature and presents the widest diversity of forms like, tuber shape, size, colour, taste, etc.The areas of origin and domestication of the cultivated diploid potatoes was in the high plateau of Peru-Bolivia in the region of the lake Titicaca. Cultivated tetraploids are also found in this region. Wild potato tubers are all bitter in taste and contain potentially toxic steroidal alkaloids. So, the first step in the evolution of the crop is the emergence of alkaloid free diploids which might have happened in the period 2000 to 5000 B.C. Among the cultivated potatoes, the two most important ones are the autotetraploides, *Solanum tuberosum* subsp. *tuberosum* and *Solanum tuberosum* subsp. *andigena* and two diploids, *Solanum phureja* and *Solanum stenotomum*.

There have been two views regarding the origin of cultivated potato i) it might have arisen from the diploid ancestral form *Solanum stenotomum* by simple chromosome doubling. *Solanum stenotomum* was also the first domesticated tuber bearing species originated from the wild *Solanum leptophyes*, ii) it might have arisen as a spontaneous amphidiploid hybrid from the ancestor, *Solanum stenotomum* and a diploid weed *Solanum sparsipilum*.

In another school of thought, *Solanum vernei*, a wild diploid species occurring in the north-western Argentina, gave rise to the *andigena* forms of tuber bearing *Solanum* following autopolyploidy and further human selection led to the evolution of the species of commerce, *tuberosum*.

Potato belongs to the genus *Solanum*, series Tuberosa of the family Solanaceae. More than 200 tuber bearing *Solanum* species are grouped under section Petota, sub-section Potatoe and 19 series.Of the two most important cultivated autotetraploid species, *Solanum tuberosum* subsp. *tuberosum* is wide spread in Europe, USA, Canada and Asia while the other one, *Solanum tuberosum* subsp. *andigena* is grown only in the Andes of South America.

***Solanum tuberosum* subsp. *andigena*:** Thin and long stem, small and narrow leaflets. Flowers are produced more profusely. They have long stolons and mostly coloured deep-eyed tubers. Tuberization takes place only under short-day conditions at high altitudes (above 2000m).

***Solanum tuberosum* subsp *tuberosum*:** Shorter and thicker stem, larger and wider leaflets. Tubers are generally white and oval. Tuberization takes place both under long days in temperate climate as well as under short day conditions in the tropics.

The important diploid cultivated species used as source of resistance to some forms of virus and fungal diseases are *S. demissum*, *S.stenotonum* and S. *phureja*. There are several wild species of the series Tuberosa such as, *S. sparsipilum, S. leptophyes, S. berthaultii, S. spegazzinii, S. vernei,* etc. However, all potato cultivars cultivated for commercial purposes outside South America belong to the species *Solanum tuberosum*. The ploidy levels of potato species varies from 2x to 6x with 73% of these being diploid, 4% triploids, 15% tetraploid, 2% pentaploid, and 6% hexaploid.

Adaptation in India

The most possible introduction of potato to India was with Portugese in Bengal. Firstly, it was introduced at the then Saptagram, Hooghly in West Bengal in 1535 A.D and later it was introduced to Burdwan, part of Hooghly and Midnapur as commercial crop in 1570 A.D. Second evidence proposed that potato was introduced in India from Scotland by East India Company and distributed to Burdwan, Midnapore and Chitagaon (now in Banladesh).

However, the earliest reference of potatoes in India is from the account of the voyage of Edward Terry, who was chaplain to Sir Thomas Roe, British Ambassador to the court of Mughal Emperor Jahangir from 1615-1619. It was recorded to have grown in Surat and Karnataka around 1675 and later in the

Nilgiri hills in 1822. Its cultivation started in the hills by the English during 1787. The early potato introductions in India did not belong to *S. tuberosum* ssp. *tuberosum* but to *S. tuberosum* ssp. *andigena*. Those initial introductions eventually got established in India and further selections by Indian farmers resulted in several indigenous cultures known as *desi* varieties. Among these, Phulwa, Darjeeling Red Round and Gola were most popular. By the end of 19th century, the potato had spread throughout North India. Now potato is a ubiquitious crop in India.

Potato cultivation in India started with these *desi* varieties of *Solanum andigena* background. Later, long duration European varieties of *Solanum tuberosum* background like, Magnum Bonum, Royal Kidney, Great Scot, etc. were introduced. These long duration varieties suitable for European conditions did not perform well under comparatively short growing seasons of the tropical and sub-tropical plains of India where high temperature prevails both at the time of planting and harvesting. Systematic breeding of potato suitable for Indian condition started with the establishment of Central Potato Research Institute at Shimla. All the present day potato varieties have been bred through hybridization involving different exotic and indigenously bred varieties along with different tuber bearing *Solanum* species. However, despite the adoption of hybridization between many tuber-bearing *Solanum* species in breeding potato the present varieties are only being put under *Solanum tuberosum.*

Improved varieties

Early maturing group (Varieties of this group requires 70-80 days after planting to give economic yield in the Indo-Gangetic plains): Kufri Chandramukhi, Kufri Ashoka, Kufri Lauvkar, Kufri Kuber, Kufri Jawahar, Kufri Alankar, Kufri Khyati.

Medium maturing group (Varieties of this group requires 90-100 days after planting to give economic yield in the Indo-Gangetic plains): Kufri Jyoti, Kufri Bahar, Kufri Lalima, Kufri Sutlez, Kufri Arun, Kufri Pushkar, Kufri Sadabahar, Kufri Girdhari, Kufri Pukhraj, Kufri Chipsona 1, Kufri Chipsona 2, Kufri Anand, Kufri Himalini, Kufri Sheetman, Kufri Moti, Kufri Kundan, Kufri Red, Kufri Alankar, Kufri Sherpa, Kufri Swarna, Kurphi Megha, Kufri Thenmalai.

Late maturing group (Varieties of this group requires 110-130 days after planting to give economic yield in the Indo-Gangetic plains): Kufri Sindhuri, Kufri Badshah, Kufri Kanchan, Kufri Giriraj, Kufri Swarna, Kufri Jeevan, Kufri Dewa, Kufri Khasigaro, Kufri Naveen, Kufri Chamatkar, Kufri Kisan, Kufri Kumar, Kufri Safed, Kufri Neela, Kufri Neelamani.

Varieties suitable for processing: Kufri Chipsona-1, Kufri Chipsona-2, Kufri Chipsona-3, Kufri Clipsona-4, Kufri Frysona, Kufri Surya, Atlantic, Diamant, Fritolay Hybrid 1333.

Varieties moderately resistant to late blight disease: Kufri Alankar, Kufri Jyoti, Kufri Swarna, Kufri Megha, Kufri Naveen, Kufri Neelamani, Kufri Muthu, Kufri Jawahar, Kufri Sutlej, Kufri Pukhraj, Kufri Giriraj, Kufri Jeevan, Kufri Shailga, Kufri Girdhari.

Varieties moderately resistant / resistant to early blight disease: Kufri Chamatkar, Kufri Sinduri, Kufri Jyoti, Kufri Badshah, Kufri Lalima, Kufri Jeevan, Kufri Khasigaro, Kufri Sherpa.

Varieties resistant to wart disease: Kufri Jyoti, Kufri Sherpa, Kufri Naveen Kufri Kanchan.

Varieties resistant to PVX virus: Kufri Badshah, Kufri Sherpa.

Varieties moderately resistant to PVY virus: Kufri Sherpa.

Varieties resistant to cyst nematode: Kufri Swarna.

Red-skinned varieties: Kufri Sindhuri, Kufri Lalima, Pimpernel, Kufri Kanchan.

Heat torerant varity: Kufri Surya.

Cultural requirements

Potato is a cool season crop but is susceptible to frost. At low temperature, vegetative growth is restricted and at temperatures nearing the freezing point there is permanent and often irrecoverable frost injury to the plant. For these reasons, potato is grown as a summer crop in the hills and as a winter crop in the plains of the tropical and subtropical regions. At higher temperature, the respiration rate increases and the carbohydrates produced by photosynthesis are consumed rather than stored in the tuber. For this reason, in the vegetative growth period the temperature may be higher but during tuberization phase the temperature should be lower. Long day coupled with high temperature conditions promote haulm growth without formation of tuber and short day with low temperature condition induce tuberization. The optimum temperature for good plant growth is 15°C to 20°C. Potato should be grown when maximum day temperatures are below 34°C and night temperatures are not above 20°C. Most potato genotypes do not tuberize when the prevailing night temperature is above 21°C. Well drained light soils like, loam and sandy-loam rich in organic matter are ideal for growing potato. Potato can be grown in all types of soil, alluvial, black, red and laterite having pH ranging from 6.5 to 7.5 and EC: < 0.5%. Alkaline or saline soils are not suitable for potato cultivation. However, Gangetic plain with almost neutral soil is most suitable for growing potato.

Land Preparation

Deep ploughings followed by 2-3 cross and shallow ploughings are necessary for making the soil well pulverized, loose, friable ,well levelled and good tilth. Well rotten organic manure preferably farm yard manure or compost @ 30-35 t/ha need to be added during land preparation. Land is levelled properly to provide good drainage. For amelioration of acid soil, 1.5 tonnes dolomite or 1.0 tonne slacked lime/ha (depending upon the soil test) need to be applied. Application of Phorate 10 G @ 15 kg/ha is advocated to control soil inhabiting insects like cut worm, mole cricket, weevil, etc. Pre-planting irrigation is necessary if enough moisture is not available at the time of planting.

Planting time

1. Planting in the plains

Early crop: Planting should be done within 10-20 September in Punjab and western Uttar Pradesh; First week of October in Central Uttar Pradesh; Second and third week of October in Bihar and 2nd fortnight of October in West Bengal.

Main crop: Planting should be done in November in north-western plains, Gangetic plains and West Bengal and October – December in Plateau region of peninsular India.

Late crop: Planting should be done during last week November to middle of December in Gangetic plains.

Spring crop Planting should be done during December- January in North-western plains.

2. Planting in Hills

North-eastern hills: May- June in the valleys and high altitudes (3000-3500m); April in the high hills (2500- 3000 m); January – February in Mid hills (1000 – 1800 m) as spring crop and August – September for autumn crop; September in Ayodhya hills of West Bengal as autumn crop; January – February in the lower hills and valleys for spring crop.

Nilgiri hills: April and August.

Planting of tuber

Potato is traditionally propagated from tubers which have a dormancy of nearly 8-10 weeks after harvest. Multiple sprouting stage of the tuber is the best for planting because yield depends on the number of main stems. It is recommended to plant the certified seed from Government approved organizations. Whole seeds tubers should be taken out from cold storage 10-15 days before planting and kept in shady place under diffused light condition in thin layer on the floor which helps in sprouting the tubers. Use of at least 30 g sized (3-5 cm diameter), disease free healthy whole or cut tubers having 2-3 sprouts is the best. Cutting instruments must be cleaned and disinfected with 1% potassium permanganate solution each time after cutting any infected tubers. Potato is mainly infested with seed borne diseases hence, it is essential to treat the seed before planting. Seed tuber should be treated with Mancozeb @ 2.5 g / litre of water for 10 minutes or *Trichoderma viridi* @ 4-5 g/ litre of water for 10 minutes or *Pseudomonas fluorescence* @ 2.0 g/ litre of water for 10 minutes. For treating one quintal of seed tubers, 50 litres solution is required and the same solution can be used for 3-4 times after which fresh solution need to be prepared. Treated tubers should be dried under shade before planting. Average seed rate is about 25-30 quintals per hectare.. Seed tubers may be planted either on ridges or on flat beds. Planting should be done at a depth of 3-5 cm keeping the sprouts upward. Planting in rows in flat beds is suitable in light soils. Tubers are planted and covered with soil to make ridge. During planting care should be taken so that the seed tubers do not come in direct contact of the fertilizers that have been applied as basal.

Planting of seedlings from TPS

True potato seed (TPS) technology saves considerably the large quantity of seed tubers (25-30 quintals/ha) and their storage, space and transport. 100-150 g TPS is sufficient to produce seedlings for one hectare area. TPS can be stored safely at ordinary room temperature and under low humidity condition for 2-3 years.

Spacing

A spacing of 60 x 20 cm is recommended for medium sized tubers (40-50 cm) giving the plant population of 83,000/ha; 60 x 25 cm for larger sized tubers (50-55 cm) with the plant population of 67,000/ha and 60 x 15 cm for small sized tubers (35-40 cm) with the plant population of 1, 11, 000/ha.

Nutrient Management

Potato is very responsive to the application of nutrients. Nitrogen is the primary limiting nutrient in potato production. Potassium increases the size of tubers. A general fertilizer dose of 180-200 Kg N, 90-100 Kg P_2O_5 and 120-150 Kg K_2O per hectare is recommended. However, nutrient requirement depends on the type and nutrient status of the soil. Calcium ammonium nitrate (CAN) and ammonium sulphate are better sources of nitrogen than urea for potato. Similarly, di-ammonium phosphate (DAP) and super phosphate is effective source of phosphorous and the better source of potassium in the potassium sulphate than the muriate of potash for potato. Application of half the dose of nitrogen and potash and full dose of phosphorus in furrows at the time of planting gives the best results. Rest half dose of nitrogen and potash should be top-dressed in two split doses at 25-30 days after planting and 40-45 days after planting followed by earthing-up. In case of micro-nutrient deficiency, foliar application of 0.5% Zinc sulphate and 0.4% Borax is recommended.

Irrigation

Adequate irrigation is essential for proper growth and yield of potato. Stolon formation, tuber initiation and tuber development stages are the critical stages for irrigation. One light irrigation, if necessary, is given before planting to ensure uniform germination. Depending on the soil moisture condition, splash irrigation at 3-4 days interval up to 2-3 days before first earthing up need to be applied. The crop needs to be irrigated at an interval of 7-10 days after 1st earthing up. Level of irrigation water must not exceed 3/4th part of the ridges. Total water requirement for the crop varies between 350-550 mm. Irrigation should be stopped 10-15 days before harvesting the crop to allow the tuber skin to become firm.

Interculture and weed control

Earthing up is necessary in potato cultivation because proper development of tubers depends upon aeration, moisture availability and proper soil temperature. First earthing up should be done 3-4 weeks after planting and the second 10 days after first earthing up. Earthing up also prevents greening of tubers which occurs due to exposure of the tubers to sunlight. Mulching in the ridges with paddy straw, wheat husk, maize or jowar stalks is effective in inducing quick germination and conserving soil moisture. Two to three hand weedings will make the plot weed free and well pulverization of the soil. Application of herbicides like, Pendimethalin @ 1.0 kg a.i./ha 1-2 days after planting of tubers is recommended and post-planting application of Simazine @ 0.5 to 1.0 kg/ha over the ridges just after planting of potatoes is also effective.

Harvesting the crop

Potato should be harvested before the temperature rise above 30°C to avoid rotting of tubers due to high temperature. It is better to cut the haulms 10-15 days before harvesting when irrigation has been stopped and Mancozeb @ 2.5 g/ litre of water is applied on cut portion of haulms. The crop should be harvested when haulms start yellowing and falling on the ground manually with the help of khurpi or spade or mechanically with the help of 1-4 row potato diggers.

Post harvest handling and storage

The tubers should be cured in a stack under shade for 10-12 days for hardening of tubers. The height of the stack should be 1-1.5 m and width 3-5 m. The cut, rotten, insect damaged and rubbed tubers should be discarded after harvesting. As per market demand, the tubers need to be graded into small (<25 g), medium (25-50 g) large (50-75g) and extra large (> 75 g) on the basis of their size. The graded potatoes are generally packed in gunny bags. The table purpose varieties are stored in cold storage at temperature of 2°C-4°C and 80-85% relative humidity. The low temperature checks sprouting and rot of tuber and high relative humidity reduces weight loss from the tubers. The processing varieties are stored in cold storage at temperature of 10-12°C and 80-85% relative humidity. Potato stored at 10-12°C generally sprout and it is necessary to check sprout growth to reduce storage losses. To check sprout gowth, a sprout suppressant should be used. Most commonly used sprout suppressant chemical is CIPC (chrolo isopropyl carbamate). For treating small quantity of potatoes, powder form of CIPC (25 mg dust/ t of tuber) is convenient. For a large scale treatment, liquid formulation of CIPC (Chloropropham 50% a.i marketed in the name of Oorja) of 30-35 ml is required for fogging one ton of tuber. However, the dose may vary depending upon the loss of fog during treatment due to leakage, pore size of bags used, the quantity of potatoes stored, etc. After CIPC treatment, the store house should be kept air tight for 24-48 hours and 10-12°C temperature is maintained. The treated potatoes are safe for consumption 3-4 weeks after the treatment.

Plant Protection Measures

Physiological disorders

Uneven sprouting in the field

Adequate crop stand cannot be maintained for this disorder which happens due to i) planting soon after their removal from cold storage, ii) soil moisture deficiency and iii) untreated cut tuber pieces if infected by the fungus.

Control measures

- Holding of the tubers in storage till end of the dormancy period.
- Treating the cut tuber pieces with 0.2% Diathane M-45 for 10 minutes.
- Placing the tubers somewhat deep, particularly in light soil with the cut surface facing downwards.

Dormancy

Dormant or even semi-dormant tubers do not sprout readily after planting causing delayed and erratic crop stand. This physiological hindrance becomes a problem when two potatoes are taken in rotation and tubers produced in the hills are used as seed soon after harvest.

Control measures

- In advanced stage of dormancy, cutting of tubers may terminate dormancy.
- Immersing the cut tuber pieces immediately in 1% thiourea solution for one hour and then plant them as soon as possible.
- Triple treatment: Spraying the whole tubers with 3% ethylene chlorohydrin solution @ 20 ml/kg of tuber in an air tight chamber for 4 hours; thereafter dipping the cut pieces of tuber in 1% thiourea solution for one hour and finally dipping the tubers in 1 ppm gibberellic acid solution for 10 minutes. The treated tuber pieces are dried in shade before planting.

Diseases

Late blight *(Phytophthora infestans)*

Late blight is the worst disease of potato. It was first reported in the 1830s in Europe and in the US. It is famous for being the cause of the 1840s Irish Potato Famine, when a million people starved and a million and a half people emigrated. Late blight continued to be a devastating problem until the 1880s when the first fungicide was discovered. In the recent years, it has reemerged as the most serious problem. This fungal disease is prevalent in almost all potato growing areas of the country. Seed tuber is the primary source of inoculums. Cool weather (10-20°C) and relative humidity of 80% and above induce the disease development. The dead areas appear at the tip or margin of the leaves and spread downward and inward. Decaying leaves emit offensive odour. In present day condition, stem infection is also very common, where dark brown lesions appear on the stem or on the branches and the whole plant die within few days.

The sign of tuber infection is brown to purple discolouration of the skin followed by a brown rot which extends to about half an inch below the surface of the tubers.

Control Measures

- Cultivation of resistant varieties like Kufri Alankar, Kufri Jyoti, Kufri Swarna, Kufri Megha, Kufri Naveen, Kufri Neelamani, Kufri Muthu, Kufri Jawahar, Kufri Sutlej, Kufri Pukhraj, Kufri Giriraj, Kufri Shailga, Kufri Girdhari, etc.
- Use of disease-free, healthy and certified seed tubers.
- Restriction of water supply particularly when weather condition is favourable for disease development.
- Keeping the compost piles away from potato growing areas.
- Keeping the tubers covered with soil throughout the season to prevent tuber infection.
- Removal of infected tubers before storing to prevent the spread of disease in storage.
- Dehaulming before harvest to avoid inoculation of the tubers during harvest.
- If the late blight causing weather condition prevails, spraying should be started with protectant fungicides like, Copper-oxychloride (0.3%) / Mancozeb (0.25%) / Chlorothalonil (0.2%) / Metiram (0.3%) at 10 days interval.
- If late blight is present in field in pockets, spraying should be started with Metalaxyl-mancozeb (0.25%)/Cymoxanil-Mancozeb (0.25%)/ Fenamidone – Mancozeb (0.25%)/Dimethomorph (1g) + Mancozeb (2g)/ l of water immediately and alternate spray schedule with protectant and systemic fungicides at 7-10 days interval depending upon weather condition.
- Careful irrigation or avoidance of top dressing and irrigation under late blight prevailing condition.

Early blight *(Alternaria solani)*

This disease, also known as target spot, rarely affects young, vigorously growing plants. It is found on older leaves first. It is a common soil borne, fungal disease. Warm-moist weather favours the occurrence of the disease. Small, isolated, scattered, pale brown spots first appear on the leaflets. Later the spots grow in size and in the necrotic tissues, concentric rings develop.

Control measures

- Crop rotation and field sanitation.
- Growing of resistant/tolerant varieties like Kufri Chamatkar, Kufri Sinduri,

Kufri Jyoti, Kufri Badshah, Kufri Lalima, Kufri Jeevan, Kufri Khasigaro, Kufri Sherpa.

- Spraying the crop with 0.3% Mancozeb or 0.1% Bavistin or Chlorothalonil 0.2% at the occurrence of intermittent rains in dry wheather 3-4 times at 10 days interval.
- Two sparys of 0.2% Chlorothalonil or Difenconazole (0.5g /l) or Metiram (0.3%) is very effective.

Charcoal Rot (*Macrophomina phaseolina*)

It is fairly wide spread soil and tuber-borne fungal disease causing mostly tuber rot in storage. Dry and warm weather (28 -35°C) favours this disease infection. The disease may appear as stem blight, wet charcoal coloured tuber rot or as dry tuber rot. Small black spots appear at the site of lenticels and eyes of the infected tubers and foul smell emits from the rotten mass.

Control measurers

- Use of disease free tubers.
- Growing early variety so that the crop may be harvested earlier before the temperature reaches 28°C.
- Frequent irrigation to lower the soil temperature.

Black scurf (*Rhizoctonia solani*)

Characteristic symptoms of this disease are brownish-black sunken lesions on the underground stems and stolons. The disease may cause non-uniform stands of weak, spindly-looking plants. Early-season infections often result in the pruning of young stolons where lesions girdle them completely. Dark stem lesions occurring below the soil line may girdle the main stem resulting in yellowish or purplish leaves that curl upwards. On relatively healthy-looking plants, aerial tubers may form. During mid-season, the fungus may develop a white powdery mold growth on the stems that extends just above the soil line. This often is associated with stem lesions below the ground. The fungus forms sclerotia (survival structures) on the tubers. The sclerotia vary from netted or scurfy residues to individual black masses on the tuber surface. Tubers may be misshapen, cracked or may develop a russet like skin.

Control measurers

- Use of disease free seed tubers.
- Warming of the seed tubers prior to planting.

- Any practice that promotes rapid emergence of sprouts will reduce the attack.
- Use proper crop rotation, preferably grasses or cereals.
- Spraying the crop with Validamycin (0.2%) is highly effective.

Soft rot (*Erwinia carotovora, E. carotovora* ssp. *atroseptica, E.carotovora* ssp. *carotovora*)

It is a common bacterial disease where infection in the field is often carried over to storage. In the field, striking brown-black or jet-black slimy areas on the stem just above the ground level appear which is called 'black leg'. In storage, the surface of the affected tuber becomes discoloured, sometimes wrinkled, depressed and the internal tissues turn soft and mushy resulting in disintegration of the whole tuber.

Control measures

- Dipping of the seed tubers in 0.01% Streptocycline solution before planting.
- Soil application of stable bleaching powder @ 15 kg/ha before planting.
- Dipping of the tuber before storage in 0.3% Boric acid solution for 5 minutes.
- Storage of the healthy tubers in dry cool (4°C temperature) and well-ventilated condition.
- Spraying the crop with 0.01% Streptocycline + 0.3% Copper-oxychloride.

Common scab (*Streptomyces scabis*)

Two types of scab-like roughened lesions on tubers are caused by these bacteria viz., shallow scab and deep scab. In shallow scab, superficial rough areas, sometimes raised above and often slightly below the healthy skin are caused. It consists of corky tissues which arise from abnormal proliferation of the cells of tuber periderm. In deep scab, 1-3 mm deep lesions which are darker than shallow scab is caused and the tissue around turns corky.

Control measures

- Dipping the seed tubers in the solution of 0.25%Agallol-6 or Emisan-6 for 30 minutes before planting.
- Keeping the field in well-irrigated condition from the fifth week of planting up to the ninth week.
- Green manuring minimizes the incidence of the disease.

Leaf Roll (Potato leaf roll virus)

This virus is transmitted by several species of aphids.Upward and inward rolling of leaflets and leaves become leathery and brittle. The infected plant produce short stolons and small tubers get clustered around the base of the plant.

Latent Mosaic (Potato virus X (PVX) and Potato virus S (PVS)

This sap transmitted virus cause mild mottling of leaves or light green discolouration. Severe infection leads to stunting of plants and cause degeneration in the performance of the varieties.

Control measures

- Use of disease free certified seed stock.
- Uprooting the infected plants along with tubers showing the disease symptom from the field.
- Avoidance of frequent movement in the field.
- Washing of all tools and implements with 3% solution of trisodium phosphate before use them.
- Spraying the crop with systemic insecticides like, Imidacloprid (1ml/3l) or Fipronil (2ml/l) or Monocrotophos (2ml/l) to control the aphids and leaf hoppers.

Insect Pests

Potato tuber moth (*Phthorimea operculella*)

This serious pest both in storage and field was introduced in India through seed material imported from Italy. This pest completes 10-13 generations in a year under favourable environmental conditions. The moth lays eggs singly near the eyes of the exposed tubers and sometimes on the ventral surface of the leaves. The caterpillars after hatching bore into the tubers, petioles and terminal shoots. The caterpillars burrow into the tuber flesh, often tunneling it freely. Infected tubers become useless and ultimately rot by secondary infection.

Control measures

- Deep planting of healthy tubers and proper earthing up leaving no open cracks.
- Trapping the adults by light trap and pheromone trap.

- ❒ Spraying the crop with 0.2% carbaryl (Sevin WP 4 g/litre of water).
- ❒ Spraying the crop with *Bacillus thuringiensis* (0.05%).
- ❒ Treating the gunny bags with 10% neem seed kernel extract.

Hadda beetle (*Epilachna vigintioctopunctata, E. occellata*)

Adults and grab of these polyphagous pests feed voraciously on the leaves by scraping in a characteristic manner leaving the veins intact.Characteristic skeletonised patches are developed on the leaves which later dry out.

Control measures

- ❒ Spraying the crop with 2 g carbaryl or 1.5 ml endosulfan or 2.5 ml triazophos or 1.25 ml profenophos per litre of water.
- ❒ Collection and destruction of initial egg masses are useful to suppress the pest population.

Cut worms (*Agrotis ipsilon* and other *Agrotis* spp.)

The larvae after hatching feed gregariously on the foliage for few days and upon development, the caterpillars disperse and hide in the cracks and crevices of the soil.At night the caterpillars come out and cut the young potato plants near the ground level.

Control measures

- ❒ Clean cultivation and uprooting the stubbles and breaking big soil clots.
- ❒ Flood irrigation in the infected plots to expose the larvae for bird predation.
- ❒ Drenching the soil around the plants and the ridges with 2.5 ml chloropyriphos or 2 g carbaryl per litre of water.

Mole cricket (*Grylotalpa africana*)

In some cases the mole cricket causes extensive damage by feeding on the developing tuber which reduces the market value of the crop.

Control measures

- ❒ Application of 2.5 ml chloropyrifos/ litre of water along with rice bran in the infected plots during evening hours reduce the pest population drastically.

Root Knot Nematode (*Meloidogyne incognita*)

These plant parasitic nematodes infest root and during feeding process, the cortical cells are stimulated into giant cells which cause root galls.The affected plants fail to draw nutrients and water from the soils which make them pale green and stunted.

Control measures

- Summer ploughing to reduce the load of the nematodes.
- Growing of non-host crop like marigold or wheat in rotation.
- Application of Furadon 3 G @ 20 kg/ha (1/2 during planting and rest part during earthing up).

Yield

Tuber yield depends on the variety, climate and cultural practices. In plains, under well managed conditions yield may be 25-35 t/ha and in the hills except valleys expected yield is 20 t/ha.

Solanaceous Fruit Vegetable Crops

- **Tomato**
- **Brinjal**
- **Chilli**
- **Sweet pepper**

Tomato

Tomato, *Solanum lycopersicum* L under the family Solanaceae (Chromosome No.: 2n = 2x = 24) is one of the major vegetable crops for fresh consumption as well as for processing purpose in India and the world as a whole. It is one of the most versatile crops in the world because of its fast and wide climatic adaptation. Tomatoes have been used as food by the inhabitants of central and South America since pre-historic time. Long before the discovery of the Americas, the pre-Columbian and Aztecs developed larger fruited and new colours of tomato from the wild cherry tomato.

The crop

Two types of growth habits are basically found in tomato: determinate and indeterminate. Determinate tomatoes grow to a certain height, then flower and set their fruit within a short time. Indeterminate tomatoes continue to grow and produce flowers and fruits until senesce naturally. Leaves are more or less hairy, strongly odorous and pinnately compound in nature.

The fundamental inflorescence type of tomato is regarded as a "scorpioid cyme" of a monochasinm. In cyme inflorescence, the shoot apex differentiates into a flower and subsequent growth occurs due to activity in an axillary branch which will eventually terminate in a flower. The cyme may be simple or branched depending on the genotype. The five-petaled flowers are yellow, 2 cm across, pendant, and clustered. Ovary is superior (above the attachment of the petals, sepals and stamens, and is free from the receptacle) and with 2-9 compartments. It is mostly self- but partly cross-pollinated. Bees and bumblebees are most important pollinators. Fruits are berries that vary in diameter from 1.5 to 7.5 cm or more. They are usually red, scarlet, or yellow, though green and purple varieties do exist, and they vary in shape from almost spherical to oval and elongate to pear-shaped. Each fruit contains at least two cells of small seeds surrounded by jellylike pulp.

The tomato fruit is classified as climacteric in terms of ripening which is accompanied by a peak in respiration and a concomitant burst of ethylene.

Exposure to exogenous ethylene accelerates ripening of green tomatoes. Fruit ripening culminates in dramatic changes in colour, texture, flavour, and aroma of the fruit flesh. Chloroplasts are transformed into chromoplasts, chlorophyll is degraded and carotenoids accumulate. The characteristic pigmentation of red tomato fruit is due to the deposition of predominant carotenoid pigment, lycopene. The characteristic flavour of tomato fruits results from the volatile compounds produced within the fruit during the ripening. Over 400 volatile compounds are found in tomato fruit. Among them seven are the most important contributors to the aroma: hexanal, hexenal, hexenol, 3-methylbutanal, 3-methylbutanol, methylnitrobutane, and isobutylthiazole.

Importance and use

Tomato is the world's most popular and widely cultivated vegetable crop and grown in all kinds of climate viz., temperate, subtropical and tropical. Tomato is basically a warm season crop and has the potential to grow all the year round in mild climatic condition. It is universally treated as 'protective food' and provides almost all types of vitamins and minerals in quite fair amount. It is eaten directly as raw vegetable in sandwiches, salad etc. and used in a number of ways for culinary purposes. Its ripe fruits are utilized on a large scale in the preparation of a variety of processed products such as, puree, paste, powder, ketchup, sauce, soup, syrup, juice, drinks and canned whole peeled fruits. The mature green fruits are used for making pickles and preserves and are also consumed after frying in oils. Fruit firmness, thick pericarp (0.5 cm and above), low locule number (2-3) and slow fruit ripening character is preferred for good transportability and better shelf-life. High total soluble solids (°Brix 5.5 and above), low pH (below 4.5), solid /acid ratio of 15 and sugar/ acid ratio of 8.5 and high viscosity of fruit pulp is ideal for processing purpose. Tomato fruits contain various types of flavoring compounds which enrich the taste and colour thus, it is used in a number of ways to improve the flavour and characteristics of other foodstuffs. Red colour of tomato is due to the carotenoid pigment lycopene which is regarded as the powerful antioxidant.

Nutritional value

Tomato fruit contains intermediate levels of vitamin C, carotenoids and provitmin A. It is rich source of minerals particularly potassium and organic acids particularly citric and maleic acid. Ascorbic acid content ranges from 16-75 mg /100g of fruit weight. Average total sugar content is 2.5% in ripe fruit. Free sugars generally representing more than 60% of the total soluble solids are mainly glucose and fructose with traces of sucrose. The ripe fruits contain important amino acids like, glutamic acid, aspartic acid, triptophan and tyrosine. The attractive red colour of tomato is due to carotenoid pigment lycopene which is the potent non-nutrient bioactive substances and act as an anti-oxidant. It contains a steroidal glycoalkaloid, tomatine to the range of 130-150 mg per 100g fresh tomatoes. The strong odour of the leaves, stem and young fruits is attributable to this alkaloid and for a long time tomato was not consumed by human due to persistent superstitions of its poisonous nature.

Medicinal value

Several epidemiological studies have shown that consumption of tomato protects against cancers of digestive tract, stomach, colon, rectum, gastrointestinal neoplasm and prostate and this beneficial effect is due to presence of antioxidant compounds such as lycopene, β carotene and vitamin C. Among all the antioxidants lycopene, the major carotenoid without pro-vitamin A activity is the most efficient quencher of singlet oxygen. Tomato is said to be useful in curing cankers and sores in the mouth. The secondary metabolites i.e., colorless glycoalkaloids a-tomatine and dehydrotomatine significantly reduce plasma cholesterol. Feeding high tomatine green tomatoes induce a greater reduction in plasma LDL (bad) cholesterol and triglyceride levels than feeding low tomatine red tomatoes. Pickled green tomatoes are widely consumed in many countries have high tomatine content. The tomato pulp and juice are mild aperients (laxative), a promoter of gastric secretion and blood purifier and is considered to be the intestinal antiseptic. It is a very good appetizer and remedy for the patients suffering from constipation if used as soup. It stimulates torpid (sluggish, dull, inactivated) liver and is good in chronic dyspepsia. Tomato poultice is used in the treatment of edema during pregnancy. The people suffering from hyperacidity and from stones in gall bladder should avoid the excessive use of tomato in daily diet.

Origin and taxonomy

For several years tomato was known as classified under *Lycopersicon* although initially Linnaeus (1753) classified it in the genus *Solanum*. Recently, molecular studies provided new data on the relationships of tomato-potato and based on these results, tomato has been again assigned to the genus *Solanum*. *Solanum* section *Lycopersicon* includes the cultivated tomato (*S. lycopersicum*) and its wild relatives, *S. lycopersicum* being the only domesticated species. The wild cherry tomato, *Lycopersicon esculentum* (*Solanum lycopersicum*) var *cerasiforme*, the ancestor of the modern cultivated tomato has originated in Peru-Equador-Bolivia region of the Andes. The Veracruz-Puebla region of Mexico is the centre of domestication of the cultivated tomato. From Mexico, tomato was taken to Italy, Spain, Portugal and other

European countries, Africa and Middle East by the explorers in the 16th century. It was taken to Philippines after its discovery by Magellan in 1521, and from there it was taken to other Asian countries. During the early years of its introduction tomato was considered to be poisonous in Britain, France, Spain and other European countries. Tomato moved to the USA from northern Europe around 1781 where it was first grown by Thomas Jefferson in Virginia.

In total, the cultivated tomato, *Solanum lycopersicum* (*Lycopersicon esculentum*) has 12 wild relatives including and its weedy escaped forms that are distributed worldwide viz., *Solanum lycopersicoides* (*Lycopersicon lycopersicoides*), *Solanum sitiens* (*Lycopersicon sitiens*), *Solanum juglandifolium* (*Lycopersicon ochranthum*), *Solanum ochranthum* (*Lycopersicon juglandifolium*), *Solanum pennellii* (*Lycopersicon pennellii*), *Solanum habrochaites* (*Lycopersicon hirsutum*), *Solanum chilense* (*Lycopersicon chilense*), *Solanum huaylasense* (Part of *Lycopersicon peruvianum*), *Solanum peruvianum* (*Lycopersicon peruvianum*), *Solanum corneliomuelleri* (Part of *Lycopersicon peruvianum*), *Solanum arcanum* (Part of *Lycopersicon peruvianum*), *Solanum chmeilewskii* (*Lycopersicon chmeilewskii*), *Solanum neorickii* (*Lycopersicon parviflorum*), *Solanum pimpinellifolium* (*Lycopersicon pimpinellifolium*), *Solanum cheesmaniae* (*Lycopersicon cheesmaniae*) and *Solanum galapagense* (Part of *Lycopersicon cheesmaniae*).

Adaptation in India

Britishers believed to have introduced tomato in India in 1828 through the Royal Agri-Horticultural Society, Calcutta and afterwards it spread to other parts of the country. In India tomato improvement was initiated around 1950 with the introduction of exotic cultivars. By this time, tomato was spread throughout India because of its wide adaptability in tropical environmental condition and during the course its spread, some selections from the introduced materials have also been made in different parts of the country. Few such old selections like, Annaji of Tamil Nadu, Meeruti of Uttar Pradesh, Patharkuchi of West Bengal are still being cultivated commercially as local cultivars.

Open pollinated improved varieties

Determinate type : Pusa Early Dwarf, HS 101, HS 102, Hisar Arun, Punjab Chhuhara, Sonali, Roma, Pusa sheetal, Pusa Uphar, Pusa Rohini, Pusa Gaurav, CO-1, CO-3, Punjab Keshri, La Bonita, Pusa Red Plum, Hisar Lalima, Narendra Tomato-2, Arka Meghali, Arka Alok, Arka Ashish, KS-2,Sakthi, TLB-111, Pusa Rohini, Azad T-2,Pusa Sadabahar, Pant Bahar, Punjab NR-7, Utkal Pallavi, Kashi Hemant, Gujarat Tomato-2, Utkal Pragyan, etc.

Indeterminate/ Semi-determinate type: Pusa Ruby, Arka Vikas, Arka Saurabh, Arka Ahuti, Arka Abha, Punjab Tropic, Pusa 120, Sioux, Marglobe, Best of All, Keck-Ruth, PKM-1, Arka Meghali, Utkal Kumari, Pusa Uphar, Bhagyashree, Dhanashree, NDT 5, Pant T-1, Pant T 3, Hisar Lalit, Solan Gola, Kalianpur Angoorlata, Junagadh Ruby, Parbhani Yashashri, Kashi Sharad, etc.

For fresh market: Pusa Early Dwarf, Pusa Ruby, PantT3, Arka Vikas, Arka Saurubh, CO-3, BT 12, Punjab Keshri, Pant Bahar , HS 110.

For long distance market: Pusa Gaurav, Roma, Punjab Chhuhara, Pusa Uphar, Arka Saurabh, Pant T3, NDT 120, Pantharkuchi.

Rootknot nematode resistant: Pusa 120, Hisar Lalit, Punjab NR-7.

Bacterial wilt resistant : BT-1, BT-10, BT-12, Arka Alok, Arka Abha, Sakthi, Swarna Naveen, Swarna Lalima, Utkal Raja,

Leaf curl virus resistant: Hisar Anmol, TLB-111, Kashi Vishes, Kashi Amrit, Kashi Anupam.

Spotted wilt and leaf curl resistant: Dhanashree.

Abiotic stress resistant: Pusa Sheetal (low temp regime, upto 8°C night temperature), Pusa Sadabahar (both low and high temp regime, above 29°C average daily mean temperature).

Round fruit: HS 102, Hisar Arun, Co-1, Hisar Lalima, NDT 120, HS 110, Hisar Lalit, Pant T3, Arka Saurabh, Marglobe, Best of All, Pant T1, Utkal Kumari, Pusa Uphar, Bhagyashree, Pusa Rohini.

Flattish -round fruit: Pusa Sheetal, Sweet 72, Azad T-2, Hisar Anmol (H-24), Pusa Ruby, Pant Bahar, PKM-1, Patharkutchi.

Pear shaped fruit: Punjab Chhuhara,

Elliptical fruit: Roma, Pusa Gaurav, Punjab Keshri, Arka Ashish, Pusa Sadabahar, Sonali.

Public sector hybrids: Pusa Hybrid-1, Pusa Hybrid-2, Pusa Hybrid-4, Pusa Hybrid-8, Pusa Divya Pant hybrid-1, Pant Hybrid-2, Arka Shreshta, Arka Vishal, Arka Vardan, Arka Abhijit, Rajashree, TH-2312, TH-802, COTH-1, Bidhan Tomato Hybrid-4, Bidhan Tomato Hybrid-62, Swarna Baibhav, Swarna Sampada, Solan Shagun, Arka Vardhan.

Cultural requirements

Tomato is a highly adaptive and warm loving crop and can be grown in the plains as well as in the hill condition. The optimum daily mean temperature is

21°C to 24°C. Average daily mean temperature below 16°C and above 28°C is harmful for the crop. Synthesis of lycopene pigment in the fruits is highest at 21° to 24°C daily mean temperatures and drops off rapidly above 27°C. Night temperature is critical for fruit set, the optimum range being 15°C to 20°C and fruits generally do not set when night temperature is either below 13°C or above 20°C. There are some varieties and hybrids which can set fruits at low temperature (around 8°C night temperature) and high temperature (around 29-30°C average daily mean temperature). Well drained fertile, organic matter rich soils are ideal for tomato cultivation however, sandy loam soil is best for early crop and clay loam soil for getting higher yield. The ideal soil pH is between 6.0 and 7.0 although it can tolerate a little soil acidity upto pH 5.5 but not below it.

Nursery management

Seed beds are prepared finely, well drained, 15-20 cm raised, 1.0m wide and of convenient length. Fine and fully decomposed farmyard manure or compost @ 3-4 kg/m^2 should be well mixed to the beds. The beds are drenched with formaldehyde (4.0%) and covered with polythene sheet for 5-7 days for soil disinfection to avoid damping off disease. The seed bed soil can also be treated with Captan @ 2-3g/litre 3-4 days before seed sowing. 400 to 500g seeds of open pollinated varieties and 125 to 175g seeds of hybrids are required per hectare. The seeds are treated with Captan or Thiram @ 2-3 g per kg of seeds before sowing. Seeds are sown at shallow depth at 5.0 cm apart and covered with finely sieved leaf mould or sterilized cow dung manure. After sowing, the beds are covered with straw or long dry grass with regular sprinkling of water till seeds are germinated. Seedlings may be raised under poly tunnels having side open provision to protect the young seedlings from heavy rain or scorching sunlight. The seedlings are watered daily or in alternate day with rose can. Hardening of the seedlings can be done by withholding water at least 4 to 6 days before transplanting. Stocky, 25-35 days old, healthy seedlings of 10-15 cm height with 3-4 leaves are ideal for transplanting.

Sowing time

Tomato is grown in three distinct seasons viz., early autumn, winter-spring and spring-summer in the plains of India although sowing and transplanting season vary with the region. In the hills, tomato is sown in February-March or March-April. For early autumn crop in Maharashtra, Gujarat, Karnataka, Tamil Nadu, Andhra Pradesh and Rajasthan sowing is done in May-July. For winter-spring crop in Maharashtra, Gujarat, Karnataka, Tamil Nadu and Andhra Pradesh, West Bengal, eastern Uttar Pradesh, Bihar, Orissa, Assam sowing is done in September- October. For spring-summer crop in Punjab, Haryana, Delhi, Western Uttar Pradesh, Madhya Pradesh, West Bengal sowing is done in October-December. For spring-summer crop in Maharashtra, Gujarat, Karnataka, Tamil Nadu, Andhra Pradesh and Rajasthan sowing is done in December-February.

Land preparation

The land is prepared in well advance with repeated ploughings (at least 4-5 ploughing) to a fine tilth. All stubbles, weeds, etc need to be removed from the land. Well rotten organic manure preferably farm yard manure or compost @ 20-25 t/ha are added during land preparation and the land is levelled properly.

Transplanting

Healthy seedlings of 28-35 days old, 12-15 cm height are ready for planting. After hardening the seedlings for 4-6 days, the seed bed is irrigated before lifting to facilitate easy pulling of young seedlings. Seedlings should be planted in the field preferably during afternoon hours for better field establishment. The field should be irrigated just after transplanting.

Spacing

Spacing varies with the growth habit of the variety or hybrid. Spacings administered for indeterminate hybrids are 90 cm x 90 cm, 90cm x 75cm; for determinate hybrids is 75 cm x 75 cm; for indeterminate varieties are 90 cm x 60 cm, 75 cm x 75 cm and for determinate varieties are 60 cm x 60 cm and 60 cm x 45 cm.

Nutrient management

Adequate nutrition is necessary not only for obtaining higher yield but also for better fruit quality. A general fertilizer dose of 100 kg N, 80 kg P_2O_5 and 80kg K_2O per hectare is recommended for open pollinated improved varieties.One-third nitrogen along with other fertilizers should be given as basal and rest N should be top dressed in two split doses at 30 days interval after transplanting. A general fertilizer dose of 200 kg N, 100 kg P_2O_5 and 100kg K_2O per hectare is recommended for the hybrids. One-fourth nitrogen along with other fertilizers should be given as basal and rest N should be top dressed in three split doses at 30 days interval after transplanting. Two foliar sprays of 0.5% $ZnSO_4$ is beneficial for both yield and quality. Two foliar sprays of 0.2% calcium chloride and two sprays of 0.3% borax check blossom end rot due to calcium deficiency and cracking of fruits due to boron deficiency. Calcium ammonium nitrate can be used as the source of nitrogen in calcium deficient soil. Vermicompost @ 4-6t /ha as broadcasting can be applied during final land preparation. Azotobactor @ 800g/ha or Phosphate solublizing bacteria @800g/ha can also be applied as seedling root dip. Suspension of the bacterial inoculants is prepared in sufficient water and the seedlings are dipped in the bacterial suspension in shade for 2-3 hours in the afternoon hours. The treated seedlings should be transplanted immediately.

Irrigation

The plants require adequate moisture throughout their growth period. During summer irrigation should be given at 4-5 days interval and during winter at 10-15 days interval. About 95.0 cm total water is needed for successful tomato crop and 2.5 cm irrigation water need to be applied at every fortnight. Critical stages of irrigation are flower development, fruit set and after each harvest. Most widely adopted irrigation method is furrow method of irrigation. Nutrients can also be applied through drip irrigation. Irrigation should be stopped at fruit ripening stage during winter.

Interculture and weed control

In indeterminate varieties and hybrids, fruit size can be increased and at the same time disease problem can be checked by training and pruning of the plants. Two or three strong basal stems are retained and all axillary branches on them are removed by disbudding, allowing only the main stems to grow and produce fruits. The plants are staked to avoid damage of fruits, the common practice being to stake individual plant or provide support to plants in a row. Shallow inter- cultivation is practiced during first 4 weeks to remove the weeds and conserve soil moisture. Two hand hoeings at 1st and 3rd fortnight after transplanting and earthing up at 2nd fortnight after transplanting is generally recommended. Intercultural operations should be stopped when the plants have started flowering. Foliar application of 50 ppm PCPA increases fruit set at high and low temperature condition. Pre-planting application of Pendimethilin @1.0 Kg a.i. /ha or Fluchloralin @ 1.25 Kg ai/ha along with one hand hoeing 45 days after transplanting is effective in controlling the weeds.

Harvesting the crop

Tomato fruits are harvested at different stages of maturity depending on the distance of the market and the purpose viz., mature green, pink and breaker, hard ripe and fully ripe. Breaker or pink stage is the ideal harvest stage which ensures good quality, better shelf-life and proper development of colour. The fully ripe fruits are harvested for processing purpose. Tomato fruits are generally harvested at an interval of 4-5 days in summer and 6-7 days of in winter. Tomato fruits should not be pulled from the vine rather is picked with a twisting motion of hand to separate the fruits from the stem. Pre-harvest sprays of 0.3% Difolatan at 10 days interval are effective in controlling post harvest diseases.

Post harvest handling and storage

Pre cooling of tomato after harvest at 12-13°C on the farm prolongs their storage-life. Application of 0.2% Benomyl is effective in preventing growth of most of the damaging fungi. Mature green tomatoes can be stored in polythene bags of 100 gauge thickness at 12-14°C and 90-95% relative humidity for 18-21 days. Fully ripe tomatoes store well at 4-5°C and 85-90% relative humidity for 7 days in cold storage. Intact pedicel with fruits increases shelf life by 3-4 days. Evaporative cool storage (Zero energy cool chamber) extends shelf life of tomato harvested at breaker stage by 4-5 days. Tomato fruits treated with GA_3 at 10 ppm or BA at 25 ppm show minimum physiological weight loss and incidence of decay. Fruit dip in 0.5% calcium nitrate or calcium chloride extends the shelf life by checking CO_2 and ethylene production. Usually bamboo basket and wooden boxes of various size and shapes are used for packing of tomatoes. Wooden boxes are generally used for packing for long distance markets.

Plant Protection Measures

Physiological disorder

Blossom end rot

Brown, water-soaked discolouration appears at the blossom end of the fruit where senescent petals are attached while the fruit is still green. The spots enlage and darken rapidly and the affected portion becomes sunken, leathery and dark coloured. This disorder is caused due to i) sudden change in the rate of transpiration specially in moisture stress condition, ii) continuously high evapo-transpiration regime and a large leaf area, iii)increasing level of nitrogen content in the fruits and iv) fall in the level of calcium content in the fruit.

Control measures

- Increasing the frequency of irrigation.
- Increasing the level of phosphate which decreases the incidence of this disorder.
- Two foliar sparys of 0.2% calcium chloride at the time of fruit development.

Puffiness

Partially filled fruit become light in weight and lack in firmness. Cross-section of the affected fruits shows emptyness or pockets. This disorder is caused due to i) non fertilization of ovules, ii) embryo abortion after normal fertilization, iii) necrosis of vascular and placental tissue after normal development of the fruit, iv) high temperature and v) high soil moisture.

Control measures

- Avoidance of over irrigation.
- Three sparys of 0.3–0.4% borax solution at initial plant growth, fruiting and fruit development stages.

Fruit cracking

In the severe form of cracking, surface of the mostly fully matured and ripe fruits cracks radially from the stem end. In another form of cracking, surface of the mature green fruits cracks concentrically around the shoulder of the fruit. Fruit cracking may be a genetically controlled character. However, it may be caused due to i) irrigation or rainfall after a long dry spell ii) exposure of the fruits to sun iii) boron deficiency.

Control measures

- Irrigation at regular interval.
- Avoidance of pruning during summer season.
- Three sprays of 0.3-0.4% borax solution at initial plant growth, fruiting and fruit development stage.

Sunscald

Exposed portion of either green or nearly ripe fruits get blistered and water soaked. Rapid desiccation leads to sunken area, grey in green fruits and yellow in pink or red fruits. It is generally caused due to exposure of the fruits to extreme heat of scorching sunshine.

Control measures

- Avoidance of training and pruning in the summer months.
- Protecting the plants from defoliation caused by diseases and insect pests.

Gold fleck

Tiny yellow spots appear as gold flecks around the calyx and fruit shoulder due to deposition of calcium oxalate. With high incidence of this disorder the fruits become unattractive. This disorder is caused due to i) higher supply of phosphatic fertilizers, ii) higher supply of calcium fertilizers and iii) increased magnesium concentration in the fruit

Control measures

- Summer shading lowers the incidence of this disorder.

Diseases

Damping off (*Pythium, Rhizoctonia, Phytophthora,* etc.)

It is very destructive fungal disease in nursery beds, especially during early season. In pre- emergence damping off, the growing points are killed in the initial stages of seed germination before they come out through the soil. In post-emergence damping off, the seedlings topple over the ground due to collar rotting and rapid shrinking of the cortical tissue of the hypocotyls.

Control measures

- Hot water treatment of seeds (at 51.7^0 for 30 minutes).
- Treating the seeds with Thiram/ Captan / Metalaxyl-Mancozeb @ 2.5-3.0g / kg of seeds.
- Drenching the nursery beds 7 days before sowing with Thiram/ Captan or any copper fungicide @ 3g / litre of water.
- Covering the beds with transparent polythene sheets before sowing and left to open sun for at least15 days which is known as soil solarization.
- Spraying the young seedlings with 0.2% Blitox or 0.25% Metalaxyl-Mancozeb or 0.25 % Fosetyl-Al.
- Provision of proper drainage in the beds.
- Maintenance of plant population in the seed bed.
- Removal of the affected seedlings from beds as soon as the symptoms are visible.
- Avoidance of flooding the beds to check spread of the disease.

Late blight (*Phytophthora infestans* (Mont) de Bary)

It is a serious fungal disease when temperature lay between 10° to 20°C with relative humidity of 85% and above for 18 hours continuously for at least two consecutive days. Brown to purple-black lesions appear on leaflets, petiole or stem, advanced rapidly to cause a severe blight in favorable condition white fructification of fungus appears on the lower side of the leaves. Grey water-soaked spots enlarged to indefinite size and shapes appear at stem end of the fruit.

Control measures

- If the late blight causing weather condition prevails, spraying should be started with protectant fungicides Copper-oxychloride (0.3%) / Mancozeb (0.25%) / Chlorothalonil (0.2%) at 10 days interval.
- If late blight present in field in pockets, spraying should be started with Metalaxyl-mancozeb(0.25%)/Cymoxanil-Mancozeb(0.25%)/Dimethomorph (1g) + Mancozeb(2g)/ litre of water immediately and alternate spray schedule with protectant and systemic fungicides at 7 – 10 days interval depending upon weather condition.
- Careful irrigation or avoidance of top dressing and irrigation under late blight prevailing condition.

Early blight (*Alternaria solani* (Ellis and Mart) Jones and Grout)

It is serious fungal disease when temperature lay between 18^0 C to 30^0 C and soil moisture is abundant. Circular to angular and dark brown to black spots appear on the leaflets. Usually, a narrow chlorotic halo is found around the necrotic spots. Fruits may be infected both at the green and ripe stage. At stem end, black or dark brown lesions usually develop with sunken spots which become leathery later.

Control measures

- Spraying the crop with 0.3% Mancozeb or 0.1% Bavistin or Chlorothalonil 0.2% at 10-15 days interval starting form 45 days after transplanting as a prophylactic measure.
- Two sparys of 0.2% Chlorothalonil or Difenconazole (0.5g /l) is very effective.

Buckeye rot (*Phytopthora parasitica*)

This fungal disease becomes severe when soil moisture is plenty and soil temperature is around 18°C. The disease symptoms appear on the lower fruits. Brownish spot appears in the fruit which enlarges and the surface of lesion assumes a pattern of concentric rings of narrow, dark brown and wide, light brown bands.

Control measures

- Spraying the crop 3-4 times with 0.3% Blitox or Mancozeb or Metalaxyl-Mancozeb (0.25%) / Cymoxanil-Mancozeb (0.25%) at 10 days interval starting from disease appearance.

Bacterial wilt (*Ralstonia solanacearum* E.E. smith)

It is a serious bacterial disease particularly in the sandy loom and lateritic soil. Drooping of the lower leaves followed by wilting of the entire plant. Leaf epinasty, marginal leaf necrosis and development of adventitious roots along the stem take place. A slimy grey material oozes out when the cut stem is dipped in clear water.

Control measures

- Treating the seeds with hot water at 52°C for 20 minutes or with 0.01% streptomycin solution for 30 minutes.
- Proper crop rotation with non-solanaceous crops for at least 2 to 3 years reduces the infestation. Effective rotations are maize- okra-cabbage/ cauliflower; maize-beans-maize, or okra-rice-mustard.
- Application of bleaching powder @ 15-20 kg /ha to the soil at least 3-4 week before planting.
- Maintenance of proper drainage in the field.
- A combination of Dhaincha as green manure crop and chemical soil amendment with lime and bleaching powder and application of bioagents as soil application is most promising in management of bacterial wilt of solanaceous vegetable crops.

Tomato leaf curl (Tomato leaf curl virus)

Tomato leaf curl virus is transmitted by white fly (*Bemisia tabaci*). Severe stunting of the plans with downward curling of leaves and excessive branching are the characteristic symptoms. Older leaves turn leathery and brittle and the newly emerging leaves look pale yellow and smaller in size. In severe infection, the plants remain unfruitful.

Control measure

- Destruction of all weeds around the field which serve as alternate host of the virus.
- Covering the nursery bed with 40 mesh fine nylon net and spray the seedlings with Imidacloprid (3.5 ml/10 litre).
- Pre-sowing treatment of nursery beds with Furadon 3G @15g/100 sq.m.

- Application of Thimet 10G @ 10-15 kg/ha 10 days after transplanting followed by 2-3 foliar sparys with 0.05% dimethoate (Rogor 1.5 ml/l) or Imidachlorpid or Acetameprid or Thiomethoxam (3.5 ml/10 litre) at 10 days interval and alternate the spray schedule with NSKE.
- Application of 2% power oil to the plants to prevent acquisition and inoculation of the virus by white flies.
- Sowing of 5-6 rows of border crops of maize, jowar and bajra all round the tomato plot at least 50-60 days before transplanting of tomato.
- Mixed cropping of tomato, French bean and brinjal.
- Use of Resistant varieties/hybrids *viz*., H-24, Bidhan Tomato Hybrid-4, TLBRH-6, Kashi Vishes, Kashi Amrit, Kashi Anupam.

Tomato mosaic (Tomato mosaic virus)

The tomato mosaic virus (TMV) is transmitted by contact with clothes, man, animal and machinery during cultivation or by contact of infected plants, plant debris and through infected seeds. Yellow-green mottling appears on leaves and occasionally on fruits. Infected leaflets are usually distorted, puckered and smaller than normal. Number of branches, inflorescence and flowers are substantially reduced.

Control measure

- Soaking the seeds in trisodium phosphate solution (90g/l of water) for 15 minutes a day or two before sowing.
- Crop rotation excluding solanaceous crops.
- Washing all tools and implements with 3% solution of trisodium phosphate before using them again.

Insect pests

Tomato is infested by a numbers of insects and non insect pests of which some of them attained major pest status. The insect pests those are considered as major pests are discussed here.

Fruit borer (*Helicoverpa armigera*)

It is a widely distributed polyphagous insect pest causing devastating damage by boring the young developing fruits and feed inserting their head within it. Young larvae first feed on tender leaves and later infest the fruits by mining circular holes and feed therein.

Control measures

- Collection and destruction of initial invading larval population at early stage keeps the crop free from severe damage.
- Planting 14 rows of tomato alternated with one row of African marigold as trap crop.
- Inundative release of egg parasitoid *Trichogramma chilonis* and *Trichogramma brasiliensis* @ 50,000/ha/week suppress the pest population.
- Application of selective NPV is also very effective against this pest.
- Provision of birds perching site in tomato field suppress larval population as the predatory birds like drongo, starling, etc. successfully predate on them.
- Spraying the crop with 0.05% Chlorpyriphos (Dursban 2.5 ml/l) or 0.2% Carbaryl (Sevin 4g/l) at 7-10 days interval after fruit initiation.

White fly (*Bemisia tabaci*)

It is a dangerous pest of many important vegetable crops. Though it directly causes very little damage however, indirectly it causes profuse damage by transmitting virus to the healthy crops. The tiny nymphal stages of the whitefly directly suck sap from the ventral surface of leaves and the affected portion becomes yellowish, leaves wrinkle and curl downwards. Severe infestation causes stunted growth of the crops.

Control measures

- Inundative release of nymphal-pupal parasitoids, *Encarsia formosa* to suppress population outbreak of the whitefly.
- Spraying the crop with Azadirectin @ 1.0% (Neemazol 1ml/l).
- Spraying the crop with Diafenthiuron @ 100 ppm/l (Pegasus 0.5gm/l) or 0.03% Phosphamidon (Demecron 1.5 ml/l), 3-4 round at regular 7 days interval targeting only young apical parts of plant canopy.

Tobacco caterpillar (*Spodoptera litura*)

The larvae feed gregariously during early instars and voraciously during fourth onward stages preferably at night on the foliages. Severe infestation of this pest in nursery bed causes entire seedlings to be defoliated overnight.

Control measures

- Collection and destruction of leaves along with the egg masses and early gregarious larval stages are one of the best ways to suppress the pest.

- Application of selective NPV is also very effective against this pest.
- Inundative release of egg parasitoids, *Trichogramma chilonis, T. brasiliensis* and larval parasitoids, *Bracon hibator, Apantales* sp. to suppress the population outbreak of the pest.
- Spraying the seedlings with 0.05% Endosulfan (Thiodan 1.5 ml/l) or 0.2% Carbaryl (Sevin 2 g/l) at 10 days interval.

Epilachna beetle (Epilachna vigintioctopunctata)

The grubs and adults feed on the leaves along with the chlorophyll bearing tissues leaving the veins and cuticle intact resulting in the development of characteristic skeletonised patches on the leaves.

Control measures

- Collection and destruction of egg masses along with the grubs and adults in the initial invading population is very effective to check destructive form of this pest.
- Spraying the crop with 0.2% Carbaryl (Sevin 4g/l) or 0.05% Endosulfan (Thiodan 1.5ml/l) at 10 days interval.

Root knot mematode (*Meloidogyne incognita* and *M. javanica*)

The nematodes attack the roots and produce swellings or galls there. Infested plants become stunted and yellow in colour. Plants wilt in hot and dry weather and in severe cases, die.

Control measures

- Ploughing the field in summer months to expose the soil to high temperature to reduce the nematode population.
- Application of neem cake or neem leaf @ 500 g/m^2.
- Growing trap crops like marigold, wheat, etc. in rotation or intercropped with tomato.
- Application of Nemagon, Nematex or Vapam to the soil.
- Application of Thimet 10G (25 kg/ha) or Furadon 3 G (33.3 kg/ha) in the field at transplanting.

Yield

Ripe fruit yield varies with the type of variety and hybrid. Open pollinated determinate varieties may yield 20-25 tonnes/ha and open pollinated indeterminate varieties may yield 25-35 tonnes/ha. Fruit yield of the hybrids may be as high as 50-80 tonnes/ha in open condition. In protected condition under poly house, fruit yield of indeterminate hybrids may go up to to 300 tonnes/ha.

Brinjal

Brinjal, also known as aubergine or eggplant, *Solanum melongena* L under the family Solanaceae (Chromosome No.: 2n = 2x =24) is one of the important vegetables of the tropical countries particularly India, Bangladesh, Pakistan, China, Japan, Indonesia, Thiland, Malaysia and the Philippines. It is also commonly grown in France, Italy, Spain, Bulgaria, the USA and several countries of Africa. It was known in India from far ancient times. In India, it is one of the most common, popular and principal vegetable crops grown throughout the country except higher altitudes. It is a versatile crop adapted to different agro-climatic regions and can be grown throughout the year. It is a perennial but grown commercially as an annual crop.

Besides the commonly grown species, *Solanum melongena* L as brinjal or eggplant, other species of *Solanum* are also grown for vegetable purpose in different parts of the world *viz*., *Solanum muricatum, Solanum quitoense* and *Solanum sessiliflorum* in central and South America, *Solanum macrocarpon* in south-east Asia, *Solanum aethiopicum*, *Solanum gilo* and *Solanum macrocarpon* in Africa.

The crop

Brinjal is a delicate, ligneous, tropical perennial plants often cultivated as a tender or half-hardy annual vegetable crop. The stem is often spiny. It grows 40 to 150 cm tall, with large, coarsely lobed leaves that are 10 to 20 cm long and 5 to 10 cm broad. Semi-wild types can grow much larger, to 225 cm with large leaves over 30 cm long and 15 cm broad.

At the level of each flower, there is dichotomous branching that grows more or less regularly, depending on the species and variety. The sympodia generally consist of two leaves and the axillary bud of the leaf below each flower frequently gives rise to a new branch. Growth and flowering are continuous throughout the life of the plant. Brinjal flowers are large, white to violet coloured and solitary or in clusters of two or more. Flower consists of calyx: sepals 5, united, persistent; corolla: petals 5, united, usually cup shaped; Androecium : stamens 5, alternate with corolla; Gynoecium: carpels are united, ovary superior. The hypogynus gynoecium is syncarp located obliquely in relation to the median. In most varieties the perfect flowers are borne singly and opposite the leaves. In brinjal, heterostyly is a common feature. Four types of flowers have been reported depending on the length of styles, *viz*. (i) long-styled with large ovary, (ii) medium-styled with medium size ovary, (iii) Pseudoshort-styled with rudimentary ovary and (iv) true short-styled with very rudimentary ovary.

The long and medium-styled flowers produce fruits whereas pseudo-short and short-styled flowers do not set any fruits. The percentage of long and medium styled flowers is a varietal character. Brinjal is usually self-pollinated with about 10% natural cross pollination. However, the extent of cross-pollination may go up to even 48% in predominantly long styled flowers when population of pollinators are high hence, it is classified as often cross-pollinated crop.

The fruit is botanically classified as a berry, white flesh with a meaty texture. The fruit contains numerous small, soft seeds that, though edible, taste bitter because they contain nicotinoid alkaloids like the related crop, tobacco.

Brinjal display a wide range of fruit shapes and colours, ranging from oval or egg-shaped to long club-shaped; and from white, yellow, green through degrees of purple pigmentation to almost black. Purple skin colour of fruits is due to anthocyanin pigment and black colour is due to high contents of both anthocyanin and chlorophyll. On an average, the oblong-fruited eggplant cultivars are rich in total soluble sugars, whereas the long-fruited cultivars contain a higher content of free reducing sugars, anthocyanin, phenols, glycoalkaloids (such as solasodine), dry matter, and amide proteins. Bitterness in eggplant is due to the presence of glycoalkaloids which varies between 0.37 to 4.83 mg/100g fresh in the commercial cultivars of India. Generally, high glycoalkaloids content to the level of 20mg/100g fresh weight produce a bitter taste and off flavour. The cut surface of the flesh rapidly turns brown when the fruit is cut open due to to high polyphenol oxidase activity.

Importance and use

Brinjal is the most common, popular and principal vegetable crop grown widely in India for its various shape, size and colour of fruits. High productivity, wide adaptation and ease in availability make the crop to find its place as the common man's crop. In India, it is cooked in vegetable and fish curry, sambar, mashed, fried, grilled and stuffed with spices and then fried (varta).

Nutritional value

Brinjal fruits are a fairly good source of calcium, phosphorus, iron and vitamins particularly 'B' group. Fruit analysis revealed that it contains 1.4 g protein, 18 mg calcium, 47 mg phosphorus, 0.11 mg riboflavin, 0.9 mg niacin, 124 IU vitamin A, 12 mg vitamin C per 100 g of edible part. Purple coloured brinjal contains more vitamin C than green one. High amount of glyco alkaloids (20 mg/100g fresh weight) in the fruits impart bitter taste and off-flavour. There is much variation in chemical constituents in the fruits of different cultivars. Copper content and polyphenol oxidase activity are the highest in purple colour fruits and lowest in white coloured fruits. Similarly, Fe content and catalase activity are highest in the green and lowest in the white cultivars. Green cultivars have the better processing qualities than the more popular purple cultivars.

Medicinal value

The first use of brinjal in India was probably medicinal rather than culinary. Fruits are excellent remedy for those suffering from liver troubles and high blood cholesterol problem. Fresh or dry product of fruit markedly drops blood cholesterol level. The de-cholesterolizing action is attributed to the presence of higher amount of (65%) ploy-unsaturated fatty acids (linoleic and lenolenic acids) in the flesh and seeds of the fruit. Magnesium and potassium contents in the fruits further help in reduction of cholesterol. It is rich in bioflavonoids which can prevent stomach cancer. Piercing brinjal fruits with a needle and fried in sesame oil is a remedy for toothache. White brinjals are said to be a remedy for diabetic patients. Aqueous extracts of fruit juice inhibits choline esterase activity of human plasma and dry fruit is reported to contain goitrogenic principles.

Origin and taxonomy

Brinjal is a native of India, believed to be Indo-Burma region and perhaps China is the secondary centre of origin of this crop. *Solanum incanum*, a wild species and having wide distribution in atleast 10 habitats in India is the progenitor of the cultivated species, *Solanum melongena*. The first record of brinjal in India was during 300 BC to 300 AD. Atleast 33 Sanskrit names for brinjal have been mentioned in the ancient literatures of India and the most common being Varttaka, Bhantaki and Vattingan. It was cultivated in Africa around 9th centuryA.D. It was known in Italy at the end of the 14th century and introduced into southern Europe in the 15th century and the name" eggplant" was probably derived from the white egg like fruits.

Brinjal, the non-tuberiferous *Solanum* (2n = 2x = 24) belongs to sub-genus *Leptostemonum,* section Melongena of the family Solanaceae, sub-family Solanoideae and Tribe Solaneae.

There are three botanical varieties of *Solanum melongena* characterized by fruit shape and size and growth habit of plant.

S. melongena var. esculentum: Fruit round or oval.

S. melongena var. serpentinum: Fruit long, slender.

S. melongena var. depressum: Plant short and dwarf.

Adaptation in India

India is the primary centre of origin of brinjal and rich biodiversity exists in Eastern Ghats, north-eastern region, central India, eastern India particularly in Orissa and West Bengal. There are more than 100 local cultivars are grown in India in different names like, Muktakeshi, Bhangar, Patakata, Elokeski, Nababganj Local, Makra, Soyla, Kanta Makra, Nurki, etc. in West Bengal; Singhnath, Bholanath in Tripura; Ratan Mani, Burupada, Borigumma, Kutiguda, Dhenkanal, Batparsi, Banmalipur Local, Narayanpur Local, Nisan Local, Radha, Green Rocket, Chitrakonda, etc. in Orissa; Hydarpur, Improved Muktakeshi, Sharanpur Long, Kathvaia, etc. in Bihar; Muktakeshi, Ramnagar Giant, Batia, Sheonath, Bundelkhand, Dudhya baingan, Jafrabadi, etc. in Uttar Pradesh and north-central region; Manjari Gota, Bourad Local, Borgona Local, Khandesh, Khed Shivpur, Malkapuri, Shivpur Khandala, Chikhalgaon Local, Hingana, Dorla, etc. in Maharashtra; Junagadh Bhattu, Surati Gota, Gulabi Dorla, etc. in Gujarat; Coorg Local, Irangeri, West Coast Green Round, Udupi Gulla, etc. in Karnataka; Nurki Round, Nurki Long, Banjari, etc. in Madhya Pradesh; Wynad Giant, Pattabiram, Annamalai, etc. in Tamil Nadu; Gola Baingan, Ludhiana Local, Batia, etc. in Punjab.

Improved open pollinated varieties

Long fruited variety: Pusa Purple Long, Mysore Long, Hisar Pragati, Pusa Bhairav, KS-331, JB-15, Green Long, Punjab Barsati, Punjab Chamkila, PH-4, Pant Samrat, Azad Kranti, Arka Sheel, Arka Shirish, Arka Keshav, Punjab Sadabahar, NDB-25, Annamalai, Haritha, Swarna Abhilamb, Junagadh Long, Kashi Taru, Narendra Brinjal-2, Pusa Shyamala, etc.

Round or oval fruited variety: MDU-1, BR-112, Surya, CHBR-1, DBR-8, Azad B-1, Pusa Uttam, Pusa Upkar, Pusa Purple Round, Pant Rituraj, KalyanpurT-3, Punjab Jamuni Gola, Punjab Neelam, Hisar Shyamal, Azad Kranti, Punjab Bahar, Swarna Shyamali, Swarna Shree, Swarna Pratibha, Black Beauty, Bidhan Super, etc.

Oblong fruited variety: CO-1, CO-2, Bhagyamati, Utkal Tarini, Hisar Jamuni, Junagadh Oblong, Kashi Prakash, Utkal Madhuri, Utkal Keshari, Pusa Kranti, Pusa Anupam, Bidhan Superme.

Longish-oblong or cylindrical fruited variety: Utkal Madhuri, Utkal Keshari, Pusa Kranti, Pusa Anupam.

Small fruited variety: PKM-1, Shyamala, Aruna, Pusa Ankur, Pusa Bindu, Manjari Gota, Jawahar Brinjal-64.

Cluster bearing variety: PLR-1, Arka Neelkanth, KKM-1, ARU-1, Arka Nidhi, Utkal Jyoti, Gulabi, Vaishali, Pragati, Pusa Purple Cluster, ARU-2C, Arka Kushumkar, Punjab Barsati.

Bacterial wilt tolerant variety: Utkal Tarini, Utkal Madhuri, Utkal Keshari, Arka Nidhi, Utkal Jyoti, Bidhan Super.

Public sector hybrids: Arka Navneet, Pusa Hybrid-5, Pusa Hybrid-6, Pusa Hybrid-9, Krishna, COBH-1, Neelam, Kashi Ganesh, Kashi Sandesh, Pant Brinjal Hybrid-1, Anand Brinjal Hybrid-1, BH-1, BH-2, Narendra Hybrid Brinjal-1, Narendra Hybrid Brinjal-2, Narendra Hybrid Brinjal-3, Swarna Ajay, Gujarat Brinjal Hybrid-2, Anand Brinjal Hybrid-2, Krishna.

Cultural requirements

A long and warm growing season with average temperature range of 21°C –27°C is most suitable. Extreme temperature (below 10°C and above 35°C) adversely affects the reproductive growth. Warm and humid climate favours luxurious growth, while poor growth is exhibited in cool climate. It grows on all types of soil ranging from light sand to heavy clay however, well-drained, fertile, silt or clay- loam soils are ideal. Light soils are good for early crop, while clay-loam soils are well-suited for getting high yield. It can tolerate slightly acidic soil however, the ideal soil pH range of the soil is 5.5 to 6.6. In the crop rotation, brinjal should follow a leguminous crop to enrich soil fertility.

Nursery management

Fine, well drained, 15-20 cm raised, 1.0m wide and of convenient length seed bed should be prepared. Fine and fully decomposed farmyard manure or compost @ 3-4 kg/m^2 should be well mixed to the beds. The beds may be drenched with formaldehyde (4.0%) and covered with polythene sheet for 5-7 days for soil sterilization. 375 to 500g seeds of open pollinated varieties and 150 to 250g seeds of the hybrids are required per hectare. The seeds are treated with Captan or Thiram @ 2-3 g per kg of seeds before sowing. Seeds are sown at shallow depth 5.0 cm apart in the row and covered with finely sieved leaf mould or sterilized cow dung manure. After sowing the beds are covered with straw and water is sprinkled regularly till seeds are germinated. Seedlings may be raised under poly tunnels to protect the young seedlings from heavy rain or scorching sunlight. The seedlings are watered daily or in alternate day with rose can. Hardening of the seedlings can be done by withholding water at least 4 to 6 days before transplanting. Stocky, 25-35 days old, healthy seedlings of 10-15 cm height with 3-4 leaves are ideal for transplanting.

Sowing time

Sowing time for the autumn – winter crop i.e., main season crop is May-July; for spring – summer crop is October-November and for summer-rainy crop is February-March. However, the crop can be grown round the year in mild climate. In the hills, seeds are generally sown in April.

Land Preparation

The land is prepared well in advance with repeated ploughings (at least 4-5 ploughing) to a fine tilth. All stubbles, weeds, etc. are removed and the land is leveled. Well rotten organic manure preferably farm yard manure or compost @ 20-25 t/ha should be applied during land preparation followed by proper levelling of the land.

Transplanting

28-35 days old, healthy seedlings of 12-15 cm height are ready for planting. After hardening the seedlings for 4-6 days, the seed bed is irrigated before lifting to facilitate easy pulling of young seedlings. The seedlings should be planted in the field preferably during afternoon hours for better field establishment and the field is irrigated just after transplanting.

Spacing

Spacing varies with the type of the varieties and hybrids. Spacing for the round fruited improved varieties and hybrids is 90 cm x 90 cm; for vigorous growing cultivars is 75 cm x 60 cm; for semi vigorous and mid season cultivars is 60 cm x 60 cm; for long fruited cultivars is 60 cm x 45 cm and for early and dwarf types is 45 cm x 45 cm.

Nutrient management

Brinjal is long duration crop and is good feeder of macro-nutrients. A general fertilizer dose of 125 kg N, 80 kg P_2O_5 and 60kg K_2O per hectare is recommended for open pollinated improved varieties.One-third N along with other fertilizers should be given as basal and rest N should be top dressed in two split doses at 30 days interval after transplanting. A general fertilizer dose of 200 kg N, 100 kg P_2O_5 and 80kg K_2O per hectare is recommended for the hybrids.One-fourth N along with other fertilizers should be given as basal and rest N should be top dressed in three split doses at 30 days interval after transplanting. Two foliar sprays of 0.5% $ZnSO_4$ and single spray of 0.15% $CuSO_4$ increase yield and quality of fruits. Vermicompost @ 4-6t/ha may be applied as broadcasting during final land preparation. Azotobactor @ 800g/ha or Phosphate Solublizing Bacteria @800g/ha is applied as seedling root dip. A suspension of bacterial inoculants is prepared in sufficient water and the seedlings are dipped in the bacterial suspension in shade for 2-3 hours in the afternoon followed by their immediate transplanting.

Irrigation

Brinjal being a shallow rooted crop needs irrigation at frequent interval. During summer, irrigation is given at 4-5 days interval and during winter at 10-15 days interval. About 100 to 110 cm total water is needed for successful brinjal crop. Most adopted method of irrigation is furrow method however, drip irrigation is beneficial for reducing water requirement and weed control. Nutrients can also be applied through drip irrigation.

Interculture and weed control

Shallow inter- cultivation is generally practiced particularly a few days after every irrigation to remove the weeds and conserve soil moisture. 3-4 hoeings are normally needed to check the weeds. Earthing up after second top dressing of nitrogen fertilizer is helpful. Pre-planting application of Pendimethilin1.0 kg a.i. /ha is effective for controlling the weeds. Use of 2-5ppm 2,4-D or 20 ppm 4CPA or 60 ppm NAA as whole plant spray at flowering promote fruit set and increase yield. Spraying of the plant protection chemicals should be avoided in the morning hours to facilitate free movement of the pollinating insects like, bees.

Harvesting the crop

The fruits are harvested when they develop good colour, marketable size but still immature and tender. The good marketable quality fruits are bright, glossy appearance having freshness and optimum size without any fading or change in fruit colour. Fruits are harvested with a portion of stalk. Generally 15-19 days fruits are optimally matured for harvest in big fruited varieties.

Post harvest handling and storage

After harvest, dipping the fruits in the solution containing 200 ppm NAA in combination with 90 ppm prochloraz retards fruit senescence and decaying. Under ordinary condition, the normal storage life of brinjal fruits is 1-2 days in the winter months however, 2-3 weeks storage life can be obtained when fruits are stored in cold storage at 8° -10°C temperature and 85-90% relative humidity. Waxing the fruits in 3% concentration help to reduce weight loss and retard the loss in acidity, starch, sugar and ascorbic acid contents. For long distance transport, fruits should be harvested towards evening hours and thereafter cooled them with sprinkling of water. The fruits should be stored in perforated polythene bags or PVC stretch film.

Plant Protection Measures

Diseases

Damping off (*Pythium* spp., *Phytophthora* spp.)

It is very destructive fungal disease in nursery beds, especially during rainy season. Affected seedlings turn pale green and brownish lesion is found at the basal portion of the stem. The lesion girdles the stem and finally the seedlings topple over.

Control measures

- Hot water treatment of seeds (at 51.7° for 30 minutes).
- Seed treatment with Thiram/Captan/Ceresan or Agrosan G.N @ 2.5g/kg of seeds.
- Drenching the nursery beds 7 days before sowing with Thiram/ Captan or with any copper fungicide @ 3g/litre of water.
- Covering the beds with transparent polythene sheets before sowing and left to open sun for at least15 days which is known as soil solarization.
- Spraying the young seedlings with 0.2% Blitox or 0.25% Metalaxyl-Mancozeb or 0.25 % Fosetyl-Al.
- Provision of proper drainage to the beds.
- Removal of the affected seedlings from beds as soon as the symptoms are visible.
- Avoidance of flooding the beds to check the spread of the disease.

Phomopsis blight (*Phomopsis vexans*)

This is a serious fungal disease of brinjal. Infected leaves show small, circular spots which become gray to brown with irregular blackish margin. The affected leaves may turn yellow and die. These lesions may also develop on the petiole and stem. On the fruits, soft, water decayed portion are observed which later turn black. The whole or major portion of the fruit may rot. The disease is a major problem in seed crop.

Control measures

- Use of seeds from disease-free plants.
- Seed treatment with Thiram or Captan @ 2.5 g/kg of seed.
- Keeping the fields neat and clean all the time.
- Hot water treatment of seeds before sowing at 50ºC for 30 minutes.
- Spraying the crop with Diathane M-45 @ 0.2.5% or Bavistin @ 0.1% at 7-10 days interval.
- Collection and burning the diseased plants and fruits.
- Growing of the tolerant varieties like Pusa Bhairab, Pant Samrat, etc.
- Crop rotation with leguminous or cucurbits may also help to reduce the disease incidence.

Bacterial wilt *(Ralstonia solancearum)*

This is a severe bacterial disease of brinjal. The characteristic symptoms of the disease are yellowing and dropping of the foliage, stunting and wilting of the plant gradually, followed by collapse of the entire plant.

Control measures

- Treating the seeds with hot water at 52ºC for 20 minutes or with 0.01% streptomycin solution for 30 minutes.
- Proper crop rotation with non- solanaceous crops like, maize- okra-cabbage/ cauliflower, maize-beans-maize or okra-rice-mustard for at least 2 to 3 years reduces the incidence.
- Application of bleaching powder @ 15-20 kg /ha to the soil atleast 3-4 week before planting.
- Maintenance of proper drainage in the field.
- A combination of Dhaincha as green manure crop and soil amendment with lime and bleaching powder and application of bioagents as soil application is promising in management of wilt of solanaceous vegetable crops.
- Growing of the tolerant varieties like, Utkal Tarini, Utkal Madhuri, Utkal Keshari, Arka Nidhi, Utkal Jyoti.

Little leaf

This disease is caused by a phytoplasma. The affected plants are shorter in stature but produce larger number of branches, leaves and roots than the healthy one. The leaves are malformed to small and yellowish in colour. Internodes get shortened and give the plans a bushy appearance. Infected plants normally do not bear any fruits. The disease is transmitted by jassids.

Control measures

- Removal of the infected plants from the field and burn them to check the spread of the disease.
- Spraying the crop with Rogor @ 1.5 ml/litre or Carbaryl @ 0.3% to control the vector population.
- Early sown crop (mid June) become less affected due to prevalence of low leaf hopper vector population.

Insect Pests

Shoot and fruit borer (*Leucinodes orbonalis*)

This is the most dangerous pest of brinjal causing devastating damage. The warm seasoned crops are the worst affected than the winter crops. The larva initially attacks the terminal shoots and bores inside, as a result of which the shoots wither, drooping down and dry. In the bearing stage, it prefers to bore into the young fruits by making holes and feeds inside. The drooping shoots of the plant is an indication of its presence. Affected fruits get rot and fall down.

Control measures

- The withering shoots should be nipped off and the bored fruits are picked up regularly and bury them deep into the soil to reduce the pest load.
- The affected plant parts and fruits along with the larvae may be collected into bag prepared by mosquito net to conserve the larval-pupal parasite like *Trathela flavoorbitalis* and *Bracon* sp.
- Application of Carbofuran 3G one and half teaspoon full per plant 2-3 weeks after transplanting.
- Spraying the crop with Spinosad (0.15ml/l), 0.25% Carbaryl or 0.05% Endosulfan at the early flowering stage and after harvesting of fruits during bearing stage is very effective.
- Spraying the crop with Profenophos or Triazophos 1.25ml/l of water.
- Crop rotation excluding brinjal.
- Use of pheromone trap to destroy the adult males.

Epilachna beetle (*Epilachna vigintioctopunctata*)

It infects most of the solanaceous and cucurbitaceous crops. The adults and grubs are very destructive, feeding voraciously on the leaves and tender plant parts. Characteristic skeletonized patches are developed due to their feeding which turn Ladys'lace like appearance on leaves and later dry away.

Control measures

- Spraying the crop with 0.2% Carbaryl (Sevin 2 g/l) or Endosulfan 0.05% (Thiodon 1.5 ml/l) at the early stage of infestation.
- Mechanical control by hand picking and destruction of gregarious population along with egg masses and adult are very effective to keep suppress the destructive population of the pest.

Jassids (*Amrasca biguttula biguttula*)

The adult and nymphs are destructive which colonize on the under surface of the leaves and cause damage by sucking saps. The infested leaves curl upwords along the margins which may turn yellowish, and show burnt up patches. Jassids also transmit little leaf disease.

Control measures

- Spraying the crop with Azadirectin 1% periodically as prophylactic measure is very effective to prevent the pest outbreak.
- Application of Imidacloprid @ 0.3ml/l is very effective against the pest.
- Application of Aldicarb granules @ 1.0 kg a.i./ha at 5 days after transplanting, followed by spraying of Carbaryl 0.2% or Diafenthiuron (Pegasus 0.5g/l) after fruit-set provide satisfactory result.

Thrips (*Scirtothrips dorsalis*)

Both the adults and nymph of this tiny insect cause damage by sucking sap from both the surfaces of the leaves. Continuous sucking of huge population along the veins makes silver shiny and pale appearance of leaves. The severely infested leaves gradually wither.

Control measures

- Application of Starthane (Acephate) 1ml/l or acetamiprid @ 0.4ml/l is very effective against this pest.

Red spider mite (*Tetranychus* spp.)

Four spider mite species, *Tetranychus urticae* Koch, *T. macferlanei* Baker & Pritchard, *T. ludeni* Zacher and *T. neocaledonicus* Andre are commonly found to attack vegetable crops. They are tiny, microscopic and morphologically alike without visual difference among these four species. However, *T. urticae* is found everywhere to attack the crops. Brinjal is severely infested by these mite species and found to feed from the dorsal surface of the leaf. All the active stages of the mite suck sap resulting whitish appearance of the crop. Severely infested crops are covered by their heavy webbing that affect the photosynthetic activity of the plants, the leaves gradually dry up and fall off prematurely. Summer crops are worst affected.

Control measures

- Use of balanced fertilizer and proper irrigation keeps the plants healthy and reduce mite attack.
- Spraying the crop with neem based acaricides during first occurrence of mite in the crop.
- Release of predatory mite, *Amblyseius longispinosus* is very effective against these mites.
- Heavy infestation requires acaricidal application (Dicofol @ 2.5 ml/l; Propergite @ 1.5 ml/l or Diafenthiuron @ 1ml/l) to save the crop.

Stem borer (*Euzophera perticella*)

The caterpillars attack stems and often kill the young plants. The growth of infested plant is stunted or withers.

Control measures

- Removal and destruction of the affected plant parts.
- Spraying the crop with 0.03% Diazinon or 0.2% Carbaryl.

Root-knot nematodes (*Meloidogyne* spp.)

The nematodes attack the roots and produce swellings or galls there. The affected plants become stunted in growth, and leaves show chlorotic symptoms and become rough in texture.

Control measures

- ❒ Application of Nemagon or Nematex to the soil.
- ❒ Planting of 6 week old *Tagetes erecta* (Marigold) seedlings in the alternate rows of brinjal which acts as trap crop for nematode.
- ❒ Deep ploughing 2-3 times in the summer months is beneficial in minimizing the nematode population.
- ❒ Application of Thimet 10 g (12 kg/ha) or Furadon 3 G (15 kg/ha) in the field at transplanting.

Yield

Yield of brinjal varies according to the region, cultivar and duration of the crop. Fruit yield may be 20-30t/ha for early crop; 35-40 t/ha for long duration crop and 40-80 t/ha for the hybrids.

Chilli

Chilli or hot pepper, *Capsicum* species under the family Solanaceae (Chromosome No. 2n =2x =24) is a tropical vegetable crop commonly used throughout the world as spice for its pungency and red colour of ripe dried fruits and also for its green fruits for pungency and flavour. The genus name *Capsicum* derived from the Latin word 'capsa' meaning chest or box because of the shape of fruit which encloses seeds very neatly, as in the box. In 2014, world production of fresh green chillies and peppers was 33.2 million tonnes, led by China with 48% of the global total. Global production of dried chillies and peppers was about nine times less than for fresh production, led by India with 32% of the world total. India is one of the leading chilli producing and exporting countries in the world. Other major chilli growing countries are China, Mexico, Turkey, Indonesia, Japan, Ethiopia, Uganda, Nigeria, Thiland, and Pakistan.

By the turn of the 15th century when the Spanish and Portuguese discovered South America, chilli peppers were widely cultivated for human consumption. The chilli pepper has been cultivated in South America for nearly 10,000 years, so it has had plenty of time to evolve. The Portuguese brought the stock of *Capsicum* species form Brazil to India during 1584. Chilli or hot pepper or chilli pepper and sweet pepper or bell pepper belongs to the same species, *Capsicum annuum* L. Besides this most commonly grown species, *Capsicum annuum* as chilli of commerce, the other species of *Capsicum* grown as chilli in different South American countries are *Capsicum baccatum* var. *pendulum* in southern Columbia, Equador, Peru, Bolivia, southern Brazil, northern Chile and Argentina; *Capsicum chinense* in West Indies, low land south America, southern Bolivia and southern Brazil; *Capsicum frutescens* in Mexico, central America, lowland south America, southern Bolivia and southern Brazil. *Capsicum chinense* and *Capsicum frutescens* are also grown as chilli in India also particularly in the Northeastern states.

The crop

Capsicum annuum encompassing the varieties of chilli or hot pepper, bell, sweet and paprika peppers is a perennial herbaceous plants but often cultivated as an annual. It is a many-branched plant, growing up to 75-80 cm in cultivated varieties and often shrubby in appearance. The leaves are simple and alternate, elliptical to lanceolate, with smooth margins (entire). The small flowers (around 1.5 cm diameter) are borne singly or, rarely, in pairs in the axils; they are white or occasionally purple, campanulate (bell-shaped), often with 5 lobes, and contain 5 bluish stamens.

The fruits are many-seeded berries and pod-like, but with no sutures are there. The fruits vary considerably in size and shape, ripening to yellow, orange, red or purple. Some *Capsicum* species synthesize and accumulate anthocyanins in different tissues and organs.

The numerous varieties that have been developed are categorized in five major groups: 1) Cerasiforme (cherry peppers); 2) Conoides (cone peppers); 3) Fasciculatum (red cone peppers); 4) Grossum (bell or sweet peppers); and 5) Longum (chili or cayenne peppers).

The substances that give chili peppers their pungency (spicy heat) when ingested or applied topically is Capsicin ($C_{18}H_{27}NO_3$) which is a condensation product of 3 – hydroxy 4 methoxy benzylamine and decylenic acid (8-methyl-*N*-vanillyl-6-nonenamide). Capsicin and several related chemicals, collectively called *capsaicinoids*. Capsaicin is present in large quantities in the placental tissue, the internal membranes, and to a lesser extent, the other fleshy parts of the fruits. The seeds themselves do not produce any capsaicin. The quantity of capsaicin varies with variety and growing conditions. Water stressed peppers usually produce stronger pods. When chilli peppers are consumed, capsaicin binds with pain receptors in the mouth and throat, potentially evoking pain *via* spinal relays to the brainstem and thalamus where heat and discomfort are perceived. Human affection for chili heat — a form of irritation is mysterious. However, taste is not merely a sensation, pleasing or not, in food. It is a biological force. Chili heat taps into the body's heat and metabolic regulation systems. Capsaicin - ironically, a close chemical relative of vanilla triggers receptor proteins in our mouths, nose, skin and eyes that detect heat, tricking the body and mind with a sensation that the temperature is rising.

The intensity of the "heat" of chili peppers is commonly reported in Scoville heat units (SHU). Historically, it was a measure of the dilution of an amount of chili extract added to sugar syrup before its heat becomes undetectable to a panel of tasters; the more it has to be diluted to be undetectable, the more powerful the variety, and therefore the higher the rating. The modern method is a quantitative analysis of SHU using high-performance liquid chromatography (HPLC) to directly measure the capsaicinoid content of a chili pepper variety. Pure capsaicin is a hydrophobic, colorless, odorless, and crystalline-to-waxy solid at room temperature, and measures 16,000,000 SHU.

Chief distinguishing characters of five cultivated *Capsicum* species are presented below :

C. annuum Blue or blue-purple anthers, milky white corolla without corolla throat spot, calyx teeth present but inconspicuous, generally

	one and occasionally two flowers per node, fruit become red, orange, yellow, brown, violet and black at maturity, seeds tan or straw in colour.
C. baccatum	Anther colour white before dehiscence and brown afterwards, cream coloured corolla with yellow, brown, and dark green corolla throat spot, calyx teeth present and conspicous, generally one flower per node, fruit become red, orange, yellow at maturity, seeds tan or straw in colour."Distinguished from *C. annuum* by the presence of yellow, tan or brown spots on corolla and prominent calyx teeth"
C. frutescens	Anther colour blue, pale greenish-yellow corolla without corolla throat spot, calyx teeth absent, generally 2-5 (usually 2) flowers per node, fruit become red at maturity, seeds tan or straw in colour
C. chinense	Anther colour blue, white corolla without corolla throat spot, calyx teeth present, generally 2-5 (usually 3) flowers per node, fruit become red, orange, yellow, white at maturity, seeds tan or straw in colour."Calyx of the matured fruit is having angular constriction at junction with pedicel is the only difference from *C. frutescens*"
C. pubescens	Anther colour purple, corolla white or purple with a white basal zone without corolla throat spot, calyx teeth present, generally one flower per node, fruit become red, yellow at maturity, seeds black in colour."It is distinct from other cultivated peppers because of black and rugose seeds".

Different "International cultivar groups" of hot chillies under *Capsicum* species can be classified on the basis of fruit shape and pungency.

Ancho	*C. annuum*, large, heart shaped ,thin-walled fruits, mildly pungent
Cayenne	*C. annuum*, fruits tapered, slender, wrinkled with thin wall, highly pungent.
Cherry	*C. annuum*, spherical, small, slightly flattened, thin-walled fruits, sweet and pungent.
Chiltepin	*C. annuum*, very small, egg-shaped fruits, thin walled, highly pungent.
Cuban	*C. annuum*, fruits of irregular blunt shape, mildly pungent, thin walled.

Jalapeno	*C. annuum*, fruits small, almost cylindrical with rounded ends, thick walled, skin russeted in some varieties, highly pungent.
New Mexican	*C. annuum*, long, slender with a tapered end, thin walled, moderately pungent, some varieties have sweet fruits.
Serrano	*C. annuum*, fruit short, elongated, thin walled with blunt end,very pungent
Squash	*C. annuum*, fruits flattened scallop-shaped, medium to thick walled, pungent
Tabasco	*C. frutescens*, fruits small, slender, erect, thin walled, extremely pungent.
Habanero	*C. chinense*, fruits small, thin, puffy-walled, extremely pungent.

Importance and use

Chilli is grown throughout the country and is used in almost all dishes. Chilli of commerce, *Capsicum annuum* is classified into several groups of cultivars on the basis of fruit shape and the manner in which it is prepared and used. They can be eaten fresh as biting chilli or in curries, sambar, rasum and other meat, fish and vegetable dishes to impart pungency, colour and flavour to food items. Dried fruits are used to make universal curry powder and curry paste. Fresh green and ripe chillies are used to make all kinds of pickles, different sauces and paste. The red colour, capsanthin is used in high quality cosmetic preparations like, lipstick. The essential oil, oleoresin is used in the food and beverage industries. Chilli extractions are used in the preparation of ginger beer and other beverages. Capsaicin, which is obtained from *C. annuum* and other *Capsicum* species, is an intense skin and eye irritant, and is the ingredient used in pepper sprays sold for self-defense. This non-conventional use of chilli as self-defense sprays is gaining popularity in different countries.

Nutritional value

The nutritional value of chilli lies in their vitamin (A and C) and phosphorus contents otherwise, chillies add more than just flavour to the diet. Chilli is particularly rich in vitamin C (ascorbic acid content 120-130 mg/100g fresh). Chilli also contains appreciable amount of vitamin A (280-290 IU/100 g fresh). It has high crude fibre content (25.0-27.0 %). Chilli peppers lose their vitamin content when they are cooked or pickled or if they are allowed to ripen too much.

Medicinal value

The stimulation of the hot chilli causes the heart to beat faster thus causing the body to sweat and act as body's natural air conditioner. The warmth of chillies not only stimulates the taste buds but also the appetite and digestion through mouth watering effects of increasing saliva flow. Chilli has the protective effect against peptic ulcer disease. Chilli peppers get rid the body of enough fats to lower the blood cholesterol level and reduce the chances of heart attack. Capsaicin, the pungent principle has substantial antigenotoxic and anticarcinogenic effects. The carotenoides, predominantly, capsanthin and â carotene have anti oxidant property. The flavonoides present in red pepper particularly, quercetin, kaempferol, myricetin and luteolin are responsible for induction of enzymes that detoxify carcinogens. Capsicin, once ingested, cause the brain to release endorphins into the blood stream which can induce a natural feeling of well being similar to that achieved by long distance runners. Capsaisin also has numerous medical uses, including pain balms, vapour linements, skin ointments for cold, sore throat, chest congestion, etc. Capsaicin is also used in the preparations of prickly heat powder. The oil content of the capsicum, if dissolved in ether and applied with cotton wool is considered to be very useful in relieving rheumatic pains.

Origin and txonomy

It is generally accepted that the *Capsicum* genus originated in Bolivia and consists of 25–30 species. Five of these *Capsicum* species were domesticated: *C. annuum*, *C. baccatum*, *C. chinense*, *C. frutescens*, and *C. pubescens*. The largest group of varieties is found among the *C. annuum* spp. which is grown worldwide. Both hot pepper and sweet pepper belongs to the family Solanaceae under the same species *Capsicum annuum*. The wild progenitor of *C. annuum* is thought to be the bird pepper, whose domestication occurred in Mexico.

The archaelogical excavations at Tehuacan in Mexico have indicated the existence of *C. annuum* around 7000B.C. On his return from Americas, Columbus brought chilli, *Capsicum annuum* to Spain in 1493. It spread to other parts of the Mediterranean region. Later it went to north in England in 1548 and moved to central Europe in the 16th century. It was introduced to India by Portuguese in 16th century.

Hot chillies and sweet peppers, *Capsicum annuum* var. *annuum* have been originated from the wild and weedy *Capsicum annuum* var. *minimum* distributed from the southern United States to the northern South America. Another cultivated species, *Capsicum baccatum* var. *pendulum* was originated from its wild progenitor, *Capsicum baccatum* var. *baccatum* in South America.

The species *Capsicum frutescens* is wide spread as a wild, weedy or semi-domesticated species in low land tropical America including Mexico, central South America and lowland South America. Perhaps, *Capsicum chinense* has been originated from the wild type of *Capsicum frutescens*. No ancestral type could be identified for the other cultivated species *Capsicum pubescens*.

Domestication, human selection and breeding have produced hundreds of varieties in the chilli growing areas of the of world from the great natural diversity of the *Capsicum* species differing greatly in fruit size and shape, pungency and ranging in colours from orange, red, yellow, violet to green.

Adaptation in India

Mixed stock of *Capsicum* species was introduced in India by the Portuguese about 425 years ago into Kerala. Christian missionaries also introduced *Capsicum* species in the North-eastern states long ago separately. High adaptation in the tropical Indian condition, wide spread almost through out the country and farmers' selections particularly for pungent and red types have resulted in the evolution of huge genetic diversity of chilli in India. Huge number of local cultivars of chilli are grown in India in different names like, Suryamukhi, Bhangar Local, Suti, Sili, Beldanga Local, Jateswar Local, Bullet, etc. in West Bengal; Oosimilagi, Lavang Mirche, Golgonda Mirche, Sadasivpet, Shivali, Mundu, Nallapadu, Warangal, Samba, *etc*. in Andhra Pradesh; Salem, Ramanand, Samba Kodakal, Sattur Samba, Kovilpatti, Nambiyur Local, Kandangadu Local, Ramanathapuram Gundu, *etc*. in Tamil Nadu; Javari, Javageal, Byadgi, Dubba Bayadgi, Kadda, Sankeshwar, Chincholi, Kollegal, Arsikera, Madhugiri, etc. in Karnataka; Kandhari, Achar, Vella Kandhari, etc. in Kerala; Muslawadi, Sankeshwar, Sindhur, Dondicha, Achalpuri, Malkapuri, etc. in Maharashtra; Gholar, Patta, *etc*. in Gujarat; Mathania, Mandoria, etc. in Rajasthan; Elichip, Kalipeeth, etc. in Madhya Pradesh; Sirhindi, Sanauri Red, Sanauri Yellow, Rajpura Long, Jullundhur, Shokati, Longi, *etc*. in Punjab; Kandhari, Suryamukhi, etc. in Uttar Pradesh; Rahri, Patna Red, Sabour, Angon, Longi, *etc*. in Bihar; Puri Red, Puri Orange, *etc*. in Orissa; Latabh, Baghi, Gauhati Black, etc. in Assam. Bhoot Jalokia and Naga Jhall cultivars of Assam, Nagaland and other north-eastern states belong to *Capsicum chinenese*. The cultivars of small and very pungent chilli of upright fruit orientation cultivated in different parts of India including north-eastern states belong to *Capsicum frutescens*.

Improved open pollinated varieties

Andhra Jyoti (G-5), Bhagyalaxmi (G-4), K-2, K-1, J-218, Muslawadi, LCA 206b, JCA 283, Phuley Jyoti, Phule Suryamukhi, Bhaskar, CO-1, CO-2, MDU-

1, NP 46-A, Punjab Lal, Pusa Jwala, Punjab Surkh, Punjab Guchhedar, Pant C-1, Pant C-2, JCA-154, Kalyanpur Selection, Pusa Sadabahar, X-235, Kalyanpur Red, Agnirekha, Arka Lohit, Sindhur, Aparna,

Mild to medium pungent varieties: (capsaicin content 0.20 to 0.55 mg/g fruit) : Kovilpatti-1, Kovilpatti-2, Bhagyalakshmi (G-4), Pusa Jwala, CO-3, Jwala Mukhi, Jwala Sakhi, Ujwala, RHRC Pendant Cluster, Andhra Jyothi (G-5), PMK-1, PLR-1 Muslawadi, CA 960, NP 46 A, Prakash, Sabour Arun, Agnirekha, Phule Jyoti, Phule Suryamukhi, Arka Abir, Jayanti, Kashmir Long-1, Arka Suphal, Kashi Anmol, Prashnath, Anugraha, Gujarat Vegetable Chilli-111, Gujarat Vegetable Chilli-121, RCH-1, LCA-353.

Pungent to highly pungent varieties: (capsaicin content more than 0.55 mg/g fruit): CO-1, CO-2, Pant C-1, Pant C-2, PKM-1, Arka lohit, Pusa Sadabahar, Jawahar Mirch 218 (J –218), Punjab Lal, Hisar Vijay (H-28), Hisar Shakti (H-44), Bhaskar, G-2 , G-3, MDU-1, Sindhur, X-197, Phule Sai, Jawahar Mirch 283, Hot Portugal, Chanchal, Punjab Guchhedar, Punjab Surkh, Utkal Ava, Azad Mirch-1, BCC-1, Bidhan Chilli-4.

Samba type varieties: (long and pendant): Kovilpatti-2, Bhaghalakshmi (G-4) , Pusa Jwala, Co-3, Jwala Mukhi, Muslawadi, NP 46 A, Sabour Arun, CO-1, Hisar Vijay (H-28)

Gundu type varieties: (short, conical, somewhat round or oblong in shape): PMK-1, PLR-1, CO-2, Andhra Jyothi (G5).

Variety suitable for colour extraction: KTPL-19, Arka Abir

Variety high in oleoresin content: Hisar Shakti, Punjab Lal, Bhaskar (LCA-235), Ujwala, CO-3, RHRC Pendant Cluster

Varieties for pickling: GCA 154, PLR-1

Varieties tolerant to leaf curl virus: Pusa Jwala, Jwala Sakhi, Pusa Sadabahar, Jawahar Mirch 218 (J-218), Punjab Lal, Hisar Vihay (H-28) Hisar Shakti (H-44)

Varieties tolerant to thrips: Bhaskar, Kovilpatti-2, Kovilpatti-1, Andhra Jyothi, Phule Suryamukhi.

Varieties suitable for export: Bhaskar, Bhagyalakshmi, Andhra Jyoti, G-3, Pant C-1, Arka Lohit, X-235, Panvel, Ansari, Musalwadi

Public sector hybrids: CH-1, CH-3, Kashi Surkh, HCH-9646, Arka Harita, Arka Sweta, Arka Meghna.

Cultural requirements

Chilli performs well in warm humid tropical and sub-tropical regions. A temperature range of 20-33°C is optimum for chilli. Extreme temperatures, below 15°C and above 35°C significantly reduce reproductive growth and pollen viability. Humidity favours growth but frost is injurious. Low temperature at fruit ripening stage may delay colour development in the fruits. Chilli grows best in a well-drained loam soil rich in organic matter. Chilli can also be grown in heavy soil rich in organic matter particularly for dry chilli production. The pH should be around 6.5-7.5. It can tolerate salinity to a considerable extent. Chilli performs well with a rainfall of 600-1200 mm spread over 4 to 5 months. Enough water should be available, but the soil should also have good drainage facility. The crop prefers moderate sunshine, high relative humidity and drizzling rains during early establishment.

Nursery management

Seed beds are prepared finely, well drained, 15-20 cm raised, 1.0m wide and of convenient length. Fine and fully decomposed farmyard manure or compost @ 3-4 kg/m^2 should be well mixed to the beds. The beds are drenched with formaldehyde (4.0%) and covered with polythene sheet for 5-7 days for sterilization of seed bed soil. 500-600 g seeds of open pollinated varieties and 250 to 300 g seeds of hybrids are required per hectare. Seeds are treated with Captan or Thiram @ 2-3 g per kg of seeds before sowing. Seeds are sown at shallow depth 5.0 cm apart in the row and covered with finely sieved leaf mould or sterilized cowdung manure. After sowing the beds are covered with straw and water is sprinkled till seeds are germinated. Seedlings may be raised under poly tunnels to protect the young seedlings from heavy rain or scorching sunlight. Thinning of the seedlings should be done at least 20 days before transplanting for proper growth of the seedlings. The seedlings are watered daily or in alternate day with rose can. Hardening of the seedlings can be done by withholding water at least 4 to 6 days before transplanting. Stocky, 40-45 days old, healthy seedlings with 4-5 leaves are ideal for transplanting. Clipping the terminal bud of the seedlings 10 days prior to transplanting helps in better establishment of the seedlings and also accelerates the growth of the axilary buds.

Sowing time

Chilli can be grown in West Bengal, Maharashtra and South India almost throughout the year. However, chilli is grown in three distinct seasons viz., early autumn-winter, spring-summer and summer-rainy in the plains of India although sowing and transplanting season vary with the region. For early autumn-winter crop sowing is done in May-July; for spring-summer crop in October-November and for summer- rainy crop in February. In the hills, chilli is sown during March-April.

Land Preparation

The land is prepared well in advance with repeated ploughings (at least 4-5 ploughing) to a fine tilth. All stubbles, weeds, etc are removed and the land is leveled. Well rotten organic manure preferably farm yard manure or compost @ 20-25 t/ha are applied during land preparation followed by proper levelling of the land. Vermicompost @ 4-6t/ha may be applied as broadcasting during final land preparation.

Transplanting

40-45 days old, healthy and clipped seedlings of 12-15 cm height are ready for planting. In open pollinated varieties, approximately 50-60 thousand seedlings are required for one hectare land. After hardening the seedlings for 4-6 days, the seed bed is watered before lifting to facilitate easy pulling of young seedlings. The seedlings are planted in the field preferably during afternoon hours for better field establishment. The field should be irrigated just after transplanting.

Spacing

Spacing depends on type of varieties and hybrids. Spacing for dwarf and medium tall varieties for green chilli is 45-60 cm x 30-45 cm; for tall varieties or hybrids for green chilli is 75 cm x 60 cm and for early winter-winter season varieties for dry chilli is 75 cm x 75 cm or 90 cm x 90 cm.

Nutrient management

Chilli is long duration crop and is good feeder of macro-nutrients. A general fertilizer dose of 100 kg N, 60 kg P_2O_5 and 60kg K_2O per hectare is recommended for open pollinated improved varieties.One-third N along with other fertilizers should be given as basal and rest N should be top dressed in two split doses at 21 and 42 days after transplanting. A general fertilizer dose of 150 kg N, 80 kg P_2O_5 and 80kg K_2O per hectare is recommended for the hybrids.One-fourth N along with other fertilizers should be given as basal and rest N should be top dressed in three split doses at 30 days interval after transplanting. Vermicompost @ 4-6 t/ha should be applied as broadcasting during final land preparation. Azotobactor @ 800g/ha or Phosphate solublizing bacteria @800g/ha can be applied as seedling root dip. A suspension of bacterial inoculants is prepared in sufficient water and the seedlings are dipped in the bacterial suspension in shade for 2-3 hours in the afternoon hours followed by immediate transplanting.

Irrigation

Chilli needs judicious irrigation for proper growth and yield. Chilli cannot withstand excess moisture. Frequent and heavy irrigations induce lanky growth and cause flower shedding. During summer, irrigation should be given at 4-5 days interval and during winter at 10-12 days interval. About 18 acre-inch of total water is needed for successful chilli crop. Critical stages of irrigation are tenth leaf to flowering, fruiting and after periodical harvests. Drip irrigation is beneficial for reducing water requirement and weed control. Nutrients can be applied through drip irrigation.

Interculture and weed control

Shallow inter-cultivation is needed particularly a few days after every irrigation to remove the weeds and conserve soil moisture. 3-4 hoeings are normally needed to check the weeds. Light hoeing at the base of the seedlings should be done 21 days after transplanting and after first topdressing with nitrogenous fertilizer. Pre-planting treatment of Fluchloralin @1.0-1.5 kg a.i. / ha followed by one hand weeding 30 days after transplanting is effective in controlling the weeds.

Harvesting the crop

Harvesting of chillies depend on the type and purpose for which they are grown. Harvesting of green chillies may be started 60 to 70 days after transplantation. Ripe fruits are picked at an interval of 1-2 weeks and harvesting continues over a period of about 3 months. The number of picking varies from 6-10 depending upon the season, cultivar and cultural practices.

Post harvest handling and storage

The packaging used for transportation should have ventilation and alternatively, jute bag can be used. Green chillies can be stored in the cold storage up to 40 days at 0°C and 95-98 % relative humidity. Green chillies treated with 200 ppm GA3 significantly increase the shelf life. For dry chilli, harvesting should be done when the fruits are well ripened and partially withered in the plant itself. The harvested fruits are kept in heaps either indoor or in shade away from direct sunlight for 2-3 days to develop uniform red colour. The red ripe fruits are dried in the sun on clean dry polythene sheets, cemented/

concrete drying yards, etc for 8-15 days depending upon the weather condition. The red ripe fruits can also be dried in hot air oven at about 54.4°C for 2-3 days. The red ripe fruits are spread in thin layers for uniform drying with frequent stirring to prevent the growth of mould and discoloration. The dried pods are heaped and then covered by clean gunny bags or polythene sheets. The moisture content of the dry pods is to be kept at 8-10%. The well-dried fruits after removing the extraneous matters like plant parts, etc are kept in clean and dry gunny bags in the storage well protected from dampness. The bags are stacked 50 to 60 cm away from the wall of the cold storage. The product may be stored for 8-10 months in cold storage however, storing dry chillies for longer period may lead to deterioration in colour. Insects, rodents, etc are to be effectively prevented from getting access to the premises where chilli is stored.

Yield

Fruit yield varies with the variety and hybrid and it may be 7.0-10 t/ha for green chillies of open pollinated variety; 12-15 t/ha for green chillies of hybrids; 1.5-2.5 t/ha for dry chillies of open pollinated variety and 3.0-3.5 t/ha for dry chillies of hybrid.

Sweet pepper

Sweet pepper or capsicum or bell pepper is a cultivar group of the species *Capsicum annuum* under the family Solanaceae (Chromosome No. 2n =2x =24). Sweet pepper requires slightly cooler climate than hot chillies and is grown worldwide for its its thick and fleshy fruits having delicate taste, pleasant flavour and colour. The major countries producing sweet peppers are China, Indonesia, Sri Lanka, Pakistan, Turkey, Korea, Spain, Bulgaria, Hungary, Romania, Italy, Yugoslavia, Nigeria, Ghana, Tunisia, Mexico, USA, Argentina and Peru.

Sweet pepper is also known as bell pepper, pimento, Simla mirch or paprika. In India it is generally grown in the cooler weather particularly in the periphery of cities. Major capsicum growing areas are situated around Bangalore, Belgaun and Mysore of Karnataka, Pune and Thane in Maharashtra, Nilgiri hill areas of Tamil Nadu, Ranchi of Jharkhand, Darjeeling hill and some part of 24 Parganas of West Bengal, hill regions of Uttar Pradesh and Uttaranchal, Himachal Pradesh and Jammu and Kashmir.

The crop

The fruits are many-seeded berries, but no sutures are there. Different cultivars produce fruits of different colours including red, yellow, orange, green, chocolate/brown, vanilla/white and purple. The most common colours of bell peppers are green, yellow, orange and red. More rarely, brown, white, lavender, and dark purple peppers can be seen, depending on the variety. Bell peppers are sometimes grouped with less pungent pepper varieties as "sweet peppers".

The bell pepper is the only member of the *Capsicum* genus that does not produce capsaicin, a lipophilic chemical that can cause a strong burning sensation when it comes in contact with mucous membranes. The lack of capsaicin in bell peppers is due to a recessive form of the gene that eliminates capsaicin and, consequently, the "hot" taste usually associated with the rest of the *Capsicum* genus producing hot peppers or chilli. Sweet pepper is internationally classified into several groups of cultivars on the basis of fruit shape and pungency.

The characteristic aroma of green peppers is caused by 3-isobutyl-2-methoxypyrazine (IBMP). Its detection threshold in water is estimated to be 2 ng/litre.

Different cultivar groups of sweet peppers are classified primarily on the basis of fruit shape and size.

Bell	Large, blocky fruits with a blunt end of 3 to 4 lobes, thick walled, sweet, fruit colours, green, red, yellow, orange, violet.
Cheese	Mostly globose shaped fruits, small to medium in size and of various shape, mild, yellow to green maturing to red.
Cuban	Fruits of irregular blunt shape, mildly pungent, thin walled.
Long Wax	Fruits long, tapered but in some cases blunt, mildly pungent.
Pimiento	Fruits large, cone or heart shaped, thick walled, not pungent.

Importance and use

The fruits of nonpungent (sweet) varieties are eaten raw in salads or cooked, mixed and stuffed vegetable. It is also widely used in the preparations of pickles, sauces and soups. In stews and other preparations sweet pepper imparts a novel flavour. The dried fruits are ground to a powder (paprika) and used as an ingredient in curry powder.

Nutritional value

Sweet pepper is particularly rich in Vitamin A and C otherwise, it add novel and delicate flavour to the diet. It contains very high amount of vitamin C (175 mg/100 g fresh) and is also particularly rich in vitamin A (683 mg/100g fresh). Red peppers have twice the vitamin C content of green peppers. It has high crude fibre content. Sweet peppers lose their vitamin content when they are cooked or pickled or if they are allowed to ripen too much.

Medicinal value

The delicate flavour of capsicum stimulates not only the taste buds but also the appetite. Appreciably high Vitamin A and C contents in sweet pepper act as potential antioxidant. The carotenoides, predominantly, capsanthin *and* â carotene have anti oxidant property. The flavonoides present in red pepper particularly, quercetin, kaempferol, myricetin and luteolin are responsible for induction of enzymes that detoxify carcinogens. Different preparations are used to counter irritants, in lumbago, neuralgia and rheumatic disorders. It has tonic and carminative action especially in dyspepsia.

Adaptation in India

In India, mixed stock of *Capsicum annuum* including both hot and sweet peppers was introduced in Kerala by the Portuguese from Brazil in 1584. Christian missionaries also introduced *Capsicum* species in the North-eastern states long ago separately. However, chillies or hot pepper spread much more widely and rapidly than sweet pepper in India. High adaptation in the tropical Indian condition, wide spread, extensive course of cultivation almost throughout the country and farmers' selections particularly for pungent and red types have resulted in the development of huge numbers of local cultivars of chilli under different names in India. In contrast, very few local cultivars have been developed in sweet pepper suggesting that Indian farmers were in favour of pungent and red hot pepper. Hence, most of the non-pungent or mildly pungent types, if al all have been introduced in India as mixed stock, possibly got eliminated. Very few local cultivars of sweet pepper having pungent or non-pungent, small, roundish fruits are still being grown by the farmers in Karnataka (Srirangapatna area of Mandya district), Goa and coastal areas of Kerala. However, leading present day varieties in India are important introductions although, some pure line selections have also been done from exotic genotypes.

Origin and taxonomy

Capsicum annuum var. *annuum* under which both hot and sweet pepper belong was domesticated in Mexico. Great natural diversity of the *Capsicum annuum* var. *annuum* exists differing greatly in fruit size and shape, pungency and ranging in colours from orange, red, yellow and violet to green. Domestication, human selection and breeding have produced many varieties of sweet pepper from the highly diversified genotypes of this species. Mexico is probably the centre for morphological diversity for this important species on a world scale since it includes all the commercially important sweet peppers and many hot peppers.

Improved open pollinated varieties

California Wonder, Yolo Wonder, World Beater, Early Bountiful, Chinese Giant, Giant Bull Nose, Arka Basant, Arka Gaurav, Arka Mohini, Nishat-1,

Public sector hybrids: Pusa Deepti, Solan Hybrid-1, Solan Hybrid-2, DARL-202, KTCPH-3.

Cultural requirements

Sweet pepper performs well in mild climate hence, it is considered as a cool season crop. A temperature ranging from 26-28°C during daytime and 16-18° C at night is ideal. It cannot withstand extreme temperatures, below 15°C and above 30°C significantly reduce growth and fruit set. It is susceptible to frost. High relative humidity is good for growth and yield, fruit set and yield however, this condition is favourable for the development of foliar and fruit diseases. It is highly sensitive to environmental factors. Hence, its cultivation under poly-tunnels or poly-houses is preferred. Growing sweet pepper inside the poly house in winter months increases fruit set and yield because of slightly higher relative humidity and comparatively higher night temperature due to green house effect. Providing shade through poly or net house during summer and only shade net in open during other seasons is beneficial. Well-drained red-loam and alluvial soil with slightly acidic pH is considered ideal for sweet pepper cultivation. On sandy loam soil, it can successfully be grown with proper manuring and timely irrigation. It can stand acidity to a certain extent. The pH should be around 5.5 - 6.8. Enough water should be available, but the soil should also have good drainage facility. The plant has bi-directional root system and it is important that there is free unhampered root growth and root aeration for successful crop growth.

Nursery management

Seed beds should be prepared finely, well drained, 15-20 cm raised, 1.0m wide and of convenient length. Well decomposed farmyard manure or compost @ 3-4 kg/m^2 and a fertilizer dose of 0.5 kg NPK of 15:15:15 per bed, 1.0 m wide and 7.5 m long should be mixed 10 days before sowing the seeds. The beds are drenched with formaldehyde (4.0%) and then covered with polythene sheet for 5-7 days for soil sterilization. The seed beds may also be covered with transparent polyethylene sheet for 15-20 days (solarization) for soil disinfection. 500 g seeds of open pollinated varieties and 250 g seeds of hybrids are required per hectare. The seeds are treated with Captan or Thiram @ 2-3 g per kg of seeds before sowing. The seeds are sown in rows 8-10 cm apart and more than 1 cm deep and covered with finely sieved leaf mould. After sowing, the beds are covered with straw and water is sprinkled till seeds are germinated. Seedlings may be raised under poly tunnels to protect the young seedlings from heavy rain or scorching sunlight. The nursery should be covered with 40 mesh nylon net to protect the seedlings from feeding by virus carrying vector, white fly. Thinning of the seedlings should be done at least 20 days before transplanting for proper growth of the seedlings. The seedlings are sprayed with 0.25% Dithane M-45 or Difolaton at weekly interval to reduce the chance of post emergence damping off. The seedlings are watered daily or in alternate day with rose can. Hardening of the seedlings can be done by withholding water at least 4 to 6 days before transplanting. Stocky, 35-40 days old, healthy seedlings with 4-5 leaves are ideal for transplanting. Clipping the terminal buds of the seedlings 10 days prior to transplanting helps in better establishment of the seedlings and also accelerates the growth of the axilary buds.

Sowing time

Sowing is done in July-September for autumn-winter crop; in October-November for winter-spring crop in poly-tunnel in north Indian condition and February- March for summer crop in the hills.

Land Preparation

The land is prepared in well advance with repeated ploughings (at least 4-5 ploughing) to a fine tilth. All stubbles, weeds, etc are removed and the land is levelled properly. Well rotten organic manure preferably farm yard manure or compost @ 20-25 t/ha is applied during land preparation and the land is levelled properly. Vermicompost @ 4-6t/ha can be applied as broadcasting during final land preparation.

Transplanting

Healthy and clipped seedlings of 35 -40 days old and 12-15 cm height with 4-5 leaves are ready for planting. It is better to transplant the seedlings in one side of the ridges. After hardening the seedlings for 4-6 days, the seed bed is irrigated before lifting to facilitate easy pulling of young seedlings. The seedlings are planted in the field preferably during afternoon hours for better field establishment. The field should be irrigated just after transplanting.

Spacing

Spacing for open pollinated varieties is 45 x 45 cm and 60 cm x 30 cm. Very high plant population of 55,000 plants/ha can be maintained with 60 cm x 30 cm spacing. Spacing for the hybrids is 75-80 cm x 30-45 cm.

Nutrient management

It is a good feeder of macro-nutrients. However, it is a shy rooted crop hence, dipping the seedlings in 1.5 % superphosphate solution before transplanting gives better result. A general fertilizer dose of 120 kg N, 80 kg P_2O_5 and 60kg K_2O per hectare is recommended for open pollinated improved varieties. One-third N along with other fertilizers should be given as basal and rest N should be top dressed in two split doses at 21 and 42 days after transplanting. A general fertilizer dose of 200 kg N, 150 kg P_2O_5 and 100kg K_2O per hectare is recommended for the hybrids. One-fourth N along with other fertilizers should be given as basal and rest N should be top dressed in three split doses at 30 days interval after transplanting. Nitrate form of nitrogen is more preferable than ammoniacal form of nitrogen. Vermicompost @ 4-6 t/ha can be applied as broadcasting during final land preparation. The VAM fungus, *Glomus itraradices* can also be applied in the seed bed. Azotobacter @ 800g/ha or Phosphate solublizing bacteria @800g/ha or Azospirillium @ 2 kg/ha is applied as seedling root dip. A suspension of bacterial inoculants is prepared in 10 litres of water and the seedlings are dipped in the bacterial suspension in shade for 30 minutes in the afternoon hours followed by immediate transplanting.

Irrigation

Sweet pepper needs judicious irrigation for proper growth and yield. It is sensitive to both soil moisture deficit and excess. Frequent and heavy irrigations induce lanky growth and cause flower shedding and reduce yield. In sandy-loam soil, it is best to provide irrigation at 60% soil moisture depletion level in the top 30 cm soil profile. Generally irrigation is given during winter at 12-15 days interval. Average consumptive use is 446 mm. Critical stages of irrigation are flowering, fruiting and after periodical harvests. Drip irrigation is beneficial for reducing water requirement and weed control preferably under polyhouse condition. Nutrients can also be applied through drip irrigation.

Interculture and weed control

Shallow inter-cultivation is generally practiced particularly a few days after each irrigation to remove the weeds and conserve soil moisture. 3-4 hoeings are normally needed to check the weeds. Earthing up the seedlings 25 days after transplanting should be done after first topdressing with nitrogenous fertilizer. Pre-planting treatment of Fluchloralin @1.0-1.5 kg a.i. /ha followed by one hand weeding 30 days after transplanting is effective. Basalin @ 0.48 kg a.i. / ha applied 5 days after transplanting followed by one hoeing 35 days after transplanting effectively controls weeds. The plants are sprayed twice at 10 days interval with 25 ppm GA or 50 ppm NAA at flowering stage to increase fruit set. In determinate varieties which are mostly cultivated in tropical and sub-tropical regions, judicious pinching of the flowers is needed to produce 8-10 big fruits per plant of more than 100 g weight. The plants should be staked properly to produce blemish less fruits.

Pruning

Some sweet pepper varieties, particularly grown under poly house in the temperate climate are indeterminate in growth habit, that is, they continually grow new stems and leaves. For this reason, the plants have to be pruned and trained on a regular basis in order to ensure a balanced growth for maximum fruit production. Generally, the plants are managed with two main stems per plant, resulting in a density of 6 stems/m² from an initial planting density of 3 plants/m². Pruning also improves air circulation around the plant which helps to reduce disease. Plants are generally pruned every two weeks. As new leaves and lateral side shoots develop from the axils of the new nodes on the growing stems, they have to be pruned to maintain the two main stem architecture of the plant. These two stems are managed to carry the full production of the plants throughout the year. Twine hung from the overhead support wires is used to support each stem.

Harvesting the crop

The fruits are harvested when they are still green and full grown, firm and crisp. Red and yellow pepper varieties can be harvested after the development of respective colour. Pimento and paprika cultivars are picked at dark red ripe stage. Harvesting of sweet peppers starts 35-50 days after flowering depending on the variety. Harvesting can be done once in 10-12 days with 5-6 pickings in open pollinated varieties and 8-10 pickings in the hybrids having determinate growth habit. The fruits should be picked with an upward twist with a piece of peduncle attached.

Post harvest handling and storage

In the open field grown crop, the harvested fruits should be kept in shade to avoid sunscald. Immediate pre-cooling by immersing preferably in ice-cold water enhances the shelf-life. Bell pepper fruits are wiped, cleaned in fresh water, graded in size and packed in wooden boxes, paper cartons or plastic crates before marketing. It can be stored up to 14 to 21 days at 7-10°C and high RH of 90-95% and for at least 40 days at 0°C and RH of 95–98% with minimum shrinkage (only 4%). Dipping the harvested fruits for 2 minute in 1 or 2% potassium bicarbonate solution reduces decay development after storage and increases shelf life. The fruits should be transported in cold temperature condition of 7-10°C.

Yield

Fruit yield varies not only with the varieties/hybrids but also with the growth habit and growing condition. Open pollinated variety of determinate growth habit grown in open field condition may give yield of 10.0-15.0 t/ha; open pollinated variety of determinate growth habit grown in poly house may give double yield of 20.0-25.0 t/ha. Fruit yield of the hybrids of determinate growth habit grown in poly house is 35.0-45.0 t/ha. Fruit yield of the hybrids of indeterminate growth habit grown in poly house may be as high as 100.0 t/ha.

Plant Protection Measures For Chilli and Sweet Pepper

Physiological disorder

Flower and fruit drop

It is one of the major constraints in both chilli and sweet pepper cultivation. It may be caused due to i) low humidity and high temperature condition, ii) decreasing light intensity and iii) short day during early flowing stages.

Control measures

- Irrigation at flowering and fruit set stage reduces blossom and fruit drop.
- Foliar application of 50 ppm NAA at full bloom stage effectively controls the drop.

Skin cracking

In this physiological disorder of sweet pepper, cracking occurs around the shoulder of the fruits. Fluctuations in temperature and humidity are responsible for this disorder. High day temperature and average relative humidity increase the incidence of cracking.

Sunscald

It is a physiological disorder of sweet pepper. Soft, light coloured area appears on the fruit, which becomes slightly wrinkled, slightly sunken and papery later. It occurs when the fruits are exposed to scorching sunlight.

Blossom end rot

In this physiological disorder of sweet pepper, water soaked spots appear at the blossom end which become light brown and papery as the lesions dry out. Heavy irrigation after a period of low soil moisture condition, heavy application of nitrogenous fertilizers and calcium deficiency causes this disorder.

Control measures

- Judicious and timely irrigation.
- Application of judicious amount of P and K fertilizers along with N fertilizer.
- Two foliar sparys of 0.2% calcium chloride the time of fruit development.

Disease

Damping off (*Pythium*, *Rhizoctonia*, *Phytophthora*, etc.)

It is very destructive fungal disease in nursery beds, especially during early season. The attack usually starts on the germinating seed, spreading to the hypocotyls, basal stem and developing taproot. In pre-emergence damping off, the growing points are killed in the initial stages of seed germination before they come out through the soil. In post-emergence damping off, the seedlings topple over the ground due to collar rotting and rapid shrinking of the cortical tissue of the hypocotyls.

Control measures

- Treating the seeds with Thiram/Captan/Mancozeb @ 3.0g/kg of seeds.
- Drenching the nursery beds 7 days before sowing with Thiram/Captan or any copper fungicide @ 3g/litre of water.
- Covering the beds with transparent polythene sheets before sowing and left to open sun for at least15-20 days which is known as soil solarization.
- Application of bioagents like *Trichoderma* @ 5g/m^2 reduces the pathogen load in seedbed.
- Thin sowing and proper thinning of the seedlings at early stage.
- Spraying the young seedlings with 0.2% Blitox or 0.25% Metalaxyl-Mancozeb or 0.25 % Fosetyl-Al.
- Provision of proper drainage in the seed beds.
- Removal of the affected seedlings from the beds as soon as the symptoms are visible.
- Avoidance of flooding the beds to check spread of the disease.

Die-back or Anthracnose (*Colletotrichum capsici*)

It is the most damaging and seed borne fungal disease. Elongated ash coloured, angular spots develop on veins and ventral surface of the leaf. Leaves fall down and the branches die back from the tip. Usually circular and sunken lesions with black margin appear on ripe fruits, which spread forming concentric markings with dark fructifications as the disease advances. The disease spread rapidly under hot and humid condition.

Control measure

- Seed treatment with 0.2% Bavistin or 0.3% Thiram.
- Spraying the seedlings in the seed bed before transplanting with 0.3% Captan or 0.1% Bavistin or 0.2% Chlorothalonil or 0.1% Hexaconazole.
- Spraying the crop with 0.1% Hexaconazole followed by 0.3% Blitox after removal of the infected twigs at 10 days interval.

Fruit rot or Phytophthora blight (*Phytopthora capcisi*)

This fungal disease is the most destructive disease of sweet pepper. The disease mostly appears in comparatively high temperature and high relative humidity condition and is spread through rain splash and irrigation water. Initial symptoms in the field occur as crown rot from soil inoculum. In the early stage, wilting of the plants occur with purplish-black lesions at the collar region.Dark green water soaked spots later appear on the fruits which enlarge rapidly to cover the entire fruit causing rotting.

Control measures

- Crop rotation excluding sweet pepper and chilli.
- Avoidance of excessive irrigation and provision of proper drainage in the field.
- Spraying the crop with 0.3% Blitox or 0.25% Mancozeb or 0.25% Metalaxyl-Mancozeb or 0.25% Fosetyl-Al at 10 days interval.

Bacterial leaf spot (*Xanthomonas campestris* pv. *vesicatoria*)

These seed borne bacterial disease of both chilli and sweet pepper favours 21-31°C temperature and 85-90% relative humidity coupled with continuous drizzling and cloudiness for its development. Yellowish green spots appear on young leaves while dark coloured water soaked areas on older leaves. The spots later appear with straw-coloured centers and dark margins. Small blisters like watery spots appear on infested fruits.

Control measures

- Hot water treatment of seeds at 50°C for 20 minutes.
- Spraying the crop with a combination or 100 mg Agrimicin 100 + 3.0g Blitox per litre of water during favourable weather condition.

Soft rot (*Erwinia carotovra*)

It is a common post harvest bacterial disease of particularly sweet pepper causing fruit decay which usually starts at wounds in the carpel or peduncles.

Control measures

- Avoidance of injury to the fruits during harvesting.
- Storage of fruits in low temperature condition.

Chilli leaf curl (Chilli leaf curl virus)

It is a destructive virus disease of both chilli and sweet pepper. This virus is transmitted by white fly (*Bemisia tabaci*). Upward and downward curling of the leaves, yellowing of leaves accompanied by puckering of interveinal areas and thickening of veins are the characteristic symptoms. In advanced stage of the disease, axillary buds are stimulated producing clusters of small leaves, and the plant appears bushy with stunted growth.

Control measures

- Destruction of all weeds around the field which serve as alternate host of the virus.
- Covering the nursery bed with 50 mesh fine nylon net and spraying the seedlings with Imidacloprid (3.5 ml/10 litre).
- Pre-sowing treatment of nursery beds with Furadon 3G @15g/100 sq.m.
- Application of Thimet 10G @ 10-15 kg/ha 10 days after transplanting followed by 2-3 foliar sparys with 0.05 % Dimethoate (Rogor 1.5 ml/l) or Imidachlorpid or Acetameprid or Thiomethoxam (3.5 ml/10 litre) at 10 days interval and spraying alternately with NSKE.
- Application of 2% power oil to the plants to prevent acquisition and inoculation of the virus by the white flies.
- Sowing 5-6 rows of border crops of maize, jowar and bajra all round the chilli plot at least 50-60 days before transplanting of chilli.

Insect pests

Thrips (*Scirtothips dorsalis*)

Both nymphs and adults lacerate the tender leaf tissues and feed on the oozing sap resulting in curling and crumpling of the leaves. Severe infestations cause shortening of internodes, thickening and reduced size of leaves with corrugated leaf lamina. Buds and flowers are also attacked. Infestation becomes severe during warm and dry weather conditions.

Control measures

- Seedling root dipping before transplanting in the aquous solution of Fipronil (2ml/l) or Imidacloprid (1ml/3l) or Monocrotophos (2ml/l).
- Spraying the crop with Acetamiprid (0.4ml/l), Fipronil (1.5ml/l) or Imidacloprid (5ml/15 l) or Thiomethoxam (1g/5 l) or Profenophos (1ml/l) at fortnightly interval.

Aphid (*Aphis* spp., *Myzus* spp.)

The aphid species *Myzus persicae* and *Aphis gossypii* attack chilli crop in early season of crop growth during January-February. They multiply rapidly and besides continuous sucking from the leaves and young apical twigs, they also secret honeydew on the plant canopy resulting formation of sooty mould. Severely infested plants may wither gradually.

Control measures

- Release of lady bird beetles and hover fly reduce the aphid population to some extent.
- Application of Imidacloprid @ 0.3 ml/l or Chlorpyriphos @ 2.5 ml/l control the aphids.

Yellow mite (*Polyphagotarsonemus latus*)

Among the different mite species attacking vegetable crops, the yellow mite, *Polyphagotarsonemus latus* is one of the most destructive causing enormous yield losses of chilli. Though it has hundreds of host plants but chilli is one of the most preferred one which is attacked by the mite throughout the warm and humid period of the year. The mite attack young apical leaves producing characteristic curling and crumpling symptoms which might be confused with that produced due to infestation of thrips and leaf curl virus. In case of mite attack, the young leaves usually curl downward in an inverted boat shaped manner, the texture of the leaves become leathery, lower surface turn shiny silvery-whitish appearance which gradually become brownish and curl completely. At the early stages of crop growth the infested leaves get thickened and later crack. The apical shoot gets crumpled, become rossetted and the growth of the plant is stunted. Infestation at the flowering stage causes bud and flower shedding resulting complete loss of yield.

Control measures

- Prophylactic spray with Azadirectin 5% @ 1ml/l at 7day interval.
- Planting of chilli and sweet pepper during September- October to obtain maximum desynchronizing in the occurrence of mite.
- Application of Carbofuran @ 1 kg a.i/ha 15 days after transplanting.
- Spraying the crop with acaricides like Dicofol @ 2.5 ml/l; Propergite @1.5 ml/l or Diafenthiuron @ 1 ml/l during early budding stage of the crop at 15 days interval.

Cole Crops

- Cauliflower
- Cabbage
- Broccoli

Cauliflower

Cauliflower, *Brassica oleracea* L var. *botrytis* under the family Brassicaceae (Cruciferae) having chromosome No. 2n = 2x =18 is the most popular vegetable among the cole crops, particularly in Asia. It is grown in all continents of the world, of which Asia is the leading one followed by Europe. It covers small acreages in North America, South America, Africa, Australia and New Zealand. India is the largest producer of cauliflower in the world and other major producers are China, France, Italy, UK, USA, Spain, Poland, Germany and Pakistan. Cauliflower is a major vegetable crop of India and grown in almost all states particularly Bihar, Uttar Pradesh, Odisha, West Bengal, Assam, Haryana, Punjab and Maharashtra.

The crop

The name cauliflower has originated from the Latin words 'Caulis' meaning stem and 'Floris' meaning flower. The edible part of cauliflower is 'curd' which is botanically the pre-floral fleshy apical meristem. Morphologically, the curd is made up of numerous divided hypertrophic branches which terminate the main stem of the plant. It is highly suppressed and extremely ramified hypertrophied flower stalk.

The pattern of development has been modified in various ways during the domestication of cole vegetable crops. The leaves on the main stem have a different shape in both kales (wrinkled and curly) and cabbages (packed into a head). In Brussels sprouts, the many secondary shoots arising along the main stem generate small heads of leaves. In knolkhol or kohlrabi, the region of the main stem that produces leaves is itself grossly swollen.

In both cauliflower and broccoli, the flowering stem is modified. In broccoli, the stems are much shorter, and many buds are densely packed. In cauliflowers, flower buds are not usually produced at all. Instead, the inflorescence meristem continuously generates replicas of itself in a spiral on its flanks. Each new meristem can in turn produce more, and so on until the tenth order of branching and beyond may be generated. This results in a closely packed, geometric cluster of undifferentiated inflorescence meristems (curd) that has a novel form and texture not present in wild-type plant, *Brassica oleracea*. The causes of all these striking variants are likely to be genetic. A gene with a major role in generating the cauliflower growth pattern has been identified and cloned from the model laboratory plant *Arabidopsis thaliana,* a relative of the cole vegetables. A variant of *Arabidopsis* was described in 1993 that resembled cauliflowers, albeit on a miniature scale in a strain of *Arabidopsis* originally from Wassilewskija in Belarus. By itself, the recessive mutant allele of the *cauliflower (cal)* gene has no effect. However, when combined with the recessive flower mutant *apetala 1 (ap1),* all floral properties are lost and the inflorescence meristem generates copies of itself indefinitely, as in the cauliflower of commerce. It was already known that the wild-type *AP1* gene of *Arabidopsis* plays a role in ensuring that the primordia that grow on the flanks of the inflorescence meristem develop as flowers.

In India, two separate groups of cauliflower are grown namely, Indian or tropical cauliflower and temperate annual or Snowball type of cauliflower. The typical tropical cauliflowers form curd and mature at average daily temperature around 20°C to 30°C and can tolerate high humidity. These are long stalked plant with broadly wavy and loosely arranged leaves producing highly flavoured, flat to semi-spherical , uneven and some what loose and yellow to cream colored curds. The morphological characters of Indian cauliflower are very close to the characters typical of Cornish while some leaf and curd character can also be Roscoff and Italians like. Cultivars of maturity group I and II (early and mid-early or Kanwari, Katki, Aghani group) belong to typical tropical cauliflower. Cultivars of maturity group III (mid-late or Poosi or Pousi group) also show most of the characters of tropical cauliflower group. There are several local cultivars in varying maturities, commonly named after the season of curd maturity. Seeds of these local varieties are marketed by different private companies of Bihar (particularly Hazipur), Uttar Pradesh (particularly Faizabad), Punjab and Haryana.

The annual temperate or Snowball group of cauliflower is not adaptive to high temperature condition. They generally form curd within the temperature range of 10°C and 16°C. This cauliflower group shows similarities with Erfurt and Alfa types of cauliflower originated in Germany and the Netherlands. These are dwarf plant with dark green, wavy, short, erect (sometimes spreading) leaves and producing solid, very deep, knobby, bright white, less flavoured and high quality curds which are less sensitive to riceyness. Cultivars of maturity group IV (late or Maghi group) come under this type in strict sense.

Importance and use

Cauliflower is the most important crop among the Cole crops in India, China and many other countries. India is the largest producer of cauliflower. With the development of tropical or Indian cauliflowers and tropical F_1 hybrids in addition to the temperate or Snowball type, it has now become possible to grow cauliflower almost throughout the year particularly in the northern and central part of India. All the F_1 hybrids of the tropical types have basically been developed from the Indian materials. In India there is hardly any house where it is not used regularly as vegetable. It is generally used as cooked vegetable either singly or mixed with potato, peas, capsicum or other vegetables as fried or in curry form or in sambar. Cauliflower is also used in fish curry as a delicacy. It is also used in the preparation of cauliflower pickle or mixed pickle with other vegetables. Boiled cauliflower is also consumed in continental form applying only salt and black pepper powder. Grated cauliflower is used for the preparation of stuffed food items. Cauliflower is generally processed in frozen form.

Nutritional value

It is not a rich source of nutrients however, a substantial amount of protein, carbohydrate, phosphorus, calcium, iron and ascorbic acid is present. Protein content of cauliflower is much higher (2.4 g/100g fresh) than cabbage. It contains fairly good amount of Vitamin C (75 mg/100 g fresh). It is a good source of calcium (30 mg), iron (17 mg) and phosphorus (76 mg) per 100 g fresh. It also contains fair amount of vitamin A (70 IU/100 g fresh).

Medicinal value

Cauliflower is fairly high source of glucosinolates (40-80 mg/100g fresh), predominantly sinigrin and glucobrassicin which have prominent anti-carcinogenic property. The products of hydrolysis of these glucosinolates formed by enzymatic or non-enzymatic action are biologically active compounds with diverse effect on human health. The product of glucosinolate breakdown, particularly the isothiocyanates act as anti-carcinogens. Cauliflower also contains S-methylcysteine sulfoxide (average 14 mg/100 g fresh) which also show anti-carcinogenic property. The carotenoid, β-carotene present in cauliflower act as potential anti-oxidant.

Origin and taxonomy

The cole crops have been originated mainly by mutation and introgression from wild species during evolution, human selection and adaptation. It has descended through mutation and selection from a common ancestor *Brassica oleracea* var. *sylvestris*, the leafy kale like wild cabbage. Probably kale was the first cole crop to be selected and adapted by man. The cultivated species selected by man had biennial habit with yellow flowered racemose inflorescence unlike the ancestral kale or broccoli of the Mediterranean region which had cymose inflorescence with white flowers. Natural crossing between ancestral broccolis (*Brassica cretica* ssp. *nivea*), the wild annual with cymose white flowers and the biennial wild types have resulted into many types of *Brassica oleracea* including various forms of broccoli and cauliflower several thousand years ago. However, cultivation of broccoli was not documented till 1660.

Cauliflower is comparatively of a later origin, about 500 years ago, probably through introgression within broccoli genepool having a close genetic affinity. It was probably originated in the island of Cyprus from where it moved to other areas like, Syria, Turkey, Egypt, Italy, Spain and north-western Europe. It was first introduced in Italy by the Genoese from Levant or Cyprus around 1490. In the middle of the 16th century, the first illustration and description of cauliflower was presented by the herbalist Dodoens (1544). The crop became more common particularly in the beginning of the 18th century.

Italy is the centre of genetic diversity for both cauliflower and broccoli where several land races of these two crops are still available even today. At the beginning of 17th century cauliflower reached Germany, France and England although, the crop was cultivated more commonly in Europe from the beginning of 18th century. Many types of cauliflowers were developed in Europe in 16th to 17th century. All the cauliflowers grown in the Europe were either temperate annuals (eg. Cornish, Alpha, Erfurt, Danish, Lecerf, Mechelse, etc.) or biennial types with winter hardiness and vernalization requirement for flowering (Old English, Walcheran, Roscoff, Angers, etc.).

Many types of cauliflower have been developed in Europe particularly in France, U.K., Sweden, Denmark and the Netherlands from 16th century onwards in addition to different land races developed in Italy being the centre of diversity. Some of the cauliflower groups of Europe are Originals or Italians in Italy (Jezi, Naples, Romanesco and Florablanca), Effurt, Alfa and Snowball in Germany and the Netherlands, Cornish and Northerns in England and Roscoff and Angers in France. The European bianneals include Old English, Walcheran, Roscoff, Angers and Sent Malo and the annuals include Alfa, Erfurt, Danish, Lecerf and Mechelse.

The Tropical or Indian group of cauliflowers have been originated as a result of intercrossing among European annuals and biennial groups and simultaneous selections. The Cornish types of cauliflower were first introduced in India in 1822 from England and later other biennial European types like, Roscoffs, Italians and Northerns were also introduced. Natural crossing between different European annual and biennial types during attempts of seed production followed by selection by farmers and adaptation through the course of more than 100 years resulted in the development of Tropical or Indian cauliflower that are distinctly different from the temperate European types. The first four varities listed by Sutton and Sons during 1929 were Early Patna, Main crop Patna, Early Benaras and Main crop Benaras. The early and heat tolerant cultivars were introduced to different countries like, Japan, South Korea, Taiwan, Sri Lanka, Isreal, Hawaii, Brazil, etc. Many heat tolerant improved varieties and hybrids have been developed particularly in Japan, South Korea and Taiwan from this genetic base originally developed in India.

The *Brassica oleracea* cytodeme is a polymorphic aggregate species with chromosome number, 2n= 18. Cauliflower is a monogenomic species having genomic constitution C and chromosome number n = 9. It belongs to the Order Cruciferae, Tribe Brassicae, Sub-Tribe Brassica under the family Cruciferae or Brassicaceae.

Adaptation in India

Cauliflower was introduced in India from England in 1822 by Dr. Jemson, a botanist from Royal Kew garden who was the incharge of the Company Bagh, Saharanpur, Uttar Pradesh (erstwhile United Provinces). The Royal Agri-Horticultural Society, Calcutta also introduced cauliflower in 1824 from South Africa. The cauliflower type introduced in India more than 190 year ago was Cornish cauliflower of England. Later other European types like, Roscoffs, Italians and Northerns were also introduced. During the period of 100 years (1822–1929) these introduced cauliflower varieties underwent selection by the local growers when seed production was attempted by them in north Indian plains. Selections were exercised for early maturity and wider adaptability in hot and humid weather conditions. Natural crossings between Cornish and other European types followed by selection by the farmers and adaptation in high temperature condition resulted in the development of a separate group of cauliflower, the Indian or Tropical cauliflower. These tropical cauliflowers are distinctly different from the temperate European types and are adapted to early sowing and early harvest. Thus a new development in cauliflower took place in India in the form of tropical cauliflower resistant to hot weather and high humidity. These tropical or Indian types are grown in the plains of India during June to December and normally followed by the temperate types known as snowball or late cauliflower.

Open pollinated improved varieties

Cauliflower varieties grown in India can be grouped according to their temperature requirement and time of curd maturity with respect to the plains of northern part of India.

Maturity group		Temperature requirement for curd initiation and proper curd development
Early crop	Mid September to mid October	20-27°C
	Mid October to mid November	20-25°C
Mid -early	`Mid November to mid December	16-20°C
Mid-late	Mid December to mid January	12-16°C
Late	Mid January to mid February	10-16°C

Early group (Mid September to mid October maturity): Early Kunwari, Pant Gobi-3, Pusa early Synthetic, Synthetic 78-1, Pant Gobi-3, Kashi Kunwari.

Early group (Mid October to mid November maturity): Pusa Deepali, Pusa Katki, RC-JOB-1, Pant Gobi-2.

Mid -early group (Mid November to mid December maturity): Improved Japanese, Pusa Sharad, Pant Gobi-4, Pant Shubra, Pusa Meghna.

Mid-late group (Mid December to mid January maturity): D-96, Pusa Synthetic, Pusa Himjyoti, Pusa Shubhra, Punjab Giant-35, Punjab Giant-26, Hisar 1.

Late group (Mid January to mid February maturity): Dania, Pusa Snowball-1, Pusa Snowball-2, Pusa Snowball K-1, Snowball-16, Ooty-1.

Public sector hybrid: Pusa Hybrid-2 (Mid November to mid December maturity), Pusa Kartik Sankar (Mid October to mid November maturity).

Cultural requirements

Cauliflower is a thermo-sensitive crop. Both tropical and temperate varieties are grown in India. The tropical heat tolerant or Indian cauliflowers have been classified into four maturity groups depending on the average temperature requirements for curd initiation and development under north Indian condition. The maturity groups of Indian cauliflowers are September (20-27°C), October (20-25°C), November (16-20°C) and December (12-16°C). Well drained, fertile, sandy- loam and clay-loam soils rich in organic matter are ideal. Light soils are good for early crop, while clay- loam soils are well suited for getting high yield. The ideal soil pH is 6-7. Cauliflower is sensitive to soil acidity, boron and molybdenum deficiency.

Nursery management

Fine, well drained, 15-20 cm raised, 1.0 m wide and of convenient length seed bed should be prepared. Fine and fully decomposed farmyard manure or compost @ 3-4 kg/m^2 should be well mixed to the beds. The beds may be drenched with formaldehyde (4.0%) and covered with polythene sheet for 5-7 days for soil sterilization. 500 to 750g seeds of open pollinated varieties of early and mid-early maturity group and 300-500 g seeds of mid-late and late maturity group are required per hectare. Generally one gram contains 300-350 seeds. The seeds are first treated by soaking for half an hour in Streptocycline solution (250mg/5 litre of water) and then after drying in shade with Captan or Thiram @ 2-3 g per kg of seeds before sowing. Seeds are sown at shallow depth at 5.0 cm apart, 1-1.5cm depth and covered with finely sieved leaf mould or vermicompost. After sowing, the beds are covered with straw and water is sprinkled regularly till seeds are germinated. Seedlings of the early and mid-early varieties, in particular, should be raised under poly tunnels having sides open to protect the young seedlings from heavy rain or scorching sunlight. The seedlings are watered daily or in alternate day with rose can. Hardening of the seedlings can be done by withholding water at least 4 to 6 days before transplanting. Stocky, healthy seedlings of 6-8cm height, 4-6 leaves of 35-40 days old are ready for transplanting for the early and mid-early varieties and seedlings of 28-30 days old are ready for transplanting for the mid-late and late varieties.

Sowing time

The time of sowing depends on the variety, season and location. Sowing time is May - June for the varieties of September maturity group; June – first fortnight of July for the varieties of October maturity group; July-August for the varieties of November maturity group; July-August for the varieties of December maturity group; September for the varieties of December maturity group and October for the varieties of January-February maturity group.

Land preparation

The land is prepared well in advance with repeated ploughings (at least 4-5 ploughing) to a fine tilth. All stubbles, weeds, etc. are removed and the land is leveled. Well rotten organic manure preferably farm yard manure or compost @ 25-35 t/ha should be applied during land preparation followed by proper levelling of the land. The land should have adequate drainage facility particularly for the varieties of early and mid-early maturity groups.

Transplanting

Healthy seedlings of 35-40 days old for early and mid-early varieties and 28-30 days old for mid-late and late varieties are ready for planting. After hardening the seedlings for 4-6 days, the seed bed is irrigated before lifting to facilitate easy pulling of young seedlings. The seedlings should be planted in the field preferably during afternoon hours for better field establishment and the field is irrigated just after transplanting. Planting in ridges is ideal particularly for early and mid-early varieties when rainfall is high. In one hectare of land, 35-37 thousand seedlings can be accommodated. Spraying the 15 days old seedlings with the starter solution of ammonium sulphate (50g/10litres of water) is beneficial for better establishment. Spraying the seedlings with 0.1ppm IBA or 10ppm NAA gives better establishment, enhance vegetative growth and ensure more yield.

Spacing

The spacing of 60 cm x 30 cm is given for early and mid-early varieties and 60 cm x 45 cm for mid-late and late varieties.

Nutrient management

Adequate nutrition is necessary for obtaining higher yield and better curd quality. A general fertilizer dose of 125 kg N, 80 kg P_2O_5 and 60kg K_2O per hectare is recommended for open pollinated improved varieties. Half N along with other fertilizers should be given as basal and rest N should be top dressed in two split doses at 35 days interval after transplanting. Mid-late and late varieties require comparatively more nutrients than the early and mid-early varieties. A general fertilizer dose of 200 kg N, 100 kg P_2O_5 and 80kg K_2O per hectare is recommended for the hybrids.Half N along with other fertilizers should be given as basal and rest N should be top dressed in two split doses at 20 days interval after transplanting. Judicious application of phosphatic fertilizer increase curd yield and dry matter content. Cauliflower is responsive to sulphur application hence, it is better to apply ammonium sulphate as the source of nitrogen. Weekly spray of 1-2% urea 20 days after transplanting gives better growth and increases yield. Three foliar sprays of 0.3% borax after 20, 35 and 50 days after transplanting is beneficial in increasing yield and checking browning disorder.

Irrigation

Cauliflower is a shallow rooted vegetable crop and requires proper soil moisture throughout the cropping period. The early and mid-early crop requires only protective irrigation depending on the rain. Over-watering should be avoided. Irrigation is generally required at 10-12 days interval during cool season. Total water requirement about 350mm.

Interculture and weed control

Curd yield could be reduced to the tune of 36-69% if weeds are not controlled in the early stages of growth. The crop needs at least 4 weeks weed free period after transplanting to prevent yield loss. Pre-planting application of Basalin (Alachlor) 2-2.5 litre/ha or Stomp (Pendimethalin) 3.3 litre /ha in the finally prepared field followed by one hand weeding 40-45 days after is the best. Clear plastic mulching results in faster growth and advances the harvest upto 10 days and suppresses weed growth. Mulching with 10 cm thick paddy straw is also beneficial. Earthing up should be done about 4-5 weeks after transplanting after topdressing with nitrogen fertilizer. Blanching i.e. covering the curd through tying the outer leaves up over the curd improve curd colour and quality.

Harvesting the crop

The curds are harvested when they reach prime condition. It is better to harvest early than late. Late harvest causes ricy or fuzzy and loosening of curds. It is better to harvest above the soil surface with a large and sharp knife or sickle.

Post harvest handling and storage

Large leaves are trimmed leaving only sufficient jacket leaves to protect the curd from bruising and other mechanical injury in transport. More number of jacket leaves is kept when cauliflowers are transported loose in gunny bags or basket. When they are transported in crates, the jacket leaves are trimmed leaving a fringe of leaves projecting 2-3 cm above the curds. Tight packing is essential to prevent shifting and bruising. Cellophane and transparent films are also used to pack cauliflower which is hydro-cooled to remove field heat. Use of polythene packing provides modified atmosphere and consequently reduces decaying, softening, loss of total solids and weight. Consumers' package should be ventilated to prevent CO_2 injury. It can be stored for 30 days at 0° to 1.7°C with 85-90% RH. Post-harvest spray of 5-15ppm BA (N6-benzyladenine) checks yellowing of curd in storage.

Yield

Yield may be 10-15 tonnes/ha for open pollinated early and mid-early varieties; 15-20 tonnes/ha for mid-early hybrids; 18-20 tonnes/ha for open pollinated mid-season varieties; 20-25 tonnes/ha for mid-season hybrids; 25-30 tonnes /ha for open pollinated late varieties and 35-38 tonnes /ha for the hybrids.

Cabbage

Cabbage, also known as white cabbage, *Brassica oleracea* L var. *capitata* under the family Brassicaceae (Cruciferae) having chromosome No. 2n = 2x = 18 is one of the most popular vegetables in the world. Cabbage was one of the oldest cultivated vegetables, used by the ancient Greeks and Romans. It is also called white cabbage or headed cabbage. It is grown widely in Europe, Russia, Asia, USA, Canada, South America, Japan and East Africa. Traditionally cauliflower occupies more area in India however, after the development of tropical hybrids it is now produced almost round the year widely throughout the country. India is the second largest producer of cabbage in the world. In India the major cabbage producing states are Uttar Pradesh, Bihar, Orissa, West Bengal, Assam, Maharashtra and Karnataka and other important growing states are Haryana, Rajasthan, Gujarat, Uttarakhand, Himachal Pradesh and Nilgiri hills of Tamil Nadu.

The crop

The plants have fibrous and shallow root systems and about 90 percent of the root mass is found in the upper 20–30 cm of soil. The edible part of cabbage is 'head' which is formed by thickening of apical bud with thick tightly packed overlapping leaves. It is a biennial plant, grown as an annual vegetable crop for its dense-leaved heads. Cabbage heads generally range from 0.5 to 4.0 kg in weight, and can be green, purple and white. Smooth-leafed firm-headed green cabbages are the most common. Some varieties with smooth-leafed purple and crinkle-leafed "Savoy" cabbages of both green and purple colour are also found. Most cabbages have thick, alternating leaves with margins that range from wavy or lobed to highly dissected; some varieties have a waxy bloom on the leaves. Many shapes, colours and leaf textures are found in various cultivated varieties of cabbage. Leaf types are generally divided between crinkled-leaf, loose-headed "Savoy" cabbage and smooth-leaf firm-head cabbages. Different head shapes viz., oblate, round and pointed shape are found in different classes and varieties of cabbage.

Genetic diversity in cabbage is mainly found in north-west Europe. The cultivated types of cabbage show great variation with respect to size, shape and colour of leaves and size, shape, colour and texture of the head. Three cultivated forms of cabbage have been recognized.

- *Brassica oleracea* var. *capitata* f. *alba* (white cabbage)
- *Brassica oleracea* var. *capitata* f. *rubra* (red cabbage)
- *Brassica oleracea* var. *sabauda* (savoy cabbage)

White cabbage: Cultivars of this group are the commercial cabbage grown in the world. Leaves are generally smooth and light green in colour. Inside colour of the head is creamy white. Varieties of this cabbage group are available in three different head shape namely, round, flat or drum head and pointed.

Savoy cabbage: Leaves of this cabbage group are blistered and puckered in texture and light green or green in colour. Varieties of this cabbage group are available in three different head shape namely, round, flat or drum head and pointed.

Red cabbage: Leaves of this cabbage group are smooth, red due to anthoeyanin pigment and having distinct coating of wax. Varieties of this cabbage group are available in round head shape.

All the tropical cabbage hybrids of "White cabbage" group have basically been developed in Japan. Cabbage cultivation to the tropical regions of South-east Asia was spread due to the classical "KK" and "KY" crosses developed during 1950's. Japanese, Korean, Vietnamise and Taiwan Seed trade still hold the monopoly for these tropical cabbage hybrids. The present day hot weather hybrids form compact heads under tropical conditions of day temperatures of even 30-35^0 C and have been accepted widely by the farmers for growing in warmer conditions of east, south and western India. However, from the seed production point of view, these hot weather cabbage types are also biennial in nature although the span of cold temperature requirement for vernalization is comparatively brief.

Importance and use

Cabbages are prepared in many different ways for eating. They can be cooked, pickled, fermented for dishes such as sauerkraut, steamed, stewed, sautéed, braised, or eaten raw as salad. It is generally used as cooked vegetable either singly or mixed with potato, peas, or other vegetables as fried or in curry form. It is used as stewed or boiled items in continental form. A well known lactic acid fermented product called "sauerkraut" is prepared from cabbage. Shredded cabbage after blanching in steam or boiled water is used for the preparation of dehydrated product by freeze drying or hot air drying. Blanched cabbage is canned in 1% salt solution as covering liquid. Cabbage after shredding in small pieces or grated is mixed with different vegetable oils or vinegar, spices and creams, called "cole slaw" and is taken as salad however, this preparation cannot be stored for long period.

Nutritional value

Cabbage is a good source of vitamin K, vitamin C, Vitamin A, Ca, P, Na, K, S and dietary fibre. It contains good amount of Vitamin C (average 100 mg/ 100 g fresh). It is a good source of calcium (average 46 mg) and phosphorus (average 38 mg) per 100 g fresh. Good amount of phosphorus content is helpful in utilization of calcium and assimilation of carbohydrates and fats in human body. It also contains fair amount of vitamin A (average 80 IU/100 g fresh).

Medicinal value

Cabbage has remarkable healing properties. Hippocrates prescribed a dish of boiled salted cabbage for violent colic. Ancient Romans and Greeks used cabbage to alleviate headaches, gout, and ingestion of poisonous mushrooms. Roman soldiers applied cabbage leaves to their wounds. There are many anecdotal observations of cabbage leaves to be effective in curing severe inflammations and infections. Cancer preventive properties of cabbage include antioxidant, anti-mutagenic, and detoxification activities exhibited by substances and enzymes contained in cabbage juice.

Cabbage is fairly high source of glucosinolates (40-90 mg/100g fresh), predominantly sinigrin and glucobrassicin which have prominent anti-carcinogenic property. The products of hydrolysis of these glucosinolates formed by enzymatic or non-enzymatic action are biologically active compounds with diverse effect on human health. The product of glucosinolate breakdown, particularly the isothiocyanates act as anti-carcinogens. Cabbage also contains S-methyl cysteine sulfoxide (18 mg/100 g fresh) which also show anti-carcinogenic property. Cabbage protects against bowel cancer due to presence of indole-3 carbinol in it. The carotenoid, β-carotene present in cabbage act as potential anti-oxidant. Rich vitamin C and fairly high fibre and calcium contents reduce the risk of colon cancer. Cabbage is effective against ailments like gout, diarrhoea, stomach and colic troubles. Cabbage juice is a remedy against poisonous mushrooms and as a gargle against hoarseness. The leaves are traditionally used to cover ulcers and wounds. Fermented product of shredded leaves of cabbage (sauerkraut) has curative effect on scurvy disease due to high content of Vitamin C in it.

Origin and taxonomy

Cabbage is an important vegetable known to mankind for over 4,000 years. Cabbage was most likely domesticated somewhere in Europe before 1000 BC, although savoy cabbage were not developed until the 16th century. By the middle Ages, cabbage had become a prominent part of European cuisine.

All cole crops have been originated mainly by mutation and introgression from wild species during evolution, human selection and adaptation. It has descended through mutation and selection from a common ancestor *Brassica oleracea* var *sylvestris*, the leafy kale like wild cabbage. Probably kale was the first cole crop to be selected and adapted by man. The wild diploid species *Brassica oleracea* var. *sylvestris*, leafy kale-like plant which occurred in the northern shore of the Mediterranean sea and on the lime cliffs in the north-eastern England and along the North sea is the progenitor of the cultivated kale and non-heading cabbage. Such cabbage was probably in general use, around 2000-2500 BC by the Greeks and Romans. It is said that cabbage was held in high esteem by the ancient Greeks and even worshipped by the Egyptians. The wild non- heading cabbage was originated in eastern Mediterranean region and Asia Minor which was probably brought into Western Europe by the Celts.

The hard-headed cabbage developed through selection in the North Europe descended from this wild leafy non-heading cabbage, originating in the eastern Mediterranean region and Asia Minor which was probably brought into Western Europe by the Celts. The wild perennial forms occur in south-western Britain, western France, northern Spain and the Mediterranean region. The present day hard headed white cabbage evolved in Germany around 1150 A.D. by human selection descended from the wild non-heading cabbage which was perennial in nature and, later it moved to England in the 14th century. Savoy cabbage was originated earlier in Italy, probably from Portuguese kale (*Brassica oleracea* var. *costata*).

The *Brassica oleracea* cytodeme is a polymorphic aggregate species with chromosome number, 2n = 18 having cabbage, cauliflower, Brussels sprouts, broccoli, knolkhol and kale. Cabbage has a somatic chromosome number of 18 and its genomie constitution is C. It is a secondary polyploid with basic chromosome number 6. Three basic chromosomes are present in duplicate and remainder are single i.e. ABBCCDEEF = 9. The order of duplicate loci is often different on different chromosomes. Considerable chromosome rearrangement and restructuring has occurred during the evolution of the species. It might be difficult to identify the possible structure of a lower chromosome number progenitor or to determine if *B. oleracea* evolved by duplication of loci and complex rearrangements.Cabbage is a monogenomic species and belongs to the Order Cruciferae, Tribe Brassicae, Sub-Tribe Brassica under the family Cruciferae or Brassicaceae.

Adaptation in India

Cabbage was introduced in India by the Portuguese during the Mughal period in the 16th century however, it became popular during the British period.The exact date of introduction of cabbage into India is not certain but it is known to be cultivated during 16th century in the Mughal period as has been mentioned in the Ain-i-Akbari. However, it is not known whether cabbage seed production had been attempted during the Mughal period. The only recorded evidence of organized seed production of temperate vegetables including cabbage is in the early forties of twentieth century, when experiments were conducted at Quetta in Baluchistan and later in Kashmir valley, which later extended to Himachal Pradesh where its chilling requirement to enter into the reproductive phase after the vegetative phase is fulfilled, being biennial crop in nature. This was necessitated due to stoppage of seeds import from Europe during World War II. Whether Mughal rulers regularly imported cabbage seeds and seeds of other temperate vegetables like carrot, turnip, beet root etc. or these were being presented by the European traders during their visits to the Mughal courts, is neither known nor could be established.

Although cabbage was introduced in India much earlier than cauliflower yet, unlike tropical or Indian cauliflowers no such tropical cabbage types evoved in India.

Open pollinated improved varieties

White cabbage varieties grown in India can be grouped on the basis of days required for maturity of heads after transplanting.

White cabbage

Early (60-75 days) : Golden Acre, Pride of India, Copenhagen Market, Express, Pusa Mukta, Pusa Agethi, Pusa Synthetic, Express, Drumhead Early, Palampur Green

Mid season (80-95 days) : Glory of Enkhuizen, Kalimpong English Ball

Late season (90-130 days) : September, Pusa Drumhead, Late Large Drumhead, Kalimpong Eclipse Drumhead

Pointed headed white cabbage : Jersey Wakefield, Charleston Wakefield

Red cabbage : Mammoth Rock Red, Red Acre, Kinner Acre

Savoy cabbage : Chieftain, Perfection

Tropical hybrids : Most of the high temperature tolerant tropical hybrids are early maturing and are produced by different private seed companies of Japan, Korea, Taiwan, Vietnam, etc.

Cultural requirements

Cabbage is basically a cool season crop and thrives best in cool and moist conditions. Cabbage is hardier vegetable crop than cauliflower and can withstand frost and extreme cold weather. The minimum temperature for growth of cabbage is just above 0°C. 15-20°C temperature is optimum for growth and head formation and growth is checked when temperature exceeds 25°C. However, the tropical heat tolerant hybrids form tight heads even at 25° to 35°C temperature regimes. Cabbage loses its flavour when grown under warm and dry conditions. Cabbage is grown on a wide range of soil. Well drained, fertile, sandy-loam and clay-loam soils rich in organic matter are ideal. Light soils are good for early crop, while clay- loam soils are well suited for getting high yield. Due to slow growth on heavy soils, keeping quality of heads is improved because of compactness. The ideal soil pH is 6.0-6.5 since phosphorus availability is more in this pH range. Cabbage does not grow well in highly acidic soils and pH should not be below 5.5. Most of cabbage types are moderately tolerant to soil salinity. Plants grown in saline soils are more susceptible to diseases like black leg.

Nursery management

Fine, well drained, 15-20 cm raised, 1.0 m wide and of convenient length seed bed should be prepared. Fine and fully decomposed farmyard manure or compost @ 3-4 kg/m^2 should be well mixed to the beds. The beds may be drenched with formaldehyde (4.0%) and covered with polythene sheet for 5-7 days for soil sterilization. 500 g seeds of open pollinated early varieties and 375 g seeds of late varieties are required per hectare. Generally one gram contains 300-350 seeds. The seeds are first soaked for half an hour in Streptocycline solution (250mg/5 litre of water) and then after drying in shade with Captan or Thiram @ 2-3 g per kg of seeds before sowing. Soaking the seeds in 0.1% boric acid before sowing enhance yield. Seeds are sown at shallow depth at 5.0 cm apart, 1-1.5cm depth and covered with finely sieved leaf mould or vermicompost.

After sowing, the beds are covered with straw and water is sprinkled regularly till seeds are germinated. Seedlings may be raised under poly tunnels to protect the young seedlings from heavy rain or scorching sunlight. The seedlings are watered daily or in alternate day with rose can. Hardening of the seedlings can be done by withholding water at least 4 to 6 days before transplanting. Stocky, 30-35 days old, healthy seedlings of 6-8 cm height with 4-6 leaves are ideal for transplanting.

Sowing time

The time of sowing depends on the variety/ hybrid, season and location. With the selection of proper variety and high temperature tolerant tropical hybrid, cabbage can be grown almost round the year in mild climate of Karnataka and Maharashtra. Generally sowing time is June– August for high temperature hybrid in eastern India; mid to late September for Early variety in eastern India; August-September for early variety in northern India; September-October for late variety in northern India; March-June for growing in high hill condition and almost throughout the year in mid-hill condition.

Land Preparation

The land is prepared well in advance with repeated ploughings (at least 4-5 ploughing) to a fine tilth. All stubbles, weeds, etc. are removed and the land is leveled. Well rotten organic manure preferably farm yard manure or compost @ 25-30 t/ha should be applied during land preparation followed by proper levelling of the land. The land should have adequate drainage facility particularly for the high temperature hybrids and early varieties.

Transplanting

Healthy seedlings of 35-40 days old for early and mid-early varieties and 28-30 days old for mid-late and late varieties are ready for planting. After hardening the seedlings for 4-6 days particularly for early and mid-early varieties, the seed bed is irrigated before lifting to facilitate easy pulling of young seedlings. The seedlings should be planted in the field preferably during afternoon hours for better field establishment and the field is irrigated just after transplanting. One hectare land can accommodate 40-45 thousand seedlings. The field should be irrigated just after transplanting. Planting in ridges is ideal particularly for high temperature hybrids and early varieties when rainfall is high. Spraying the 15 days old seedlings particularly of the high temperature tolerant hybrids with the starter solution of ammonium sulphate (50g/10litres of water) is helpful for better establishment.

Spacing

Spacing of 60 cm x 30 cm, 45 cm x 45 cm is generally given for high temperature tolerant hybrids and early varieties; 60 x 45 cm for mid season varieties and 60 cm x 45 cm, 60 cm x 60 cm, 75 cm x 30 cm for late varieties.

Nutrient management

Cabbage is shallow rooted heavy feeder and requires adequate nutrient supply for profitable yield. A general fertilizer dose of 150 kg N, 60 kg P_2O_5 and 80kg K_2O per hectare is recommended for open pollinated improved varieties. Half N along with other fertilizers should be given as basal and rest N should be top dressed in two split doses at 25 days interval after transplanting.

A general fertilizer dose of 200 kg N, 125 kg P_2O_5 and 150kg K_2O per hectare is recommended for the hybrids. Half N along with other fertilizers should be given as basal and rest N should be top dressed in three split doses at 25 days interval after transplanting. Weekly spraying of 1-2% urea 20 days after transplanting gives better growth and increases yield. Two sprays of sodium or ammonium molybdate 20mg/litre of water at 25 and 40 days after transplanting increase the volume and weight of head and enhance the maturity. Application of non-symbiotic nitrogen fixing bacteria *Azotobacter* is helpful in increasing the yield. Combined application of 60 kg nitrogen / ha as inorganic source and *Azotobacter* as biofertilizer helps in reducing the fertilizer requirement and increasing yield.

Irrigation

Cabbage is a shallow rooted vegetable crop and requires proper soil moisture throughout the cropping period. The optimum growth occurs at moisture content of the soil varying from 60-100% field capacity with an average of 80 per cent. Irrigation should be given when the moisture content of soil had dropped below 50% of field capacity. Irrigation should be given immediately after transplanting and thereafter is generally required at 10-12 days interval during cool season. Critical stages of irrigation are at head formation and enlargement stage. Total water requirement about 350-400 mm. Excessive and frequent irrigations have an adverse effect on growth and yield.

Interculture and weed control

Yield is reduced significantly if weeds are not controlled in the early stages of growth. Critical crop weed competition period is 35-40 days after planting and two weedings during this period is recommended. Due to shallow root system, deep hoeing should always be avoided as it does harm to the feeder roots. Pre-planting application of Basalin (Alachlor) 2-2.5 litre/ha or Stomp (Pendimethalin) 3.3 litre /ha in the finally prepared field followed by one hand weeding 40-45 days after is the best. Black polythene mulch is effective in controlling weeds, conserving moisture and ensure satisfactory growth and higher yield of cabbage. Mulching with 10 cm thick paddy straw is also beneficial. Earthing up should be done about 4-5 weeks after transplanting after topdressing with nitrogen fertilizer.

Harvesting the crop

The heads are harvested at suitable size, firm and tender and before they start cracking. Delay in harvesting results in cracking of heads. High temperature hybrids stay long in the field without the problem of head cracking. Heads are cut with a sharp knife attached with some wrapper leaves. Large non-wrapper leaves are trimmed away leaving only sufficient jacket leaves to protect the head from mechanical injury during transport. Proper handlings of cabbage at the time of harvesting prevent post-harvest loss by mechanical, physio-biochemical and microbial loss.

Post harvest handling and storage

Cabbage can be stored at its best for 4-6 weeks at 0° to 1.7°C temperature and 90-95% relative humidity. However, late varieties with very firm heads can be stored up to 12 weeks at 0° to 1.7°C temperature and 90-95% relative humidity. It can be stored for 265 days under control atmosphere storage (3-4%CO_2+2-3%O_2 and 93-95%RH) with a 17% total mass loss. Cabbage can also be stored for some period in zero energy cool chambers applying the principle of evaporative cooling. Use of polythene packing provides modified atmosphere storage. Tight packing in sacks is essential to prevent bruising particularly for the early varieties having loose heads however, transportation in crates is the best.

Yield

Cabbage yield ranges widely depending on the variety or hybrid and climate and it is 25-35 tonnes/ha for open pollinated early varieties; 30-45 tonnes/ha for open pollinated late varieties and 50-80 tonnes /ha for the hybrids.

Broccoli

Broccoli, *Brassica oleracea* L var. *italica* under the family Brassicaceae (Cruciferae) and having chromosome No. 2n = 2x =18 produces fleshy, flower heads arranged in a tree-like fashion on branches sprouting from a thick, edible, sturdy and meaty stalk. The word "broccoli" comes from the Italian plural of broccolo, which means "the flowering crest of a cabbage" and is the diminutive form of *brocco*, meaning "small nail" or "sprout".

China is the largest producer of broccoli in the world followed by India. Other major producers are United States, Spain, Mexico and Italy. Broccoli cultivation in India has gained momentum in the recent times especially around big cities. Initially cultivation of broccoli started in India mainly to cater the demands of the big hotels for the delicacy of the foreigners. Now it has become quite popular particularly among the city dewellers of India mainly because of its high nutritive values and cancer prevention properties.

The crop

Broccoli resembles cauliflower, which is a different cultivar group of the same species. It is classified in the Italica cultivar group of the species *Brassica oleracea*. Broccoli has large flower heads, usually green or purple in colour, arranged in a tree-like structure branching out from a thick, edible stalk. The mass of flower heads is surrounded by lavish leaves.

There are three commonly grown types of broccoli. The most familiar is Calabrese broccoli, often referred to simply as "broccoli", named after Calabria in Italy. It has large (10 to 20 cm) green heads and thick stalks. Sprouting broccoli has a larger number of heads with many thin stalks. Purple cauliflower is a type of broccoli grown in Europe and North America. It has a head shaped like cauliflower, but consisting of tiny flower buds. It sometimes, but not always, has a purple cast to the tips of the flower buds.

Importance and use

Broccoli is well known not only for its delicate taste but also for its high nutritive, antioxidant and cancer prevention properties. Besides the main head, long slender smaller heads which are called spears are developed in the axils of the leaves. Both heads and fleshy stems and spears are eaten either as cooked vegetables single or mixed with other vegetables particularly potato or as salad.

Nutritional value

Among the cole crops, broccoli is the richest source of nutrients. It is a good source of vitamin C, iron, fibre, potassium, vitamin A, calcium, zinc, magnesium, β carotene, and vitamin B. It has about 50 times more vitamin A contents than cauliflower and cabbage. It is a rich source of protein, minerals and vitamins particularly, vitamin A (average 3500 IU/ 100g fresh). Broccoli also contains other carotenoid compounds like, lutein and zeaxanthin. Protein content of broccoli is much higher (average 3.3 g/100g fresh) than cauliflower and cabbage. It contains much higher amount of Vitamin C (average 137 mg/ 100 g fresh) than cauliflower and cabbage. It is a good source of calcium (average 80 mg), iron (average 17 mg) and phosphorus (average 79 mg) per 100 g fresh. It also contains fair amount of riboflavin (average 0.12 mg/100 g fresh). Broccoli has low content of carbohydrates, fat and dietary fibre.

Medicinal value

Broccoli is now marketed as a health-promoting food because it naturally has high content of bioactive phytochemicals such as glucosinolates, phenolic compounds, vitamin C, and mineral nutrients. Thus, a diet rich in broccoli plays a role in the prevention of chronic diseases, such as cardiovascular and carcinogenic pathologies, and breast and prostate cancers. Broccoli has also been found to exhibit antioxidant activity that prevents oxidative stress related to many diseases.

It is thought to be important in the prevention of diabetes, osteoporosis and high blood pressure. Broccoli has high antioxidant activity mainly because of the presence of considerable amount of flavonoids like kempherol, β-carotene and ascorbic acid. Broccoli is rich source of glucosinolates (40-80 mg/100g fresh), predominantly sinigrin and glucobrassic in which have prominent anti-carcinogenic property. The products of hydrolysis of these glucosinolates formed by enzymatic or non-enzymatic action are biologically active compounds with diverse effect on human health. The product of glucosinolate breakdown, particularly the isothiocyanates act as anti-carcinogens. Broccoli is also a rich source of sulphoraphane, a compound associated with reducing the risk of cancer.

Origin and taxonomy

Broccoli is a result of careful breeding of cultivated *Brassica* crops in the northern Mediterranean starting in about the 6th century BC. Since the time of the Roman Empire, broccoli has been considered a uniquely valuable food among Italians. The greatest genetic diversity of cole crops including broccoli is the Mediterranean gene centre including Greece, Syria, Cyprus, Sicily, Italy, Spain and Portugal. The cole crops have been originated mainly by mutation and introgression from wild species during evolution, human selection and adaptation. It has descended through mutation and selection from a common ancestor *Brassica oleracea* var *sylvestris*, the leafy kale like wild cabbage. Probably kale was the first cole crop to be selected and adapted by man. The cultivated species selected by man had biennial habit with yellow flowered racemose inflorescence unlike the ancestral kale or broccoli of the Mediterranean region which had cymose inflorescence with white flowers. In was thought that many types of *Brassica oleracea* including various forms of broccoli and cauliflower might have originated by natural crossing between ancestral broccolis (*Brassica cretica* ssp. *nivea*), the wild annual with cymose white flowers and the biennial wild types, followed by human selection several thousand years ago. As all the flower buds open in broccoli compared to prefloral fleshy apical meristem of cauliflower, it is thought that broccoli must have preceded cauliflower in the evolution of species.

Broccoli was developed in eastern Mediterranean region and Italy is the main centre of its diversification. Broccoli was brought to England from Antwerp in the mid-18th century by Peter Scheemakers. Broccoli was first introduced to the United States by Southern Italian immigrants, but did not become widely popular until the 1920s.

Adaptation in India

In India, broccoli is comparatively a recent introduction particularly from USA and Japan for commercial cultivation by the private seed companies around 1990. All the varieties developed in India are selections from the exotic materials.

Improved varieties

The green sprouting broccolis are classified in accordance to their maturity, e.g. "early", "medium" and "late" cultivars. There is more demand for green sprouting broccoli having green, firm and compact crown heads. The side shoots or heads are less preferred in the Indian market. Important Indian varieties include Pusa Broccoli, Palam Samridhi, Palam Haritika, Palam Vichitra, Punjab Broccoli-1, etc.

Early season type (maturity 65-70 days after transplanting): Early Danish Giant, Decicco, Green–Bud, Spartan Early, Atlantic.

Mid season type (maturity 85-90 days after transplanting): Green Sprouting Medium, Pusa Broccoli, Palam Samridhi, Punjab Broccoli-1.

Late season type (maturity 115-130 days after transplanting): Waltham 29, Green Mountain, Coastal, Atlantic, Palam Haritika, Palam Vichitra, Palam Kanchan.

Cultural requirements

It is a cool season crop. Some cultivars of broccoli exhibit resistance to frost to a certain extent however, it depends on the size of the plant. In the hills it is grown as spring-summer crop. The average temperature of 20°–25°C is optimum for its proper growth, while 15°–20°C is the best for heading stage. The heads become loose with rise in temperature. High temperature also gives rise bracts during developmental stage of head. Broccoli can be cultivated in a wide range of soils. The soils should be well-drained and sufficiently fertile. The crop is best suited in deep loamy soil. It requires moist soil for fast and appropriate growth. The shoots become more fibrous under dry soil. It is slightly tolerant to acidic soil but the optimum pH of soil for its cultivation is 6.0-6.5.

Nursery management

Fine, well drained, 15-20 cm raised, 1.0m wide and of convenient length seed bed should be prepared. Fine and fully decomposed farmyard manure or compost @ 3-4 kg/m^2 should be well mixed to the beds. The beds may be drenched with formaldehyde (4.0%) and covered with polythene sheet for 5-7 days for soil sterilization. For cultivation of broccoli 400-500 g seed is needed for one hectare. The seeds are first soaked for half an hour in Streptocycline solution (250mg/5 litre) and then after drying in shade, treated with Captan or Thiram @ 2-3 g per kg of seeds before sowing. Soaking the seeds in 0.1% boric acid before sowing enhance yield. Seeds are sown at shallow depth at 5.0 cm apart, 1-1.5cm depth and covered with finely sieved leaf mould or vermicompost.

After sowing, the beds are covered with straw and water is sprinkled regularly till seeds are germinated. Seedlings may be raised under poly tunnels to protect the young seedlings from heavy rain or scorching sunlight. The seedlings are watered daily or in alternate day with rose can. Hardening of the seedlings can be done by withholding water at least 4 to 6 days before transplanting. Stocky, 30-35 days old, healthy seedlings of 6-8 cm height with 4-6 leaves is ideal for transplanting.

Sowing time

The time of sowing depends on the variety, season and location. Sowing is generally done from mid July to first week of August if the early varieties are grown in mid hill region and the crop matures by October. In high hill situation sowing of the early varieties is done during March – April which matures by July-August. The early varieties are sown in mid September for growing in the plains. The mid season varieties are sown from first week of September to last week of October which matures by December – January in plain condition. The late varieties are sown up to first week of November (suitable for the plain of northern India) which matures by February- March.

Land preparation

The land is prepared well in advance with repeated ploughings (at least 4-5 ploughing) to a fine tilth. All stubbles, weeds, etc. are removed and the land is leveled. Well rotten organic manure preferably farm yard manure or compost @ 25-30 t/ha should be applied during land preparation followed by proper levelling of the land. The land should have adequate drainage facility particularly for the early varieties. Planting should be done on ridges to avoid water logging problem.

Transplanting

Healthy seedlings of 28-30 days old are ready for planting. After hardening the seedlings for 4 – 6 days, the seed bed is irrigated before lifting to facilitate easy pulling of young seedlings. The seedlings should be planted in the field preferably during afternoon hours for better field establishment and the field is irrigated just after transplanting. One hectare land can accommodate 40-45 thousand seedlings. The field should be irrigated just after transplanting. Planting in ridges is ideal. Spraying the 15 days old seedlings with the starter solution of ammonium sulphate (50g/10litres of water) is helpful for better establishment. Spraying the seedlings with 0.1ppm IBA or 10ppm NAA gives better establishment, enhance vegetative growth and ensure more yield.

Spacing

The optimum spacing for broccoli in the main field is 45 cm x 45 cm. In very rich soils, spacing can be reduced to 45 cm x 30 cm to avoid stem hollowness due to rapid plant growth. At a wider spacing, plants produce more laterals. The closer spacing is preferred for mechanical harvesting of the central head. However, closer spacing delays maturity.

Nutrient management

Use of optimum doses of fertilizers is important for its proper growth since both rapid and slow growth is undesirable. The bud clusters become loose and hollow-stem results from rapid growth however, slow growth affect yield adversely. A general fertilizer dose of 100-125 kg N, 80 P_2O_5 and 50 kg K_2O per hectare is recommended for open pollinated improved varieties. The doses differ from place-to-place depending upon the fertility status of the soil. The full dose of P, K and half of N are applied at the time of preparation of land. The remaining dose of N should be top dressed in 2 equal split doses. The first is applied 4–5 weeks after transplanting, whereas second before head formation. A high yield of side shoots can be obtained by liberal use of N after harvesting central bud cluster. Micronutrient requirement of broccoli is fairly high. Molybdenum and Boron may be supplied by soil application or foliar sprays. Soils with pH above 7.0 affect the availability of boron. Three foliar sprays of 0.3% borax after 20, 35 and 50 days after transplanting is beneficial in increasing yield and checking browning disorder.

Irrigation

Broccoli requires sufficient moisture in the soil for uniform and continuous growth. Frequent irrigation at 10-15 days interval is generally given depending on weather conditions. The dry conditions adversely affect the quality and yield of shoots by being more fibrous. Water logging conditions should always be avoided as it depresses plant growth. It is better to follow furrow system of irrigation.

Interculture and weed control

The crop should be kept weed free in early stage. Hoeing is necessary in order to provide good aeration as well as to make the plot weed–free before irrigation. Since it is a shallow rooted crop, hoeing should not be done beyond the depth of 5-6 cm close to the plant to avoid injuries to the roots. A light earthing–up at final hoeing is beneficial.

Harvesting the crop

The heads of broccoli should be harvested iimediately after attainment of proper size when the bud clusters are compact and the buds have not yet been opened. Usually, only the central head along with 15 cm long fleshy stem is harvested. Generally harvesting continues for 4-6 weeks. After cutting, part of the foliage is removed from the harvested shoots. The head may be 15-20 cm in diameter and weighs 250-600 g depending on the vatiety.

Post harvest handling and storage

After harvesting, the heads should be immediately sorted, graded, packed in baskets and sent to markets. The side sprouts or spears are tied together after harvesting to give them a shape of single head before marketing. A high rate of respiration results in quick deterioration of its quality. They should be cooled at 4°C and then packed with ice in crates and stored in refrigeration. They can be stored well for 7-10 days at 4°C.

Yield

Yield varies greatly with the the cultivars, maturity period and method of harvest. Open pollinated varieties generally give 10-15 t/ha while yield of the hybrids may be 18-20 t/ha.

Plant protection measures of cauliflower, cabbage and broccoli

Physiological disorder

Browning

It is a boron deficiency disorder and cauliflower is most sensitive to this micro-nutrient deficiency disorder followed by broccoli. In cauliflower and broccoli, water–soaked lesions first appear in the stem, leaf and on the surface of the curd and head which later turn rusty brown in color. Browning of curd is sometimes associated with hollow stem symptom. The affected curd or head is bitter both in fresh and cooked condition.

Cabbage is less sensitive to this boron deficiency disorder in which water–soaked lesions first appear in the stem and leaf. Browning of the stem also occurs along with thickening and brittleness of leaves and top portion of the seedlings.

Control measures

- Correction of soil reaction and salinity.
- Soil application of borax @ 10-15kg/ha once in two years in deficient soil.
- Three sprayings of 0.25-0.50% borax solution @ 1-2kg/ha along with a sticker 20, 35 and 50 days after transplanting.

Whip tail

It is a molybdenum deficiency disorder particularly in the soil pH below 5.5. In cauliflower and broccoli, young plants become chlorotic and may turn white, particularly along the leaf margins and finally become cupped and wither. Leaf blade of the older plants does not develop properly and may be strap like and severely savoyed. In extreme cases only mid rib develops. Growing point of the plant is usually deformed and no marketable curds or heads are produced. In cabbage, growing point gets distorted and leaf area is reduced.

Control measures

- Correction of the soil pH upto 6.5 by proper liming
- Soil application of 1.5kg/ha sodium or ammonium molybdate mixed with fertilizer or irrigation water when the plants are set in field.
- Three sprayings of 0.1% ammonium molybdate solution along with a sticker 15, 30 and 45 days after transplanting.

Buttoning

It is a disorder of cauliflower in which the curd does not develop to full size and remain as a button. It is caused due to i) nitrogen deficiency, ii) planting with over aged seedlings, iii) planting early variety late in the season and iv) exposure of the seedlings to poor light conditions

Control measures

- Right selection of the varieties for different temperature regimes.
- Transplanting 35-40 days old seedlings.
- Supply of adequate nutrients.

Blindness

It is also a disorder of cauliflower in which curds are not formed due to distortion of terminal bud and only large, dark green, thick and leathery leaves develop in these plants. This disorder is caused due to i) mechanical or insect injury of the terminal bud, ii) planting late variety early in the season and iii) exposure of the seedlings to very low temperature.

Control measures

- Careful handling of the plants and good protection against insects
- Avoidance of very low temperature exposure

Riceyness

It is a disorder of cauliflower and marked by velvety or granular appearance on the surface and premature initiation of the flower bud. This disorder is caused due to i) exposure to the temperature higher or lower than the optimum required for particular variety in curd development stage, ii) temperature fluctuation during curd development and iii) poor seed stock.

Control measures

- Selection of proper variety for the particular growing season.
- Use of good seed stock and growing under favourable weather condition.

Diseases

Damping off (*Pythium*, *Rhizoctonia*, *Phytophthora*, *Fusarium*, *Sclerotium*, *Botrytis*, *Sclerotinia*)

It is very destructive fungal disease of cauliflower, cabbage and broccoli in nursery beds, especially during early season. In pre-emergence damping off, the growing points are killed in the initial stages of seed germination before they come out through the soil. In post-emergence damping off, the seedlings topple over the ground due to collar rotting and rapid shrinking of the cortical tissue of the hypocotyls.

Control measures

- Hot water treatment of seeds at 50^0 for 30 minutes.
- Treating the seeds with Thiram/ Captan @ 3.0g /kg of seeds.
- Drenching the nursery beds 7 days before sowing with Thiram/ Captan or any copper fungicide @ 3g / litre of water.
- Covering the beds with transparent polythene sheets before sowing and left to open sun for at least15 days, known as soil solarization.
- Spraying the young seedlings with 0.2% Blitox or Carbendazim (1g) + Mancozeb (2g) per litre of water.
- Provision of proper drainage to the beds.
- Removal of the affected seedlings from beds as soon as the symptoms are visible.
- Avoidance of flooding the beds to check spread of the disease.

Downy mildew (*Peronospora parasitica*)

It is fungal disease of cauliflower, cabbage and broccoli where discoloration occurs in the young seedlings and in severe cases the whole plants die. In mature plants, purplish or yellow-brown spots appear on the upper surface of the leaf and white fluffy downy growth occurs on the lower leaf surface.

Control measures

- Hot water treatment of seeds at 50^0 for 30 minutes.
- Crop rotation excluding cruciferous crops.
- Spraying the crop with 0.25% Mancozeb, Metalaxyl-Mancozeb (0.25%) and Cymoxanil-Mancozeb (0.25%) with a sticker at weekly intervals.

Yellowing (*Fusarium oxysporum* f. sp. *conglutinans*)

It is a serious fungal disease of particularly cauliflower and cabbage where the fungi attack the seedlings through the rootlets. Yellowing starts from the lower leaves and progress towards the upper leaves very fast. Leaves drop prematurely and the plants remain stunted.

Control measures

- Application of adequate organic matter in the soil.
- Application of lime to raise the soil pH to 7.2
- Seed treatment with *Trichoderma viride* @ 5g/ kg of seeds.
- Treating the seeds with Thiram/ Captan /kg of seeds.
- Spraying the crop with 0.1% Bavistin at 10 days interval

Curd blight (*Alternaria brassicae*)

It is primarily a seed borne fungal disease of cauliflower where small, brown to black circular to slightly elongated spots appear on the leaves, petioles and stem and later blight patches occur on the curd.

Control measures

- Treating the seeds with Thiram/ Captan @ 3.0g /kg of seeds.
- Spraying the crop with Copper-oxychloride (0.3%) or Mancozeb (0.25 %)/ Chlorothalonil (0.2%) or Difenconazole (0.5g/l) at 10 days interval.

Black leg (*Phoma lingam*)

It is a seed borne fungal disease of particularly cabbage which attacks mostly the seedlings. Seedlings topple over due to distortion of the vascular bundle and root system. Stem of the affected plants when splits open vertically, shows severe black discoloration of the vascular bundles.

Control measures

- Hot water treatment of seeds at 50°C for 30 minutes.
- Seed treatment with Thirum or Captan @ 3g/kg of seeds.
- Spraying the plants with 0.2% Captan or Chlorothalonil (0.2%) or Thiophanate-methyl (0.1%)

Club root (*Plasmodiophora brassicae*)

It is a soil borne fungus of particularly cabbage which causes swelling of roots and rootlets like the shape of a club and xylem vessels are plugged. It declines the vigour and imparts pale green coloration to the plants. The plants tend to wilt in hot sunny condition.

Control measures

- Solarization of the field during summer by ploughing and covering with transparent polythene film.
- Liming the soil to raise the soil pH (7.2) as the disease appears more in acidic soil condition.
- Disinfection of the seed bed with 4% formalehyde.
- Seed treatment with Thiram @ 3g/kg of seeds.
- Seedling root treatment with 4% Calomel (mercurous chloride) before transplanting.

Black rot (*Xanthomonous campestris* pv. *campestris*)

It is a seed borne bacterial disease of cabbage, cauliflower and broccoli. The bacteria infect the plants in the nursery as well as in the main field. The infected plants become stunted, cotyledonary leaves turn yellow to black and drop prematurely which results in killing of the seedlings. In the main field, first sign of the disease appears near the leaf margin, characterized by chlorosis which progress towards the centre of leaf blades in V – shaped lesions. In the affected portion, veins and veinlets turn brown and finally black. In severe cases leaves wither, curds become discoloured and subsequently soft rot may set in.

Control measures

- Growing of resistant varieties of cauliflower like, Pusa Subhra, Pusa Snowball K–1.
- Seed treatment with hot water at 50-52°C for 30 minutes followed by a dip in 0.01% streptocycline solution for 30 minutes.

- Spraying the crop with 0.01% streptocycline or Agrimycin-100 at transplanting and curd initiation stage.
- Crop rotation excluding cruciferous crops.
- Spraying the crop with Copper-oxychloride (0.3%) + Streptomycin sulphate (100 ppm) for management of black rot. Micronutrient (B+Mo) spray is beneficial after application of antibiotics.

Soft rot (*Erwinia carotovora*)

This bacterial rot occurs after the occurrence of black rot or after mechanical injury to the matured curd. Water soaked patches appear on the upper surface of the curd which expands and cover the whole curd. The water soaked portion turns to light brown and ultimately dark brown within a week. Sticker and spreader need to be added with spray mixture for spraying on cabbage and cauliflower.

Control measures

- Spraying the crop with Copper-oxychloride (0.3%) + Streptomycin sulphate (100 ppm) for management of soft rot.

Insect pests

Cabbage borer (*Hellula undalis*)

This insect pest attacks all the cole crops. The caterpillars after hatching mine into the leaves along the side veins and make it a white papery structure filled with their excreta and later they bore into the curd and head. The crop loss potentiality is very high in heavily infested crops.

Control measures

- Collection and destruction of larvae at early invading population by wire devices is one of the best ways to reduce the destructive form at later stage of crop.
- Application of DDVP @ 0.75 ml/l may control the pest.

Diamond back moth (*Plutella xylostella*)

This insect pest causes serious damage to cabbage followed by cauliflower. The caterpillars feed on lower side of the leaves producing typical whitish patches and later they bore small holes in the leaves giving a shot hole appearance all over.

Control measures

- Picking and destruction of the larvae at the early stages of the crop.
- Growing mustard as a trap crop in the field.
- Spraying the crop with 1.25 ml Profenophos or 1.0 ml Cypermethrin per litre of water.

Aphid (*Lipaphis erysimi*)

The aphid species attack all the cruciferous crops during seedling as well as later stage of crop growth. Severe infestation reduces the vigour of the crop due to their continuous sucking.

- Encouraging the population growth of lady bird beetles and hover fly in the field.
- Spraying the crop with 0.03% phosphamidon or 0.05% Malathion or 0.03% Methyl demetion or neem-based pesticides along with sticker to control the pests.

Cucurbits

- **Watermelon**
- **Muskmelon**
- **Bottle gourd**
- **Ridge gourd**
- **Pointed gourd**
- **Cucumber**
- **Pumpkin**
- **Bitter gourd**
- **Wax gourd**

Watermelon

Watermelon, *Citrullus lanatus* under the family Cucurbitaceae (Chromosome number: 2n = 22) is an important and common summer season crop in India. China is the largest producer growing over two thirds of the world's watermelons and other major producers include Turkey, USA and Iran. In India, Punjab, Haryana, Karnataka, Assam, West Bengal, Odisha, Himachal Pradesh, Uttar Pradesh, Tamil Nadu and Rajasthan are the major watermelon growing states. It is a major river-bed crop of Uttar Pradesh, Rajasthan, Gujarat, Maharashtra and Andhra Pradesh.

The crop

It is a trailing annual with stems, as long as 400 cm. The roots are shallow (40-50 cm) and extensive (60-90 cm), with taproot and many lateral roots. Leaves are ovate (oval but broader towards the base) to obovate (oval but broader at the apex), 8-20 cm long, scabrid (rough to touch) and deeply pinnatified. The lobes are pinnately divided into three or four pairs of lobes. The stems are hairy, rounded to angular in cross-section, and have branched tendrils at each node. The tendrils are pinnately divided into three or four pairs of lobes.

The plant bears separate staminate and pistillate flowers on the same plant (monoecious). The flowers are unisexual, solitary, axillary and pedicillate. The pedicel of male flower is slender and 12-30 cm long; the female flower pedicel is slightly stouter. Flowers are small, 4-5 cm across, sulphur yellow in colour, and less showy than other cucurbits.

The fruit, botanically called "pepo" consists of a firm outer rind, a layer of white inner rind flesh of 0.5-1.4 cm thick, and an interior coloured edible pulp embedded with seeds. The edible part of the fruit is the endocarp (placenta), which contrasts with muskmelon (*Cucumis melo*), where the edible part of the fruit is the mesocarp. The fruit is round to oblong. The skin is smooth with colour ranging from yellow, orange, light green to almost black, either solid or striped with a paler green or marbled. The flesh may be white, cream yellow, pale red, red or dark red. In Asia, smaller fruit in the range of 1-4 kg is popular. Fruit rind varies from thin to thick and from brittle to tough.

The seeds are embedded in the edible pulp, small in size (4-6 cm), long and flattened. Seed colour can be white, tan, brown, black, red, green or mottled.

Importance and use

Watermelon is primarily consumed fresh for their sweet and juicy fruits and are often eaten as desserts. The fruit can be prepared and eaten in many ways like, drinks, appetizers and dessert. It is relished as an excellent dessert fruit. The fruit juice makes an excellent refreshing and cooling beverage after adding a pinch of salt and black pepper.

The rind can be used as aphrodisiac and may be consumed in pickled or candied form. The seeds are also used as a source of oilseed, eaten as snack or can be added to other dishes. Roasted seeds of this crop are eaten throughout Asia and the Middle East, and watermelon seeds are ground into flour and baked as bread in some parts of India. In some African cuisines, however, immature fruits are served as a cooked vegetable. In parts of the former Soviet Union and elsewhere watermelon juice is fermented into an alcoholic beverage. Watermelons are also sometimes used as feed for livestock.

Nutritional value

Watermelons are mostly water - about 92 percent, but this refreshing fruit is soaked with nutrients. Each juicy bite has significant levels of vitamins A, B6 and C, lots of lycopene, antioxidants and amino acids. There's even a modest amount of potassium. Fruit is a rich source of lycopene, the powerful antioxidant, which is the scavenger of free radicals. In fact, today watermelon is top of the lycopene list much ahead of tomato.

Medicinal valsue

Fruit is used in cooling, strengthening, aphrodisiac, astringent to the bowels, indigestible, expectorant, diuretic, and stomachic. It purifies the blood, allays thirst, cures biliousness, well for sore eyes, scabies and itches. The seeds are tonic to the brain.

Origin and taxonomy

Watermelon was originally domesticated in central and southern Africa region. Watermelon emerged as an important cultigen in northern Africa and south-western Asia prior to 6,000 years ago. Archaeological data suggest that they were cultivated in ancient Egypt more than 5,000 years ago, where representations of watermelons appeared on wall paintings and watermelon seeds and leaves were deposited in Egyptian tombs. Watermelons are the important part of the "most widespread and characteristic African agricultural complex adapted to savanna zones" in that they are not only a food plant but also a vital source of water in arid regions. Indeed, watermelons refer to as "botanical canteens." From their African origins, watermelons spread *via* trade routes throughout much of the world, reaching India by 800 and China by 1100 A.D. In both of these countries, as in Africa, the seeds are eaten and crushed for their edible oils. Watermelons became widely distributed along Mediterranean trade routes and were introduced into southern Europe by the Moorish conquerors of Spain, who left evidence of watermelon cultivation at Cordoba in 961 and Seville in 1158 A.D. Then, watermelons spread slowly into other parts of Europe, perhaps largely because the summers are not generally hot enough for good yields. However, they began appearing in European herbals before 1600 A.D. Their first recorded appearance in Great Britain dated to 1597A.D. and by 1625 A.D.

Watermelons reached the New World with European colonists and African slaves. Spanish settlers were producing watermelons in Florida by 1576 A.D. and by 1650 A.D., they were common in Panama, Peru, and Brazil, as well as in British and Dutch colonies throughout the New World. The first recorded cultivation in British colonial North America dates to 1629 A.D. in Massachusetts. Like cucumbers and melons, watermelons spread very rapidly among Native American groups. Prior to the beginning of the seventeenth century, they were being grown by the tribes in the Ocmulgee region of Georgia, the Conchos nation of the Rio Grande valley, the Zuni and other Pueblo peoples of the Southwest, as well as by the Huron of eastern Canada and groups from the Great Lakes region. By the mid-seventeenth century, native Americans were cultivating them in Florida and the Mississippi valley, and in the eighteenth and early nineteenth centuries the western Apache of east-central and southeastern Arizona were producing maize and European-introduced crops including watermelons as they combined small-scale horticulture with hunting and gathering in a low rainfall environment. This fact is ethnographically significant because other transitional foraging -farming groups, such as the San people of the Kalahari Desert of southern Africa had parallel subsistence practices involving watermelons. Watermelons and muskmelons were also rapidly adopted by Pacific Islanders in Hawaii and elsewhere as soon as the seeds were introduced by Captain James Cook (1778 A.D.) and other European explorers.

The important role of watermelons in American popular culture was in numerous areas including-folk art, literature, advertising and merchandising and the large number of annual summer watermelon festivals throughout the country with "parades, watermelon-eating contests, seed spitting contests, watermelon queens, sports events, and plenty of food and music". Growing and exhibiting large watermelons is an active pastime in some rural areas of the southern United States. Closely guarded family "secrets" for producing large watermelons and seeds from previous large fruit are carefully maintained. According to The Guinness Book of Records, the largest recorded watermelon in the United States was grown by B. Carson of Arrington, Tennessee, in 1990 which weighed a phenomenal 119 kg.

Watermelon has 22 chromosomes in the diploid form. The genus *Citrullus* belongs to the sub-tribe *Benincasinae*. Other members of similar genera in Cucurbitaceae are *Acanthosicyos* and *Eureiandra* whilst those with 22 chromosomes include *Gymnopetalum*, *Lagenaria, Momordica, Trichosanthes* and *Melothria*. The genus *Citrullis* has now been revised to include *C. lanatus, C. ecirrhosus, C. colocynthis*, and *C. rehmii*. It was originally thought that the wild-growing colocynth (*Citrullus colocynthus*), also known as bitter apple the ancestor of the watermelon. But DNA analysis from chloroplasts appears to indicate that the *Citrullus colocynthus* (colocynth) and *Citrullus lanatus* (watermelon) evolved from a common ancestor found in Namibia, *Citrullus ecirrhosus*.

Adaptation in India

The crop has reached India in prehistoric times which resulted evolution of extensive variability and this designated India as the secondary centre of diversity of this crop. Cultivation of this crop began in ancient Egypt and India and then spread to different countries *via* Mediterranean, Near East and Asia. As a result of prolonged cultivation and selection, new forms of table watermelon have evolved. The varieties grown today have little resemblance to the ancestral African forms. Watermelon cultivars grown in India comprise different locally adapted types under region specific names, improved varieties developed either through selection over indigenous types or combination breeding approaches and different introduced varieties.

Improved varieties

Open pollinated varieties : Ashahi Yamato, Improved Shipper, New Hampshire Midget, Dixielce, Charleston Grey, Durgapura Meetha, Durgapura Kesar, Sugar Baby, Arka Manik, Fuken, PKM-1, Special No.-1.

Public sector Hybrids : Durgapura Lal, Arka Jyothi.

Cultural requirements

Watermelon is a warm season crop and cannot withstand frost or very low temperatures. It requires a long growing season in the subtropics, but fast growing in the tropical regions. Flowering and fruit development are promoted by high light intensity and high temperature. Hot, dry climate and a long growing season preferably with warmer days and cooler nights is the best for growing watermelon. For seed germination, optimum moisture and a soil temperature between 25°–30°C is ideal. Plant growth is optimum under 28°–30°C, while fruiting is better at 24°–27°C. Higher temperatures are beneficial during ripening. Arid regions of Rajasthan, Haryana, and Punjab are best suited for production of quality fruits. Watermelon may be grown on a wide variety of soils. Sandy loams are best for early crop, while loams have high-yielding potential. Alluvial river-beds are also good for watermelon. Heavier soils do not permit perfect root growth and hence only short duration varieties with smaller fruits are suitable. The soil should be well-drained and should have ample organic matter. It is generally cultivated in river-beds by making trenches and sowing in hills or pits. Soil with a pH of 6.5–7.0 is ideal.

Land preparation

The field is prepared with 4-5 deep ploughings to make into a fine tilth before seed sowing. All the weeds, stubbles are removed and the land is levelled by laddering. Manures should be mixed with the soil at the time of last ploughing. Generally furrows are opened at 2.0 to 3.0 m spacing and sowing is done on the top of the sides of furrows and the vines are allowed to trail on the ground. The crop is also grown on flat or raised beds, about 2.0 to 2.5 m wide and seeds are sown on both sides of the bed. In rainy season, raised beds or mounds are prepared to facilitate drainage. Mounds in the furrow, flat bed method or in river bed system of planting are prepared with 15-20 t FYM/ha and basal dose of fertilizers mixed with the soil. Growing watermelon with polyethylene mulching is the best.

Sowing time

Sowing time is late-February to mid-March in north Indian plains; November–January in north-eastern and western India; mid-November in West Bengal; December–January in central and south India and August–September in Rajasthan. However, good quality hybrids can be grown any time between February and September in North and Eastern India with polyethylene mulch.

Seed sowing

Seeds are sown by hand dibbling in the mound. Sowings can be done following flat bed or furrow method or in river bed system depending on the season and convenience. The seed requirement for planting depends on spacing between plants and rows which again depend on vining habit of the variety/ hybrid. In general, a seed rate of 3.0–3.5kg for small-seeded types and 5.0kg for large-seeded types is sufficient for one hectare. Seedlings can be transplanted with ball of earth after raising seedlings in polyethylene bags. For transplanting, 1–2 seeds/ bag are sown in perforated polyethylene bags of 150–200 micron thickness and 8cm × 10cm size. Seedlings at 2–3 leaves stage with ball of earth should be transplanted with care so that roots are not disturbed. Seeds should be soaked overnight and kept at moist warm place for 48 hour for initiation of germination. Pre-germinated seeds could be sown in hills on the raised sides of furrows in upland and in trenches or in pits in the river-beds. Two plants/ hill in trenches or furrows and 4 plants/ pit are retained for growing. In pit sowing, pits of 60cm × 60cm size are dug up and filled with soil mixed with sufficient quantity (equal) of organic manure and fertilizer mixture before sowing.

Spacing

Spacing is 2.0-3.0 m x 90-150 cm for spring-summer crop; 2.0-3.0 m x 100-150 cm for river bed system of cultivation and 2.0-2.5 m x 60-90 cm for early autumn crop.

Nutrient management

Watermelon responds well to manuring and fertilizer application. The dose of manures and fertilizers depend on the soil type, climate and system of cultivation. Well rotten farm yard manure at the rate of 15-20 t/ha should be applied at the time of land preparation. A general fertilizer dose of 100 kg N, 60 kg P_2O_5 and 60 kg K_2O per hectare is recommended for the open pollinated improved varieties. Half N along with full P and K fertilizers and farm yard manure should be given as basal and rest N should be top dressed in two split doses at the time of vining (25-30 days after sowing) and at blooming stages (40 – 45 days after sowing). A general fertilizer dose of 200 kg N, 100 kg P_2O_5 and 100kg K_2O per hectare is recommended for the hybrids. One-fourth N along with full P and K fertilizers and farm yard manure should be given as basal and rest N should be top dressed in three split doses at the time of vining (25 - 30 days after sowing), at blooming (40-45 days after sowing) and after first harvest stages (80-85days after sowing).

Crop regulation

In general, high N nutrition under high temperature condition promotes staminate flowers and lower the number of pistillate flowers/ vine, resulting in low fruits-set and yield. The best way to control the vine growth within reasonable limits is by adjusting nitrogen fertilizer doses and frequency of irrigation and pruning. If apical shoot is pinched and 2–4 side shoots are allowed to grow, it gives significantly higher fruit yield than un-pruned plants. In archway system of training, plants are spaced at 45 cm to accommodate 9,500 plants/ha, whereas in inclined cordon system, plants are spaced at 54 cm in rows alternately at 2.14 m and 76 cm bed centre. In vertical cordon system, plants are spaced 54 cm apart in rows at 1.52m to accommodate 12,000 plants/ha. Application of different growth regulators, secondary and micro-nutrients viz., 50 ppm 2,3,5 tri-iodobenzoic acid (TIBA), 50 ppm maleic hydrazide (MH), and 2 ppm boron twice, first at two true leaves of the plants i.e. 15-20 days after sowing and subsequently repeated after 7-10 days helps in increasing the number of pistillate flowers/plant and thereby fruit yield.

Irrigation

Interval and amount of irrigation depends upon the season and type of soils. Soil moisture is important for good germination. Watermelon responds very much to irrigation but it cannot withstand waterlogged conditions. It is generally cultivated as a spring-summer crop and irrigation is very important and the crop should be irrigated at 10-15 days intervals during summer season. Frequent irrigations also promote excessive vegetative growth, especially in heavy soils. Soil moisture stress during pre-flowering, flowering and fruit development stages drastically reduces yield. However, irrigation should be checked during ripening of the fruits as it adversely affects fruit quality and promotes fruit cracking. Water application should be restricted to the base of the plant and root zone so that water may not wet the vines, especially when flowering, fruit-set and fruit development is in progress. Frequent wetting of leaves, stems and developing fruits promote disease incidence. River-bed crop needs watering only in initial stages of crop growth. Later when roots go deep up to the water table, no irrigation is required.

Interculture and weed control

Watermelon does not require much attention on interculture operation although it is sensitive to weeds in initial stages of plant growth. Yield losses up to 30% have been observed due to weeds. In early stage, the beds and ridges should be kept weed-free. Intercultural operations need to be started 15–20 days after sowing. Depending upon soil and environmental factors, 2–3 weedings would be required. At the time of topdressing of nitrogenous fertilizer, weeding and earthing-up are done. When the vines start spreading, weeding in between rows or ridges is neither necessary nor feasible since vine growth can smother the weeds. Vigorously growing weeds should be manually pulled out, without disturbing vines at later stages. Pre-emergence herbicidal treatments of Butachlor @ 2.0 kg a.i./ha and Trifluralin @ 1–2 kg a.i./ha are effective.

Harvesting the crop

Watermelon should be harvested at proper stage of maturity. Irrigation in preceding one week of harvest may give high turgidity to fruits and the skin of the fruit may crack while transporting them and hence irrigation before harvest should be avoided. Size of fruits and colour of the skin are not the good indicators to know the proper stage of fruit maturity. The crop is ready for harvesting 90–120 days after sowing depending upon cultivar and season. Change in the colour of the portion of the fruit which rests on the ground from white to creamy in case of light skin colour, or to yellow in case of dark skin colour is a useful guide of maturity. A metallic sound when the fruit is tapped with the back of hand or with fingers denotes immaturity, whereas a heavy dull sound indicates ripeness. The drying of tendril at the base of the fruit is also a sign of maturity. However, the knack of recognizing a ripe fruit comes with experience. The fruits generally become fit for harvesting 30–40 days after anthesis. Fruits should be separated from the vine with the help of a knife.

Post harvest handling and storage

Improper harvesting, handling, transportation and distribution result in significant losses. Jerk to fruits while loading and transport should be avoided. They can be shipped to long distances, even without refrigeration. Flesh colour, sugar content and flavour do not improve/deteriorate appreciably after harvesting but there is a gradual softening or granulation of pulp around the seeds.

Watermelons generally keep well for 1–3 weeks at 2.2°–4.40°C and 80% relative humidity. Shilf of fruits may be reduced when exposed to ethylene gas prodncing fruits such as apple, stone fruits, grapes, etc.

Yield

Fruit yield may vary between 25-30 tonnes/hectare for open pollinated improved varieties and 50-60 t/ha for the hybrids.

Cucumber

Cucumber, *Cucumis sativus* under the family Cucurbitaceae (Chromosome No: 2n = 14) is one of the oldest cultivated vegetable crops in India. *Cucumis* in Latin means 'cucumber', the word was derived from Greek for cucumber, *kykyon*. The epithet *sativus*, also Latin, means 'that is sown'. Cucumber is the second most widely grown cultivated cucurbit in the world after watermelon. In the temperate countries it is extensively grown in the green house. It is grown throughout the world to be consumed as fresh fruits, as slicing cucumber, and as pickles in immature stage. In 2014, world production of cucumbers and gherkins was 75 million tonnes, led by China with 76% of the total followed by Turkey, Russia, Iran and Ukraine.

It is one of the important and popular vegetable crops extensively grown during spring-summer, summer-rainy and early autumn seasons in all parts of India including the hills of the southern and northern regions. Gherkin (*Cucumis anguria*) and small sized pickling cucumbers have become important processed quality vegetables for export from Karnataka, Andhra Pradesh and Tamil Nadu.

The crop

The plant is a morphologically variable annual herbaceous climber. The stems are prostrate, angular, and covered in white pubescence. Stipules are absent, and the plant bears unbranched axillary tendrils up to 30cm long. The pubescent leaves are alternately arranged on 10-16 cm long petioles, simple, basally cordate, and apically acute with 3-7 palmate lobes. The palmately-veined leaves are nearly orbicular, 7-20cm long and broad. The vine use their simple tendrils to climb over structures or other vegetation.

It is basically monoecious in sex form and its axillary flowers are actinomorphic and rarely bisexual. Both staminate and pistillate flowers have a pubescent 5-parted calyx composed of 0.5-1cm long white pubescent sepals. The sepals are long, narrow, and acute; on pistillate flowers, the calyx is fused to the ovary and forms a hypanthium. The corolla is approximately 2cm long, yellow, fused less than half of its length, campanulate, 5-parted with oblong to lanceolate lobes. Staminate flowers are solitary or 3-7 on pubescent pedicels (0.5-2cm long), bearing 3 stamens, of which two have 2-celled anthers (0.3-0.4cm long) and one has 1-celled anther. Pistillate flowers are solitary or paired on pedicels shorter than the staminate (< 0.5cm long) before fruit development, whereas during fruit development the pedicels can elongate up to 5cm long. The pistilate flowers have 3 staminodes, a pubescent, ellipsoid, unilocular ovary 2-5cm long, with a short style and 3 stigmas. The pistillate flowers are pollinated by bees that are attracted by the nectar produced by the flowers; honey bees are more effective than wild bees. The pollen is sticky, which makes wind pollination improbable.

Botanically, the fruit is classified as a pepo with no internal divisions. The fruit is indehiscent and cylindrical with many seeds. The fruits are glabrous, and can be smooth or warty, yellow or green, ranging from 5-100cm long and weigh 50g to 4kg.

Cucumis melo, also known as melon, has leaves that are apically round, and the fruits are smooth to reticulate and ovate to oblong in shape. In comparison, *C. sativus* has apically acute leaves, and the fruits are warty and cylindrical.

Depending on the variety, cucumbers may have a mild melon aroma and flavour, in part resulting from unsaturated aldehydes, such as (E,Z)-nona-2,6-dienal, and the *cis*- and *trans*- isomers of 2-nonenal. The slightly bitter taste of cucumber rind results from cucurbitacins.

Importance and use

Immature fruits are universally used as salad and for pickling. Immature fruits are also used with curd for the preparation of "Rayata". Mature fruits are also used as vegetable in India. Off-season cucumber particularly during winter is sold at high price. Cucumber particularly gherkin has great export potential to South-East Asia and Gulf and European countries. Varieties or hybrids of gherkin having green, small sized (7.5 cm x 2.4 cm) fruits of about 200 in number per kg is considered as premium quality. The green leaves are used as potherbs. Edible oil can be extracted from the seeds and used for cooking.

Nutritional value

Edible portion of cucumber contains 0.4% protein, 2.5% carbohydrates and 0.1% fat and 7.0 mg vitamin C per 100 g fresh edible portion. Tender fruits provide about 25 mg phosphorus, 10mg cacium and 1.5 mg iron per 100 g edible portion.

Medicinal value

The fruits and seeds have cooling effect. Fruits are good for people suffering form constipation, jaundice and indigestion. Fruits are demulcent and diuretic. Fruits contain photolytic enzymes, ascorbic acid oxidase, succinic and malic dehydrogenases. Fruits are also used as an astringent and antipyretic. The seeds can be used to expel parasitic worms. Seed oil is also used as antipyretic.

Mature uncooked cucumbers bring relief for individuals suffering from celiac disease and promote skin health. Immature cucumbers can be cooked and consumed to treat dysentery. The fruit is also valued in the cosmetic industry, used to soften the skin. A poultice made from fresh cucumbers can be applied to burns and open sores. The juice from the leaves induce vomiting and aid digestion.

Origin and taxonomy

The genus *Cucumis* comprises about 30 species distributed over two distinct geographic areas like South east of Himalayan group and African group. The Asiatic species which are mostly annuals are distributed in India, China, Myanmar and Korea and the African species which include annual, perennial, monoecious and dioecious are found in different parts of Africa, Middle East, Sudan, Egypt, Ethiopia and Pakistan. The melons belong to the *Cucumis* species of the African group and cucumber belongs to the *Cucumis* species of the Asiatic group. So, cucumber (*C. sativus*) and melon (*C. melo*) are considered to be separate and unrelated species.

Cucumber has been known in history for about 3000 years and was probably originated in India from where it seems to have spread eastwards to China and west words to Asia Minor, North Africa and Southern Europe and subsequently to entire Europe. It was introduced to the new world by Columbus who planted it in Haiti in 1494 and perhaps soon afterwards it was brought to the USA. It was grown by the ancient Greeks and Romans in about 300 B.C. Cucumber was cultivated almost throughout Europe in the middle ages and was common in France in the ninth century and in England in 1327 A.D. The other *Cucumis* species of the Asiatic group occurring in India as wild include *C. sativus* var. *hardwickii, C. callosus, C. hystrix* and *C. setosus*.

Cucumis sativus var. *hardwickii* R. (Alex.), found along the southern foothills of the Himalayas crosses readily with cultivated cucumber. This has led to conclusion that var. *hardwickii* is either a feral or progenitor form of the cultivated cucumber.

Cucumber, *Cucumis sativus* L. belongs to the family Cucurbitaceae, sub-family Cucurbitoideae, tribe Melothrieae and sub-tribe Cucumerinae. The cucumber along with other Indian *Cucumis* species are diploid (2n= 14) with a basic chromosome number, x = 7. The *Cucumis* species of the African group to which melons belong are also diloid (2n = 24) but with a different basic chromosome number (x = 12).

Adaptation in India

Cucumber is believed to be indigenous to India although it is never found in wild state in India. Mentioning of cucumber as Sanskrit language equivalents

"Urvaruka" in Rigveda and "Ervaruka" in Caraka Samhita signifies its usage by the Aryan in the Vedic period. Cucumber has wide distribution in India showing large genetic variability. It is grown as spring-summer, summer-rainy and early autumn crop in the field both as sole and mixed crop or grown in different riverbeds of the country. Huge genetic diversity in cucumber exists in India because of its probable primary gene centre in Indian subcontinent (Hindustani centre). Large numbers of indigenous cultivars are mostly grown in India under different names. A number of improved varieties have also been developed through purification and subsequent improvement over the indigenous cultivars.

Improved varieties

- **Slicing type varieties**: Japanese Long Green, Straight Eight, Poinsett, China, Himangi, Phule Shubhangi, Sheetal, Khira-90, Khira 75, CO-1, Punjab-1, Pusa Uday, Swarna Sheetal, Pant Khira-1, Kalyanpur Ageti, Solan Green, Kalyanpur Green, Gujarat Cucumber-1, Swarna Ageti,
- **Public sector hybrids:** Pusa Sanyog, AAUC-1, AAUC-2, Pant Sankar Khira-1, Solan Hybrid, PCUCH-3

Cultural requirements

Cucumber is basically a warm season crop but is successfully grown in tropical, sub-tropical and temperate regions. It prefers slightly milder temperature than watermelon and muskmelon. It prefers full sun exposure in warm and humid climates. It is very susceptible to frost. Optimum temperature range for the growth and development is 18°C-24°C. The minimum temperature should not go below 10 °C and maximum above 30°C. Seeds germinate well at 25°C day temperature. Low temperature causes stunting of growth and poor germination of seeds.

Sex ratio in monoecious form is sensitive to photoperiod and temperature condition. Short days, comparatively low-night temperature and high relative humidity increase the number of pistillate flowers in the vine. Long days and high temperature above 30°C increase the number of staminate flowers and reduce the number of pistillate flowers in the vine. Excess humidity promotes some serious diseases like, downey mildew and anthracnose and increases the attack of fruit fly.

It grows well on moist, well-drained soils rich in organic matter and prefers a soil pH around 7.0 to slightly alkaline in nature. Lighter soils that warm quickly in springs are usually used for early yields. In heavier soils vine growth is greater and fruits mature late. The soils should not crack in summer and should not be water-logged in rainy season. Cucumber cannot be grown in soil having pH below 5.5. Sandy river beds have moisture beneath and warm up quickly in spring and for this reason, cucumber survives low temperature periods of winter months in river-beds and produce early crops in spring. In riverbed, cucumber is grown usually as a mixed cropping. It is sown as relay crop just before digging of potatoes in late January or early February in North India.

Land preparation

The field is prepared with 4-5 deep ploughings to make into a fine tilth before seed sowing. All the weeds, stubbles are removed and the land is levelled by laddering. Manures should be mixed with the soil at the time of last ploughing. The plant has comparatively shallow root system and is susceptible to water logging so land should be well drained. Generally furrows are opened at 1.5 to 2.5 m spacing and sowing is done on the top of the sides of furrows at 60-90 cm spacing and the vines are allowed to trail on the ground, especially during summer. The crop is also grown on flat or raised beds, about 1.5 to 2.5 m wide and seed are sown on the periphery of the beds. In rainy season, raised beds or mounds are prepared to facilitate drainage. Mounds in the furrow, flat bed method or in river bed system of planting are prepared with 15-20 tonnes of FYM per hectare and basal dose of fertilizers mixed with the soil. In river bed system of sowing, trenches of 30 cm wide, 60 cm deep and of convenient length and spaced by 2.0 to 3.0 m are made followed by filling the trenches with farm yard manure and soil.

Sowing time

Cucumber grows successfully in tropical and subtropical climate hence, can be grown year round in the southern and central Indian states due to prevailing mild winter there. However, the crop is grown in three main seasons: spring-summer, early autumn and rainy season. Sowing is done in November-December for spring-summer crop in eastern plain; February for summer crop in eastern and northern plain; June-July for rainy season crop in eastern and northern plain; August – September for early autumn crop in eastern India; December-January for winter – spring crop in southern and western regions; November-January for river bed cultivation in northern India and April-May for hill regions.

Seed sowing

Cucumber is a seed propagated crop and *in situ* sowing is generally practiced. Seeds are sown by hand dibbling in the mound or bed. Sowings can be done following flat bed or furrow method or in river bed system depending on the season and convenience. In river bed system, pre-germinated seeds are sown in the mound. Seed soaking in water for 12 hours before sowing improves germination. Seedlings raised in plastic bags (15cm x10 cm) particularly under polyethylene shed in winter month can be transplanted at 2-4 true leaf stage after 35 – 40 days with the advent of spring without disturbing the root system. This is basically practiced in north Indian condition during winter to get early harvest in spring. This seedling transplanting can be practiced in the hills to get early crop in July. 2.5 – 3.0 kg seed/ha is sufficient depending on the plant population. It is better to do seed sowing in pre-irrigated moist soil than irrigate the field after sowing. Pre-sowing soil application of 20-22 kg Furadon 3G helps in protecting the plants from root knot nematodes and other pests during initial 4-5 weeks.

Spacing

Spacing of 1.5-2.0 m x 60-90 cm is given for spring-summer crop; 2.0-3.0 m x 60-90 cm for river bed system; 2.0-2.5 m x 80-90 cm for rainy season crop; 1.5-2.0 m x 60-90 cm for early autumn crop and 2.0-2.5 m x 80-90 cm for the crop grown on bower.

Nutrient management

Cucumber is not considered as a high feeding crop. However, the dose of manures and fertilizers depend on the soil type, climate and system of cultivation. Well rotten farm yard manure at the rate of 15-20 t/ha should be applied at the time of land preparation. A general fertilizer dose of 100 kg N, 60 kg P_2O_5 and 60 kg K_2O per hectare is recommended for open pollinated improved varieties. Half N along with full P and K fertilizers and farm yard manure should be given as basal and rest N should be top dressed in two split doses at the time of vining (25-30 days after sowing) and at full blooming (50-55 days after sowing). A general fertilizer dose of of 150 kg N, 90 kg P_2O_5 and 90 kg K_2O per hectare is recommended for the hybrids. One-fourth N along with full P and K fertilizers and farm yard manure should be given as basal and rest N should be top dressed in three split doses at the time of vining (25-30 days after sowing), at full blooming (50-55 days after sowing) and after first harvest (70-75 days after sowing). Cucumber is a comparatively shallow rotted crop hence, responses well in top-dressing.

Crop regulation

Excessive vine growth due to high nitrogen nutrition coupled with high temperature and high soil moisture condition promote staminate flowers in the vine resulting low fruit set and low yield. The best way to control the vine growth within reasonable limits is by adjusting nitrogen fertilizer doses and frequency of irrigation. Application of different growth regulators *viz.,* 100-200 ppm Ethrel (1.0-1.5 ml/ l of water) twice, first at two true leaves of the plants i.e. 15 days after sowing and subsequently repeated 7 days after helps in increasing the number of pistillate flowers in the vine and subsequently yield. Growth regulators like GA @1500-2000 ppm and silver nitrate @ 200-300 ppm induce staminate flower in the gynoecious line which enables maintenance of the gynoecious line in pure form.

Irrigation

Pits, ridges or beds are lightly irrigated a day or two days prior to sowing. Preferably sprouted seeds are planted and subsequent irrigation is given 4-5 days after seed sowing. Irrigation is applied at the initiation of first true leaf during spring-summer and at its expansion stage during rainy season. Irrigation at regular interval of 4-5 days after is very important for spring –summer crop. Irrigation may not be necessary at all for rainy season crop, if rainfall is well distributed between July- September. The field is generally irrigated subsequently at 5-6 days intervals during summer and early autumn crop and as and when required for the rainy season crop depending on the rain. In light soils, the crop is irrigated at frequent intervals. Water stressed plants generally leach out large amount of amino acids and sugars from the roots during recovery upon irrigation. The crop must be irrigated, during its most critical stages i.e. flower bud development and early fruit development stages. Over irrigation during vegetative and early flowering stages may cause excessive vine growth resulting induction of more staminate flowers in the vine. Ridge and furrow method of irrigation is the best for cucumber. Drip irrigation gives more fruit yield than furrow irrigation.

Interculture and weed control

Plant grown on trellis gives better crop stand and fruit yield mainly because of better use of sunlight by maximum number of leaves and higher number of side branches. The crop needs to be trained over low trellis of 1.5 m high above the ground. Training the vines over the trellis or bower coupled with some pruning is very effective. Cucumber needs no interculture operations for successful cultivation. Deep intercultivation should be avoided as it may damage the shallow roots of the crop.

During the early stage of the crops, beds, ridges etc need to be kept free from weeds. At the time of top dressing with nitrogenous fertilizers, weeding and earthing up are done. Pruning all the primary branches after two nodes gives highest yield in green house. When the vines start spreading, weeding in between the rows or ridges become unnecessary since vine growth can smother the weeds. Application of the herbicides effectively controls weeds. Pre-emergence application of Fluchloralin @ 1.20kg a.i/ha or Fluchloralin 0.48 kg a.i. + Nitrofen @ 0.50kg a.i/ha checks weed population in the fields.

Harvesting the crop

Cucumber should be harvested at tender stage. Attainment of edible maturity is dependent on the type (slicing or pickling type) and varieties. Fruits of the pickling or slicing types generally reach optimum edible maturity stage within a week or slightly later of anthesis. If harvesting is delayed, dark green skin colour of the slicing type fruits turn brownish-yellow or russeting. In pickling type having pale-green fruit colour, turning to yellow indicate over-maturity. White spine colour is also a useful indication for edible maturity. Over matured fruits show carpel separation in transverse section of the fruits.

Post harvest handling and storage

Fruits damaged by summer rains, fruit fly and disease should be culled out before sending to the market. Crook neck and other misshapen fruits should also be separated from the normal fruits. Cucumber cannot stand much too long transportation. Cucumber can be stored for 2 week at 10.0-11.7°C temperature with 92% relative humidity in storage. Chilling injury as a physiological storage disorder occurs when the fruits are exposed to the temperature below 0°C for prolonged period. It is better to use the fruits immediately after removal from the storage.

Yield

Fruit yield may be 15 - 20t/ha for open pollinated varieties having monoecious sex form; 30-40 t/ha for the hybrids grown in open field and 60-70 t/ha for gynoecious and parthenocarpic hybrids grown in poly house.

Muskmelon

Muskmelon, *Cucumis melo* under the family Cucurbitaceae (Chromosome number: 2n = 24) is a highly relished crop because of its attractive flavour, sweet taste and refreshing effect. Muskmelon is so named because of the delighful odour of the ripe fruits. *Musk is* a Persian word for a kind of perfume; *melon is* French, from the Latin *melopepo,* meaning "apple-shaped melon" and derived from Greek words of similar meaning. Muskmelon encompasses different type of melons like netted, salmon flesh cantaloupe, smooth-skinned, green fleshed, Honey Dew, wrinkled-skinned, white fleshed, Golden Beauty and several other dessert melons. Muskmelon is one of the most economically important cucurbits, cultivated in many tropical, subtropical and temperate regions around the world. Winter production in parts of Africa (e.g. Sudan and Kenya) for export to northern Europe has increased its importance as a cash crop. It is a good cash crop in Asia and South American countries. The world production of melon in 2014 has been 29.6 million tonnes, with China accounting for 44% of the total. Other significant contributers are Iran, Egypt, Turkey, India, Spain, USA, and Mexico. In India, muskmelon is widely grown in Utter Pradesh, Rajasthan, Punjab, Bihar, some part of Andhra Pradesh, Tamil Nadu and Karnataka both under irrigated and riverbed cultivation. Depending upon local market preferences, several land races and local cultivars have established themselves in different geographical pockets and river beds of India.

The crop

Three important species of the genus *Cucumis* are distinguishable on the basis of following key characters :

Fruits spiny, muricate or echinate

- Leaves deeply lobed, fruits small, prickly lemon-yellow when mature: *C. anguria*
- Leaves shallowly lobed, fruits comparatively large, mostly cylindrical or oblong: *C. sativus*
- Fruits glabrous or pubescent, smooth or netted, mostly with sutures: *C. melo*

Muskmelon is a polymorphic species where most cultivars are andromonoecious (staminate and bisexual flowers on the same plant) however other predominant sex form is monoecious. Stem is soft-hairy to glabrous, striate or angled, leaves orbicular to ovate to reniform, usually five angled, sometimes shallowly three to seven – lobed, hairy or somewhat scrabrous, 8-12 cm across.

The staminate flowers are clustered. The pistillate or bisexual flowers are solitary on short stout pedicels.

Horticulturally important melons are classified into seven groups based on its polymorphism in leaf, flower, fruit shape, and colour

- *C. melo* var. *cantaloupensis* Naud.: Medium size fruits, round shape, smooth surface, marked ribs, orange flesh, aromatic flavour and sweet.
- *C. melo* var. *reticulatus* Ser.: Medium size fruits, netted surface, few marked ribs, flesh colour from green to red orange.
- *C. melo* var. *saccharinus* Naud.: Medium size fruits, round or oblong shape, smooth surface with grey tone sometimes with green spots, very sweet flesh.
- *C. melo* var. *inodorus* Naud.: Smooth or netted surface, flesh commonly white or green, lacking the typical musky flavour. These fruits are usually later in maturity and longer keeping than *cantaloupensis*.
- *C. melo* var. *flexuosus* Naud.: Long and slender fruit eaten immature as an alternative to cucumber.
- *C. melo* var. *conomon* Mak.: Small fruits, smooth surface, white flesh. These melons ripen rapidly, develop high sugar content but little aroma.
- *C. melo* var. *dudaim* Naud.: Small fruits, yellow rind with red streak, white to pink flesh.

Botanically, the fruit is classified as a pepo with many seeds. Usually sweet melons have a high pH value (low organic acid content). Combination of high sucrose and high organic acid, mainly citric acid leads to pleasant flavours. Forty-six flavour-related compounds identified in cantaloupe were esters and aldehydes. All esters believed to have flavour impact increased progressively after pollination, and this trend continued with increasing harvest maturity. However, compound recovery often decreased when fruits were harvested over-ripe. Most aldehydes increased during early growth stages and then tapered off with increasing harvest maturity. Genes involved in the biosynthesis of esters, the main volatile components of the Alcohol Dehydrogenase (ADH) and Alcohol Acyl-Transferase (AAT) families have been isolated but their phenotypic effect has not yet been determined.

Importance and use

Mature fruits are eaten fresh as a dessert fruit, canned or used for syrup or jam. The dessert types quench thirst and add to the nutrient contents of main diet. Immature melons are used fresh in salads, cooked or pickled in some parts of the world. Melon seeds are dietary source of unsaturated vegetable oil and protein and sometimes are lightly roasted and eaten like nuts or even used in sweets in some parts of India. Dried and ground melon seeds are used as food in some African societies. In addition to their consumption when fresh, melons are sometimes dried. Other varieties are cooked, or grown for their seeds, which are processed to produce melon oil.

Nutritional value

Muskmelon is rich source of vitamin A, C and B, and minerals like calcium, phosphorus and iron. It contains 90% water and 9% carbohydrates, with less than 1% each of protein and fat. The yellow and orange fleshed melons contain more than 350 mg of β-carotene, a precursor of vitamin A. Cantaloupes contain 45 mg and Honey Dew 32 mg of vitamin C per 100 g of edible portion. Seed kernels are edible, tasty and nutritious, since they are rich in oils and energy. Muskmelon comprises of a significant amount of dietary fibre, making it good for those suffering from constipation. The potassium present in muskmelons makes them quite helpful in the lowering of blood pressure. Being rich in Vitamin C, the fruit has been believed to be good for preventing scurvy, cancer and other chronic ailments. The fruit proves to be a good source of vitamin A and thus, helps maintain healthy skin. The folic acid present in muskmelons helps create healthy foetuses (in pregnant women) and can even prevent cervical cancer and osteoporosis. Consumption of muskmelon in hot summer helps to refresh the body and mind.

Medicinal value

Consumption of muskmelon is associated with regulating heart beat and, possibly, preventing strokes. Muskmelons might reduce the risk of developing kidney stones and age-related bone loss. Muskmelons have been found to have sedative properties, making them beneficial for the people who are suffering from insomnia. The fruit has almost zero cholesterol and thus, is safe for the people who suffer from the problem of high blood cholesterol. Regular consumption of muskmelon juice can help treat lack of appetite, acidity, ulcer and urinary tract infections. It also serves as a mild antidepressant. Since it has high water content, the fruit can help reduce the heat in the body and thus, prevent heat-related disorders. The seeds are diuretic, cooling, nutritive, and beneficial to the enlargement to prostate gland. Melon fruits, roots, leaves, and seeds play important roles in the treatment of a wide range of health problems in Chinese traditional medicine.

Origin and taxonomy

Muskmelon is native to Persia (Iran) and adjacent areas on the west and the east. Persia and the trans-Caucasus are believed to be the main centre of origin and development, with a secondary centre including the northwest part of India, including Kashmir, Afghanistan and China. Although truly wild forms of *C. melo* have not been found, several related wild species have been noted in those regions. The oldest supposed record of muskmelon goes back to an Egyptian picture of the period around 2400 B.C. In an illustration of funerary offerings of that time appears a fruit that some experts have identified as muskmelon, although others are not so sure.

The Greeks appear to have known the fruit in the third century B.C., and in the 1st century after Christ it was definitely described by the Roman philosopher Pliny, who said it was something new in Campania. The Greek physician, Galen, in the second century, wrote of its medicinal qualities, and Roman writers of the third century gave directions for growing it and preparing it with spices for eating. The Chinese apparently did not know the muskmelon until it was introduced to their country around the beginning of the Christian era from the regions west of the Himalayas.

Culture of the muskmelon spread westward over the Mediterranean area in the Middle Ages and was apparently common in Spain by the 15th century. Columbus carried the seeds on his second voyage and had them planted on Isabela Island in 1494. About this same time, Charles VIII of France introduced muskmelons into central and northern Europe from Rome.

Before the end of the 16th century, the muskmelon had been introduced by the Spaniards to many places in North America. In the first English colonies in Virginia and Massachusetts it was grown. In the mid-1600's the Indians of Florida, the Middle West and New England grew it. Later during the eighteenth century they reached the Pacific Islanders *via* British explorers. It is a polymorphic taxon comprising large number of botanical and horticultural groups, which includes both desserts as well as cooking and salad types.

Muskmelons do not cross with watermelon, cucumber, pumpkin or squash, but varieties within the species intercross frequently. There are various horticultural forms within *C. melo* based on fruit characteristics, namely, cantaloupes, nutmeg muskmelons, winter melons, white- skinned melons, snake melons, oriental pickling melons, mango melons, pomegranate melons which hybridize readily with each other and there is apparently very little sterility even among progenies from crosses involving variant types. The genome of *Cucumis melo* L. was first sequenced in 2012.

The *Cucumis* species of the African group to which melons belong are also diloid (2n = 24) but with a different basic chromosome number (x = 12).

Adaptation in India

Muskmelon was introduced in India probably during Mughal invasion around 14th century from central Asian region (some parts of southern Russia, Iran, Afghanistan). Since then, it has spread to different parts of the country as far down to Andhra Pradesh and Karnataka. Existence of huge diversity in North-West India also suggests its probable secondary centre of origin here. Dessert type muskmelon grown in India comprises different local cultivars and improved varieties developed through selection from indigenous genotypes and hybridization employing exotic introductions.

Improved varieties

- **Open pollinated varieties**: Lucknow Safeda, Pusa Sharbati, Pusa Madhuras, Pusa Rasraj, Arka Jeet, Arka Rajhans, Durgapura Madhu, Hara Madhu, Hisar Madhur, Hisar Saras, MH-10, Punjab Sunheri, Punjab Rasila, RM-43, RM-50, MHY-3, Gujarat Muskmelon-1, Gujarat Muskmelon-2, Gujarat Muskmelon-3, Kashi Madhu, Narendra Muskmelon-1, Narendra Muskmelon-2
- **Varieties resistant to downy mildew**: Hisar Saras, Punjab Rasila, Kashi Madhu
- **Public sector Hybrids**: Punjab Hybrid-1, Pusa Rasraj, Arka Rajhansa, MH-10

Cultural requirements

Muskmelon flourishes well under warm climate and cannot tolerate frost. The optimum temperature for germination of seeds is 27°–30°C. With the increase in temperature, the plants complete their vegetative growth earlier. Stormy weather particularly dust-storm during flowering reduces fruit setting. Dry weather with clear sunshine during ripening ensures a high sugar content, better flavour and high percentage of marketable fruits. High humidity increases the incidence of diseases, particularly those affecting foliage. Cool nights and warm days are ideal for accumulation of sugars in fruits. A well-drained, loamy soil is ideal. Lighter soils which warm up quickly in spring are usually utilized for early crop. In heavier soils, vine growth is more and fruit maturity is delayed. On sandy river-beds, alluvial substrata and subterranean moisture of river streams support its good growth. Its long tap root system is adapted to grow in the river beds. It is necessary that soils should be fertile and rich in organic matter. Muskmelon is sensitive to acid soils. The crop cannot be grown successfully below a pH of 5.5. A soil pH between 6.0 and 7.0 is ideal. Alkaline soils with high salt concentration are not suitable for its successful growth.

Land preparation

The field is prepared with 4-5 deep ploughings to make into a fine tilth before seed sowing. All the weeds, stubbles are removed and the land is levelled by laddering. Manures should be mixed with the soil at the time of last ploughing. Seeds should be sown on raised beds. The width of bed depends upon crop and the variety to be planted. Generally 3-4m wide raised beds are prepared for muskmelon. Planting is done on both sides of the beds. Planting can also be successfully done between long water channels. The channels can be prepared either with a tractor or with manual labour. The distance between 2 channels should be according to width of the bed. Subsequently, the vines should be raised to grow in between the cannels.

Sowing time

Early sowing is done in river-beds from November to January in north India. In northern plains, sowing is done in February and in southern plains, it is generally done in December–January.

Seed sowing

Sowings can be done in flat bed or furrow method or in river bed system depending on the season and convenience. Optimum temperature for seed germination is 24°–29°C. To get an early crop, seeds can be sown in polythene bags under protected cover. The seedlings become ready for transplanting at 2–4 true leaf stage. This practice is prevalent in Punjab, where sowing is done in last week of January or first week of February. The seedlings become ready for transplanting by February-end or first week of March. The seed rate of 1.0-1.5 kg/ha is optimum for sowing in field by dibbling method keeping 2–3 seeds/ hill. For direct sowing, a seed rate of 2.50–3.75 kg/ha is adequate. The transplanted crop matures 15–20 days earlier than the direct-seeded one. For raising seedlings, polythene bags of 15cm × 10cm size and of 100 gauge thickness is filled with a mixture of well-rotten farmyard manure, field soil and silt in equal proportion. For raising nursery for 1 hectare, 12.5–15 kg polythene bags are required. Seeds should be sown 1.5cm deep. Watering of the bags should be done daily with a sprinkling can. Watering the bags should be stopped two days before transplanting. Polythene bags are removed before transplanting the seedlings. Care should be taken that the earth-ball with the seedling does not break. The seedling with the earth ball is placed in the pits in the field, soil is put all around and irrigation is given in the channels immediately after transplanting.

Spacing

Spacing is 3.0-3.5 m x 50-60 cm for spring-summer crop and 2.0-3.0 m x 1.0-1.5 m for river bed system of cultivation.

Nutrient management

The dose of manures and fertilizers depend on the soil type, climate and system of cultivation. Well rotten farm yard manure at the rate of 25-30 t/ha should be applied at the time of land preparation. A general fertilizer dose of 125 kg N, 70 kg P_2O_5 and 70 kg K_2O per hectare is recommended for the open pollinated improved varieties. Whole P and K and one-third of N should be applied as basal in two parallel bands 45cm apart on both sides of the bed mark. The channels are opened between fertilizer bands. Rest of the N should be applied to the vines 25 - 30 days after sowing prior to earthing-up. The row-to-row and plant-to-plant distance need to be kept same as for directly sown crop. For transplanted crop, 15–20cm deep pits are dug at the sites where seedlings are to be transplanted. Each pit is filled with a mixture of 1 kg of farmyard manure, 10–15g of calcium ammonium nitrate, 40g of super phosphate and 10g muriate of potash before planting. Another dose of 10–15g of calcium ammonium nitrate (CAN) is applied to each plant after one month. A general fertilizer dose of 200 kg N, 100 kg P_2O_5 and 100kg K_2O per hectare is recommended for the hybrids. One-fourth N along with full P and K fertilizers and farm yard manure should be given as basal and rest N should be top dressed in three split doses at the time of vining (30-35 days after sowing) , at blooming (40-45 days after sowing) and after first harvest stages (80-85days after sowing).

Crop regulation

In general, high N nutrition under high temperature condition promotes staminate flowers and lower the number of pistillate flowers/ vine, resulting in low fruits-set and yield. The best way to control the vine growth within reasonable limits is by adjusting nitrogen fertilizer doses and frequency of irrigation and pruning.

Rampant vegetative growth should be avoided as it promotes maleness. If apical shoot is pinched and 2–4 side shoots are allowed to grow it gives significantly higher fruit yield than un-pruned plants. Secondary branches are pinched off up to 6-7 nodes on emergence and tertiary branches are allowed to grow. Perfect/pistillate flowers are allowed to be borne on these branches to set fruits. Tertiary branches are detopped at 2 leaves above the first perfect/ pistillate flowers. Application of growth regulator like 200 ppm maleic hydrazide (MH) or 300-350 ppm ethephon first at two true leaves of the plants (15 days after sowing) and subsequently repeated 7 days after helps in increasing the number of pistillate flowers in monoecious form and thereby fruit yield.

Irrigation

Maintainence of sufficient moisture is essential for good growth of muskmelon. Frequency of irrigation should be reduced during maturity period to get sweeter fruits. The irrigation should be as light as possible. The light sandy soils need more frequent irrigation than heavier ones. Flooding of the field should be avoided, particularly when fruits are reaching maturity. During dry summer, the crop needs irrigation at 5–7 days intervals. Water application should be restricted to the base of the plant and root zone so that water may not wet the vines, especially when flowering, fruit-set and fruit development is in progress. Frequent wetting of leaves, stems and developing fruits promote disease incidence.

Interculture and weed control

Muskmelon is sensitive to weeds in initial stages of plant growth. In early stage, the beds and ridges should be kept weed-free by hoeing. In the beginning, cultivation can be done fairly close to the plants and shallow (5–10cm) in depth. The growth of lateral roots often equals or exceeds that of the above ground parts. When vines cover the ground, inter-cultivation should be stopped and only weeds should be pulled out by hand. In Punjab and Rajasthan, during April–June severe dust storms not only hamper vegetative growth but also roll them up and in some cases even uproot the vines, if the fields are kept clear of weeds. It is, therefore, advisable to keep the area near the roots free from weeds but allow them to grow on the beds away from root zone, so that the vines during growth can get hold of them and are disturbed least by fast winds. The herbicides like, Fluchloralin or Trifluralin @ 0.75-1.0 kg/ha or Bensulide @ 5-8 kg/ha may be applied as pre-plant soil incorporation 2 weeks before sowing. Application of Butachlor @ 1 kg/ha or Chloramban @ 2-3 kg/ha as pre-emergence and Naptalam @ 2-4 kg/ha as post emergence after first weeding efficiently helps in controlling the weeds. It needs earthing up twice at one month interval to facilitate good root development and to reduce the weeds.

Harvesting the crop

The crop is ready for harvesting 75 to 90 days after sowing, depending upon the variety. Fruits maturing on vine, without becoming overripe are superior in quality to those harvested immature or left on vine after they have become mature. Harvesting at the proper stage is of major importance in marketing good quality produce. The stage of maturity in muskmelon is judged by change in the external colour of fruit. As ripening advances, a crack develops around the penduncle at the base of the fruit and when fully ripe, the fruit slip easily from the stem leaving a large scar. This is known as full slip stage. In half-slip stage only portion of the disc is removed and scar on the fruit is smaller than full slip stage. For local market, fruit should be harvested at full slip stage. The fruits for distant markets should be harvested as they reach the half-slip maturity to avoid losses from the over-ripeness and decay. The variety like, Hara Madhu however, never reaches the full-slip stage and colour of the rind can be taken as the criteria of maturity. Fruits harvested when immature as yet, never attain these desirable characteristics.The soluble solids are considerably affected by the environment, the incidence of diseases and the vigour of the plant.

Post harvest handling and storage

Improper harvesting, handling, transportation and distribution result in significant losses. Jerk to fruits while loading and transport should be avoided. They can be shipped to long distances, even without refrigeration. Flavour and texture of the flesh improve for a few days after harvesting and attain highest quality if the fruits are harvested when they have developed their maximum sugar content. Muskmelons keep well for 1.5 weeks at 1.7°–3.3°C and 85-90 % relative humidity.

Yield

Fruit yield may be 15-20 t/ha for open pollinated varieties and 30-35 t/ha for hybrids.

Pumpkin

Pumpkin, *Cucurbita moschata* Duch. ex Poir. under the family Cucurbitaceae (Chromosome No. 2n = 40) is an important tropical vegetable grown all around the world. The five cultivated species of *Cucurbita* viz., *C. pepo*, *C. moschata*, *C. mixta, C. maxima* and *C. ficifolia* were selected by American Indians long before the discovery of America. These *Cucurbita* species are important sources of minerals, vitamins and nutritional carbohydrates, which are grown and consumed worldwide. Pumpkin is the most important species of the genus *Cucurbita* and extensively cultivated in India, Africa, Latin America, Southern Asia and United States. It occupies a prominent place among the vegetables owing to its high productivity, nutritive value, good storability, long period of availability and better transport potentialities. The English name of this species "Tropical Pumpkin" is an appropriate one because the greatest diversity lies in the neotropics where the vines are grown under a wide range of ecological conditions including the hotter conditions than are tolerated by the other cultivated *Cucurbita* species. It is one of the important and popular vegetable crops extensively grown during spring-summer, summer-rainy and early autumn seasons in all parts of the country particularly in Assam, West Bengal, Tamil Nadu, Karnataka, Kerala, Madhya Pradesh, Uttar Pradesh, Odisha and Bihar.

The crop

Pumpkin is an annual dicotyledonous vegetable, with creeping or climbing stems (growing up to 5 m) bearing tendrils. The stems are strong, cylindrical or pentangular, with petioles measuring 12–30 cm. The stems and leaves are mildly hairy. The leaves are circular, kidney-shaped, heart-shaped or triangular, often deeply indented at the base, weakly lobed, wavy and toothed, more or less white spotted, up to 20 cm long and 30 cm wide. Broad leaves are attached to thick stems and, in most cultivars, are covered in trichomes that may induce dermatitis on contact.

After about 6-8 weeks of growth, the plant produces bright yellow-orange monoecious flowers. Flowering and anthesis is influenced by temperature. The flowers are large, yellow, bell-shaped, five-lobed, and up to 12 cm long. The staminate flowers produce pollen with large granules and the pistillate flowers are pollinated by honey bee and bumble bee. The peduncle is strong, with a rounded pentangular base and large apex. Fruits are round, oblate, oval, oblong, or pear-shaped, variously ribbed, 15–60 cm in diameter, and weigh up to 45 kg. The fruits are generally harvested as mature and ripe stage which has a hard rind, firm mesocarp and seeds encased in a lignified seed coat attached to the fibrous endocarp adjacent to a hollow central cavity. Their flesh is deep yellow, orange, pale green, or white, and the hollow centre contains pulpy loose fibres and numerous seeds. The seeds are oval, flat, white to brown, thin-shelled, irregularly margined, with a meaty kernel.

Chief species identifying characters of *C. moschata*, *C. pepo*, *C. maxima*, *C. ficifolia* and *C. mixata* are presented below:

- ❒ Plant perennial, seeds black or tan: *C. ficifolia*
- ❒ Plants annual; seeds white, buff or tawey: four species, *C. moschata*, *C. pepo*, *C. maxima and C. mixata*
- ❒ Peduncle hard, smoothly grooved, flared at fruit attachment; foliage non-stipulate: *C. moschata*
- ❒ Stems hard, angular; peduncle angular, grooved; peduncle hard, sharply angular, grooved, foliage stipulate: *C. pepo*
- ❒ Plants annual; seed white, buff or towey; stem soft, round; peduncles soft, terete, enlarged by soft cork: *C.maxima*
- ❒ Peduncle hard, greatly enlarged in diameter by hard cork, not flared at fruit attachment; foliage non-stipulate: *C. mixta*

Importance and use

It is the natural source of *β-carotene, the precursor of vitamin A.* Nearly all above grown parts of pumpkin viz., leaves, staminate flowers and both mature and immature fruits are utilized as vegetable in different forms. The flesh is delicious when stewed, boiled or baked. Fully matured fruits, apart from utilization as cooked vegetable can be used in preparing sweets, candy or fermented into beverages. Delicate sweet items like "halwa", other sweets and jams are prepared from meshed flesh of fully matured ripe fruit. Fully matured ripe fruits are also mixed with tomato in the preparation of sauce and ketchup. Pumpkin pulp is also used for the preparation of GTF (glucose tolerance factor) and pumpkin milk powder.Matured fruits are also used as industrial raw material for carotene production. The "Yerusseri" prepared from immature fruits is very popular in Kerala. The seeds after removing the seed coats are used in confectionary. Seed oil (50% of the seed) is a common salad oil in the southern part of Austria, Slovenia and Hungary.

Nutritional value

Average nutritive value of pumpkin (2.68) is higher than brinjal (2.41), tomato (2.09) and cucumber (1.69) mainly because of stored carbohydrate (mainly glucose) and carotenoids (74 % β-carotene) along with moderate quantities of ascorbic, nicotinic, pantothenic and folic acids and different minerals.

Edible portion of pumpkin contains 1.4% protein, 4.6% carbohydrates, and 0.1% fat and 0.6 % minerals. In pumpkin, the pH ranges from 3.9 to 4.3, total sugar content from 0.55% to 0.62%, and reducing sugar content from 0.41% to 0.47%. The young leaves, flowers and fruits are rich in β carotene and provide on an average 5.0 mg β carotene per 100 g edible portion of fruit. It provides about 30 mg of P, 10mg of Ca and 0.7 mg of Fe per 100 g edible portion of the fruit.

Medicinal value

Pumpkin is particularly important for the supply of antioxidants, especially β carotene in the foods. It helps to alleviate cardiovascular diseases and constipation. It is also a good diuretic food because of having high K : Na ratio. Its combination with sugar and honey has soothing effect to ulcers and blisters. It is used in the treatment of diabetis, rheumatism, eczema and burns and against worms and other parasites. The fruit pulp is also used in the poultice for boils, carbuncles and ulcers, etc. *Cucurbita* species are also suitable for various systems of traditional medicine such as antibacterial, anticancer, antidiabetic, antihypertensive, antiinflammation, antiparasitic, antalgic, hypocholesterolemic, immunomodulatory and intestinal. Seeds along with sugar are used to kill tape worms and are also useful in urinary diseases.

Origin and taxonomy

Cucurbits have undergone several independent domestication events beginning as early as 10,000 years ago, which pre-dates maize and bean domestication.The subtropical *C. moschata* originated in northern Colombia while the centres of origin for the temperate species *C. pepo* and *C. maxima* are located throughout Mesoamerica.

Pumpkin and squashes have been associated with man in the America for at least 10,000 years. American Indians selected the cultivated species long before and they were staple items of their diet. It was cultivated in southern Mexico in 5000 B.C. and in Peru in 3000 B.C. Archaeological evidences showed that pumpkin was widely distributed in both North and South America with the primary centre of origin being Mexico and Peru. Wild forms of pumpkin have not yet been described but primitive-appearing land races are known from Central America to northern Peru. This coupled with archaelogical evidence recently suggests that the centre of origin of pumkin is in the north-western South America.

The genus *Cucurbita* belongs to the family Cucurbitaceae, sub-family Cucurbitoideae, Tribe Melothrieae and sub-tribe Cucumerinae. There are 26 *Cucurbita* species including five cultivated species namely, *C. moschata* –

Pumpkin, *C. Pepo* – Summer squash, *C.maxima* – Winter Squash, *C. mixta* (Syn. *C. angyrosperma)* – Winter Squash and *C. ficifolia* – Fig leaf gourd or Malabar gourd. The wild form of *Cucurbita* are classified into two forms :

a) Xerophytic form: *C. digitata*, *C. palmata*, *C. cylindiata*, *C. foetodissima.*

b) Mesophytic form: *C. ficifolia*, *C. sororia*, *C.lundelliana*

Pumpkin appears to be more closely related to xerophytic forms and is considered as the axis through which the species under xerophytic forms are related to each other. The genomic structure of *Cucurbita moschata* is AABB indicate that it is an amphidiploid. Isozyme studies in pumpkin revealed that there are 28 loci conditioning the isoenzymic phenotypes and this coupled with high chromosome number (2n = 40) suggest its allotetraploid origin.

Adaptation in India

Pumpkin was introduced to India from South America long back by foreign navigators and emissaries. Its early introduction and subsequently wide range acceptability and cultivation have favoured the development of huge genetic variability of pumpkin in India. A number of vernacular names of pumpkin *viz.,* Kumra, Misti kumra, Sitaphal, Kashiphal, Kaddu, Lalbhopla, Lalkumbhara, Meetha kaddu, Pilum kohalum, Poashani, Gumbala, Kannada, Duddini, Red gourd, Saphurii kumra, etc. also suggests its wide range adaptation as well as cultivation in India. Wide range of diversity of pumpkin genotypes is found in India and all the improved varieties have been developed in India through selection followed by inbreeding from different local cultivars. In India mostly different local cultivars are grown under different name mainly after the name of the place where the particular local types are available however, a good number of improved varieties have also been developed in India from the indigenous diversity.

Open pollinated varieties

Improved varieties: Arka Suryamukhi, Solan Badami, Arka Chandan, CO-1, CO-2, Ambili, Pusa Biswas, Pusa Vikas, Kashi Harit, Sooraj, Saras, Narendra Agrim, Azad Pumpkin-1, Narendra Amrit.

Cultural requirements

Pumpkin is basically a warm season crop and susceptible to frost. The favourable temperature range for the growth and development is minimum 18.0-30°C and optimum being optimum 20°C-25°C. Short days, comparatively low-night temperature and high relative humidity is the best for pumpkin cultivation as this condition increases the propensity of pistillate flowers in the vine. High temperature and long days help to increase the number of staminate flowers and reduce the number of pistillate flowers.It can also be grown at high temperature range of 30 -35°C. Ideal range of soil temperature for seed germination is 20°C-25°C. Seed germination is hampered at soil temperature below 10°C and above 30°C. Deep well drained loamy soils with pH range 6.0 – 7.0 are ideal. Sandy loam soils are ideal for early crop but clay loam soils are suitable for getting high yield. Soil moisture should be at least 10-15% above the permanent witting point for proper growth and yield of the crop. Pumpkin is sensitive to water logging condition. Conventionally pumpkin in West Bengal is grown in relay cropping with potato and sowing is done in January- February before potato is harvested.

Land preparation

The field is prepared with 2-3 deep ploughings to make into a good tilth before seed sowing. All the weeds, stubbles are removed and the land is levelled by laddering. Manures should be mixed with the soil at the time of last ploughing. The plant has extensive root system. In spring-summer or early autumn season, pits are prepared of 60 cm diameter and 45 cm depth at a spacing of 90 to 150 cm in rows spaced 2.0-3.0 metre. Furrow can also be opened at 2.0 to 3.0 m spacing and mounds are prepared just at the side of the furrow top. In rainy season, raised beds or mounds are prepared to facilitate drainage. Soil of the pits is filled and mounds are prepared with of 20-25 tonnes of FYM per hectare and basal dose of fertilizers mixed with the soil.

Sowing time

Pumpkin grows successfully in tropical and subtropical climate hence, can be grown year round in the southern states due to prevailing mild winter there. However, the crop is grown in three main seasons: spring-summer, early autumn and rainy season.

In eastern and northern plain, sowing is done in late January-March for spring-summer crop; May-July for rainy season crop in eastern and northern plain; August-September for early autumn crop in eastern India; December-January for winter – spring crop in southern and western regions and March-April for the hill regions.

Seed sowing

Pumpkin gives little success to direct transplanting and the seeds are sown directly in the field. Seeds are sown by hand dibbling in the mound. Sowings can be done following pit, mound or furrow method depending on the season and convenience. Seedlings raised in plastic packets particularly under polyethylene shed in winter month can be transplanted after 35-40 days with the advent of spring without disturbing the root system. 3.0-5.0 kg seed/ha is sufficient depending on the plant population. On an average 4 to 5 thousand plants can be accommodated in one hectare land. It is better to sow the seeds in pre-irrigated moist soil than irrigating the field after sowing. Pre-sowing soil application of 20-22 kg Furadon 3G helps in protecting the plants from root knot nematodes and other pests during initial 4-5 weeks.

Spacing

Spacing of 2.0-3.0 m x 90-150 cm is given for spring-summer crop; 3.5 m x 1.5-2.0 m for rainy season crop and 2.0-2.5 m x 60-90 cm for early autumn crop.

Nutrient management

Well rotten farm yard manure at the rate of 15-20 t/ha should be applied at the time of land preparation. A general fertilizer dose of 100 kg N, 60 kg P_2O_5 and 60 kg K_2O per hectare is recommended for open pollinated improved varieties. Half N along with full P and K fertilizers and farm yard manure should be given as basal and rest N should be top dressed in two split doses at the time of vining (30-35 days after sowing) and at full blooming (55-60 days after sowing) stage. High nitrogen under high temperature condition promotes staminate flowers resulting in low fruit set and yield so, fertilizer dose should be adjusted according to season of cultivation.

Crop regulation

Excessive vine growth due to high nitrogen nutrition coupled with high temperature and high soil moisture condition promote staminate flowers in the vine resulting low fruit set and low yield. The best way to control the vine growth within reasonable limits is by adjusting nitrogen fertilizer doses and frequency of irrigation. Application of ethylene releasing growth regulator, Ethrel @ 250 ppm concentration (2.5 ml/10 l of water; 1ppm = 1 ml/l) four times, first at two true leaves of the plants i.e. 15 days after sowing and subsequently repeated three times at 7 days interval induces the production of pistillate flowers in the vine which helps in increasing the yield.

Irrigation

Interval and amount of irrigation depends upon the season and type of soils. Irrigation is applied at the initiation of first true leaf during spring-summer season and at its expansion stage during rainy season. The field should be irrigated subsequently at 5-6 days intervals during summer and early autumn crop and as and when required for the rainy season crop depending on the rain.

In light soils the crop needs irrigation at frequent intervals. The crop must be irrigated during its most critical stages i.e. flowering and fruit setting stages. Over irrigation during vegetative and early flowering stages may cause excessive vine growth resulting induction of more staminate flowers in the vine. Excessive irrigation at fruit maturity stage also adversely affects the storability of the fruits. Ridge and furrow or ring methods of irrigation is the best for pumpkin

Interculture and weed control

The crop may be trained over low trellis of 1.5 m high or trailed over the dried twigs spread over the ground for proper growth. Soil at the base of the plant should be loosened before top dressing with nitrogen fertilizers followed by irrigation. At the initial stages of weed growth, hand weeding and hoeing are sufficient to check the weed growth. Application of herbicides effectively controls weeds. Pre-emergence application of Besulide @ 4-6 kg a.i. /ha or Alachlor @ 2.5 kg a.i. /ha checks weed population in the fields. Application of Gramaxone @ 0.2% controls the weeds during early stage of crop growth.

Harvesting the crop

Fruits generally reach complete ripe stage 160 to 220 days after sowing depending on the variety and season of cultivation. Generally fruits are harvested at full maturity or complete ripe stage when the harvested fruits are stored. At complete ripe stage generally rind colour of the fruits turn brown or reddish-brown from green and peduncle either dries up or separates from the fruits. Periodical harvests of tender fruits increases the fruit number and thereby yield.

Post harvest handling and storage

Fruits should be harvested at complete ripe stage for long storage life and long distance transportation. Fruits can be kept for 2 to 4 months without any damage at room temperature condition however, the fruit should not be kept in heaps. It is better to keep the fruits in shelves where they can be spread out in single layer with a space between the fruits. The fruits can be kept for 6 months at 10 to 12°C temperature with 70-75% relative humidity condition.

Yield

Fruit yield is 20 to 30 t/ ha in open pollinated improved varieties.

Bottle gourd

Bottle gourd, *Lagenaria siceraria* under the family Cucorbitaceae (Chromosome No: 2n = 22) is commonly grown cucurbit vegetable crop of India, Africa, Central America and other warmer regions of the world. The bottle like shape of fruit and its use as a container for wines and spirits in the past gave it the common name of bottle gourd. It is also called Calabash gourd or Zucca melon. In India, it is extensively grown during spring-summer, summer-rainy and early autumn seasons in all parts of the country particularly Bihar, Uttar Pradesh, Madhya Pradesh, Chattisgarh, Odisha, West Bengal, Assam, Tripura, Punjab, Haryana and Telengana. It is also cultivated in different river beds of the country and also suitable for cultivation in dry areas. It might be the only plant species that was known to man in both the New and Old world from very early pre-historic times.

The crop

Bottle gourd is a vigorous, quick-growing annual, running or climbing vine with large leaves and a lush appearance. It grows fast and may begin to flower only 2 months after seeding. The thick stem is furrowed longitudinally. The vine is branched and climbs by means of long forked tendrils along the stem. The foliage is covered with soft hairs and has a foul musky odour when crushed. Size of the leaves may be up to 35 cm wide, circular in overall shape, with smooth margins, a few broad lobes or with undulate margins. Leaves have a velvety texture because of the fine hairs, especially on the under surface. The flowers are generally borne singly on the axils of the leaves, the staminate on long peduncles and the pistillate on short peduncles. The flowers are white and attractive, up to 10 cm in diameter, with spreading petals. The ovary is inferior and in the shape of the fruit. Otherwise, the staminate and pistillate flowers are similar in appearance. The anthers are borne on short filaments grouped at the centre of the flower. The stigmas are short, thickened, and branched. The brownish seeds are numerous in a whitish green pulp. Each seed is somewhat rectangular in shape with grooved notches near the attached end.

Variation in bottle gourds is sometimes spectacular. Many forms of the bottle gourd are cultivated for specific purposes, and the size of the vines, leaves, and flowers, as well as the size and shape of the fruits vary greatly. Shape of the fruits may be long, oblong, round, club, etc. in shape depending on the variety. The background rind colour of the fruit is either light green or dark green. The dark green can be distributed as a solid colour, as regular or irregular stripes, and as an irregular blotch. Size of the fruit varies from 5 to 30 cm in diameter and from 10 to 100 cm in length. The fruit can have a sterile (seedless) neck portion that varies from a few to 35 cm in length and from 2.5 to 5.0 cm in width which generally does not have seeds. In the crooked neck shaped fruit, this sterile neck portion is conspicuous. Wider necks usually contain seeds. The seed-containing portion of the fruit varies from flat to round, cylindrical, club-shaped or long and narrow.

Importance and use

Tender fruits are widely used as cooked vegetables in a number of ways. In northern India "Kofta" is most popular vegetable preparation. Immature fruits are used with milk for the preparation of different sweets like, "Kheer", "Payesh" and "Halwa". Matured fruits are also used for the preparation of different sweets like, "Petha" and "Burfi". The green leaves are used as potherbs. Tender fruits of 30 – 40 cm in length have great export potential to South-East Asia and Gulf countries. Seed and seed oils are also edible. The hard shell of the fruit is used for making wide range of articles of common use including bowls, bottles, ladles, different containers, floats for fishing nets, pipes and musical instruments.

Nutritional value

Edible portion of bottle gourd contains 0.2% protein, 2.9% carbohydrates and 0.5% fat and 11.0 mg vitamin C per 100 g fresh edible portion. It is low in saturated fat, cholesterol, high in dietary fibre. It also contains riboflavin, zinc, thiamine, iron, magnesium and manganese.Tender fruits provide about 10 mg of P, 20mg of Ca per 100 g edible portion. The young leaves are rich in carbohydrate (6.1%) and protein (2.3%).

Medicinal value

Fruits are easily digestible and have cooling effect. It also contains sodium, potassium, essential minerals and trace elements, which regulate blood pressure and prevent the risk of heart ailments.High in sodium and potassium, bottle gourd is also an excellent vegetable for people with hypertension. It is a suitable vegetable for light, low-calorie diets as well as for children, people with digestive problems, diabetics and those recovering from an illness or injury.

It is good for people suffering from biliousness and indigestion. Fruits are cardiatonic and diuretic and are good for overcoming constipation and cough. The pulp also acts as an antidote against certain poisons. The plant extract is used as a cathartic. The seeds are used for treating dropsy. Decoction of leaves mixed with sugar is given for treating jaundice. The bitter fruits are poisonous and are used as a strong purgative.

Origin and taxonomy

The bottle gourd is an ancient crop, widespread and well used, from warm parts of the temperate zone throughout the dry and wet tropics. It is the only crop known to have been cultivated in pre-Columbian times in both the Old and New World. The bottle gourd is native to tropical Africa. Wild populations of the plant have recently been discovered in Zimbabwe. Two subspecies, likely representing two separate domestication events, have been identified: *Lagenaria siceraria* spp. *siceraria* (in Africa, domesticated some 4,000 years ago) and *Lagenaria siceraria* spp. *asiatica* (in Asia, domesticated at least 10,000 years ago). The likelihood of a third domestication event, in Central America about 10,000 years ago, has been implied from genetic analysis of American bottle gourds. Domesticated bottle gourds have been recovered in the Americas at sites such as Guila Naquitz in Mexico by ~10,000 years ago.

Bottle gourd was widely distributed in pre-Columbian times, perhaps by floating on the seas. It travelled to India, where it has evolved into numerous local varieties, and from India to China, Indonesia, and as far as New Zealand. Archaeological remains show that the bottle gourd was used in Egypt about 3500 to 3300 B.C. Bottle gourd also travelled to the New World. The dried gourds with viable seeds have survived in seawater for at least 200 days.

Archaeological evidences indicated its presence in Egyptian tombs in 3000 to 5000 B.C., Spirit caves of Thailand in 6000 to 10,000 B.C., Mexico in 5000 to 7000 B.C. and Peru in 3000 to 4000 B.C. Its utilization by man is about 16,000 years old in the Old world and 12,000 years old in the New world. Recently, seeds and other remains of the bottlegourd have turned up in several archaeological digs in Florida.

The bottle gourd, *Lagenaria siceraria* (Mol.) Standl. belongs to the family Cucurbitaceae, sub-family Cucurbitoideae, tribe Benincaseae and sub-tribe Benincasinae. The genus *Lagenaria* includes six species having pantropical distribution in Africa, Madagascar, Indo-Malaysia and the neotropics. Of these, only *Lagenaria siceraria* is cultivated which is annual and monoecious while the other five species are wild, perennial and dioecious in sex form. The wild *Lagenaria* occur only in east Africa and Madagascar. The modern forms of bottle gourd are secondary polyploids derived from its progenitor with a basic chromosome number, $x = 5$ and $n = 5+5+1$ ($2n = 22$).

Adaptation in India

Bottle gourd was introduced in India in very early times because of its independent domestication in Africa and Asia. Sanskrit language equivalents signifying Vedic period and Aryan usage is available for bottle gourd as "Alabu". The bottle gourd was probably used by Proto-Australoides and cooking of this fruit "Alabu" or "Iksvaku" have been mentioned in "Yajurveda". Such mentioning of bottle gourd in ancient Indian literatures signifies that it was introduced into India in very early times and has undergone significant diversification in this country mainly in plant type, fruit shape, fruit colour and fruit weight. It has been found wild on the coast of Malabar and in the humid forests of Dehra Dun. It is grown as spring-summer, summer-rainy and early autumn crop in the field or grown in different riverbeds of the country. Bottle gourd is cultivated in India mostly with different local cultivars under different names. A number of improved varieties have been developed through selection followed by inbreeding from different local cultivars. Hybrids have become popular in India.

Improved varieties

- **Long fruited variety:** Pusa Summer Priolific Long, Pusa Naveen, Arka Bahar, Kalyanpur Long Green, Samrat, NDBG-1,Phule BTG-1, Punjab Long, Rajendra Chamatkar, Azad Nutan, Kashi Bahar, Narendra Jyoti, Pant Lauki-3, NDBG-132, Pusa Samridhi
- **Round fruited variety**: Pusa Summer Prolific Round, Punjab Round, Punjab Komal, Pusa Sandesh.
- **Bottle shaped variety:** Narendra Dharidar, Narendra Rashmi.
- **Pot shaped variety:** CO-1
- **Public sector Hybrids** : Pusa Hybrid-3 (club shaped fruit), Pant Sankar Lauki-1 (long fruited), Pant Sankar Lauki-2 (long and club shaped), Narendra Sankar Lauki-4 (long fruited), Kashi Ganga (cylindrical fruit shape), Azad Sankar Lauki-1 (long fruited)

Cultural requirements

Bottle gourd is basically a warm season crop and susceptible to cold temperature and frost. It requires a long hot growing season to mature.Low temperature causes poor germination of seeds and stunting of growth. It can be grown in warm and humid and slightly cool climate. Optimum temperature range for the growth and development is 24°C-27°C. Soil temperature of 18 to 22° C promotes good growth and ensures better yield. Sex ratio is sensitive to photoperiod. Short days, comparatively low-night temperature and high relative humidity increase the number of pistillate flowers in the vine. Long days and high temperature increase the number of staminate flowers and reduce the number of pistillate flowers. High rainfall coupled with prolonged cloudiness promote high incidence of disease and in turn drastically reduce yield. Soil moisture is important for rapid growth and it should be at least 10-15 % above the permanent wilting point. However, it can also be grown in dry areas. It is susceptible to water logging. Deep well drained, fertile, loamy or sandy-loam soils with pH range 6.0-7.0 are ideal. Sandy loam soils are ideal for early crop.

Land preparation

The field is prepared with 4-5 deep ploughings to make into a good tilth before seed sowing. All the weeds, stubbles are removed and the land is levelled by laddering. Manures should be mixed with the soil at the time of last ploughing. The plant has comparatively shallow root system and is susceptible to water logging so land should be well drained. Generally furrows are opened at 2.0 to 3.0 m spacing and sowing is done on the top of the sides of furrows and the vines are allowed to trail on the ground. The crop is also grown on flat or raised beds, about 2.0 to 2.5 m wide and seed are sown on both sides of the bed. In rainy season, raised beds or mounds are prepared to facilitate drainage. Mounds in the furrow, flat bed method or in river bed system of planting are prepared with of 15-20 tonnes of FYM per hectare and basal dose of fertilizers mixed with the soil.

Sowing time

Bottle gourd grows successfully in tropical and subtropical climate hence, can be grown year round in the southern and central Indian states due to prevailing mild winter there. Sowing time is January- February for summer crop in eastern and northern plain; June-July for rainy-autumn season crop in eastern and northern plain; August-September for autumn-winter crop in eastern India; December-January for spring crop in southern and western regions; November-January for river bed cultivation and March-April for hill regions.

Seed sowing

Bottle gourd gives little success on transplanting of seedlings hence, seeds are generally sown in the field. Seeds are sown by hand dibbling in the mound. Sowings can be done following flat bed or furrow method or in river bed system depending on the season and convenience. Pre-sowing treatment of seeds with 600 ppm succinic acid for 12 hours improves germination and seedling growth. Seed soaking in water for 12 hours also improves germination. Seedlings raised in plastic packets particularly under polyethylene shed in winter month can be transplanted after 35-40 days with the advent of spring without disturbing the root system. 4.0-6.0 kg seed is sufficient depending on the plant population for one hectare land. On an average 4 to 5 thousand plants can be accommodated in one hectare land. It is better to sow the seeds in pre-irrigated moist soil than giving irrigation after sowing. Pre-sowing soil application of 20-22 kg Furadon 3G helps in protecting the plants from root knot nematodes and other pests during initial 4-5 weeks.

Spacing

Spacing is 2.0-3.0 m x 90-150 cm for spring-summer crop; 2.0-3.0 m x 100-150 cm for river bed system; 3.5 m x 1.5-2.0 m for rainy season crop; 2.0-2.5 m x 60-90 cm for early autumn crop and 2.5-3.0 m x 1.0 m for the crop trained on the bower.

Nutrient management

Bottle gourd responds well to manuring and fertilizer application. The dose of manures and fertilizers depend on the soil type, climate and system of cultivation. Well rotten farm yard manure at the rate of 15-20 t/ha should be applied at the time of land preparation. Bottle gourd is a comparatively shallow rooted crop hence, responses well in top-dressing. A general fertilizer dose of 100 kg N, 60 kg P_2O_5 and 60 kg K_2O per hectare is recommended for open pollinated improved varieties.Half N along with full P and K fertilizers and farm yard manure should be given as basal and rest N should be top dressed in two split doses at the time of vining (30-35 days after sowing) and at full blooming (50 – 55 days after sowing). A general fertilizer dose of 200 kg N, 100 kg P_2O_5 and 100 kg K_2O per hectare is recommended for the hybrids. One-fourth N along with full P and K fertilizers and farm yard manure should be given as basal and rest N should be top dressed in three split doses at the time of vining (30-35 days after sowing) , at full blooming (50-55 days after sowing) and after first harvest (85-90days after sowing).

Crop regulation

Excessive vine growth due to high nitrogen nutrition coupled with high temperature and high soil moisture condition promote staminate flowers in the vine resulting low fruit set and low yield. The best way to control the vine growth within reasonable limits is by adjusting nitrogen fertilizer doses and frequency of irrigation and pruning. Application of different growth regulators, secondary and micro-nutrients *viz*., 150 ppm Ethrel (1.5 ml/10 litre of water), 400 ppm maleic hydrazide, 50 ppm triodobenzoic acid, 3-4 ppm boron or 20 ppm calcium twice, first at two true leaves of the plants i.e. 15 days after sowing and subsequently repeated 7 days after helps in increasing the yield.

Irrigation

Interval and amount of irrigation depends upon the season and type of soils. The crop requires frequent irrigation as high humidity is favourable for the crop. Irrigation after sowing improves seed germination because seed germination require high water supply. Irrigation should be given at the initiation of first true leaf during spring-summer and at its expansion stage during rainy season. The field should be irrigated subsequently at 5-6 days intervals during summer and early autumn crop and as and when required for the rainy season crop depending on the rain. Crop grown on light soils needs irrigation at frequent intervals. The crop must be irrigated during its most critical stages i.e. flowering and fruit setting stages. Over irrigation during vegetative and early flowering stages may cause excessive vine growth resulting induction of more staminate flowers in the vine. Ridge and furrow or ring methods of irrigation are the best for bottle gourd. Drip irrigation gives 48% more fruit yield than furrow irrigation.

Interculture and weed control

Plant grown on trellis gives better crop stand and fruit yield mainly because of better use of sunlight by maximum number of leaves and higher number of side branches. The crop needs to be trained over low trellis of 1.5 m high above the ground. Training the vines over the trellis or bower coupled with some pruning is very effective. The auxillary buds are to be removed at weekly interval till the vine attains the height of the trellis and finally the growing point of the vine is pruned 15 cm below the trellis allowing two side branches to grow over the trellis or bower. After 85 to 90 days of sowing older leaves near the bottom of the vine are pruned. Hand pollination increases yield both during summer and monsoon seasons because of enhanced fruit set. Soil at the base of the plant should be loosened before top dressing with nitrogen fertilizers followed by irrigation. In early stage of growth, the beds, ridges etc should be kept free from weeds. Hand weeding and hoeing are sufficient to check the weed growth at the initial stages. Deep intercultivation should be avoided as it may damage the shallow roots of the crop. Application of herbicides effectively controls weeds. Pre-emergence application of Linuron @ 0.5 kg/ha, Alachlor 2.5 kg/ha, Dichlormate 0.4kg/ha, Duiron 1.25kg/ha checks weed population in the fields.

Harvesting the Crop

The tenderness and edible maturity are judged by pressing the skin and little pubescence persisting on the skins marks tenderness. In the early variety or hybrid first picking of fruits may be done even 60 to 65 days after sowing.

Fruits generally take 18-20 days after fruit set to reach marketable stage when the fruits are still tender. In some big fruited variety it may takes 20-25 days after fruit set to reach marketable maturity. In small fruited hybrids, fruits can be harvested at 6-7 days intervals. Fruits are harvested by cutting the peduncle with knife.

Post harvest handling and storage

Improper harvesting, handling, transportation and distribution result in significant losses. Proper cultural operations, harvesting, transportation, storage and pre and post harvest treatments minimize post harvest losses. Under normal cool and shady conditions, fruits can be kept for 3-5 days. In cold storage at 8.0 to 10°C temperature the fruits can be stored for 2 to 3 weeks. Fruits are packed in polythene bags and kept in small boxes for distance transportation and export.For export, the fruits should be picked at edible stage and then packed in polythene bags and these bags are kept in boxes of convenient capacity.

Yield

Fruit yield is 20-25 t/ha for open pollinated improved varieties and 40-50 t/ha for the hybrids.

Bitter gourd

Bitter gourd or Bitter melon, *Momordica charantia* L under the family Cucurbitaceae (Chromosome no. 2n = 22) is widely grown in India, Indonesia, Malaysia, Singapore, Thailand, China, Japan, East Africa, the Caribbean and South America for its edible fruit. It is one of the important and popular vegetable crops extensively grown during spring-summer, summer-rainy and early autumn seasons in all parts of the country. The leading bitter gourd growing states are Uttar Pradesh, Odisha, West Bengal, Maharashtra, Gujarat, Andhra Pradesh, Tamil Nadu and Kerala. Bitter gourd has high export potential particularly to the Gulf countries. Open pollinated varieties and hybrids with complete green, lustrous, prickly, medium long and spindle shaped fruits are suitable for export.

The crop

Bitter gourd is a tropical, fast growing, climbing perennial that may grow up to 2-3 m. The stem may be either hairless or slightly hairy. The plant with thin stems and tendrils require a trellis to support their climbing vines.

There is a central taproot, from the apex of which the stems spread to climb over any available support.The well branched, slender, green stems are usually slightly 5-angled or ridged, and carry unbranched tendrils in the leaf axils. The leaves are carried singly along the stems on 3-5 cm long stalks, and each leaf is 4-10 cm long, rounded in outline, and deeply 5-9 lobed. The foliage has an unpleasant smell when crushed.

Bitter gourd is generally monoecious in sex form where staminate and pistillate flowers are borne on separate nodes. Staminate flowers have a slender basal swelling which is continuous with the base of the sepal tube, which ends in five blunt sepals. There are five oval, yellow petals 10-20 cm long, and five central stamens. Anthesis typically occurs between 3:30 and 7:30 a.m., when flowers are complete l y open and pollen viability is lost relatively rapidly. The stigma is usually receptive for 1 day before or after flower opening, after which it dries and turns brown. The pistillate flower consists of an inferior ovary and a three-lobed, wet stigma that is attached to a columnar, hollow style. Staminate flowers appear first and usually exceed the number of pistillate flowers by about 20:1. The flower opens at sunrise and remains open for only one day.

The ovary contains three carpels typical of many cucurbits, each with 14 to 18 ovules, surrounded by an ovary wall. Although the number of ovules in an ovary can be up to 60, the average is 40. Anatropous ovules are attached to parietal placenta in two irregularly aligned rows in each carpel. Unlike other cucurbits, however, no more than four ovules can be seen in ovary cross-section.

Typically, pollen tubes penetrate papillae tissue within 1 hour of pollination arriving at the ovary cavities about 6 hours after pollination, and thus fertilization is accomplished within 18 to 24 hours post pollination.

Bitter gourd comes in a variety of fruit size (medium to very big), shape (spindle, obtuse, ovate, elongated obtuse, oblong, etc.), tubercle character (deep and sharp tubercles, medium tubercles, smooth ridges, smooth surface, etc.) and fruit colour (dark green, green, light green, yellowish-white and white). The cultivar common in China is 20–30 cm long, oblong with bluntly tapering ends and pale green in colour, with a gently undulating, warty surface. The bitter melon more typical of India has a narrower shape with pointed ends, and a surface covered with jagged, triangular "teeth" and ridges. It is green to white in colour. Between these two extremes there are several intermediate forms. Cucurbitacins, the tetracyclic triterpenoid compound makes the fruit and all other plant parts bitter. Cucurbitacins are considered a type of steroid since they are derived from cucurbitane and stored as glycosides.

Bitter gourd may contain alkaloid substances like quinine and morodicine, resins, and saponin glycosides, which may be the cause of intolerance in some people. Their bitterness and toxicity may be reduced somewhat by parboiling or soaking in salt water for up to 10 minutes. Toxicity symptoms may include excessive salivation, facial redness, dimness of vision, stomach pain, nausea, vomiting, diarrhoea, muscular weakness.

Importance and use

Tender fruits are widely used as cooked vegetables in a number of ways. The fruits are more commonly used as stir-fried, boiled and stuffed with tomato, onions, green chilies, garlic and curry leaves. White fruited varieties are less bitter in taste and particularly preferred in South India. It can be canned, pickled and stored. The fruit slices can be dried and used as a vegetable as and when required and the ratio of fresh fruit: dehydrated product, in general, is 16:1. Cooked vegetable of bitter gourd remain quite fit for consumption for 2-3 days. The bitter glucoside, cucurbitacin may help in preventing spoilage of cooked vegetable of bitter gourd. Extracts of fresh fruits contain alkaloids, highly aromatic essential oil and saponin alkaloid "momordicine". Young leaves are also used as pot herb preparations. It is also grown as ornamental crop in the USA, Japan and some other countries.

Nutritional value

Bittergourd fruits are good source of carbohydrates, proteins, vitamins, and minerals and have very high nutritive value among cucurbits. The fruits contain high amounts of vitamin C, vitamin A, vitamin E, vitamins B_1, B_2 and B_3, as well as vitamin B_9 (folate). The caloric values for leaf, fruit and seed were 213.26, 241.66 and 176.61 Kcal/100g respectively. The fruit is also rich in minerals including potassium, calcium, zinc, magnesium, phosphorus and iron, and are an excellent source of folates (72µg/100g) and dietary fibre.

Edible portion of bitter gourd contains 2.0% protein, 5.0-10.0% carbohydrates and 0.2-1.0% fat and 96.0 mg vitamin C and 210 IU vitamin A per 100 g fresh edible portion. Tender fruits also provide about 2.0 mg Fe, 40-70 mg of P, 23mg of Ca per 100 g edible portion. Cultivars with small to medium sized fruits contain higher carbohydrate and fat and lower phosphorus contents than the big fruited ones.

Medicinal value

Bitter groud is a powerful nutrient-dense plant composed of a complex array of beneficial compounds which include bioactive chemicals, vitamins, minerals and antioxidants which all contribute to its remarkable versatility in treating a wide range of illnesses. Medicinal value of bitter melon has been attributed to its high antioxidant properties due in part to phenols, flavonoids, isoflavones, terpenes, anthroquinones and organosulfur compounds. Typically abundant phenolic compounds present in bitter gourd are gallic acid, gentisic acid, catechin, chlorogenic acid and epicatechin. While the concentration of phenolics varies with plant organ (fruit, leaf, root) and cultivar type, the highest content has been found in the Indian white-fruited followed by China white-fruited, China green-fruited, and the least in green-fruited cultivars of India.

The main constituents of bitter melon which are responsible for the antidiabetic effects are triterpene, proteid, steroid, alkaloid, inorganic, lipid, and phenolic compounds. Several glycosides have been isolated from the *M. charantia* stem and fruit and are grouped under the genera of cucurbitane-type triterpenoids. The fruit contains at least three active substances with anti-diabetic properties, including charantin, which has been confirmed to have a blood glucose-lowering effect, vicine and an insulin-like compound known as polypeptide-p. Charantin increases glucose uptake and glycogen synthesis inside the cells of the liver, muscle, and fatty (adipose) tissue.

Fruits are antidotal (remedy for counteracting the effects of poison), antipyretic tonic, appetizing stomachic, antibilius (serving to prevent or cure biliousness) and laxative. It is highly beneficial in the treatment of blood disorders like blood boil, scabies, itching, psoriasis, ring worm and other fungal diseases. It is also good for curing rheumatism, diabetes and asthma. Fruits and leaves are considered anthelmintic and vermifuge and useful in piles, leprosy and jaundice. Fresh leaf juice is an effective medicine in early stages of cholera and other types of diarrhoea during summer. Leaf juice is used as emetic, purgative, given in bilious affections and rubbed in burning of the soles of the feet. Roots are astringent and useful in haemorrhoids.

Origin and taxonomy

The origin of bitter gourd remains obscure. However, the centre of bitter gourd domestication likely lays in eastern Asia, possibly eastern India or southern China. Indo-Burma centre of origin of this crop is widely accepted. This proposition is amply supported by long history of its cultivation in India. It is widely distributed in India, China, Malayasia, Thailand and tropical Africa. Uncarbonized seed coat fragments have been tentatively identified from Spirit Cave in northern Thailand however, there have been no archaeological reports of hitter gourd remains in China. Bitter gourd from its native place was taken to Brazil early in the slave trade and is now wide spread throughout he tropics.

Momordica belonging to the family Cucurbitaceae, sub-family Cucurbitoideae is monophyletic and the genus can be divided inot 11 clades. The Asiatic species falls under three sects. Dioecious species like *M. cochinchinensis*, *M. dioica, M. sahyadrica, M. denticulata, M. denudata, M. clarkeana* and *M. subangulata* grouped under the sect. Cochinchinensis, and monoecious species *M. charantia* and *M. balsamina* under the sect. Momordica and *M. cymbalaria* under the sect. Raphanocarpus. The monoecious species *M charantia* and *M. balsamina* produce edible fruits, and have been widely distributed as crops becoming naturalized throughout the tropics.

There are two botanical varieties viz.; *M. charantia* var. *muricata* (syn. var. *abbreviata*) and *M. charantia* var. *charantia*, the former is mostly wild and the latter cultivated. *M. charantia* var. *muricata* is considered as the progenitor of cultivated *M. charantia* var. *charantia. M. charantia* var. *muricata* produces small and round fruits with tubercles, more or less tapering at each end. Wide variation in fruit size (medium to very big), shape (spindle, obtuse, ovate, elongated obtuse, oblong, etc.), tubercle character (deep and sharp tubercles, medium tubercles, smooth ridges, smooth surface, etc.) and fruit colour (dark green, green, light green, yellowish-white and white) are found in cultivated type under *M. charantia* var. *charantia.*

The cytological investigations have indicated that perhaps aneuploidy is involved in the evolution of the genus *Momordica*. Chromosome number of *Momordica* species ranges between 2n = 22 (*M. charantia, M. balsamina* and *M. tuberose*) and 2n = 44 (*M. foetida*) and two dioecious and vegetatively propagated species, *M. cochinchinensis* and *M. dioica* remain in between in this respect (2n = 28). Natural triploid (2n = 33) has also been reported in *M. charantia*.

Adaptation in India

Bitter gourd was mentioned in Sanskrit language equivalent "Kara vellika" in two ancient medical treatises of India, "Charaka Samhita" and Yajurveda". Bitter gourd has also been listed as a vegetable in the "Ain-i-Akbari", the detailed account of the Mughal rulers in India. Such mentioning of bitter gourd in ancient Indian scriptures signifies its very early history in India. Its cultivation throughout India has resulted significant diversification in the genotypes. Bitter gourd has been growing in this country from ancient period and undergone significant diversification mainly in fruit size, shape, colour and weight. Large number of indigenous cultivars are grown in India under different names. Improved varieties have been developed in India chiefly through selection from the wide genetic diversity present in different states and subsequent purification of the genotype by inbreeding because bitter gourd like other cucurbits does not show inbreeding depression.

Improved open pollinated varieties

Pusa Do Mousumi, Pride of Surat, Pride of Gujarat, Small Green, Arka Harit, Pusa Vishesh, Priyanka, Konkan Tara, Gangajali, Hirkani, Coimbatore Long White, MC-23, MDU-1, Coimbatore Green, Phule Green, Coimbatore Long, Preethi, and VK-1 Priya. Punjab-14, Coimbatore Long White, Pant Karela 1, Kalyanpur Baramasi, Kalyanpur Sona, C-96, NDB-1, Phule Green Gold, MDU-1, Pant Karela-2, etc.

- **Small fruited variety** (7.5 to 10.0 cm long): Pusa Do Mousumi, Pride of Surat, Pride of Gujarat, Small Green.
- **Medium long fruited variety** (10 to 15 cm long): Arka Harit, Pusa Vishesh, Priyanka, Konkan Tara, Gangajali
- **Long fruited variety** (15 to 20 cm long): Hirkani
- **Extra long fruited variety** (20 to 40 cm long): Coimbatore Long White, MC-23, MDU-1, Coimbatore Green, Phule Green, Preethi, and VK-1 Priya.
- **Spindle shaped variety**: Konkan Tara, Hirkani , Arka, Harit, Priyanka, Gangajali, CO-1
- **Green and dark green fruited variety**: Pusa Do Mousumi, Arka Harit, VK-1 Priya, CO-1, Pusa Vishesh, Coimbatore Green, Hirkani, Phule Green, Konkan Tara, Punjab 14.

- **White fruited variety:** Preethi (MC84), Priyanka (Sel.1010), Coimbatore Long White.
- **Red pumpkin beetle resistant variety:** Pant Karela 1
- **Varieties suitable for export**: Arka Harit, Pusa Vishesh, MBTH –1, Hirkani, Konkan Tara, Preethi.
- **Public sector hybrids**: Pusa Hybrid-1, Pusa Hybrid-2, Phule Priyanka.

Cultural requirements

Bitter gourd is basically a warm season crop but has a wide range of adaptability and can be grown well in subtropical climate. It is also grown in the hills in the summer season. It is susceptible to frost and does not grow well under cool conditions. Temperature below 18°C causes poor germination of seeds, stunting of growth resulting poor yield. Optimum temperature range for seed germination and growth and development is 25°C-30°C. Temperature above 36°C causes poor development of pistillate flowers leading to poor yield. Seed germination is inhibited at and below 8°C and above 40°C. Short days and comparatively low-night temperature increase the propensity of pistillate flowers in the vine as well as lowers the node number at which the first pistillate flowers appear. Long days and high temperature increase the number of staminate flowers and reduce the number of pistillate flowers. High rainfall coupled with prolonged cloudiness promote high incidence of disease particularly downy mildew and in turn drastically reduce yield. It is susceptible to water logging. It can be grown in all types of soil however, well drained fertile, sandy loam and silt loam soils with pH range 6.5-7.0 are ideal. Sandy loam soils are ideal for early crop. It is also suitable for cultivation in the partial shady condition in the coconut plantation as practiced in Kerala.

Land preparation

The field is prepared with 4-5 deep ploughings to make into a fine tilth before seed sowing. All the weeds, stubbles are removed and the land is levelled by laddering. Manures should be mixed with the soil at the time of last ploughing. Farm yard manure @ 20-25 t/ha should be mixed with the soil at the time of last ploughing. The plant has comparatively shallow root system and is susceptible to water logging so land should be well drained. Generally long furrows or channels of 60 cm width are opened at 2.0 to 2.5 m spacing and mounds are prepared at 1.5 m spacing along the channel. Seed sowing is done on the mounds by the sides of furrows and the vines are allowed to trail on the ground and it is the best method of growing bitter gourd. The crop is also grown on flat or raised beds; about 2.0 to 2.5 m wide and seed are sown on both sides of the bed. In rainy season, raised beds or mounds are prepared to facilitate drainage. Mounds in the furrow, flat bed method or in river bed system of planting are prepared with 20-25 tonnes of FYM per hectare and basal dose of fertilizers mixed with the soil.

Sowing time

Bitter gourd grows successfully in tropical and subtropical climate hence, can be grown year round in the southern and central Indian states due to prevailing mild winter there. However, the crop is grown in three main seasons: spring-summer, early autumn and rainy season. It is sown in November-December for spring-summer crop in eastern and northern plain; January- March for summer crop in eastern and northern plain; June-July for rainy season crop in eastern and northern plain; August – September for early autumn crop in eastern India; December – January for winter – spring crop in southern and western regions and April- June for the Hill regions.

Seed sowing

Bitter gourd gives little success to transplanting of seedlings. Seeds are sown by hand dibbling in the mound. Sowings can be done following flat bed or furrow method depending on the season and convenience. Pre-sowing treatment of seeds with Thiram @ 2g per kg of seeds checks damping off of the seedlings. Seed soaking in water for 6 hours also improves germination. The seed has hard seed coat and germinate slowly due to slow absorption of water. Germination takes longer time at low temperature. Seedlings raised in plastic packets particularly under polyethylene shed in winter month can be transplanted after 35-40 days with the advent of spring (January – February) without disturbing the root system. 4.0-5.0 kg seed/ha is sufficient depending on the plant population. It is better to do seed sowing in pre-irrigated moist soil than irrigate the field after sowing. Pre-sowing soil application of 20-22 kg Furadon 3G per hectare helps in protecting the plants from root knot nematodes and other pests during initial 4-5 weeks.

Spacing

Spacing is 1.2-1.5 m x 45-60 cm for spring-summer crop; 1.5-2.0 m x 50-60 m for rainy season crop; 1.2-1.5 m x 45-60 cm for early autumn crop and 1.5-2.0 m x 75-80 cm for the crop grown on the bower.

Nutrient management

Bitter gourd responds well to manuring and fertilizer application. The dose of manures and fertilizers depend on the soil type, climate and system of cultivation. Well rotten farm yard manure at the rate of 20-25 t/ha should be applied at the time of land preparation. A general fertilizer dose of 100 kg N, 60 kg P_2O_5 and 50 kg K_2O per hectare is recommended for open pollinated improved varieties. Half N along with full P and K fertilizers and farm yard manure should be given as basal and rest N should be top dressed in two split doses at the time of vining (25-30 days after sowing) and at full blooming (50-60 days after sowing). A general fertilizer dose of 200 kg N, 100 kg P_2O_5 and 100kg K_2O per hectare is recommended for the hybrids. One-fourth N along with full P and K fertilizers and farm yard manure should be given as basal and rest N should be top dressed in three split doses at the time of vining (25-30 days after sowing) , at full blooming (45-50 days after sowing) and after first harvest (70-75 days after sowing). Bitter gourd is a comparatively shallow rotted crop hence, responses well in top-dressing. High nitrogen under high temperature condition promotes staminate flowers resulting in low fruit set and yield so, fertilizer dose should be adjusted according to season of cultivation.

Crop regulation

Like other monoecious cucurbits, excessive vine growth due to high nitrogen nutrition coupled with high temperature and high soil moisture condition promote staminate flowers in the vine resulting low fruit set and low yield. The best way to control the vine growth within reasonable limits is by adjusting nitrogen fertilizer doses and frequency of irrigation. High levels of endogenous GA like substances occur in the plants between 45-60 days when the ratio of staminate: pistillate flower is low. Application of different growth regulators and micro-nutrients viz., 150 ppm ethrel, 50 ppm maleic hydrazide (0.5ml/10 litre of water), 50-100 ppm cycocel, 60 ppm gibberelic acid or 3-4 ppm boron twice, first at two true leaves of the plants i.e. 15 days after sowing and subsequently repeated 7 days after helps in increasing the yield. Soaking of seeds in 20 ppm ethrel, 3-4 ppm silver nitrate, 3-4 ppm B9 or 3-4 ppm boron induce significantly higher number of pistillate flowers in the vine and produce higher fruit yield.

Irrigation

Bitter gourd is a shallow rooted crop and roots are mostly concentrated at top 60 cm soil layer. First irrigation should be given immediately after sowing. Irrigation after sowing improves seed germination because seed germination require high water supply. Irrigation should be given at the initiation of first true leaf during spring-summer and at its expansion during rainy season. The crop should be irrigated at 3-4 days interval in summer until it flowers. Critical stages of irrigations are flower bud development and early fruit development stage when irrigation is mandatory. Over irrigation during vegetative and early flowering stages may cause excessive vine growth resulting induction of more staminate flowers in the vine. Ridge and furrow method of irrigation is the best for bitter gourd. Drip irrigation gives more fruit yield than furrow irrigation.

Interculture and weed control

In the bed system of cultivation, the vines are allowed to grow on the bed without any support. Plant grown on trellis or pandal or bower gives better crop stand and fruit yield mainly because of better use of sunlight by maximum number of leaves and higher number of side branches. Bower system of bitter gourd cultivation is very much practiced in Kerala, Tamil Nadu, Karnataka, Maharashtra, Andhra Pradesh and West Bengal. Soil at the base of the plant should be loosened before top dressing with nitrogen fertilizers followed by irrigation. In early stage of growth, the beds, ridges etc should be kept free from weeds. At the initial stages of weed growth hand weeding and hoeing are sufficient to check the weed growth. Deep intercultivation should be avoided as it may damage the shallow roots of the crop. Application of herbicides effectively controls weeds. Pre-emergence application of Glyphosphate at 4.5 kg a.i. /ha checks weed population in the fields.

Harvesting the crop

Flowering starts generally 40-45 days after sowing and the first picking can be done 60-70 days after sowing depending on the variety, sowing time, soil type and management practices. Regular harvesting at short intervals increase the number of fruits in the vine and irregular harvesting may delay the formation of successive fruit production and affect the yield adversely. The colour of the tender fruits may be light green, dark green, whitish green, yellowish-white or white depending on the variety. Fruits generally take 10-12 days after fruit set to reach marketable stage when the fruits are tender. In some big fruited variety it takes 20 days to reach marketable maturity.

Post harvest handling and storage

After harvesting, the entire insect damaged, deformed and disease affected fruits should be removed. The harvested fruits cannot be kept for long time in ambient condition so need to be sent to the market as soon as possible. Water is sprinkled over the fruits to maintain the freshness for sometime at the initial stage. The fruits can also be kept in polypropylene bags for extending the shelf life. Dipping the fruits in solution of wax emulsion (6-12%) plus sodium phenyl phenoate for 30-60 seconds prolong the shelf life up to 3 days at room temperature. Fruits can be stored in cold storage at 0.6 –1.7°C and 85–90% relative humidity for 4 weeks.

Yield

Fruit yield may be 15-18 t/ha for open pollinated improved varieties and 30-35 t/ha for the hybrids.

Ridge gourd

Ridge gourd, *Luffa acutangula* under the family Cucurbitaceae (Chromosome number: 2n = 26) is an important warm-season fruit vegetable crop, widely cultivated in India, Southeast Asia, China, Japan, Egypt and other parts of Africa. It is also known as ribbed gourd, angled loofah and Chinese okra. In India, it is cultivated both on a commercial scale and in kitchen gardens as spring-summer and rainy season crops. Cultivation cost of this crop is comparatively less which ensures good return to the farmers. The crop has also got fair export demand. In India, it is cultivated in Andhra Pradesh, Tamil Nadu, Karnataka, Gujarat, Maharashtra, Assam, West Bengal, Uttar Pradesh and Punjab.

The crop

The genus *Luffa* is monoecious with annual vines. Vine is vigorous with slender, five angled stem, Leaves subcircular, membranous, deltoid to nearly orbicular in outline, but acutely pointed at the apex, usually five to seven lobed, scrabrous, nearly glabrous and dentate margins. Tendrils are robust, often 3-fid, puberulous. The staminate flowers are borne in racemes while the pistillate flowers are solitary and short or long pedunculate. Staminate flowers: 17-20 flowers in racemes at the apex of peduncle, peduncles 10-15 cm long, pedicels 1-4 cm long, white-puberulous, probracts 3-7 x 2-4 mm, fleshy, green, ovate with 3-10 glistening glands on the upper surface, calyx tube campanulate, lobes lanceolate, 4-6 x 2-3 mm, apex acuminate, slightly reflexed, densely white-pubescent, 1-nerved, corolla pale yellow, lobes obcordate, 15-25 x 10-20 mm, both surfaces subglabrous, stamens 3, free, 1 unilocular, 2 bilocular, filaments 3-4 mm long,bearded at the base, anthers puberulous. Pistillate flowers: peduncles 5-10 cm long, ovary elongate, 10-angular, apex constricted, style short, stigmas 3, expanded, 2-lipped.

The flowers of ridge gourd open in the evening and remain open throughout the night while the flowers of sponge gourd open in early morning hours. Generally in ridge gourd, the mature unopened staminate and pistillate flower buds are used for crossing in evening and crossed flower buds are covered by paper bag. In sponge gourd, the unopened female and male flower buds are covered in the evening followed by pollination next morning.

Fruits, botanically called "pepo", are clavate-oblong, acutely 10-angled, apex obtuse or slightly acute. The fibre obtained from the mature dry fruit is xylem fibre and arranged in a cell like structure. The chemical composition of fibres is as like other lignocellulosic fibres having around 64% cellulose, 21% hemicellulose and 10% lignin.

Seeds are ovate, verrucose, 10-12 x 7-8 mm, 2 mm thick and black.

Importance and use

The young tender fruits are non-bitter which are cooked as a vegetable or used in the soups. The immature fruits are cooked in many ways and are quite commonly used as curries, stir fried, boiled and stuffed. The fibre obtained from the mature dry fruit is used in industry for filters of various sorts, good pot holders, table mats, bath room mats, slipper and shoe soles.

Nutritional value

The fruits of ridge gourd contain fair amount of minerals and fibre. Ridge gourd is good and inexpensive source of β-carotene. Ridge gourd contains 0.5% protein, 3.4% carbohydrate, 37mg β-carotene and 18 mg vitamin C per 100g edible portion.

Medicinal value

Ridge gourd has various pharmacological activities like hepatoprotective, antidiabetic, antioxidant, abortifacient and antifungal activity. Fruit is demulcent, diuretic and nutritive. It is emetic and traditionally used for the treatment of stomach ailment and fever. The ribosome-inactivating proteins (RIPs), luffins, have been isolated from seeds of ridge gourd and characterized. This abortifacient protein has been isolated from seeds of ridge gourd which posses ribosome-inhibiting properties on the replication of HIV infected lymphocyte and phagocyte cells which explain its potential as a therapeutic agent for AIDS. The seed possesses purgative and emetic properties and have been used in the treatment of asthma, sinusitis and fever.

- A recent report shows that the water extracts from fresh sponge gourds exhibited scavenging effect in terms of 1,1-diphenyl-2-picrylhydrazyl radical scavenging activity (DPPH-RSA) and butylated hydroxyanisole scavenging effect. The fresh fruits of *L. acutangula* exhibited antioxidant activity. Other than DPPH-RSA, *in vitro* antioxidant activity is also measured in terms of 2,2-azinobis-(3-ethylbenzthiazoline-6-sulphonic acid) radical scavenging activity (ABTS-RSA) and cupric ion reducing antioxidant capacity assay (CUPRAC assay).

In the developed world, the demand for *luffa* sponge products for skin care is increasing. *Luffa* sponge is a suitable natural matrix for immobilization of microorganisms and has been successful in the process of biosorption of heavy metals from waste water. The leaves are used as poultice in leprosy and in splenitis. The pounded leaves are applied locally to splenitis, haemorrhoides and leprosy. The juice of the fresh leaves is dropped in to the eyes of children in granular conjunctivitis, also to prevent the lids adhering at night from excessive meibomian gland (special kind of sebaceous gland at the rim of the eyelids inside the tarsal plate) secretion. The oil from the seeds is known to cure cutaneous problems (abnormalities of dermal fibrous and elastic tissue). The roots have laxative effects.

Origin and taxonomy

The "Loofah gourds" (*Luffa acutangula* and *Luffa cylindrica*) originated from the subtropical regions of Asia, most probably from India as a primary centre. *Luffa* is essentially an old world genus, consisting of two cultivated species, *Luffa acutangula* and *Luffa cylindrica,* and two wild species, *Luffa graveolens and Luffa echinata. Luffa graveolens* is considered as the progenitor species of the two cultivated species. *Luffa graveolens* is found in north-eastern plains, extending south to Tamil Nadu and sporadic in Eastern Himalayas. The advanced sex form of dioecism has been found only in wild primitive species *Luffa echinata*. Wild forms of the ridge gourd are extremely bitter in taste but the domesticated types are less bitter. The chromosome number of both cultivated and wild species is same (2n=26).

Adaptation in India

The genus *Luffa* under which ridge gourd belongs has been originated in subtropical Asia including India. Both ridge and sponge gourd were domesticated and known in very early times. *Luffa* was mentioned in Kautilya's Arthashastra, c. 350-300 B.C. in India. Expectedly, wide genetic variability in this crop is found in India. A large number of local cultivars varying in shape and size (one metre to a few centimetres long) are grown in India. Some of them are even slightly bitter in taste. Improved varieties have been developed in India mainly by single plant selection from the existing genetic variability followed by inbreeding to stabilize the genotype.

Improved open pollinated varieties

- Pusa Nasdar, Pusa Sada Bahar, CO-1, CO-2, Konkan Harita, Punjab Sadabahar, IIHR-8, PKM-1, Arka Sujat, Arka Sumeet, Arka Swathi, Swarna Manjari, Swarna Uphar, Deepthi, Pant Torai-1, Konkan Harita, Phule Sucheta
- Cultivar with hermaphrodite sex form: Satputia

Cultural requirements

Well drained loamy soil with good amount of organic matter is preferable for ridge gourd cultivation. Soil having pH 6.0-7.5 is suitable for its cultivation.

It requires long warm season for best production. It grows best at a temperature of 25-30°C. Rain during flowering and fruiting reduces yield considerably. Soil should have good water-holding capacity especially in summer season.

Land preparation

The land should be prepared to obtain the fine tilth to facilitate rapid and better germination. As per requirement of the crop, the raised bed, furrows or pits are prepared and field is kept readily for planting. In riverbeds, trenches or pits are prepared for sowing seeds.

Sowing time

Ridge gourd grows successfully in tropical and sub-tropical climate hence, can be grown year round in the southern and central Indian states due to prevailing mild winter there. However, the crop is grown in three main seasons: summer, early autumn and rainy season. January- February: Summer crop; June-July: Rainy season crop in eastern and northern plain; September-October: Early autumn crop in eastern India.

Seed sowing

Ridge gourd is a seed propagated crops and *in situ* sowing is generally practiced. Seeds are sown by hand dibbling in the mound or bed. Sowing can be done following flat bed or furrow method depending on the season and convenience. Seed soaking in water for 12 to 18 hours before sowing improves germination. Pre-sowing soil application of 20-22 kg Furadon 3G helps in protecting the plants from root knot nematodes and other pests during initial 4-5 weeks. 3.5-5.0 kg/ha is sufficient for open pollinated varieties. The seed before sowing should be treated with Thiram or Captan @ 3 g/kg of seed to protect against seed and soil borne diseases.

Spacing

A general spacing of 1.5-2.5 m x 60-120 cm is followed in ridge gourd.

Nutrient management

Ridge gourd responds well to manuring and fertilizer application. The dose of manures and fertilizers depend on the soil type, climate and system of cultivation. Well rotten farm yard manure at the rate of 15-20 t/ha should be applied at the time of land preparation. A general fertilizer dose of 100 kg N, 60 kg P_2O_5 and 60 kg K_2O per hectare should be applied per hectare for open pollinated improved varieties. The hybrids generally require higher fertilizer dose of 200 kg N, 100 kg P_2O_5 and 100 kg K_2O per hectare. Half the dose of N along with full P and K fertilizers and farm yard manure should be given as basal and rest N should be top dressed in two split doses at the time of vining (25-30 days after sowing) and at full blooming (50-60 days after sowing).

Crop regulation

High nitrogen under high temperature condition promotes staminate flowers resulting in low fruit set and yield so, fertilizer dose should be adjusted according to the season of cultivation. Application of NAA (200 ppm) increases pistilate flower production and in turn increases the yield significantly.

Irrigation

The first irrigation should be given immediately after the sowing. Summer crop requires more frequent irrigation than rainy season crop. It should be irrigated once in a week depending on soil moisture. During rainy season, irrigation may be required during the early growth period. Water use efficiency can be doubled with drip irrigation compared to conventional surface irrigation method.

Interculture and weed control

Plant grown on trellis gives better crop stand and fruit yield mainly because of better use of sunlight by producing maximum number of leaves and higher number of side branches.The crop needs to be trained over low trellis of 1.5 m high above the ground.During the early stage of the crops, beds, ridges etc. need to be kept free from weeds. At the time of top dressing with nitrogenous fertilizers, weeding and earthing up are done. When the vines start spreading, weeding in between the rows or ridges becomes unnecessary since vine growth can smother the weeds. Deep intercultivation should be avoided as it may damage the shallow roots of the crop. Application of herbicides effectively controls weeds. Pre-emergence application of Glyphosate @4.5 kg a.i./ha checks weed population in the fields.

Harvesting the crop

Fruits are harvested when they are still immature and the flesh should not turn fibrous. Picking is done in about 10-12 days after anthesis of pistillate flower and at an interval of 6-7 days. If there is delay in harvesting, the fruits become more fibrous and unfit for human consumption.

Post harvest handling and storage

Grading should be done according to the size of the fruit. The fruits are packed in bamboo baskets with proper cushioning to prevent injury during transit. The plastic crates are also used for packing of fruits. Fruits harvested at the marketable stage can stand 3-4 days in a cool place without any adverse effects. The fruits can be stored for 1-2 weeks at 7-10°C with 90-95% relative humidity.

Yield

Fruit yield may be 10-12 t /ha in the open pollinated improved varieties and 20-25 t/ha in the hybrids.

Ash gourd

Ash gourd, *Benincasa hispida* (Thunb.) Cogn. under the family Cucurbitaceae (Chromosome number: 2n = 24) is a monotypic genus with only one cultivated species *hispida*, synonym *cerifera*. *Hispida* refers to the hirsute pubescence on the foliage and immature fruit, whereas *cerifera* means wax bearing. It is also known as wax gourd, hairy melon, winter melon, ash pumpkin, white pumpkin, wax gourd and white gourd. Ash gourd is an important vegetable in China, India, the Philippines, Malaysia, Taiwan, Bangladesh and Sri Lanka and other parts of Asia. It is also cultivated in Latin America and the Caribbean, usually by immigrants of Chinese descent. In India, it is widely cultivated in Uttar Pradesh, Tamil Nadu, Kerala, Andhra Pradesh and Karnataka, Rajasthan, Haryana, Bihar and West Bengal.

The crop

The hairy vines of wax gourd are large and spreading. Each tendril is two or three branched, occurring opposite to the probract at the leaf axis. The large, hairy leaves are lobed and have long petioles; they emit an unpleasant odour when bruised. Wax gourd is monoecious, with solitary staminate and pistillate flowers are borne in the same vine. Above the shallow hypanthium, the petals are almost completely separate and widely spread. Sepal lobes are foliaceous.

Wax gourd, with its showy yellow flowers, large leaves, long vines and huge fruits, bears some resemblance to *Cucurbita*.

Prominent morphological difference between *Benincasa hispida* and *Cucurbita species*

Benincasa hispida	*Cucurbita species*
Three stamens of staminate flowers are separate	Three stamens of staminate flowers are more or less united
Staminate flowers are borne on long pedicels but the pistillate flowers are almost sessile	Both staminate and pistillate flowers are borne on long pedicels
Probract is present at the base of each leaf petiole	Probract is not present at the base of each leaf petiole

Anthers generally dehisce earlier before opening of the flower. Time of anthesis and anther dehiscence generally occur from 4.20 a.m. to 5.15 a.m. and 1.45 a.m. to 2.40 a.m., respectively. Pollen viability varied from 80 to 99% depending on genotype and age of pollen. Stigma became receptive 6 hour before anthesis and remained so for 12 hour after anthesis. Pollen germination starts 12 hour after pollination on the stigma.

Mature fruits are heavy, weighing even up to 40 kg although small fruit of some genotype measures about 5.5 cm in diameter. Fruits may be cylindrical or globose and rind colour is green with light-coloured speckles.

The fruit is fuzzy when young. The immature fruit has thick white flesh that is sweet and juicy when eaten. By maturity, the fruit loses its hairs and develops a waxy coating, giving rise to the name wax gourd, and providing a long shelf life. High temperature favours the formation of waxy bloom.

Immature fruits are pubescent, mature fruits may be glabrous or pubescent, some having a dense pelt of minute hairs. The hard, dry rind of the mature fruit is usually covered with a white waxy bloom. Abrasions in the waxy coating reveal the underlying green rind. The fruits are also covered with a mild prickly pubescence. Mature fruits are called winter melons because they can be stored for long period. Long as a year, the waxy coating serves to keep moisture in and insects and micro organisms out.

The major compounds identified in the ash gourds were E-2-hexenal, n-hexanal and n-hexyl formate; however, 2,5-dimethylpyrazine, 2,6-dimethylpyrazine, 2,3,5-trimethylprazine, 2-methylpyrazine, and 2-ethyl-5-methylpyrazine are the major compounds in the ash gourd beverage. The buff-coloured seeds are flat, with margins that are ridged in some cultivars and smooth in others.

Four major cultivar groups have been recognized

- Non-ridged fruit group, which has a large fruit (0.5-1m long), cylindrical dark green rind with little or no waxy bloom
- Ridged fruit group, similar to the former but with ridged seeds and leaves with shallow lobes
- Fuzzy fruit group, seeds ridged, fruits small (20-25 cm in length), narrowly cylindrical, rind green, covered with white soft hairs without waxy bloom
- Wax gourd with ridged seeds, fruits small to large (10-60 cm in diameter) round to oblong, rind light green, covered with hair or without, covered with waxy bloom, mature leaves deeply lobed.

Importance and use

Ash gourd is cultivated for its immature as well as mature fruits. Immature fruits are more often used as a cooked vegetable to make a variety of curries. Immature fruits are used to prepare soup and pickle. Young leaves, vine tips and flower buds are boiled, light fried and eaten as greens. The matured fruit are used to prepare a famous candy called "Petha". The matured fruits are also used in confectionary and ayurvedic medicinal preparations. In China, occasionally, the entire mature fruit is steamed, often stuffed with lotus seeds, vegetables, meat or other ingredients. The famous ayurvedic preparation 'Kooshmanda rasayana' is made of ash gourd fruits. A small fruited medicinal ash gourd is also grown in Kerala. Seeds are often consumed as fried, but usually as a medication instead of a food. The fruit wax, which will develop even after the fruit is harvested, is sometimes used to make candles.

Ash gourd is graft compatible with muskmelon scion. Muskmelon grafted on ash gourd root stock can be grown in areas affected by fusarium wilt caused by *Fusarium oxysporum f. sp. melonis*, as ash gourd shows resistance to wilt.

Nutritional value

The fruits have very high moisture content (96%) and are low in calories and carbohydrates. Both mature and immature fruits are easily digested when consumed. The fruits contain Vitamin B1, Vitamin B3 and Vitamin C, various minerals such as calcium, sodium, zinc, iron, phosphorus, manganese, copper, magnesium, selenium and potassium. Due to its high content in potassium, it helps to maintain blood pressure level.

Medicinal value

Fruits, seeds, leaves and roots are used in various other medications throughout southern Asia. Ash gourd is anthelmintic, antiperiodic, aphrodisiac, diuretic, epilepsy and haemophysis. It also has the capacity to cure nerve diseases and blood sugar lowering principle. It is considered good for people suffering from nervousness and debility. Juice of the ripe fruit is used for curing insanity and epilepsy, urinary infection and biliousness. It possesses anti-ulcer activity and can be used as a tonic for heart. In Malaysia, the cooling juice of the plant is rubbed on bruises. The Chinese apply rind ashes to wounds.

Origin and taxonomy

Ash gourd is indigenous to Asian tropics and believed to have originated in Java and Japan. It has been cultivated in China for more than 2300 years. Very high diversity occurs in Indo-China and India which indicated that domestication might have taken place in Southeast Asia and India. It has been mentioned in the Chinese literature of the fifth and sixth centuries and in the Athrava Veda 800 B.C. and early Buddhist Canonical works, 500 B.C. in India. Sixth century writings about northern Chinese agriculture refer to ash gourd as an introduction from the south, suggestive of an earlier origin probably in South – East Asian region.

It is monotypic genus with the only cultivated species *Benincasa hispida* (Thunb) Cogn. however, its wild species is unknown.

The ash gourd has 2n = 24 chromosomes. Its primary basic chromosome number is X = 6. It has eight pairs of medium and four pairs of sub-terminal chromosomes. There are six groups of twelve pairs of chromosomes based on their chromosome length and arm ratio. Total length of the chromosome complement is 57.36μm with an average chromosome length of 2.39μm. It is presumed ash gourd may be a stable tetraploid derived from an ancestor having a basic chromosome number, x = 6.

Adaptation in India

Ash gourd came to India from Japan through foreign navigators and missionaries. It has been introduced in India long before and its name was referred in the Athravaveda during 800 B.C. Its extensive cultivation favours development of wide genetic diversity in this crop in India. All the improved varieties have been developed in India through selection from the indigenous germplasm.

Improved open pollinated varieties

- **Globular fruited variety:** CO-1, Kashi Ujjwal
- **Oblong fruited variety:** CO-2, Pusa Ujjwal, Kashi Dhawal
- **Cylindrical fruited variety:** APAU Shakthi
- **Big fruited variety:** Mudliar

Cultural requirements

Ash gourd grows well in warm and humid climate, preferring 24-30^{0}C temperature in daytime and 19-20^{0}C during night. However, temperature above 40^{0}C has adverse effect on flowering and fruiting and yield is affected adversely when temperature exceeds 35^{0}C. It is susceptible to frost. It produces more pistillate flowers under short day, low night temperature and humid conditions.

Well-drained, loam or sandy loam soil with high organic matter content is ideal for growing ash gourd. Under rainfed conditions, it can be grown successfully in clayey soil. Optimum soil pH is 6.0-7.5 but it can be grown also on slightly acidic soil (pH-5.5-6.5).

Land preparation

The land should be prepared to obtain good tilth to facilitate rapid and better germination. As per requirement of the crop, the raised bed, furrows or pits are prepared and field is kept ready for planting. In riverbeds, trenches or pits are prepared for sowing seeds.

Sowing time

Ash gourd grows successfully in tropical and sub-tropical climate hence, can be grown year round in the southern and central Indian states due to prevailing mild winter there. However, the crop is grown in two main seasons: summer with seed sowing during January- February and rainy season with seed sowing during June-July.

Seed sowing

Seeds are dibbled on the mounds on either side of the flat or raised beds. Sowing can also be done on the mounds along the channels or furrows. Mounds are prepared by digging pits of 45 cubic centimetre dimension and soils mixed with farm yard manure are returned. Normally, 2-3 seeds are dibbled on each mound and later, only one plant is allowed to grow. In river-beds, seed sowing is done in trenches. Seed rate is 5-7 kg per hectare. The seeds are treated with *Trichoderma viride* at the rate of 4 g or *Pseudomonas fluroscens* at the rate of 10 g or Carbendazim at the rate of 2 g per kg of seeds. Ash gourd seeds soaked in water overnight before sowing helps in better sprouting. It can also be grown by transplanting 25 days old healthy seedlings raised in portrays under shade net house.

Spacing

The distance between rows is 1.5-2.5 m and 0.6-1.2 m between the mounds. Normally 2-3 seeds are dibbled on each mound and later, only one plant is allowed to grow.

Nutrient management

Ash gourd cannot be considered an exhaustive crop however, the dose of manures and fertilizers depend on the soil type, climate and system of cultivation. About 15-20 tonnes FYM or compost per hectare are applied to the soil at the time of land preparation. Fertilizer requirement for open pollinated improved varieties is about 80 kg Nitrogen, 50-60 kg phosphorus and 60-80 kg potash per hectare. Half the quantity of nitrogen along with total amount of phosphorus and potash are mixed and applied as basal dressing near the root zone at the time of furrow or pit preparations, while the remaining two-third amount of nitrogen is applied as top dressing in two equal splits about 25 days (start of vine growth) and 45 days (flower initiation) after seed sowing.

Crop regulation

Medium and low temperatures during seedling stage boost the female flower differentiation after transplanting, compared with high temperature treatment. Lower temperature increases the content of ABA and IAA in shoot apex of ash gourd. Temperature may affect pistillate flower differentiation of ash gourd by changing the contents of ABA and IAA in shoot apex. High nitrogen under high temperature condition promotes staminate flowers resulting in low fruit set and yield so, fertilizer dose should be adjusted according to the season of cultivation.

Irrigation

Ash gourd has extensive root system and responds well to irrigation. Irrigation during summer should be given at an interval of 3-4 days depending upon weather and soil type. Irrigation should be given in channels in ridge-sown crop and flood irrigation is generally recommended during summer to create the warm and humid micro-climate which favours the growth, flowering and fruiting. Vine growth, flowering and fruit development are the most critical stages for irrigation requirement since water stress during these phases may reduce plant growth, flowering and fruiting. Moisture deficit during flowering and fruiting may cause wilting and drying of apical portion of the developing fruits, however, frequent and heavy irrigation should be avoided, especially in heavy soils, as it promotes excessive vegetative growth. In rainy season, proper drainage is equally important to drain out the excess rainwater since water logging reduces the photosynthetic rate by reducing stomatal conductance and chlorophyll content of leaf.

Interculture and weed control

The vines may be trained over low trellis of 1.5 m high or trailed over the dried twigs spread over the ground for proper growth. Soil at the base of the plant should be loosened before top dressing with nitrogen fertilizers followed by irrigation. At the initial stages of weed growth hand weeding and hoeing are sufficient to check the weed growth. Application of herbicides effectively controls weeds. Pre-emergence application of Besulide @ 4-6 kg a.i. /ha or Alachlor @ 2.5 kg a.i. /ha checks weed population in the fields. Application of Gramaxone @ 0.2% controls the weeds during early stage of crop growth.

Harvesting the crop

The crop matures in about 90 to 140 days after seed sowing depending on the cultivar and stage of the fruit during harvest. Immature, green, hairy fruits which are usually harvested 14-15 days after anthesis are the best for the use of cooked vegetable. The immature fruits should have uniform colour and size with partially developed seeds. Development of thick layer of wax is the index for judging the right stage for the harvesting of the fruits at full maturity.

Post harvest handling and storage

Green immature fruits can be stored for 10-14 days at 10-12.5°C and 85-90% relative humidity. However, fully matured fruits having a thick layer of wax may be stored even up to 12 months at room temperature in a cool dry place. Mature fruits are called winter melons because they can be stored for long period. Matured fruits are suitable for long distance marketing to the confectioners who prepare candy (petha) and preserve. Ash gourd fruits are quite resistant to water loss but susceptible to chilling injury and symptom of chilling injury includes widely distributed very small colourless sunken spots which become larger later followed by a darkening of external and internal tissues and watery break down.

A candy called 'Petha' from ash gourd is made by cooking mature fruit pieces in sugar syrup of increasingly heavy concentrations until it becomes tender and transparent. Cane sugar and invert sugar may be used in equal proportion for making syrup. Initially the pieces are boiled at 30° brix and allowed to stand for 24 hours and the process is repeated with 40°, 60° and 70° brix syrup.

Yield

Fruit yield depends on several factors such as variety, season, soil type and climatic conditions of the growing season. In general, average yield of open pollinated cultivars varies from 25-30 t/ha (140 days cropping span) and of the hybrid 40-60 t/ha (160-170 days cropping span).

Pointed gourd

Pointed gourd, *Trichosanthes dioica* under the family Cucurbitaceae (Chromosome No. 2n = 22) is also called in different Indian names of "Parwal", "Parmal", "Panal" and "Patal". This vegetatively propagated, dioecious and perennial cucurbit being the highly accepted vegetable holds a coveted position in vegetable markets of India, particularly during summer and rainy season. Tender fruits are available in the extensive period of eight months from February to October. This perennial vine crop survives for long time through giving rise to sprouts from the tuberous roots even left uncared. This important crop is widely grown in the eastern Uttar Pradesh, Bihar, West Bengal, Assam, Odisha, Madhya Pradesh, in some parts of Maharashtra and Gujarat and some hilly tracts of Andhra Pradesh and Tamil Nadu. It is also cultivated in different river beds of the country.

The crop

The plant is a perennial, dioecious, and grows as a vine which may extends up to 5-6 m. Vines are pencil thick in size with dark green cordate, ovate, oblong, not lobed, rigid, leaves. Roots are tuberous with long tap root system. Typically each node of the male plant bears a leaf on a long pedicel, a simple bifid or sometimes unbranched tendril, and a glandular bract. It may also have one or sometimes two solitary staminate flowers. In female plants, flowers are present in leaf axils. Flowers are tubular white with 16–19 days initiation to anthesis time for pistillate flowers and 10–14 days for staminate flowers. Stigma remains viable for approximately 14 hours and 40–70% of flowers set fruit.

The fruit, botanically called pepo, is globose, oblong and smooth, where the edible portion is mainly the pericarp with a little mesocarp. The fruits are green with white or no stripes. Size can vary from small and round to thick and long, 5 to 15 cm. Considerable variation exists in fruit shape, size and striation patterns. Fruits can be grouped mainly into four categories

- Long, dark green with white stripes, 10–13 cm long
- Thick, dark green with very pale green stripes, 10–16 cm long
- Roundish, dark green with white stripe, 5–8 cm long
- Tapering, green and striped, 5–8 cm long

Pointed gourd is usually propagated through vine cuttings and root suckers. Seeds are not used in planting because of poor germination and inability to determine the sex of plants before flowering. As a result, crop established from seed may contain 50% nonfruiting male plants. Both pre-rooted and fresh vine cuttings are used for propagation. Fresh vines used for field planting should have 8–10 nodes per cutting.

Importance and use

Great possibility exists for exploring its export to South-East Asia, Gulf countries and even to European countries. The tender fruits of pointed gourd are generally consumed as cooked and fried vegetable dishes, and also used in making curries. The immature fruits are used for preparing pickles. A famous sweet is prepared in India keeping the fruits filled with milk cake and immersed in sugar syrup. The newly emerged tender shoots with leaves are also a preferred potherb in many households in India.

Nutritional value

Pointed gourd is a good source of vitamins and minerals. It is a good source of carbohydrates, vitamin A, and vitamin C. It is the highest dietary fibre containing vegetables. It is called king of gourd because of having higher nutrient content than other cucurbitaceous vegetables. Protein content of the fruit is 10 times that of bottle gourd and four times that of snake gourd, ridge gourd and wax-gourd. Vitamin A content is about 5 times that of ridge gourd, 3 times of pumpkin and almost 100 times of bottle and wax gourd. Vitamin C content of pointed gourd is much higher than that of any gourd. It contains 2% protein, 0.3% fat, 2.2% carbohydrates, 153 IU vitamin A and 29 mg vitamin C per 100 g edible portion. Tender fruits also contain 3.0 gram fibre, calcium 30 mg, Potassium 83.0 mg, Phosphorus 40.0 mg, Mg 9.0 mg, Na 2.6 mg, Cu 1.1 mg and S 17 mg per 100 g edible portion. It also contains tannins, saponins, alkaloids, mixture of noval peptides, proteins, tetra and pentacyclic triterpenes, etc.

Medicinal value

Tender fruits are easily digestible, diuretic, laxative and cardiatomic. Fruits are particularly recommended during convalescence and for bronchitis, biliousness high fever and nervousness. It also invigorates the heart and brain and is useful in the disorders of the circulatory system. The fruits show some prospects in the control of certain cancer like conditions. It is particularly beneficial for colon problem due to very high dietary fibre content. The leaves and stems are hypocholesterolemic, hypoglyceridimic, hypoglycemic, hypophospholipemic and commonly prescribed for digestive complaints. The roots are diuretic and good medicine for ascites.

Juice of the leaves is used as tonic, febrifuge, in edema, alopecia, and in subacute cases of enlargement of liver. In Charaka Samhita, leaves and fruits find mention for treating alcoholism and jaundice. According to Ayurveda, leaves of the plant are used as antipyretic, diuretic, cardiotonic, laxative, antiulcer, etc.

Origin and taxonomy

Trichosanthes is a large genus, principally of Indo-Malayan distribution, with about 44 species of which 22 are found in India. Much earlier, De Candolle recorded in his book "*Origin of Cultivated Plants*" the species of *Trichosanthes* as being of Oid World origin, most probably India. So far, the centre of origin of *Trichosanthes* is not precisely known even though most authors agree on India or the Indo-Malayan region as its original home. The proposition of Assam-Bengal region as the primary centre of origin of pointed gourd was based on the rich diversity of this crop in this region including Bangladesh. Several wild forms of pointed gourd occur particularly in Assam plains and Brahmaputra valley of India. However, wild forms of pointed gourd are also available through out North India.

Pointed gourd (*Trichosanthes dioica* Roxb.) of the family Cucurbitaceae belongs to the sub-family Cucurbitoideae, tribe Trichosantheae and sub-tribe Trichosanthinae. It is one of the two cultivated species, *Trichosanthes dioica* and *Trichosanthes anguina* (snake gourd) among the 22 species occurring in India.Some of the wild species found throughout the world are *Trichosanthes cucumerina*, *T. palmata, T. cordata, T. nervifolia, T. cucumerina, T. wallichiana, T. cuspida, T. incisa, T. laciniosa, T. kirilowii*, etc. The basic chromosomes number is 11 (2n=22). A heteromorphic pair of chromosome, "an incipient sex chromosome" present in all the clones is suggested to carry the sex determining genes.

Adaptation in India

Pointed gourd was known in India in very early times and Sanskrit language equivalents signifying Vedic period and Aryan usage is available for pointed gourd as "Patol". Cooking of this fruit "Patol" have been mentioned in two medical treatises "Caraka Samhita" and Yajurveda". Several wild forms of pointed gourd occur particularly in Assam plains and Brahmaputra valley of India. Extensive clonal variation in this crop exists in West Bengal, Assam, Tripura, Bihar, and eastern part of Uttar Pradesh and asexual propagation perpetuate large number of diverse cultivars on farmers' field. Only fruit characters have been found contributed significantly to the separation of the female clones of pointed gourd. The clones could be grouped under four major types depending on fruit shape and size viz. spindle shaped fruits, oval fruits, nearly cylindrical fruits and plant bearing small fruits of different shape with dark, light green or pale-white colour with or with out stripes.

Important cultivars

Fruits of different cultivars are marketed under different local names without any standardization in nomenclature. Some of such popular local cultivars in three important pointed gourd-growing states are given below:

- **West Bengal:** Damodar, Kajli, Kajli Bombai, Kajli Damodar, Sandhyamani, Hilly, Guli, Haibathkhali, Shampuria, Dhapa, etc.
- **Uttar Pradesh:** Dandali, Kalyani, Guli, Bihar Sharif, Guthaliae, etc.
- **Bihar:** Dandali, Nimia, Hilly, Santokhwa, Bihar Sharif, Niria, etc.
- **Madhya Pradesh:** Green Oval, Green Long Striped, White Oval
- **Improved clonal selection:** Swarna Rekha, Swarna Alaukik, Rajendra Parwal-1, Rajendra Parwal-2, FP-1, FP-3, FP-4, FP-5, BCPG-3, BCPG-4, BCPG-5.

Cultural requirements

Pointed gourd is a warmth-loving crop thus, thrives well under hot or moderately warm and humid climate. Abundance of sunshine and fairly high rainfall favour good crop yield. Optimum temperature for proper growth ranges from 25° to 35°C. Regeneration of new sprouts is generally impaired below 20°C and severe cold below 5°C is bullying for the crop. Vine growth becomes highly restricted during winter, which starts again along with sprouting from fleshy root at the onset of spring. Pointed gourd can be forced during harsh winter months (November to February) in different river-beds or river basins, familiarly called 'Diara' lands in Uttar Pradesh and Bihar, through planting of rooted vine cuttings in sand of riverbeds with required watering till they drive roots up to the water level below. Such moisture laden sandy beds get warm up quickly and wipe low temperature effect of the winter. High rainfall coupled with prolonged cloudiness promote high incidence of fruit rot disease and in turn drastically reduce yield. It can be grown in a wide variety of light textured soils having good drainage facility. Well-drained sandy-loam to loam soil with slightly acidic reaction (pH 6.0-7.0) is ideally suited for this crop. The crop can withstand water stress. It is susceptible to water logging hence, grown successfully in upland condition. Deep well drained fertile, loamy or sandy-loam soils with pH range 6.0-7.0 are ideal.

Land preparation

The land with good drainage and plenty of sun shine is preferred for this crop. The field is prepared with 4-5 deep ploughing to make fine tilth before planting of vine or root. All the weeds and stubbles are removed and the land is levelled by laddering. Manures should be mixed with the soil at the time of last ploughing. The plant is susceptible to water logging so land should be well drained. Raised beds of 15-20 cm high with 3.0 m width and convenient length are prepared. A spacing of 60-75 cm is maintained between two beds which serves the irrigation-cum-drainage channel. Mounds are prepared on both sides of the bed at close proximity of the channel at 60 cm spacing. Mounds in the bed method or in river bed system of planting are prepared with the mixture of 15-20 tonnes of FYM per hectare and basal dose of fertilizers, mixed with the soil.

Planting time

Early planting can be done in upland situation but it is not beneficial in medium land situation as frequent heavy downpour hampers establishment of the cuttings while, planting of the vines after the onset of winter delays sprouting and establishment of the plants. Planting time is late August to mid September in upland situation; first to second week of October in medium land situation in high rainfall areas and late October to November in the river beds.

Planting

Unlike other cucurbits, the seed is not recommended as commercial propagating material because of poor seed viability and germinability, production of about 50% of male plants from the seeds due to segregation of male and female plants and long, even two years time for initiation of flowering. Vine cutting of 50- 60 cm having 4-5 nodes should be taken from the matured vine. The rooted cuttings should be of 15-20 cm in length and of pencil thickness. Pointed gourd is a long duration crop of about 8 months so, a very high-density planting affects its growth and yield adversely. Generally 5000-6000 cuttings are required for one hectare land in 2.0 x 2.0 m spacing. 6000 plants per hectare in 3.0 x 0.60-0.75 m spacing also give satisfactory yield. Undeveloped fruit due to lack of pollination is a common problem in pointed gourd cultivation. In the field 8-10 percent male plant population is considered enough for getting maximum fruit set in female plants.

Spacing

Spacing is 2.0 m x 2.0m; 3.0 m x 0.60-0.75 cm for planting in bed; 2.0 m x 2.0m in river bed system and 2.0 m x 60 cm for the crop on the bower.

Planting methods

Vine cutting: The defoliated vine cuttings are planted in the field following different methods.

- **Ring Method**: This planting method is widely followed where the cutting are coiled into a spiral or ring shape and planted directly on the mounds, covering one-half of the ring under the soil.
- **Lachhi method**: The cuttings are folded into a figure of eight and placed flat in the mound and pressed 3-5 cm deep in the middle into the soil.
- **Moist lump method**: The cuttings are encircled over a lump of moist soil leaving both ends of 15 cm free and the lump are buried 10 cm deep leaving the ends of the cuttings above the soils.
- **Straight vine method**: The cuttings are planted horizontally, 5 cm deep leaving both the ends above the soil.
- **Rooted cutting method**: Cutting previously rooted in the nursery under sand medium are planted late particularly in the diara land of Uttar Pradesh and Bihar.
- **Root cutting:** The tuberous roots from the matured vines are collected in September-October and planted on the mounds.

Nutrient management

Pointed gourd responds well to manuring and fertilizer application. Farm yard manure should be applied @ 15-20 t/ha at the time of preparation of mounds. In September-October planting, a general fertilizer dose of 150 kg N, 60 kg P_2O_5 and 40 kg K_2O per hectare is recommended. One-third N along with full P and K fertilizers and farm yard manure should be given as basal and rest N should be top dressed in two split doses, at 80-90 days after planting and the rest, one month later. In case of late planting, half of nitrogen and full dose of phasphatic and potassic fertilizers are applied as basal and rest half of nitrogen is side dressed 110-120 days after planting. Excessive nitrogen application under frequent irrigation promotes excessive vegetative growth, especially in heavy soils which reduce the propensity of flowering in both male and female plants.

Irrigation

The field should be irrigated once in 5 or 6 days in summer depending upon the soil, location, temperature, etc. During rainy season, irrigation is generally not required rather drainage is important. Frequent irrigation also promotes excessive vegetative growth, especially in heavy soils. Irrigation water should be restricted to the base of the plant or root zone without wetting the vines, especially when flowering, fruit set and fruit development are in progress. Frequent wetting of stems, leaves and developing fruits promote rotting disease of the vines and fruit.

Training over trellis

The crop is generally grown on the bed without any support to trail the vines. This crop can also be grown by training the vines over low trellises or bowers of 60 cm height, made up with bamboos and ropes or wires. The planting distance is reduced to 2.0 m x 60 cm to accommodate more plants per unit area. Main advantages of such types of training vines are i) lower incidence of disease particularly vine and fruit rot, ii) fruit become attractive in appearance, shape and size, iii) longer availability of fruits, iv) increase yield per unit area due to accommodation of more plants and higher return from the quality product.

Interculture and weed management

Regular weeding should be done to ensure proper growth of the vines. Mulching with straw, water hyacinth, sugarcane trash, dried grass, etc upto 8-10 cm thickness or black polythene helps in conserving moisture, suppressing weeds and protecting the fruits from rotting on contact with wet soil. Pointed gourd grown on beds are intercropped with different vegetable crops, like palak, radish, coriander, fenugreek, cauliflower, pea, etc. during early stages of growth (October-January) for better land use and greater economic return however, success of the companion crop largely depends on proper weeding. Application of weedicides like Gramoxone @ 1 litre a.i. or Fernoxon @ 0.8 litre a.i. per hectare as post-emergent spray along with mulching helps in checking the weeds effectively.

Ratooning

Consecutively three successful crops can be taken from a single planting. In ratoon cropping, the vines are pruned 15 cm from the ground level during October after the fruiting is over. Recommended dose of farmyard manure and fertilizers are applied by loosening the soil around the mound by the end of winter. Half the recommended nitrogen is applied as top dressing one month after flowering.

Harvesting the crop

Harvesting generally starts 90-120 days after planting depending on the time of planting. In October planted crop of Gangetic alluvial zone of West Bengal, harvesting of fruits generally starts from middle of February and continues to July at frequent intervals and even up to November if new flushes come. Harvesting should be done when the fruits are immature and tender before seeds become hard. Harvesting of the fruits at 12 to 15 days after fruit set depending on the cultivar is ideal for both quality consideration and yield. Picking should be done frequently so that maximum fruits could be harvested from a vine. Delay in harvesting makes the seeds mature in the fruits which reduces fruit quality and fruiting capacity of the vine.

Post harvest handling and storage

Fruits remain marketable for 2-3 days under ordinary storage conditions. After harvesting, placing of the fruits in cold water (10°C) sanitized with sodium hypochlorite solution (100 mg/ litre) for 20 minutes and then dipping in Carnauba wax solution (1 part wax + 10 part water) for one minute is very effective for increasing shelf life. Fruit dipping in 250 ppm sodium benzoate or 100 ppm citric acid solution for 10 minutes extends the shelf life of fruits up to eight days. Dipping of freshly harvested tender fruits in solution of growth substances, like kinetin (50 ppm), GA_3 (20 ppm), CCC (100 ppm) or NAA (20 ppm) for 10 minutes also increases the shelf life of the fruits upto 8 days without shrinkage and yellowing. The fruits have short storage life for 10 days at 8°C and 90% relative humidity. Fruits packed in 1 kg capacity polyethylene bag remain fresh up to 25 days at 8°C and 90% relative humidity. Because of thick skinned nature, pointed gourd can be transported to distant markets. Fruits are packed in basket or gunny bags for sending the market in a crude way. Pointed gourd must be collected in plastic crates while harvesting and packed in CFB boxes and staking properly in trucks during transit.

Yield

The fruit yield varies widely from 10.0 to 50.0 t/ha depending on the cultivar, time of planting, crop husbandry, training on trellises, pollination management and ratooning.

Plant protection measures for the cucurbits

Physiological disorders

Preponderance of staminate flowers

Predominant sex form of all the cucurbits excepting pointed gourd is monoecious. The staminate and pistillate flowers are borne separately in the same plant and fruit yield depends on the number of pistillate flowers. Preponderance of staminate flowers is caused due to i) excessive nitrogen application, ii) high temperature condition , iii) long day length and iv) over irrigation.

Control measures

- Avoidance of rampant vegetative growth through control in fertilizer application, checking in irrigation and training and pruning.
- Application of nitrogenous fertilizer at a proper dose.
- Avoidance of excess irrigation
- Application of growth substance or micronutrients for altering the sex ratio.
  - In watermelon, application of triiodobenzoic acid (TIBA) @ 50 ppm or maleic hydrazide (MH) @ 50 ppm twice, first at two true leaves of the plants i.e. 15 days after sowing and subsequently repeated after 7 days.
  - In muskmelon, application of ethephon/ethrel @ 200 ppm twice, first at two true leaves of the plants i.e. 15 days after sowing and subsequently repeated after 7 days.
  - In bitter gourd, application of ethephon/ethrel (2-Chlorethyl phosphonic acid) @ 150 ppm or 6-benzylaminopurine (BA) @ 25 ppm twice, first at two true leaves of the plants i.e. 15 days after sowing and subsequently repeated after 7 days.
  - In bottle gourd, application of ethephon/ethrel @ 150 ppm twice, first at two true leaves of the plants i.e. 15 days after sowing and subsequently repeated after 7 days.
  - In cucumber, application of ethephon/ethrel @ 150 ppm twice, first at two true leaves of the plants i.e. 15 days after sowing and subsequently repeated after 7 days.
  - In pumpkin, application of ethephon/ethrel @ 250 ppm four times, first at two true leaves of the plants i.e. 15 days after sowing and subsequently repeated three times at 7 days interval
  - In ridge gourd, application of ethephon/ethrel @ 300 ppm twice, first at two true leaves of the plants i.e. 15 days after sowing and subsequently repeated after 7 days.

Unfruitfulness

It is major disorder of pointed gourd caused due to lack of pollination, being dioecious in sex form. Pistillate flowers in the female plants are shed if pollination and fertilization does not occur. In some cases, ovary of the unfertilized flower may grow a bit due to parthenocarpic stimulation which also abscises after a few days.

Control measures

- Male plants must be grown in the field with the female plants in the ratio of 1:10 to ensure adequate pollination and fruit set.
- Plant protection chemicals should not be used in the morning hours as it hinders the visit of pollinators.
- Hand pollination should be done early in the morning (before 6.00 a.m) to achieve high fruit set. Pollens of one male flower can pollinate 7-8 female flowers satisfactorily.

Delay in fruit ripening

Delay in fruit ripening disorder is found in watermelon and muskmelon. This disorder is sometimes associated with less sweetness and cracking of fruits which occur due to high moisture level and temperature fluctuation at ripening stage.

Control measures

- Irrigation should be stopped at the ripening stage to hasten ripening.
- Sowing time should be adjusted in such a way that fruits ripe in hot and rainless condition which hastens ripening and at the same time, improves sweetness of the fruits.

Blossom end rot

This characteristic disorder is found in watermelon. Brown, water-soaked discolouration appears at the blossom end of the fruit where senescent petals are attached at the immature stage of the fruit. The spots enlarge and darken rapidly and the affected portion becomes sunken, leathery and dark coloured. This disorder is caused due to i) sudden change in the rate of transpiration specially in moisture stress condition, ii) continuously high evapo-transpiration regime and a large leaf area, iii)increasing level of nitrogen content in the fruits and iv) fall in the level of calcium content in the fruit.

Control measures

- Increase in the frequency of irrigation.
- Two foliar sprays of 0.2% calcium chloride the time of fruit development.

Disease

Powdery mildew (*Sphaerotheca fuliginea* and *Erysiphe cichoracearum*)

These are the two most commonly recorded fungi causing powdery mildew in most of the cucurbits. White, powdery fungal growth develops on leaf surfaces, petioles, and stems which are primarily asexual spores called conidia. It usually develops first on crown leaves, shaded lower leaves, and leaf under surfaces. Yellow spots may form on upper leaf surfaces opposite to powdery mildew colonies. Older plants are affected first. Severely attacked leaves become brown and shrivelled leading to premature defoliation. Plants may senesce prematurely. Fruit infection is rare on watermelon and cucumber. Cleistothecia are dark brown, minute structures (about 0.003 inches in diameter), barely discernible without a hand lens, that develop late in the growing season and protect the sexual spores within from adverse conditions. Fruits of the affected plant do not develop fully.

Control measures

- Collection and burning of all infected leaves
- Spraying the crop with Bavistin or Benlate or Topsin M @ 1g/l of water 3 to 4 times at 15 days interval.
- Two sprays of 0.1% Bavistin followed by two sprays of 0.1% Karathane or Tridemorph effectively control the disease.

Downy mildew (*Pseudoperonospora cubensis*)

Almost all the cucurbits particularly bitter gourd, cucumber, muskmelon and ash gourd are attacked by this fungus. This disease is prevalent in high humidity, especially when summer rains occurs regularly. The disease is characterized by formation of yellow, more or less angular spots on the upper surface of the leaves. White-purplish spores appear on the under surface of the leaves. The disease spreads rapidly killing the plant quickly through defoliation.

Control measures

- Growing the crop at wider spacing on well-drained soil
- Removal of infected leaves and spraying the crop with 0.25% Mancozeb 3 to 4 times at 10 days interval effectively control the disease.
- Two sprays of 0.25% Fosetyl Al or Cyamoxanil- Mancozeb or Metalaxyl-Mancozeb at 10 days interval effectively control the disease.

Anthracnose (*Colletotrichum lagenarium*)

Anthracnose is a destructive seed borne fungal disease of most of the cucurbits, particularly watermelon, muskmelon, cucumber, bottle gourd and ash gourd. This disease mostly occurs during warm and moist season and appears more during the period of heavy summer rains. All aboveground plant parts can be infected. Symptoms vary among the four cucurbits which are particularly susceptible to this disease. Leaf lesions begin as water soaked and then become yellowish circular spots.

Watermelon: Foliage the spots are irregular and turn dark brown or black.

Cucumber and muskmelon: The spots turn brown and can enlarge considerably. The lesions are dotted with pink conidia in moist condition. Stem lesions on muskmelon can girdle the stem and cause vines to wilt. Stem cankers are less obvious on cucumber and in severe cases, fruits detach from the pedicel.

Ash gourd: Symptoms are found on all the plant parts above gorund from cotyledonary leaf to first true leaves, stem, tendrils and fruits. Light brown circular spots which later turn to deep brown appear on the leaves. Elongated lesions are observed on the stem and circular to oval sunken lesions appear on the fruit which subsequently shrivels, darkens and finally dries up.

Control measures

- Use of disease-free seeds.
- Seed treatment with Thiram or Carbendazim @ 2.5 g /kg of seed.
- Rotation of cucurbits with unrelated crops in at least three-year rotation.
- Practice of good sanitation by ploughing under fruits and vines at the end of the season.
- Spraying the crop with Carbendazim (0.15%) or Chlorothalonil (0.2%) or Mancozeb (0.25%) or Hexaconazole (0.2%) at 10 days interval for 3-4 times for the management of the disease.

Damping off and Fruit rot (*Fusarium, Rhizoctonia and Phytophthora* spp.)

This fungal disease is particularly serious in bitter gourd. The fungi can infect the young seedlings causing the rot of the seedlings. However, fruit rot is very much prevalent which may occur in the field itself as well as in after harvest in storage or transit. Infected fruits are enveloped by luxuriant cottony mycelial growth of fungus.

Control measures

- Use of disease free seeds and its treatment with Thirum @ 2.5 g/kg of seeds or Carbendazim@ 2.5 g/kg.
- Crop rotation and proper drainage in the field.
- Spraying the crop with 0.25% Mancozeb or Carbendazim (1g) + Mancozeb (2g) or Metalaxyl-Mancozeb (0.25%) effectively check fruit rot in the field.

Rot disease (*Pythium* spp. and *Phytophthora*)

It is a serious fungal disease of pointed gourd. The incidence of shoot and fruit rot caused by soil borne fungi is mostly seen in rainy season, particularly when water stagnation occurs. The fungi attack the tender shoots, leaves and fruits and as a consequence, white cottony growth develops on the affected plant parts. The spread of fungi is aggravated through rain splash and irrigation water.

The other soil borne fungus, *Macrophomina phaseolina*, causing collar rot attacks root and collar region of the vines and the affected portion gradually rot and finally become dry. This disease becomes rampant in the crop preceded by jute.

Control measures

- Growing the crop on raised beds with good drainage facility to avoid water stagnation.
- Spraying the crop with Copper oxychloride @ 3-4 g/litre of water or Mancozeb @2.5g/ litre of water or Metalaxyl-Mancozeb @2.5g/ litre of water or Cyamoxanil-Mancozeb @2.5g/ litre of water at 7-10 days interval.

Fusarium wilt (*Fusarium solani* and *Fusarium oxysporum* f.sp. *lagenariae*)

This soil-borne as well as seed-borne fungus attacks the young seedlings of most of the cucurbits particularly bottle gourd, cucumber and muskmelon. This disease cause withering of cotyledons. Older plants wilt suddenly and vascular bundles in the collar region become yellow or brown. Ash gourd generally show resistance to this disease.

Control measures

- Seed treatment with Thiram or Carbendazim @ 2.5 g /kg of seed.
- Seed treatment with *Trichoderma viride* along with neem cake application is effective to control the disease.

- Application of lime to raise the soil pH to 7.0- 7.4
- Planting in well drained soils and judicious irrigation.
- Spraying the crop with 0.2% Captan or Bavistin at weekly interval checks further spread of the disease.

Fruit Rot (*Pythium, Phytophora, Rhizoctonia*, etc.)

The fungi can infect the seedlings, young plants and even young fruits of many cucurbits particularly bottle gourd and cucumber. Initial infection is noticed as small spots on the leaves, which rapidly increases in number and size. After coalescing of adjoining spots and on severely affected leaves, a burning effect or blight symptoms is seen. Fruit rot may occur in the field or it may start after harvest.

Control measures

- Use of disease free seeds
- Seed treatment with Thiram or Carbendazim @ 2.5 g /kg of seed.
- Crop rotation and proper drainage in the field.
- Spraying the crop with 0.25% Mancozeb or 0.25 % Metalaxyl-Mancozeb or 0.25 % Cyamoxanil-Mancozeb effectively check fruit rot in the field.
- Borax wash (2.5%) at 45°C for 30 seconds or at 40°C for 2 minutes before packing of the fruits prevents fruit rots in transit.

Angular leaf spot (*Pseudomonas syringae* pv. *lachrymans*)

It is an important bacterial disease of cucumber where angular brown spots appear on the leaves as first symptom. On these spots, bacterial ooze dries out to a white incrustation. The centre of such spot on leaves later dry and fall out giving ragged appearance. Spots on fruits are circular which crack open to exude amber-coloured liquid containing bacteria that subsequently infect the fruits and cause severe rotting.

Control measures

- Use of disease free seeds.
- Hot water treatment of seeds at 50°C for 30 minutes.
- Seed treatment with 0.01% Streptocycline solution for 30 minutes.
- Spraying the crop with 0.01% Streptocycline or Agromycin -100 immediately after the appearance of the disease checks the secondary spreads.

Mosaic (Bottle gourd mosaic virus, cucumber mosaic virus (CMV), Watermelon Mosaic Virus (WMV), Pumpkin mild mosaic virus)

These mosaic virus diseases are particularly severe in bitter gourd, cucumber, watermelon, bottle gourd, ash gourd and pointed gourd. These virus diseases are transmitted by sap, seed and by several species of aphids (*Aphis gossypii* and *Myzus persicae*) but it does not persist in vectors. The virus is carried over through seasons by infected crops, weeds and contaminated planting materials. The young leaves develop small greenish-yellow areas which are more translucent than the remaining parts of the leaf. Leaves show characteristics mottling with yellowish-green and blistering and in severe infection, growth of the plant is stunted. In cucumber, infected fruits are mottled with yellowish green colour and blistered. In pointed gourd, it causes mosaic mottling, irregularly distributed yellowish patches and green vein banding with deep brown to blackish vein areas on the lamina.

Cucumber green mottle mosaic virus (CGMMV)

Pointed gourd is seriously infected by this virus. The causal virus (*Tobamo virus*), an undeveloped rod shaped virus, causes irregularly distributed chlorotic spots on leaf lamina. It is transmissible by contact and mechanical inoculation. Vectors of the virus are not confirmed.

Control measures

- Spraying the crop with 0.05% Dimethoate or 0.05% Monocrotophos or 0.05% Phosphamidon at weekly interval to control aphid vectors.
- Elimination of weed hosts and infected plants from the field.
- Crop rotation with non-cucurbitaceous crops.
- Growing 5-6 rows of border crops like sunflower, sorghum or maize all round the crop sown 50-60 days before sowing the cucurbits.

Insect pests

Fruit fly (*Bactrocera cucurbita, Bactrocera dorsalis* synonym *Dacus cucurbita, Dacus dorsalis*)

All the cucurbits are attacked by this insect pest. Adult fly of this polyphagous pest having a wide host range is 4 – 5 mm long, ferruginous-brown in colour with hyaline wings. On hatching, the maggots feed inside the fruits and infested fruits can be identified by the presence of resinous liquid which oozes out of the punctures made by the flies for oviposition. The infested fruits start rotting due to secondary infection of different micro-organisms.

Control measures

- Collection and destruction of the infested fruits along with the maggots to prevent the carry over of the pest.
- Stirring the soil under the vines frequently and plough the infested field after harvest the crop to kill pupae.
- Spraying the crop with 0.05% Endosulfan.
- Application of poison bait. Bait is prepared by mixing 20 g Malathion 50% WP or 50 ml Diazinon 20% EC with 500 g molasses + 20 g yeast hydrolysate. This mixture is diluted with 2 litres of water for poison baiting and 20 liters of water for bait spray.
- Adoption of male annihilation technique using 0.1% methyl eugenol + 0.025% Malathion in a bait trap during fruiting season. Methyl eugenol is a sex attractant and males are attracted towards the bait and killed by Malathion. This trap solution is applied at 7-10 days interval.
- Use of fish meal trap with 5 gm of wet fish meal and 1 g of Dichlorvos in cotton; 50 traps are required per hectare. Fish meal and Dichlorvos impregnated cotton are to be renewed once in 7 days.

Red pumpkin beetle (*Aulacophora foveicollis*)

This insect pest attacks all the cucurbits. Shiny yellowish-red adults of this polyphagous pest lay eggs in moist soils around the plant. On hatching, the grubs feed on the roots and underground portion of host plant and also on the fruits touching the soils. Adult beetles feed voraciously on leaf lamina making irregular holes. Young seedlings and tender leaves are mainly preferred by the adults and damage at this stage may kill the seedlings.

Control measures

- Clean cultivation
- Collection and destruction of the beetles during cool hours of early morning when they remain sluggish.
- Deep ploughing of infested fields just after harvesting to kill the grubs in the soil.
- Dusting the crop with 5% Carbaryl or spraying the crop with 0.2% Carbaryl.
- Repellent dusting of ash mixed with kerosene oil but heavy dusting should be avoided as it hampers plant growth.

Epilachna beetle or Hadda beetle *(Henosepilachna vigintioctopunctata)*

It is a very destructive pest of most of the cucurbits particularly bitter gourd and ridge gourd causing damage by defoliating the entire crop of the field. Both adult and grabs are destructive and they feed by scraping the leaf tissue, leaving behind the skeletonized leaf. The late season crops are badly affected.

Control measures

- ❐ Application of carbaryl @ 2g/l is very effective against this pest.
- ❐ Clean cultivation, mechanical collection and destruction of adult beetles during cool hours of early morning when they remain sluggish.
- ❐ Collection and destruction of yellow egg masses, grabs and the invading adult population is very effective.
- ❐ Spraying the crop twice with 1.5ml Endosulfan or 2.5ml Chloropyriphos or 2.0 g Carbaryl per litre of water at 8-10 days interval effectively control the pest.

American bollworm (*Helicoverpa armigera*)

Sometimes it appears as a dangerous pest of watermelon and muskmelon causing devastating damage by feeding on the ovary and developing fruits.

Control measures

- ❐ Collection and destruction of larvae mechanically at the early infestation can reduce the destructive pest population.
- ❐ Installation of bird pertching site on the field bund helps in controlling invading population.
- ❐ Spraying the crop with Pyridalyl + Fenpropathrin @ 1.25 ml/l has been very effective.

Thrips (*Scirtothrips dorsalis*)

This insect pest cause serious damage to watermelon and muskmelon. This small insect attack on the leaves and young shoots causing silvery curling appearance. Severe attack reduces the growth of the plants. They also indirectly damage by transmitting viral diseases in the plants.

Control measures

- Spraying the infested crops with Acephate @ 1g/l or Acetamiprid @ 0.2g/l is very effective against the pest.

Aphids (*Aphis gossypii, Myzus persicae*)

All the cucurbits are attacked by the both nymph and adults which suck sap from tender twigs and ventral surface of leaves. Affected parts turn yellow, get curled, wrinkled and ultimately die away. The aphids also transmit several viral diseases in bitter gourd, cucumber, watermelon, bottle gourd and ash gourd.

Control measures

- Clipping off and destruction of the affected plant parts along with the crowded aphids thereon in the initial stage of attack.
- Spraying the crop with 0.03% Phosphamidon or 0.05% Malathion or 0.03% Methyl demeton.

Root knot nematode (*Meloidogyne incognita*)

These plant parasitic eelworms incite root galls in most of the cucurbits particularly bottle gourd, cucumber, pointed gourd, ash gourd and pumpkin. Nematode infestation cause poor and stunted growth of the infested plants. Root knot nematodes that harbour pointed gourd incite profuse galls on roots, resulting hamper in the supply of food and water inside the plant system. In severe attack, the growth remains stunted and plants may even die.

Control measures

- Long crop rotation with non-host crops like, cereals, marigold, etc.
- Ploughing the fields in the summer months.
- Application of oil cakes of mahua, neem and mustard, neem cake in particular, @ 250 kg/ha significantly suppress the incidence of *Meloidogyne incognita.*
- Growing of marigold as companion crop with pointed gourd significantly lowers gall formation and egg mass development of *M .incognita* in roots.
- Application of chopped leaves of *Ricinus communis, Calotropis procera, Leucaena leucocephala* or *Melia azedarach* in soil also helps in decreasing infection by nematode.
- Treatment of the vines of pointed gourd with chemical pesticide, like Carbosulfon @ 0.05% for 6 hours followed by shade drying before planting significantly reduces nematode incidence.
- Apply granular insecticides like Aldicarb 10 G @ 25 kg/ha or Thimet 10G @ 20kg/ha or Furadon 3G @40kg/ha.

Red spider mite (*Tetranychus urticae*)

Spider mites cause significant damage attacking on the foliages of the many cucurbits particularly pumpkin. Numerous populations continuously suck cell sap from both the surface of mature leaf which is expressed in form of numerous whitish specks on the dorsal surface of leaf. Hot and dry weather is very congenial for growth and development of red spider mite species. The plant of infested field loss vigour and dry up quickly.

Control measures

- Prophylactic application of Azadirectin 1% @ 1ml/l of water reduces the pest population.
- Spraying the crop with acaricides like, Dicofol @ 2.5 ml/l or Propergite @ 1.5 ml/l.

Brown scale insect (*Saissetia coffeae*)

It is still not a serious insect of pointed gourd except in some parts of India. This insect sucks cell sap remaining within the shelter of scale.

Control measures

- Spraying the crop with 0.2% Carbaryl or 0.05% Quinalphos
- Release of *Eublemma scitula* being an important predator to controls this insect biologically.

Root Crops

- Radish
- Carrot
- Garden beet
- Turnip

Radish

Radish, *Raphanus sativus* under the family Brassicaceae (Crucifereae) having chromosome No. 2n = 2x =18 is grown for fleshy roots and leaves in both tropical and temperate regions throughout the world. It is cultivated widely in Asia (India, China, Japan, and Korea), Europe and USA. It is grown through out India and more extensively in Uttar Pradesh, Bihar, West Bengal, Punjab, Haryana, Assam, Himachal Pradesh and Gujarat.

The crop

Radishes are annual or biennial *Brassica* crops grown for their swollen tap roots which can be globular, tapering, or cylindrical. The root skin colour ranges from white through pink, red, purple, yellow and green to black, but the flesh is usually white. The colour of pink or black skin is due to anthocyanin pigments. Numerous varieties are available, varying in size, flavour, colour, and length of time they take to mature. Smaller types have a few leaves about 13 cm long with round roots up to 2.5 cm in diameter or more slender, long roots up to 7 cm long. Both of these are normally eaten raw in salads. A longer root form, including oriental radishes, daikon or mooli, and winter radishes, grows up to 60 cm long with foliage about 60 cm high with a spread of 45 cm. The flesh of radishes harvested timely is crisp and sweet, but becomes bitter and tough if left in the ground too long.

Radish is the crop of tropical, sub-tropical and temperate climate. The tropical, high temperature tolerant types mainly developed in India are more pungent than both Oriental and European types. Large number of tropicalized cultivars with heat tolerance has been evolved in India. The European radishes are basically early, juicy and mild type with crispy roots. True Oriental radishes are also mildly pungent and crispy in nature.

Radishes owe their sharp flavour to the various chemical compounds produced by the plants, including glucosinolate, myrosinase, and isothiocyanate. Glucosinolates are very stable water-soluble precursors of isothiocyanates. The relatively non-reactive glucosinolates are converted to isothiocyanates on wounding of the radish. The tissue damage releases myrosinase, a glycoprotein that is physically segregated from its glucosinolate substrates.

Salted radish roots have a characteristic yellow colour, which generates during storage. 4-Methylthio-3- butenyl-glucosinolate (4-MTBG) is the substrate of the main pungent principle of radish and is one of the essential factors for the formation of the yellow pigment. The yellow compound 1-(2′-pyrrolidinethion 3′-yl)-1,2,3,4-tetrahydro-β-carboline-3-carboxilic acid is presumed to have been the condensation product from the degradation of 4-methylthio-3-butenylisothiocyanate and L-tryptophan, which carboline compound is considered to play an important role in the formation of the yellow pigment in salted radish roots.

Leaves are arranged in a rosette. They have a lyrate shape, meaning they are divided pinnately with an enlarged terminal lobe and smaller lateral lobes. The white flowers are borne on a racemose inflorescence. The fruits are small pods which can be eaten when young.

Importance and use

Radish is mostly eaten raw as a crunchy salad or cooked as a vegetable. Roots are also used for preparation of pickles. Young leaves are cooked as leafy vegetable particularly in northern and eastern India. The rat tail radish (*Raphanus sativa var. caudatus*) is extensively grown for its long slender pods and is cooked as a vegetable or eaten raw as salad. The oilseed radish *Raphanus sativa var. oleifer* is used for fodder and vegetable oil.

Nutritional value

Radish contains 0.7% protein, 3.4-6.8% carbohydrates and 0.2% fat and 50 IU vitamin A, 15-40mg vitamin C (ascorbic acid) per 100 g fresh edible portion. Tender roots provide about 50 mg calcium, 22mg phosphorus and 0.5 mg iron per 100 g edible portion. It also contains some trace elements including aluminum, manganese, silicon and iodine upto 18μg/100 g fresh. Radishes are low in calories and high in vitamin C, folate, and potassium. Pink skinned radish is generally richer in carbohydrate and ascorbic acid than the white skinned varieties. Radish contains glucose as major sugar and smaller quality of fructose and sucrose. Radishes are rich in dietary fibre content. Salted radish roots are essentially one of the traditional Japanese foods. The salted radish roots have a characteristic yellow colour, which generates during storage.

Medicinal value

Radish was considered as a medicinal plant in the Puranas like, Ramayana and medical treatise like "Susutra Samhita". In the Sutra period (800 B.C to 300 B.C) radish root was mentioned as aiding digestion when munched after meals. Radish roots are good for liver and gallbladder trouble, stomach and intestinal disorders, bile duct problems, loss of appetite, pain and swelling (inflammation) of the mouth and throat, tendency towards infections, inflammation or excessive mucus of the respiratory tract, bronchitis, fever, colds, and cough. In homeopathy they are used for neuralgic headaches, sleeplessness and chronic diarrhoea. The roots are useful in urinary complaints, piles and problems of spleen. According to Ayurveda, radish is believed to have a cooling effect on the blood. Radishes contain sulfurous compounds, such as sulforaphane, which have anti-cancer properties, and are expectorant.The seeds are said to be peptic, expectorant, diuretic and carminative. The juice of fresh leaves is used as diuretic and laxative.

Origin and taxonomy

It most likely originated in the area between the Mediterranean and the Caspian Sea. However, wide variability in the cultivated forms for root morphology, ecology and temperature sensitivity amply indicates multicentre origin of radish. Varieties of radish are now broadly distributed around the world, but almost no archeological records are available to help determine their early history and domestication. It is possible that radishes were domesticated in both Asia and Europe. India, central China, and Central Asia appear to have been secondary centers where differing forms were developed.

Based on recent studies using chloroplast single sequence repeats (cpSSRs), three independent domestication events was postulated which include black Spanish radish and two distinct cpSSR haplotype groups. One of the haplotype groups is geographically restricted to Asia, presenting higher cpSSR diversity than cultivated radish from the Mediterranean region or wild radish types. This implies that Asian cultivated radish cannot be traced back to European cultivated forms which spread to Asia, but might have originated from a still unknown wild species that is different from the wild ancestor of European cultivated radish.

Two distinct types of radish are available on the basis of their origin viz., European or Occidental radish and Oriental (Chinese and Japanese) radish. Maximum diversity in *Raphanus* species is available in the Mediterranean region. A few important wild species viz., *Raphanus raphanistrum, R.maritimus, R. landra, R. microcarpus and R. rostratus* available in the Mediterranean region are considered to be the probable progenitors of the small rooted European radish. However, its wild progenitor does not exist. This European radish maintains its distinctness as temperate race.

Inscriptions on the walls of pyramids show that radish was an important crop in Egypt about 4000 years ago. It was also cultivated by the ancient

Babylonians, Greeks and Romans. It was spread to China nearly 2000 years ago and to Japan some 1000 years ago.

Asiatic races of radish of widely varying root size and suiting to sub-tropical to temperate conditions have been evolved in South East Asia. Radish was introduced into China more than 2400 years ago from the eastern Mediterranean through ancient silk route and then to Japan more than 1250 years ago. The Chinese radishes are native to western region of Central Asia and south-west China or Central Asia. The Japanese radish has close relationship with *Raphanus sativus* var. *raphanistroides* which occurs wild in the coastal regions of Japan, Korea and South China. Even though China and Japan have a large range of radish varieties suiting to sub-tropical to temperate conditions, the tropical radish with heat tolerance and capable of forming roots even under day temperature of 35°C is the significant feature of adaptation of Asiatic races in India. The Indian group of radishes including the rat-tail radish evolved in the area of their present distribution in north-western region or west Asia but their ancestors do not exist.

Radish, *Raphanus sativus* belongs to the section Raphanis and family Brassicaceae (Cruciferae). There are five botanical varieties of radish having the same chromosome number ($2n = 2x = 18$) which are commonly cultivated in various regions of the world.

- *R sativus* var *radicula*: Small, cool season radish
- *R sativus* var. *niger*: Black or Spanish radish; large radish with wider range of temperature adaptation.
- *R sativus* var. *caudatus*: Rat tail radish or "Mougri" radish forming no fleshy roots but produce large slender (20-60cm) pods.
- *R sativus* var. *oleifera*: Oilseed radish; fodder radish producing no fleshy roots.
- *R sativus* var. *raphanistroides*: Japanese and Chinese winter radish.

All the five botanical varieties intercross freely among each other and also with related wild species.

Adaptation in India

Radish is one of the most ancient vegetables grown in India. Radish was cultivated in India from ancient times during 300-800 B.C. Although, introduction of radish in India has not been chronicled, mentioning of radish in different ancient literatures points its introduction before Christian era and "Mulaka" was the Sanskrit equivalent of radish in ancient India. In "Susrut Samhita", the medical trreatise of 3rd to 4th century A.D. radish had been recorded having medicinal properties. Development of large range of tropicalized cultivars suitable for growing in subtropical and tropical condition with heat tolerance and capable of forming roots under day temperature of even 35°C is the significant feature of adaptation of Asiatic races in India. These tropicalized Asiatic types flower and produce seeds freely in the plains with out vernalization or low temperature requirement. These adaptable Asiatic types of India (Desi type) have been utilized in breeding either in straight selection or in combination breeding scheme. The cultivars grown in India can be categorized under two broad groups viz., European or temperate type and tropicalized Asiatic type depending on their temperature sensitivity, root characters and seed production behaviour. The European radish maintains its distinctness as temperate race and most of the cultivars of temperate European radish race grown in India are exotic introductions mainly at the time of British regime.

In India there are many local cultivars of radish grown by the farmers under different names. Some of such cultivars are Jaunpuri Giant (very big, about 1 metre long, 50-60 cm in girth and 10-15 kg in weight) in Jaunpur, Uttar Pradesh,; Kontai Long (big, red skinned, pungent, juicy, 2-4 kg), Bombay Long Red (big, red skinned, pungent, juicy, 2-4 kg), Bombay Long White (big, white skinned, pungent, juicy,1-3 kg), Aus Mula, Baramasi (small rooted, grown at very high temperature condition) in West Bengal; Newari, Kannauji, Baramasi, Bombay Red, Nadauni in Himachal Pradesh, etc.

Improved varieties

Tropicalized Asiatic type: Pusa Desi, Pusa Reshmi, Pusa Chetki, Japanese White, Punjab Safaid, Punjab Ageti, CO-1, Kalyanpur No-9, Kashi Sweta, Arka Nishant, Chinese Pink, Kashi Hans, Punjab Pasand

European or Temperate type: White Icicle, Pusa Himani, Rapid Red White Tipped, Scarlet Globe, Scarlet long.

Cultural requirements

Radish is predominantly a cool season crop and grows best in mild and cool climate. These temperature requirements will be of higher side for tropicalized Asiatic cultivars and lower side for European cultivars. The tropicalized Asiatic types can tolerate high temperature than the temperate types although it performs at its best in cool growing condition. The high temperature tolerant Asiatic cultivars when grown at high temperature condition generally produce small root and comparatively profuse top. The roots become tough during hot weather condition. Relatively high temperature of 20-25°C for early root growth and low temperature of 10-18°C in the later part of root development is optimum for best flavour, texture and size of the roots.It can be grown in all types of soil but best results are obtained in light, friable loam soil that contain ample organic matter. For early crop, sandy or sandy loam soils are preferred. For summer crop a cool, moist soil gives the best result. Heavy soil or those with rocks, stones or pebbles below the surface produce rough, misshapen roots with a number of small fibrous laterals, and such soils should be avoided. Radish can be grown in slightly acidic soil also with soil pH ranging between 5.5 and 6.8.

Land preparation

The soil should be thoroughly prepared so that there are no clods to interfere with the development of the root. The soil should not contain any undecomposed organic matter, which may result in forking or misshapen roots. All stubbles, weeds, etc are removed from the land. Well rotten organic manure preferably farm yard manure or compost @ 10-15 t/ha is applied during land preparation and the land is levelled properly. It is better to prepare 20-25 cm high ridges as it facilitates irrigation in furrows made along the ridges.

Sowing time

The time of sowing depends on the type, variety and location. The tropical types when sown later than November in the plains of India bolt early. On the other hand, the European types when sown before October in the plains produce poorly developed roots. In places where the climate is mild, radish can be grown all the year round.

Northern, eastern and central Indian plains: Sowing between June-January for Asiatic types and October – February for European types

South Indian plains: Sowing between July-January for Asiatic types and October- December for European types.

North-eastern hills: Sowing starts from first fortnight of March till late October or beginning of November

South Indian Hills: Sowing is done between April- June.

Sowing of seeds

Radish is a direct seeded crop because swelled primary roots are economic plant parts. Roots may be misshaped in the transplanted seedlings. The seeds generally take 5-6 days for germination under optimum environmental conditions. Seed germination in field can be improved when seeds are treated with 5-10 ppm GA_3 before sowing. The seeds are sown on ridges or on flat beds. The seeds are sown as far as possible on the surface of the soil and covered by thin layer of soil. For long rooted cultivars, the seeds have to be sown at a depth of 1.5-3cm. Varieties of Asiatic types require 18-10 kg seeds and varieties of European types require 10-12 kg seeds for sowing in one hectare land. Seeds are sown either by broadcasting or line sowing. After a fortnight of sowing, the seedlings are thinned out maintaining the requisite spacing.

Spacing

Spacing of 30-45cm x 8-10cm is given for Asiatic cultivars and 20-25cm x 5-6 cm for European cultivar.

Nutrient management

Adequate nutrition is necessary for obtaining higher yield and better root quality. Well decomposed FYM @ 10-15 t/ha should be applied at the time of land preparation, 20-30 days before sowing of seeds. The FYM to be applied should be well decomposed because fresh or undecomposed manure induce forking of roots with many fibrous roots. A general fertilizer dose of 100 kg N, 60 kg P_2O_5 and 80 kg K_2O per hectare is recommended for open pollinated improved varieties. FYM along with the whole fertilizers should be given as basal before sowing seeds. For better growth and yield, the fertilizer should be ploughed to a depth of 25 cm. Two foliar sprays of 0.3% borax after 25, 40 days after sowing is beneficial for enhancing root growth.

Irrigation

Irrigation immediately after sowing is the best to get vigorous crop growth and production of tender roots. Next irrigation should be on the 3rd day and subsequently once in 5-7 days. Moist condition was found better in increasing the yield than wet condition. During summer months frequent irrigation is necessary otherwise the growth will be checked and the roots will be pungent and tough. The irrigation requirement for radish is about 210-250 mm.

Interculture and weed control

The plants should be thinned as soon as the seedlings establish. Shallow weeding helps for better root development. Regular hand weeding is necessary to check the growth of weeds. Radish normally tend to bulge out as it grows, so proper earthing up after 30-35 days of sowing is necessary to get quality roots. Root yield is high from the plants raised from seeds treated with GA_3 at 10 ppm. Pre-sowing application of Tok E-25 (Nitrofen 25%) control weeds effectively in the field.

Harvesting the crop

Radish roots become ready for harvesting in 25-55 days after sowing depending upon the type and variety. Varieties of the European types are quite early and become ready 25-30 days after sowing. Roots should be harvested at tender stage as the delay in harvesting cause pithiness of roots. Varieties of the European types become bitter and pithy much quickly if left unharvested after attaining full maturity. A light irrigation need to be given before harvest to facilitate easy lifting of roots. Roots are pulled out by hand ensuring that the roots do not break or get damaged during harvest.

Post harvest handling and storage

The harvested roots are washed, graded and tied in bunches before transported to market. In advanced countries commercial radish growers use a single row harvester that pulls the plants from the soil, cuts the roots from the tops then places them in burlap bags for transportation to a packing shed where the roots are washed, graded and bagged in plastic bags for market. Radish roots can not be stored for more than 2-3 days under room temperature without impairing its quality. Roots can be stored for about 2 months in cold storage at 0°– 1°C and 90-95 percent relative humidity.

Plant protection measures

Physiological disorder

Hollow root

High temperature during 16-30 days after sowing inhibits the formation of secondary meristem in the centre of roots leading to development of intercellular spaces which would otherwise have been filled with parenchyma. This results in the hollow root formation.

Wart

Wart is a physiological disorder which is a protrusion of white inner root tissue through splits in skin which mainly occur due to soil moisture deficiency.

Akashin

This physiological disorder is caused due to boron deficiency coupled with exposure of high day and night temperature of 30°C/20°C.

Pithyness

It occurs more in summer than spring or autumn crop. Pithiness in root is caused due to excess application of fertilizers, soil moisture stress and high temperature condition 3 weeks before harvest.

Control measures

- Keeping proper moisture condition in the field.
- Avoidance of growing sensitive varieties during summer.
- Two foliar sprays of 0.3% Borax after 25, 40 days after sowing.

Root forking

Roots get forked in poorly prepared soil having clods. Roots in touch of the lumps of undecomposed organic matter also get forked.

Control measures

- Land should be prepared finly so that the primary root may grow freely.

Diseases

Alternaria blight (*Alternaria raphani*)

It is an important fungal disease particularly for the seed crop. Small, yellowish, slightly raised lesion first appears on the leaves of seed stem. These enlarge many times as they become older. Lesions appear later on the stems and seed pods. Infection spreads rapidly during rainy weather, and the entire pod may be so infected that the style end becomes black and shriveled. The fungus also penetrates the pod tissues and ultimately infects the seeds.

Control measures

- Use of disease-free seed
- Seed treatment with hot water at 50°C for 30 minutes
- Seed treatment with Captan or Thirum @ 2g/kg of seed
- Spraying the crop with Mancozeb (0.25%) or Propiconazole (0.1%) or Difenconazole (0.05%)

White rust (*Albugo candida*)

Disease symptoms due to infection of this fungus appear on the leaves and flowering shoots which become deformed and bear only malformed flowers.

It produces white powdery substances in patches on the undersurfaces of the leaves.

Control measures

- Spraying the crop with 0.2-0.3% Zineb or 0.25% Metalaxyl-Mancozeb

Insect pests

Aphids (*Myzus persicae, Brevicoryne brassicae*)

Aphids are the most serious pests of radish. It attacks both seedlings and mature crops. Cloudy and humid conditions favour the spread of their infestation. In case of heavy infestation the plant are completely devitalized, leaves and shoot curl up, become yellowish and finally die.

Control measures

- Spraying the crop with Rogor @ 1.5ml/l of water.

Mustard saw fly (*Athalia lugens proxima*)

The insect attacks the crop in both vegetative and flowering stage. The blackish larvae are damaging which feed on the foliage. The leaves of severely infested crops are eaten up completely. The larvae are very shy in nature which roll and fall down with little disturbance.

Control measures

- Spraying the crop with Nuvacron @ 1 ml/l or Dursban (Chlorpyrifos) @ 2.5 ml/l or Malathion @ 2 ml/l or Sevin W. P. @ 4 g/l of water.

Yield

Root yield is15-20 t/ha in tropicalized Asiatic types and 5-7 t/ha in temperate types.

Carrot

Carrot, *Daucus carota* L. under the family Apiaceae (Umbelliferae) having chromosome No. 2n = 18 is a globally important root crop whose production has quadrupled between 1976 and 2016 through development of high-value products for fresh consumption, juices, and natural pigments and cultivars adapted to warmer production regions. It is a cool season crop and grown all over the world in spring, summer and autumn in temperate countries and during winter in tropical and sub-tropical climate. Carrots are one of the ten most economically important vegetable crops in the world. World production of carrots (combined with turnips) in 2014, was 38.8 million tonnes, with China producing 45% of the world total. Other major producers were Uzbekistan , Russia , the United States and Ukraine. Main carrot growing states of India are Uttar Pradesh, Assam, Karnataka, Andhra Pradesh, Punjab and Haryana.

The crop

The tap root system develops from the hypocotyl with secondary lateral roots branching from the xylem. Together, the hypocotyl and the tap root form the 'Carrot Root'. Most of the taproot consists of a pulpy and sugar loaded outer cortex (phloem) and an inner core (xylem). High-quality carrots have a large proportion of cortex compared to core. Although a completely xylem-free carrot is not possible, some cultivars have small and deeply pigmented cores; the taproot can appear to lack a core when the colour of the cortex and core are similar in intensity. Taproots are typically long and conical, although cylindrical and nearly-spherical cultivars are also available. The majority of the carotenoid pigments are deposited in the outer cortex. The periderm skin is composed of suberin and other waxy substances. Optimum root growth occurs at 15-21°C. The sweetness of carrots depends on the proportion of sugar still present.

Flower development begins when the flat meristem changes from producing leaves to an uplifted, conical meristem capable of producing stem elongation and a cluster of flowers. The cluster is a compound umbel, and each umbel contains several smaller umbels (umbellets). The first (primary) umbel occurs at the end of the main floral stem; smaller secondary umbels grow from the main branch, and these further branches into third, fourth, and even later-flowering umbels. A large, primary umbel can contain up to 50 umbellets, each of which may have as many as 50 flowers; subsequent umbels have fewer flowers. Individual flowers are small and white, sometimes with a light green or yellow tint. They consist of five petals, five stamens, and an entire calyx. The stamens usually split and fall off before the stigma becomes receptive to receive pollen. The stamens of the brown, male sterile flowers degenerate and shrivel before the flower fully opens. In the other type of male sterile flower, the stamens are replaced by petals, and these petals do not fall off. A nectar-containing disc is present on the upper surface of the carpels. the stamens release their pollen before the stigma of the same flower is receptive. The arrangement is centripetal, meaning the oldest flowers are near the edge and the youngest flowers are in the center. Flowers usually first open at the outer edge of the primary umbel, followed about a week later on the secondary umbels, and then in subsequent weeks in higher-order umbels. The usual flowering period of individual umbels is 7 to 10 days, so a plant can be in the process of flowering for 30–50 days. The distinctive umbels and floral nectaries attract pollinating insects. The fruit that develops is a schizocarp consisting of two mericarps; each mericarp is a true seed. The paired mericarps are easily separated when they are dry.

Asiatic carrots : This class of carrot might have been domesticated around 10th century in the areas of Afghanistan, Russia, Iran and India. The greatest diversity of these carrots is also found in these regions. These carrots are often called anthocyanin carrots because of their purple roots due to accumulation of primarily anthocyanin pigment. However, assortment of colours, such as yellow and red are also found in this class of carrot. Red and yellow carrots of this group are produced due to accumulation of carotenoid pigments, lycopene and lutein, respectively. They have pubescent leaves giving them a gray-green colour, and bolt easily.

European carrots : These carrots originated more recently during 16-17th century from the Asiatic carrots either by mutation or selection among hybrid progenies of yellow Asiatic carrots probably in Turkey. Adaptation to northern latitudes has been accompanied by change in photoperiod response. These carrots are also called carotene carrots because of their orange roots due to accumulation of primarily β-carotene pigment. Orange carrots were first cultivated in the Netherlands and the present day cultivars seem to have been originated from long orange varieties developed there. European carrots have also red, yellow or white roots. White carrots have also been originated by mutation from Asiatic carrots. White carrots and wild subspecies are grown in the Mediterranean regions.

The popularity of orange carrots is fortuitous for modern consumers because the orange pigmentation results from high quantities of alpha- and beta-carotene, making carrots the richest source of provitamin A. Lycopene and lutein in red and yellow carrots, respectively, are also nutritionally important carotenoids.

Asiatic carrots	European carrots
Roots branched, reddish-purple to purple-black or yellow	Roots un-branched, orange, red, yellow, occasionally white
Slightly dissected leaves	Strongly dissected leaves
Greyish-green pubescent foliage	Bright green, sparsely hairy foliage
Annual and does not require vernalization for initiation of flower stalk	Biennial and require vernalization for initiation of flower stalk

Importance and use

Carrots can be eaten in a variety of ways. It can be taken raw as well as in various cooked form. Only 3 % of the β-carotene in raw carrots is released during digestion which can be improved to 39% by pulping, cooking and adding cooking oil. Alternatively, they may be chopped and boiled, fried or steamed, and cooked in soups and stews. Together with onion and celery, carrots are one of the primary vegetables used to make various soups and stews. Roots are also used for the preparation of curries and pies. Carrot *Halwa* is a delicious and popular dish in India. Carrot jam is also popular processed item. The roots in the form of disc and slices can be dehydrated. Grated roots are used as salad, tender roots as pickles and sweetmeat. "Minimally processed" product known as 'baby' or 'cut and peel' carrots is very popular in USA and Europe. These are prepared by peeling and cutting roots into 3-7 cm length, followed by machine abrasive peeling that result in small slender carrot pieces with smooth surfaces and rounds-off cut ends. These processed roots resemble small carrots and are suitable for snack food and other fresh uses. Carrot juice is a rich source of βcarotene and is sometimes used for colouring butter and other foods. Black carrot is used for the preparation of a beverage called "kanji", considered to be a good appetizer. Carrot seed is the source of an essential oil, used for flavouring liquors and all kinds of food substitutes.

Nutritional value

Carrots contain 88% water, 4.7% sugar, 0.9% protein, 2.8% dietary fibre and 0.2% fat. It contains highest amount of β-carotene among the common vegetables. Raw fresh carrots contain on an average 8285 μg β-carotene /100 g fresh weight. Orange-coloured carrots also contain appreciable quantity of thiamine and riboflavin. The Asiatic types have more of anthocyanin pigments and less of β carotene and may be less nutritive. Potassium is the most abundant mineral in carrots, with a range from 443 to 758 mg/100 g fresh weight. Carrot dietary fibre comprises mostly cellulose, with smaller proportions of hemicellulose, lignin and starch. Free sugars in carrot include sucrose, glucose and fructose.

Carrots are also a good source of vitamin K and vitamin B6, but otherwise have modest content of other essential nutrients. Carrots also contain appreciable amount of folic acid, vitamin B6 and magnesium. Green carrot leaves are highly nutritive, rich in protein, minerals and vitamins and used as fodder and also for preparation of poultry feed.

Medicinal value

Rich β carotene content of carrot prevents night blindness which is caused due to deficiency of vitamin A. The lutein and zeaxanthin carotenoids characteristic of carrots have potential roles in vision and eye health.Carrot has been used as folk medicine for thread worms. Essential oil from carrot seeds has antibacterial properties. Consumption of carrot increases the quantity of urine and helps the elimination of uric acid. Addition of large amount of carrot to the diet has a favourable effect on the nitrogen balance in our body.

Origin and taxonomy

Undoubtedly carrot is one of the most ancient vegetables grown in the world. Carrot was originated in Afganistan in Western Asia which is having a wide diversity. The primary centres of carrot domestication are in the Middle East and Central Asia. The secondary centres of diversity are perhaps Ethiopia and North America. The first documented colours for domesticated carrot root were yellow and purple of Asiatic carrot group in Central Asia approximately 1,100 years ago, with orange carrots of European carrot group not reliably reported until the 16th century in Europe. European carrots have been originated from Asiatic carrots either by mutation or selection among hybrid progenies. Neither polyploidy nor structural changes in the chromosome seem to have played any role in the differentiation of the species.

From its centre of origin in Afganistan, it was taken by the Arabs to eastern Mediterranean and Spain in 12th century and north-western Europe in 14th century. Carrots were cultivated in Asia Minor in 10th and 11th centuries. It was introduced in India from Persia in 13th or 14th century. Carrot was first introduced into China in 13-14th century and Japan around 17th century. Carrots were first introduced to the US in the 17th century by European settlers.

Cultivated carrot belongs to the family Apiaceae (synonym Umbelliferae), subfamily Apioideae, genus *Daucus*, section *Daucus*, Species *Daucus carota* L. Based on morphological and anatomical features, the species *Daucus* is divided into five sections: Daucus L. (12 species), Anisactis DC. (three species), Platyspermum DC. (three spcies), Chrysodaucus Thell. (one species) and Meoides Lange (one species). All cultivated carrots, together with the wild ones which are known in Europe, belong to the species *Daucus carota* L. Most wild *Daucus* forms are found in the South-Western Asia and the Mediterranean, a few in Africa, Australia and America. All wild forms found in Asia, Asia Minor, Japan and U.S.A. have the same chromosome number (2n = 2x = 18) as their European relatives.

Adaptation in India

Carrot was introduced into India early in its history. The black and red types might have been introduced from the central Asian region of its origin-probably Afganistan (considered to be the main center of origin), Persia and southern Russia. These types with low carotenoid content were mostly annual, juicy and forming roots under comparatively higher temperature condition. These types showing wide variation in colour, shape and size of roots form the distinct tropical or oriental type. Colour of these types vary from absolute colourless to light lemon, light orange, deep orange, light purple, deep purple and almost black. Himachal hills and Kashmir valley still abound in its wild form where wild animals especially brown bear feed on its roots. In some parts of Kashmir, people still eat the wild carrots. Adapted cultivars of this type have been utilized in hybridization programmes to develop tropical varieties in India.

The orange coloured European types was introduced in India much later with the European traders. Most of the varieties under this group grown in India are exotic and these varieties have also been used in hybridization programme to breed temperate carrot varieties in India.

Improved varieties

Asiatic or Oriental and European or temperate carrot varieties, both are cultivated in India. Oriental or tropical varieties are mainly cultivated using different adaptable local cultivars and some improved varieties.

Asiatic or Oriental varieties: No. 29, Pusa Kesar, Pusa Meghali, Hisar Selection-1, Selection- 233, Hisar Gairic, Pusa Rudhira, Pusa Asita

European or Temperate varieties: Nantes, Chantaney, Early Nantes, Imperator, Nantes Half Long, Danvers, Pusa Yamdagini, Zeno, Early Horn, Early Gem Imperator, Ooty-1, Shalimar Carrot-1.

Cultural requirements

Carrot is a cool-season crop though some tropical types tolerate quite high temperature. A temperature range of 7.2 to 23.9°C is considered optimum for seed germination. The colour development and growth of the roots are affected by temperature. The growth is optimum at a mean temperature of 16- 18°C and the root yield and colour development is more at a temperature range of 20-25°C. Hence, cool night temperature is essential for carrot production in the tropics. Top growth is reduced at mean temperature of 28°C and roots become very strong flavoured. Carrot can be grown on all types of soils. It thrives best on a deep, loose and loamy soil. For early crop, a sandy loam soil is preferred but for high yield, deep, loose, loamy soil is desirable. The long, smooth, slender roots desired for fresh market can successfully be grown on deep, well-drained light soils. Carrots grown on heavy soils are more rough and coarse than those grown on light sandy soils. The soil pH of 6.5 is ideal and yield is reduced severely below this pH level.

Land Preparation

The land is prepared in well advance with repeated ploughings (at least 4-5 ploughings) to a fine tilth. All stubbles, weeds, etc are removed from the land. Well rotten organic manure preferably farm yard manure or compost @ 25-30 t/ha is applied during land preparation and the land is levelled properly.

Sowing time

In the northern Indian plains, the tropical types are sown during August to early-October and temperate types during October–December and in the hills during March–July.

Sowing of seeds

Carrot is direct seeded crop because swelled primary roots are economic plant parts. Roots may be misshaped in the transplanted seedlings. The seeds generally take 7- 8 days for germination under optimum environmental conditions. The seeds are sown on ridges or on flat beds. The seeds are sown as far as possible on the surface of the soil and covered by thin layer of soil. 6-8 kg seeds of the varieties of Asiatic types are sufficient for one hectare land. Seeds are generally sown in line. After a fortnight of sowing, the seedlings are thinned out to maintain the requisite spacing.

Spacing

Generally 30-45 cm x 22-25 cm spacing is given for Asiatic type and 30 cm x 22-25 cm for temperate type.

Nutrient management

Fresh cowdung manure should not be applied to a carrot field as it may result in branching of roots. About 20-25 tonnes of well rotten farm-yard manure

25 kg N, 50 kg P and 100 kg K/ha should be applied at the time of field preparation. Another dose of 25 kg N/ha need to be applied 35-40 days after sowing at first hoeing. If organic matter is not applied, the fertilizer dose of 80 kg N, 60 kg P and 100 kg K/ha gives good yield. Excess nitrogen affects sugar, dry matter, vitamin C, carotene contents and storage quality of roots.

Irrigation

After sowing, the ridges should be kept moist till the germination is completed. Irrigation is generally given at 8–10 days intervals before wilting of leaves starts. Insufficient moisture results in low yields however, excessive amount of water may decrease yield. Usually a light irrigation is given 2–3 days before harvesting.

Interculture and weed control

Thinning should be done at the early stage for proper root development. The seedlings are thinned to a distance of 22-25cm in the row. The seedlings grow slowly at first and cannot compete with weeds. All weeds should be removed especially in the early stages. The soil should be hoed time-to-time to allow proper aeration. Application of the herbicide, Stomp @ 3.5 litres/ha immediately after sowing is effective in controlling the weeds. Care should be taken that there is proper moisture in the field at the time of application of Stomp and if moisture is less light irrigation can be given. During weeding and hoeing, the crown of the roots is often exposed to light which causes greening and lowers the quality of roots. For proper development of roots one earthing-up may be done during root formation.

Harvesting the crop

The Asiatic (tropical types) carrots attain marketable stage of maturity when these are 2.5–4 cm in diameter at the upper end. Delay in harvesting makes the roots tough and unfit for consumption. For early market, the roots are pulled out manually when partly developed. They are normally dug out with a spade or khurpi when the soil is sufficiently moist.

Post harvest handling and storage

After harvesting, the roots are washed, cleaned, graded and tied in bunches of 6 to 12 roots. Carrots are sometimes marketed with their tops attached to indicate freshness. Fresh carrots cannot be stored for more than 3–4 days under ambient condition. Long-term storage is possible without appreciable change in quality in cold storage. At temperature of 0°–4.5°C with 93–98%

relative humidity carrots can be stored for 6 months. Storage at 98–100% relative humidity results in less decay than at 90–95% because at higher humidity condition moisture loss from the roots is considerably less and the produce remains crisper, firmer and of better colour. The optimum storage temperature for minimizing decay is 0°–1°C. There is no change in the nutritive and organoleptic properties of carrots stored in crates covered with perforated plastic films at 0°C and 93–96% relative humidity for 7–8 months. The stored carrots yield juice of acceptable commercial quality.

Plant protection measures

Physiological disorder

Splitting or cracking of roots

It is a major problem in many carrot-growing areas. Although the tendency of splitting seems to be controlled by genetic factors, a number of other factors may be involved. However, irrigation and application of herbicides do not significantly affect the amount of splitting of roots.

Control measures

- The splitting is reduced by low N and increases as the amount of N in the soil increases.
- High soil concentration of ammonium compounds causes more serious splitting than by other forms of N.
- Wider the spacing, the greater is the propensity of splitting and large roots are more likely to split than small ones.

Cavity spot

This disorder appears as a cavity in the cortex and in most cases the subtending epidermis collapses to form a pitted lesion. This disorder is mainly caused due to deficiency of calcium and increased accumulation of potassium.

Control measures

- Increase in calcium level in the growing medium.

Root forking

Roots get forked in poorly prepared soil having clods. Roots in touch of the lumps of undecomposed organic matter also get forked.

Control measures

- Land should be prepared finly so that the primary root may grow freely.

Disease

Alterneria blight (*Alternaria radicina, Alternaria dauci*)

This seed borne fungal disease first appears as dark brown lesion on the foliage with yellow edge and under favourable weather condition, the entire foliage dries up and withers.

Control measures

- Seed treatment with Thiram @ 3g per kg of seeds.
- Clean cultivation as infected debris provides the further source of inoculum.
- Spraying the crop with Mancozeb (0.25%) or Chlorothalonil (0.2%) or Propiconazole (0.1%) at regular intervals.

Leaf spot (*Cercospora carotae*)

The symptom of this fungal disease first appears as elongated lesions along the edge of the leaf segment, resulting in lateral curling. In dry weather, the spots are light tan in colour and in humid weather the spots are darker in colour.

Control measures

- Seed treatment with Thiram @ 3g per kg of seeds.
- Clean cultivation as infected debris provides the further source of inoculum.
- Spraying the crop with Mancozeb (0.25%) or Chlorothalonil (0.2%) or Propiconazole (0.1%) at regular intervals.

Insect pests

Carrot rust fly (*Psila rosae*).

The slender, yellowish-white maggot infests the leaves and burrows into the roots. Leaves turn rusty red to scarlet with some yellowing. Rusty-brown tunnels are seen under the outer skin of mature roots. Often the infestation causes the root to become mis-shapen and subject to decay and makes it unfit for consumption and marketing.

Control measures

- Spraying the crop with 0.1% Dimethoate (Rogar) 2-3 times.

Root knot nematode (*Meloidogyne incognita*)

These plant parasitic eelworms incite root galls causing poor and stunted growth of the infested plants and forking of the roots.

Control measures

- Long crop rotation with non-host crops like, cereals, marigold, etc.
- Ploughing the fields in the summer months.
- Application of neem cake @ 12-15 t/ha.
- Application of granular insecticides like Aldicarb 10 G @ 25 kg/ha or Thimet 10G @ 20kg/ha or Furadan 3G @40kg/ha.

Yield

Root yield is 25–30 t/ha in Asiatic type cultivars and 10–15 t/ha in European type cultivars.

Garden beet

Garden beet, *Beta vulgaris* subsp. *vulgaris* L. under the family Chenopodiaceae (Chromosome No. 2n = 18) is an important root crop in eastern and central Europe, mainly Poland, Germany, Austria, Romania, Russia and in the Middle East. It is also known as beet root, red beet, table beet, or simply beet. It is not grown as widely as radish and carrot is grown. In India, it is mainly cultivated in Haryana, Uttar Pradesh, Himachal Pradesh, West Bengal and Maharashtra.

The crop

Garden beet is the thick fleshy taproot portion of the beet plant. Upper portion of fleshy root develops from hypocotyls and basal part from tap root. Rosette of leaves is produced on the fleshy elongated hypocotyls. The root may be dark purplish red, yellow, white or striped. The taproot ranges in shape from globular to long and tapered. Skin and flesh colours are usually dark purplish red, though some are yellow, orange, nearly white or striped. Concentric rings seen in cross section of root are as a result of alternate formation of vascular tissues and storage parenchyma tissues. Dark purple red colour of beet root is due to presence of red violet glycosidic pigments, β cyanins and a yellow pigment, β xanthin. Rosette leaves develop in a close spiral, with older ones on outside.

During the first season, thick fleshy taproots are formed and in the second season after getting exposure to vernalization temperature, a tall, branched inflorescence (spike) arises which bears numerous sessile flowers in clusters of 3-4. The clusters of minute green flowers develop into brown corky fruits commonly called seed balls.

Importance and use

Garden beet is consumed as salad, cooked vegetable, pickle, steamed, canned and frozen product. It is delicious when eaten raw, but is more frequently cooked or pickled. Usually the roots are eaten boiled, roasted or raw, and either alone or combined with any salad vegetable. It is also marinated with lemon juice, herbs and olive oil.

A large proportion of the commercial production is processed into boiled and sterilized beets or into pickles. In Eastern Europe, beet soup, such as borscht, is a popular dish. In Indian cuisine, chopped, cooked, spiced beet is a common side dish. Yellow-coloured beetroots are grown on a very small scale for home consumption.

Its mild bitterness taste mixed with sweetness is highly appreciated. It is also the source of sugar. Garden beet leaves, packed with fibre, protein, vitamin K and calcium, can also be cooked and consumed as pot herbs. The leaves and stems of young plants are steamed briefly and eaten as a vegetable; older leaves and stems are stir-fried and have a flavour resembling taro leaves.

During the middle of the 19th century, wine often was coloured with beetroot juice. Betanin, obtained from the roots, is used industrially as red food colorant, to improve the colour and flavour of tomato paste, sauces, desserts, jams and jellies, ice cream, candy and breakfast cereals, among other applications. The chemical adipic acid rarely occurs in nature, but happens to occur naturally in garden beet. This industrially important dicarboxylic acid is prodneed mainly as a precursor for the production of nylon.

Nutritional value

Garden beet contains 88% water, 10% carbohydrates, 2% protein, and less than 1% fat. In a 100 gram amount providing 43 Calories, raw garden beet is a good source of folate, riboflavin, manganese and the antioxidant betaine. Garden beet is also a moderate source of fibre, magnesium, potassium, iron and vitamin C. Simple sugars, such as glucose and fructose, make up 70% of the carbohydrates in raw garden beet, and 80% in cooked materials. Garden beet has a glycemic index score of 61, which is considered to be in the medium range. The roots are low in sodium and fat which is good for health. Due to good source of folate, it supports mental and emotional health. Beet conatains high fibre of 2.8 g/100 g fresh which can help prevent heart disease, diabetes, weight gain and some cancers, and can also improve digestive health.

Medicinal value

From the middle Ages, garden beet was used as a treatment for a variety of conditions, especially illnesses relating to digestion and the blood. Bartolomeo Platina recommended taking beetroot with garlic to nullify the effects of "garlic-breath". Beet acts as a stomachic, laxative, remover of viscid mucus and bile and combatting agent of "dry emaciation." Beets are also well known for the blood promoting and liver cleansing qualities. Beets also possess the ability to treat hemorrhoids on account of its ability to kickstart the liver into producing more bile. Beetroot has liver-protecting benefits on account of its rich source of antioxidants. Drinking beetroot juice enhances the cardiovascular health. Drinking beet juice also helps to lower blood pressure. According to a study published in the *Asian Journal of Pharmaceutical and Clinical Research* in 2013, beetroot extracts showed potent anti-proliferative and immunomodulatory effects when tested against breast cancer cells. Betanin, the most common pigment in beetroots is believed to have various health benefits.

Origin and taxonomy

Garden beet has been originated from wild sea beet (*Beta vulgaris* subsp *maritima)* which is found wild in seashores of Europe and Asia. Both beet chard for leafy greens (*Beta vulgaris* subsp. *cicla*) and garden beet for fleshy roots were originated from *Beta vulgaris* subsp *maritima* in Europe through human selection and domestication. One of the earliest documents recorded that beetroot cultivation came from Babylonia circa eighth century BC. Roman and Jewish sources also mentioned a domesticated form of beetroot near the Mediterranean basin around the first century BC. Other remnants were found dating back to Neolithic aarts in north Netherlands, and Egypt's third dynasty (2650-2575 BC). It was taken to China by the Arabs through sea route.

Garden beet *Beta vulgaris* subsp. *vulgaris* L. belongs to the family Chenopodiaceae. There are two cultivated sub-species, namely, *Beta vulgaris* subsp *cicla*, the leaf beet "Chard" and garden beet, sugar beet and fodder beet all come under *Beta vulgaris* subsp. *vulgaris*.

Garden beet and all the wild species viz., *Beta patula, Beta atriplicifolia* and *Beta macrocarpa* have the same chromosome number 2x = 2n = 18 and all are crossable with the cultivated species, *Beta vulgaris* L.

Adaptation in India

Garden beet was introduced in India in remote times however, when and how the domesticated species arrived is mysterious. It is plausible that Europeans introduced garden beet strains during trade or invasion.

Improved varieties

Detroit Dark Red, Crimson Globe, Early Wonder, Ooty-1, Crosby Egyptian

Cultural requirements

Beets can survive frost and almost freezing temperatures. Beet is best adapted to cool or moderate climate although it can also be grown in warm climate. However, under high temperature condition, zoning is caused which is marked by the appearance of alternating light and dark red concentric circles in the root. It can tolerate the freezing temperature to some extent and comes up well in the plains during winter. The most favourable temperature for the development of root is 10° to 15°C and temperature below 10°C for 15 days generally induces flowering. This crop requires abundant sunshine for good development of roots. Unlike other root crops, garden beet does not have any tropical type.

Deep, well drained loam or sandy loam soils are highly suitable for cultivation of beet. Beet is highly sensitive to soil acidity and the ideal pH is 6-7. Beet root is one of a few vegetables which can be successfully grown in saline soils.

Land preparation

The soil should be thoroughly prepared so that there are no clods to interfere with the development of the root. The soil should not contain any undecomposed organic matter because it may cause misshapen roots.

Sowing time

Being a cool season crop, garden beet is raised during winter in plains and as a spring-summer crop in hills by March-April. In plains, seed sowing is done during September-November. Late sowing in the hills usually produces early seed stalks with out good root production.

Sowing of seeds

Beet is direct seeded crop and roots may be misshaped if seedlings are transplanted. For sowing in one hectare land, 6.0 – 8.0 kg seeds are sufficient. It is better to sow the crop on ridges set at 45-60 cm apart than in flat beds. The ridge should be kept moist until germination is completed. This helps in better development of roots and drainage. Water-soaked 'seed balls' which contain 2-6 seeds are drilled 2.5 cm deep in rows at spacing of 45-60 x 10 - 15 cm. Seeds are sown as far as possible on the surface of the soil and covered by thin layer of soil. Seeds should be sown thinly because each seed ball contains multiple seeds. Thinning and roguing are to be done for several times. Thinning should start when the leaves of seedlings start touching each other and continue till the plant to plant distances stands at about 10-15 cm. Staggered sowing at 1-2 weeks interval ensures steady supply of roots during the season.

Spacing

45-60 cm x 10-15 cm.

Nutrient management

Adequate nutrition is necessary for obtaining higher yield and better root quality. Well decomposed FYM @ 20-25 t/ha should be applied at the time of land preparation, 20-30 days before sowing of seeds. A general fertilizer dose of 100 kg N, 60 kg P_2O_5 and 80 kg K_2O per hectare should be applied for improved varieties. Entire farmyard manure, half of N and full P and K should be applied as basal at the time of land preparation prior to sowing and remaining half N at 30-45 days after sowing. Nitrate sources of N are preferred to ammonium sources. For better growth and yield, the fertilizer should be ploughed to a depth of 25 cm. Beets have a relatively high boron requirement and its deficiency causes internal breakdown as black rot or dry rot. Three sprays of 0.25% to 0.50% borax solution with a sticter 20, 35 says after sowing is sufficient to provide this micro-nutrient for optimum growth.

Irrigation

Irrigation immediately after sowing is the best to get good crop stand. Irrigation should be given subsequently once in 5-7 days. It is better to keep the soil moist throughout the growing period to enhance root yield. Irrigation requirement for garden beet is about 300 mm.

Interculture and weed control

Thinning is an essential operation when more than one seedling germinates from each seed. Moist soil is essential for seed germination and for further growth. Field is usually kept weed free by light hoeing at early stage of crop. Swollen roots are also to be covered with soil by earthing up 30-35 days after sowing to get quality roots.

Harvesting the crop

Generally, greens are not allowed to grow above 6 inches before harvesting. Harvesting is generally done 8-10 weeks after sowing by pulling the top with hand. Digging is an alternative way of harvesting beets. Later, tops are removed, graded and marketed. Over matured and oversized tubers become woody and crack.

Post harvest handling and storage

Clipping the tops of beets will keep them fresher for longer period. However, in European countries, where small sized bunches are in demand, root knobs are tied in bundles of 4-6 with their tops. Beet stores well at 0°C and 90% relative humidity.

Plant protection measures

Physiological disorder

Brown heart or heart rot

This disorder is caused due to boron deficiency. The deficient plants usually remain dwarf and stunted. The leaves are smaller than the normal. The young unfolding leaves fail to develop normally and eventually turn brown or black and die. The roots do not grow to full size and under severe deficiency they remain very small and distorted and have a rough, grayish appearance instead of being clean and smooth.

Control measures

- Correction of soil reaction and salinity.
- Soil application of borax @ 10-15kg/ha
- Three sprays of 0.25-0.50% borax solution @ 1-2kg/ha along with a sticker 20, 35 and 50 days after sowing.

Diseases

Leaf spot (*Cercospora beticola*)

It is a very wide spread fungal disease of garden beet. Small, circular spots of about 2 mm in diameter appear in huge numbers. These spots have definite margins usually darker in colour than the rest of the lesion.The lesion may drop and the leaf assumes a shot-hole appearance.

Control measures

- Seed treatment with Captan @ 2g/kg of seed or Thirum @ 2.5 g/kg of seed.
- Spraying with Blitox (0.3%) three times at an interval of 10 days.

Downy mildew (*Peronospora schachtii*)

It is also a serious fungal disease which may occur at any stage of growth. The fungus survives on the crop residues in the soil and also carried by seed. On the leaves, spots of various sizes upto 4.0 mm diameter appear. The affected portions become light green in colour on the upper surface while in the undersurface the fungal growth (mildew) is noticed.

Control measures

- Seed treatment with Captan @ 2g/kg of seed or Thirum @ 2.5 g/kg of seed
- Spraying the crop with 0.25% Mancozeb, Metalaxyl-Mancozeb (0.25%) or Cymoxanil-Mancozeb (0.25%) with a sticker at weekly intervals.

Insect pests

Beet leaf miner (*Pegomuia hyocyami*)

The white maggot, about 8 mm in length feeds on the tissues between the upper and lower layers of the leaf, thus causing serious injury to the leaf and consequently plant growth is checked.

Control measures

- Destruction of all fallen leaves and other plant refuse after harvesting the roots.
- Application of Imidacloprid @ 0.3 ml/l is very effective against the pest.

Yield

Average root yield varies from 25 to 30 t/ha.

Turnip

Turnip, *Brassica campestris* L. var.*rapifera* Mertz (Syn. *Brassica rapa* subsp. *rapa*) under the family Brassicaceae (Chromosome No. 2n = 20) is commonly grown in temperate climates worldwide for its bulbous taproot. The word *turnip* is a compound of *tur-* as in turned/rounded on a lathe and *neep*, derived from Latin *napus*. Small, tender varieties are grown for human consumption, while larger varieties are grown as feed for livestock. Top five turinp producing countres in the world are China, Uzbekistan, Russia, USA and Ukraine.

The crop

The swollen roots may be globular, flattish or cylindrical in shape and generally white in colour with a tinge of red, pink, purple or green. The most common type of turnip is white-skinned apart from the upper 1–6 cm, which protrude above the ground and are purple or red or greenish where the sun has hit. This above-ground part develops from stem tissue, but is fused with the root. The interior flesh is entirely white. The root is roughly globular, from 5–20 cm in diameter, and lacks side roots. The normal root below the swollen storage root is thin and 10 cm or more in length which is trimmed off before it is eaten or sold. The leaves grow directly from the above-ground shoulder of the root, with little or no visible crown or neck. Turnip roots may weigh up to 1 kg depending on the variety, growing season and length of time it has grown. Mild flavoured, small-rooted speciality varieties or baby turnips are generally eaten whole, including their leaves.

Importance and use

The swollen, tuberous tap root is consumed as salad, cooked vegetable and pickle. Young turnips should not need peeling and can be eaten raw, thinly sliced or grated into salads. Older turnips will need to be peeled and then sliced or diced before cooking. Turnips can also be steamed, roasted with other root vegetables. They can be added to soups and stews.

In Turkey, particularly in the area near Adana, turnips are used to flavour *°algam*, a juice made from purple carrots and spices served ice cold. In Middle East countries such as Lebanon, turnips are pickled. In Japan, pickled turnips are sometimes stir-fried with salt or soy sauce. In the United States, stewed turnips are eaten as a root vegetable.

Turnip leaves are sometimes eaten as "turnip greens" and they resemble mustard greens in flavour. The greens may be cooked with a ham hock or piece of fat pork meat. Stewed turnip greens are often eaten with vinegar.

Nutritional value

Turnip is a good source of minerals, antioxidants, and dietary fibre. It is also a low-calorie vegetable – a 100 gram serving only has 28 calories. The root is particularly high in vitamin C, with 21 milligrams per 100 grams. Turnip contains bitter cyanoglucosides that release small amounts of cyanide. Sensitivity to the bitterness of these cyanoglucosides is controlled by a paired gene. People who have inherited two copies of the "sensitive" gene find turnips twice as bitter as those who have two "insensitive" genes, and thus may find turnips and other cyanoglucoside-containing foods intolerably bitter.

The green leaves of the turnip top are a good source of vitamin A, folate, vitamin C, vitamin K and calcium. Turnip greens are also high in lutein (8.5 mg/100 g).

Medicinal value

Sulforaphane compound of cruciferous vegetables including turnip have been found active against some types of cancer including melanoma, esophageal, prostate and pancreatic cancer. Turnip contains high dietary fibre which helps reduce the prevalence of flare-ups of diverticulitis by absorbing water in the colon and making bowel movements easier to pass. Eating high fibre meals also helps keep blood sugar levels stable. According to a study published in the *British Journal of Clinical Pharmacology* in 2013, foods containing dietary nitrates, such as turnips and collard greens can have multiple vascular benefits which include reducing blood pressure, inhibiting platelet aggregation and preserving or improving endothelial dysfunction.

Origin and taxonomy

Turnip, an ancient crop, is known to the Greeks and the Romans since the beginning of the Christian era. Sappho, a Greek poet from the seventh century BC, calls one of her paramours *Gongýla*, "turnip". However, there are almost no archaeological records available to help determine its earlier history and domestication. Zohary and Hopf (2000) concluded that origins of turnip are necessarily based on linguistic considerations. Wild forms of the hot turnip and its relatives like, mustards and radishes are found over west Asia and Europe, suggesting their domestication took place somewhere in that area. Two primary centres of origin of turnip have been propounded: Central and Southern Europe possibly the Mediterranean region for European types and eastern Afganistan with adjoining areas of Pakistan for the Asiatic types. Asia Minor, Transcaucasus and Iran are the secondary centre of origin. Swollen rooted modern day turnip probably originated from the biennial oilseed forms in cooler part of Europe. It was introduced in England in 1590 A.D., Canada in 1540 and the USA in 1606.

Turnip, *Brassica campestris* L. var.*rapifera* Mertz (Syn. *Brassica rapa* subsp. *rapa*) is an important species of the family Brassicaceae. Turnip is closely related to Swedes, *Brassica napus* L. var. *napobrassica* Patern. Other closely related species are *B. campestris* L. ssp. *chinensis* Jusl., *B. campestris* L. ssp. *pekinensis* (Lour.) Rupr. and *B. campestris*.

Turnip is a diploid species having chromosome number, 2n = 20 with AA genome. The swede, *Brassica napus* L. var. *napobrassica* is an amphidiploid species (2n = 38, AACC genome). Turnip crosses with swede and produce sterile trilpoid hybrids (3n = 29).

Adaptation in India

Some evidence shows that turnip was domesticated before the 15th century BC and it was grown in India at this time for its oil-bearing seeds. However, information regarding introduction of turnip in India remains obscure. Some tropicalized and annual types adaptable to comparatively high temperature condition constitute the Asiatic race of turnip. Varieties under this group have been developed either by selection from the indigenous types or through hybridization with the temperate types followed by selection.

Improved varieties

Asiatic or tropical and European or temperate turnip varieties both are cultivated in India.

- **Tropical or Asiatic varieties :** Turnip L-1, Pusa Sweti, Pusa Kanchan
- **European or Temperate varieties :** Purple Top White Globe, Golden Ball, Snowball, Early Milan Red Top, Pusa Chandrima, Pusa Swarnima.

Cultural requirements

Turnip grows best in cool weather and hot weather condition cause the roots to become woody and bad-tasting. The most favourable temperature for the development of root is 10° to 15°C and temperature below 10°C generally induces flowering in Asiatic varieties. These temperature requirements will be of higher side for tropicalized Asiatic cultivars and lower side for European cultivars. The tropical or Asiatic types can tolerate high temperature than the temperate types although it performs at its best in cool growing condition. The high temperature tolerant Asiatic cultivars when grown at high temperature condition generally produce small root and comparatively profuse top.

Average rainfall of 700 to 1000 mm is ideal for growing turnip. It can be grown in all types of soil but light, friable loam soils that contain ample humus is the best. Heavy soil or those with rocks, stones or pebbles below the surface produce rough, misshapen roots. Turnip can be grown in slightly acidic soil also with soil pH ranging between 5.5 and 6.8.

Land preparation

The soil should be thoroughly prepared so that there are no clods to interfere with the development of the root. The soil should not contain any undecomposed organic matter, because they may result in misshapen roots.

Sowing time

Time of sowing depends on the type, variety and location. The tropical or Asiatic types if sown later than November in the plains of India bolt early. Hence, early September-October sowing for Asiatic types and October to December sowing for the temperate types is preferable. In the hills, seed of the European or temperate varieties are sown during April.

Sowing of seeds

Turnip is direct seeded crop because swelled primary roots are economic plant parts.Roots may be misshaped in the transplanted seedlings. It is recommended to sow the seed on the ridges set at 45 cm apart, than in flat beds which helps in better development of roots and drainage. The seeds are sown as far as possible on the surface of the soil and covered by thin layer of soil. Seeds should be sown thinly. The ridge should be kept moist until germination is completed. The average seed rate is about 3-4 kg per hectare. Plants are thinned out to a distance of 8-10 cm within the rows when the plants are 10-15 days old.

Spacing

- Tropical or Asiatic type: 30-45 cm x 10-15 cm
- Temperate type: 30 cm x 8-10 cm.

Nutrient management

Adequate nutrition is necessary for obtaining higher yield and better root quality. Well decomposed FYM @ 10-15 t/ha should be applied at the time of land preparation, 20-30 days before sowing of seeds. The FYM to be applied should be well decomposed because fresh or undecomposed manure induces missaped roots. A general fertilizer dose of 120 kg N, 80 kg P_2O_5 and 80 kg K_2O per hectare should be applied for open pollinated improved varieties. FYM along with the whole P and K fertilizers and half N fertilizer should be given as basal before sowing seeds. For better growth and yield, the fertilizer should be ploughed to a depth of 25 cm. Rest half nitrogen fertilizer should be top dressed 30 to 35 days after sowing (root bulking stage). Foliar sprays of 0.3% borax twice 25, 40 days after sowing is beneficial for enhancing root growth.

Irrigation

Irrigation immediately after sowing is the best to get vigorous crop growth and production of tender roots.Next irrigation should be on the 3rd day and subsequently once in 5-7 days.Moist condition has been found better in increasing the yield. The irrigation requirement for turnip is found to be about 230 mm. Over-irrigation and water logging cause harm to the crop.

Interculture and weed control

The plants should be thinned as soon as the seedlings establish.Shallow weeding helps in better root development. Regular hand weeding is necessary to check the growth of weeds. Proper earthing up after 30-35 days of sowing is necessary to get quality roots.Nitrogen fertilizers are top dressed at that time. Pre-sowing application of Tok E-25 (Nitrofen 25%) @ 2.0-2.5 kg a.i./ha controls weeds effectively in the field.

Harvesting the crop

Harvesting is generally done 55–60 days after sowing by pulling the top with hand. The roots should be pulled before heavy freezes or the root may crack and rot in the soil. Digging is an alternative way of harvesting the roots. Later, tops are removed, graded and marketed. Over matured and oversized roots become woody and crack.

Post harvest handling and storage

After harvest the roots are cleaned, tops are cut and roots are graded according to size and tenderness. Turnip does not store well. It can be stored for 2-3 days under room conditions. In cold storage, the topped turnips are stored at 0° C with 95% relative humidity for 8-10 weeks. Turnips are usually stored in slatted crates or bin that allows good air circulation.

Plant protection measures

Physiological disorder

Phyllody

It is the malformation of the flowering shoots and the plants usually do not produce any normal fruit. It is common and causes severe damages of the seed crops in the hills.

Brown heart

It is a boron deficiency disorder causing the core of the root turning brownish.

Control measures

- Correction of soil reaction and salinity
- Soil application of borax @ 10-15kg/ha
- Three sprays of 0.25-0.50% Borax solution @ 1-2kg/ha along with a sticker 20, 35 and 50 days after sowing.

Diseases

Alternaria blight (*Alternaria raphani*)

It is an important fungal disease particularly for the seed crop. Small, yellowish, slightly raised lesion first appears on the leaves of seed stem. These enlarge many times as they become older. Lesions appear later on the stems and seed pods. Infection spreads rapidly during rainy weather and the entire pod may be so infected that the style end becomes black and shriveled. The fungus also penetrates the pod tissues and ultimately infects the seeds.

Control measures

- Use of disease free seed.
- Seed treatment with hot water at 50oC for 30 minutes.
- Seed treatment with Captan @ 2g/kg of seed or Thirum @ 2.5 g/kg of seed.
- Spraying with Mancozeb (2.5%) or Propiconazole (1.0%) or Difenconazole (0.05%).

Downy mildew (*Peronospora parasitica*)

It is a fungal disease where discoloration occurs in the young seedlings and in severe cases the whole plants die. In mature plants, purplish or yellow-brown spots appear on the upper surface of the leaf and white fluffy downy growth occurs on the lower leaf surface.

Control measures

- Hot water treatment of seeds at 50^0C for 30 minutes.
- Crop rotation excluding cruciferous crops.

- Spraying the crop with 0.25% Mancozeb, Metalaxyl-Mancozeb (0.25%) or Cymoxanil-Mancozeb(0.25%) with a sticker at weekly intervals.

Soft rot (*Erwinia carotovora*)

This bacterial rot occurs after mechanical injury to the matured root. Water soaked patches appear on the root which turns to light brown and ultimately dark brown within a week.

Control measures

- Spraying of Copper oxychloride (0.3%) + Streptomycin sulphate (100 ppm) for management of soft rot.

Insect pests

Aphids (*Myzus persicae, Brevicoryne brassicae*)

Aphids are the most serious pests of turnip. It attacks both seedlings and mature crops. Cloudy and humid conditions favour the spread of their infestation. In case of heavy infestation the plant are completely devitalized, leaves and shoot curl up, become yellowish and finally die.

Control measures

- Spraying the crop with Chlorpyriphos @ 2.5 ml/l or Rogor @ 1.5ml/l of water
- Application of Imidacloprid @ 0.3 ml/l is very effective against the pest.

Mustard saw fly (*Athalia lugens proxima*)

The insect attacks the crop in both vegetative and flowering stage. The gray blackish larvae feed on the leaves causing defoliation.

Control measures

- Spray the crop with Nuvacron @ 1 ml/l, or Thiodan @ 1.4 ml/l or Malathion @ 2 ml/l or Sevin W. P. @ 4 g/l of water.

Yield

The average root yield varies from 20-25 tonnes per hectare depending on the variety and crop management practices.

Bulb Crops

- Onion
- Garlic

Onion

Onion, *Allium cepa* under the family Alliaceae (Amarylidaceae) having chromosome No. 2n = 16 is an important vegetable crop grown in almost all parts of the world. Top 10 onion producing countries in the world are China, India, USA, Iran, Russia, Turkey, Egypt, Pakistan, Brazil and Mexico. It is used in almost all kinds of Indian cuisine. The important onion growing states in India are Maharashtra, Karnataka, Tamil Nadu, Andhra Pradesh, Gujarat, Punjab, Haryana, Rajasthan, Uttar Pradesh, Bihar and Madhya Pradesh.

The crop

Onion also known as the bulb onion or common onion is the most widely cultivated species of the genus *Allium*. Its close relatives include garlic, shallot, leek, chive, and Chinese onion. The onion plant has been grown and selectively bred in cultivation for at least 7,000 years. It is a biennial plant, but is usually grown as an annual. Modern varieties typically grow to a height of 15 to 45 cm. The leaves are yellowish- to bluish green and grow alternately in a flattened, fan-shaped swathe. Leaves are fleshy, hollow and cylindrical with one flattened side. They are at their broadest about a quarter of the way up, beyond which they taper towards a blunt tip. The base of each leaf is a flattened, usually white sheath that grows out of a basal disc. From the underside of the disc, a bundle of fibrous roots extends for a short way into the soil. As the onion matures, food reserves begin to accumulate in the leaf bases and the bulb of the onion begins to swell when a certain day-length is reached.

The bulbs are composed of shortened, compressed, underground stems surrounded by fleshy modified scale (leaves) that envelop a central bud at the tip of the stem. In the autumn (or in spring, in the case of overwintering onions), the foliage dies down and the outer layers of the bulb become dry and brittle. The crop is harvested and dried and the onions are ready for use or storage.

Common onions are normally available in three colour varieties viz., red, yellow or brown and white. Considerable differences exist between onion varieties in phytochemical content, particularly for polyphenols, with shallots having the highest level, six times the amount found in onions. Yellow onions have the highest total flavonoid content, an amount 11 times higher than in white onions. Red onions have considerable content of anthocyanin pigments, with at least 25 different compounds identified representing 10% of total flavonoid content. The outer skin colour of the yellow or brown onion is due to the presence of a flavanoid called quercetin.

Characteristic pungency of onion results from the sulfur-rich volatile oil it contains. Freshly cut onions often cause a stinging sensation in the eyes of people nearby, and often uncontrollable tears. This is caused by the release of a volatile gas, *syn*-propanethial-S-oxide, which stimulates nerves in the eye creating a stinging sensation. This gas is produced by a chain of reactions which serve as a defence mechanism: chopping an onion causes damage to cells which releases enzymes called alliinases. These break down amino acid sulfoxides and generate sulfenic acids. A specific sulfenic acid, 1-propenesulfenic acid, is rapidly acted on by a second enzyme, the lacrimatory factor synthase, producing the *syn*-propanethial-S-oxide. This gas diffuses through the air and soon reaches the eyes, where it activates sensory neurons. Lacrimal glands produce tears to dilute and flush out the irritant.

The crop is normally harvested when the leaves die back and the outer scales of the bulb become dry and brittle. If left in the soil over winter, the growing point in the middle of the bulb begins to develop in the spring. New leaves appear and a long, stout, hollow stem expands, topped by a bract protecting a developing inflorescence. The inflorescence takes the form of a globular umbel of white flowers. The common onion has one or more leafless flower stalks that reach a height of 75–180 cm, terminating in a spherical cluster of small greenish white flowers. The seeds are glossy black and triangular in cross section.

Onions are generally classified into two broad groups.

Common onion group (*Allium cepa* var. *cepa*)

The most economically important bulb onion or common onion belongs to this group. Most of the diversity within *A. cepa* also occurs within this group. Plants within this group form large single bulbs, and are grown from seed or seed-grown sets. The range of diversity found among these cultivars includes variation in photoperiod, storage life, flavour and skin colour. Common onions range from the pungent varieties to the mild and hearty sweet onions.

Adaptations to hardy conditions of high temperature, high rainfall and short photoperiod condition have resulted in the development of wide diversity of tropicalized cultivars in western and peninsular India. Hence, varieties belonging to common onion in India are grouped according to growing season viz., winter or rabi onion and kharif onion. Rabi onions are basically intermediate type cultivars and generally require 12 hours day length for bulb formation and development. Rabi onion can broadly be classified according to the skin colour viz., red, white and yellow. Kharif onions require comparatively shorter day length of 10-11 hours for bulb formation and development.

Aggregatum group (*Allium cepa* var. *aggregatum*)

This group of onions are hardy and early maturing types grown for closely packed clusters of small sized, ovoid to pear-shaped bulblets. This group of onion generally does not produce seeds and thus are propagated vegetatively by bulblets.

Importance and use

Onions are cultivated and used around the world. It is the most common and indispensable items in every kitchen as vegetable, spice and condiment. As a food item, they are usually served cooked, as a vegetable or part of a prepared savoury dish, but can also be eaten raw or used to make pickles or chutneys. Onion is valued for its distinctive pungent flavour and is an essential ingredient of the cuisine of many regions. It is a universal flavouring agent for different vegetarian and non-vegetarian dishes all over the world. Green leaves and mature and immature bulbs are eaten raw as salad or used in preparation of various types of curry. It is used in soups, sauces and for seasoning different foods. Immature scapes are tender and consumed as green vegetable. Dehydrated onion powder is in great demand which reduces transport cost and storage losses. White onions are generally used for dehydration. Dried onion flakes can be reconstituted by cooking in water. Onions are also available in fresh, frozen, canned, caramelised, pickled and chopped forms. The dehydrated product is available as kibbled, sliced, ring, minced, chopped, granulated, and powder forms.

Nutritional value

Onions are very good source of biotin. They are also a good source of manganese, calcium, phosphorus, vitamin B6, copper, vitamin C, dietary fibre, phosphorus, potassium, folate and vitamin B1.The bulbs are fairly rich in carbohydrate (11g/100 g fresh). It also contains good amount of protein (1.2 g/ 100 g fresh) and vitamine A (35 IU/100 g fresh). Dry matter content varies from 10.6 to 14.8% and on average and white cultivars contains more dry matter than red ones.

Medicinal value

Onion bulbs contain different flavonoids like, quercetin (28.4-48.6 mg), kaempferol (0.2 mg), myricetin (0.1 mg), luteolin (0.1 mg), etc. per 100 g fresh which are potential antioxidants. The highly biologically reactive organosulfur compounds responsible for the characteristic odour and taste have potential influence on phase-I (carcinogen activation) and phase-II (carcinogen detoxification) enzymes. Consumption of onion result in fewer tendencies to form blood clots and lower the levels of cholesterol and lipoproteins associated with heart disease. Onions are diuretic. It is applied on bruises, boils and wounds. It relieves heat sensation. Bulb juice is used as smelling on hysterial faintness. It also relieves insect bites and soar throat.

Origin and taxonomy

The geographic origin of the onion is uncertain because the wild onion is extinct and ancient records of using onions span western and eastern Asia. The first cultivated onions are the subject of much debate, but the two regions that many archaeologists, botanists, and food historians point to are central Asia or Persia. It is thought to have domesticated in the mountainous regions of central Asia comprising Turkmenistan, Uzbekistan, Tadzhikistan, Afganistan, Pakistan and north Iran. They were probably and almost simultaneously domesticated by peoples all over the globe, as there are species of the onion found the world over. Food uses of onions date back thousands of years in China, Egypt and Persia.

Traces of onions recovered from Bronze Age settlements in China suggest that onions were used as far back as 5000 BC, not only for their flavour but the bulb's durability in storage and transport. Ancient Egyptians revered the onion bulb, viewing its spherical shape and concentric rings as symbols of eternal life. Onions were used in Egyptian burials, as evidenced by onion traces found in the eye sockets of Ramesses IV. In the 6th century BC, the Charaka Samhita, one of the primary works in the Ayurvedic tradition, documents the onion's use on the Asian subcontinent as a medicinal plant: "diuretic, good for digestion, the heart, the eyes, and the joints

Onion entered northern Europe through Italy and North Africa in 500 A.D. It was introduced in the USA by the Spanish explorers around 1625.

Onion is classified under the class Monocotyledon, Super order Liliflorae, Order Asparagales, Family Alliaceae, Tribe Allae and Genus *Allium*. The genus *Allium* has more than 500 species and the eight cultivated species used as vegetable are *Allium cepa* (common onion), *Allium cepa var aggregatum* (multiplier onion or potato onion), *A. cepa var viviparum* (top onion or tree onion), *A. cepa* var. *ascalonicum* (shallot), *A. sativum* (garlic), *A. schoenoprasum* (chieve), *A. fistulosum* (welsh onion or Japanese bunching onion), *A.ampeloprasam* var. *porrum* (leek), *A. tuberosum* (Chinese chieve) and *A. chinensis* (rakkyo). There are several closely related wild *Alliums* crossable with onion are *A. vavilovii*, *A. oschaninnii*, *A. pskemense*, *A. galanthum*, etc.

Chromosome number of onion 2n = 16. Chromosomes are relatively large and so very appropriate for the detection of morphological changes.

Adaptation in India

Onion is an outstanding example of adaptation carried through from the very early times before the Christian era. Onion has got a Sanskrit equivalent 'Palandu' mentioned in "Apastamba Dharma Sutra-I" during the Sutra period dated back 800 B.C. to 300 B.C. which signifies its very early introduction in India. It is also mentioned in "Charaka-Samhita", a famous early medical treatise of India, around 6th century A.D. The popularity of onion in India however, has been promoted by Mughals. Onion is originally a native of Central Asia, the temperate region, with perennial/biennial habit and long day bulbing character. In North India, rabi (winter) onion production has been developed with such biennial types. Huge genetic diversity available in the rabi onion in northern and eastern India have been utilized mainly by mass selection and other population improvement schemes. The concentration of onion cultivation in peninsular and western India resulted in the evolution of tropical and short day cultivars.

A good number of local cultivars are grown in India in different names like, Sukhsagar, Chhanchi, etc. in West Bengal; Niphad, Nasik Red, Nasik White Globe, Poona Red, Bombay Red, Bombay White, Mathewad, Shirwal, Kagar, Peth, Dhulia, Fangira, Lohru, etc. in Maharashtra; Patna Red, Patna White in Bihar; Bellary Red, Bellary White, Telagi Red, Bangalore Rose, Chikballpur Local, Krishnapuram Local, Kempgende in Karnataka; Cuddalorae, Podusa, Dindigal Red, Mutlore, Natu, Pallu Vengayam in Tamil Nadu; Rampuram Red in Andhra Pradesh; Saurastra Local in Gujarat; Kalyanpur Red, Kalyanpur Lalgol in Uttar Pradesh; Panipat Local, Rajpur Local, Hisar Local, Bahadurgarh Local in Haryana; Ludhiana Local in Punjab, etc.

Open pollinated improved varieties

True long day onion: These varieties of the temperate region require day length of more than 13 hours and lower temperature for proper development of bulbs. Example, Red Zeppelin, Walla Walla, White Spanish, Brown Spanish, etc.

Winter or rabi onion: Pusa Red, Pusa Ratnar, Pusa Madhbi, Punjab Selection, Arka Niketan, Arka Bindu, Agrifound Light Red, Udaipur 101, Udaipur 103, Sukhsagar, Punjab Red Round, Agrifound Rose, Punjab Naroya, Punjab Selection, Kalyanpur Red Round, Patna Red, VL Piaz-3, Hisar-2, etc.

Kharif onion: Niphad-53, Arka Kalyan, Agrifound Dark Red, Baswant-780, Phule Samarth.

Red skinned varieties: Pusa Red, Pusa Ratnar, Pusa Madhbi, Punjab Selection, Arka Niketan, Arka Bindu, Nasik red, Arka Pragati, Aprita, Bhima Raj, Agrifound Light Red, Udaipur 101, Udaipur 103, Sukhsagar, Punjab Red Round, Agrifound Rose, Punjab Naroya, Punjab Selection, Kalyanpur Red Round, Patna Red, VL Piaz-3, Hisar-2, Niphad-53, Arka Kalyan, Agrifound Dark red, Baswant-780, etc.

White or silver skinned varieties: Pusa White Flat, Pusa White Round, Punjab-48, Udaipur – 102, N-257-9-1, Phule Safed, Punjab White, Akola Safed, etc.

Yellow skinned varieties: Early Grano, Brown Spanish, Bermuda Yellow, Arka Pitambar, Phule Suvarna, etc.

Varieties of multiplier onion: CO-1, CO-2, CO-3, CO-4, MDU-1, Agrifound Red,

Varieties with small bulb: Agrifound Rose, Arka Bindu

Varieties suitable for export: Agrifound Rose, Arka Bindu, Niphad-53, Nasik Red, Agrifound Light red, Agrifound Dark Red, Arka Niketan, Pusa Madhbi, Pusa Red, Patna Red, etc.

Varieties suitable for dehydration: Pusa White Flat, Pusa White Round, Punjab-48, Udaipur-102, N257-9-1, etc.

Public sector hybrids: Arka Kirtiman, Arka Lalima, MOH-2.

Cultural requirements

Onion is basically a cool season crop but can be grown in a wide range of climatic regimes under both tropical and subtropical conditions. Photoperiod and temperature both influence bulb development and flowering however, photoperiod is comparatively more important for bulb formation and temperature is more important for flowering. Almost all varieties grown in India are either short day or intermediate type varieties. The rabi onion varieties require relatively high temperature and 12-13 hours day length for proper bulb development. The kharif onion varieties require relatively low temperature and 10-11 hours day length for proper bulb development. The long day onion varieties require more than 13 hours day length and lower temperature for proper development of bulbs hence, can only be cultivated in the hill region of India. Long day varieties do not bulb under short day conditions where as short day varieties if planted under long day condition will develop early bulbs. It grows well in mild climate without extreme of high or low temperatures.

The optimum temperature for good vegetative growth before bulbing is between 13-24°C with 70% relative humidity. Ideal temperature for bulb development is 15-25°C. The plants are hardy and can withstand freezing temperature. In the rabi season a sudden rise in temperature may result in early bulb maturity and smaller sized bulbs. Very low temperature at the vegetative phase results early bolting. Generally low temperature influences the development of seed stalk. It does not thrive well in places where average rainfall exceeds 75-100 cm in the monsoon periods. Onion grows well on light loam, sandy loam or clay-loam, deep friable and fertile soil rich in organic matter. In heavy soil bulb development is restricted and crops mature late compared to light soils. Onion is sensitive to high acidity and alkalinity and ideal soil pH ranges between 5.8 and 6.5.

Nursery management

Seed beds are prepared finely, well drained, 15-20 cm raised, 1.0m wide and of convenient length. Fine and fully decomposed farmyard manure or compost @ 3-4 kg/m^2 should be well mixed to the beds. The beds are drenched with formaldehyde (4.0%) and covered with polythene sheet for 7 to 10 days to disinfect the seed bed. 8-10 kg seeds/ha are required for open pollinated varieties of common onion. About 1000 square metre area is required to produce seedlings for one hectare. Before sowing the seeds are treated with Thirum @ 2-3g/kg of seeds to avoid damping off disease. Sowing of seeds in lines spaced at 5-7cm apart is recommended. The seeds after sowing should be covered with finely sieved farm yard manure or compost or vermicompost followed by light watering by rose can. The seed beds are covered with straw or dry grass till seed germination and water is then sprinkled regularly. Seedlings should be drenched with 0.2-0.3% Thiram or Captan twice, 15 days and 30 days after sowing. Side dressing with single super phosphate and muriate of potash and spraying with 0.5% urea produce healthy seedlings. Time to time hoeing, weeding and irrigation are required to raise healthy seedlings. Seedlings may be raised under low poly tunnels having side open to protect them from heavy rains. The seedlings are ready within 6-7 weeks after sowing for kharif crop and 8-9 weeks for rabi crop. Under-aged seedlings do not stand well in the field and over-aged seedlings may bolt early.

Sowing time

Onion is grown in two distinct seasons in the plains of India viz., Kharif and Rabi although sowing and transplanting season vary with the region. Sowing time of "Kharif" and "Rabi" onions in different regions of India is given below.

- Early kharif crop: April-May (Tamil Nadu, Karnataka, Andhra Pradesh)
- Kharif crop: May-June (Tamil Nadu, Karnataka, Andhra Pradesh, Maharashtra, part of Gujarat, Rajasthan, Punjab and other parts).

- Late kharif crop: August – September (Maharashtra and part of Gujarat).
- Early rabi crop: August-September (West Bengal, Odisha).
- Rabi crop: September-October (Tamil Nadu, Karnataka, Andhra Pradesh); October-November (West Bengal, Odisha and North and north-eastern region); November-December (Maharashtra and part of Gujarat).

Sowing for Rabi crop in the hills is done in September-October and for summer crop with long day varieties sowing is done in October end to early November.

Land Preparation

The land is prepared well in advance with repeated ploughings (at least 4-5 ploughing) to a fine tilth. All stubbles, weeds, etc are removed from the land. Well rotten organic manure preferably farm yard manure or compost @ 20-25 t/ha is applied during final land preparation followed by proper leveling. The field should be divided into beds and channels.

Transplanting / direct planting

The seedlings are transplanted in the main field when they are 6-7 weeks old for kharif crop and 8-9 weeks old for rabi crop. The variety with small bulb like Agrifound Rose can be directly seeded by broadcasting or in rows in the main field with 20-25 kg seeds per hectare. Multiplier onions are directly planted in the main field with 1.0 1.2 tonnes of bulblets of 1.5 to 2.0 cm size per hectare. The seed beds need to be irrigated before lifting to facilitate easy pulling of seedlings. Late afternoon hours are best for transplanting the seedlings in the main field for better establishment. The field is irrigated lightly just after transplanting or direct seeding.

Spacing

Spacing of 15cm x 10cm is given for common big size onion; 8cm x 5cm for small pickling onion and 30 x 15cm for multiplier onion.

Nutrient management

Soil should be liberally manured and fertilized for getting good yield. Organic manure should be applied atleast 15 days before transplanting the seedlings. A general fertilizer dose of 100 kg N, 60kg P_2O_5 and 100kg K_2O per hectare is recommended for open pollinated improved varieties.Half N along with entire phosphate and potash fertilizers should be given as basal and rest N should be top dressed in two split doses at 30 and 45 days after transplanting. Onion responses well to sulphur application and it helps in enhancing biosynthesis of allyl propyl disulphide which increase pungency and storability. Application of single super phosphate as basal and ammonium sulphate as top dressing which contains sulphur fulfil the major demand of sulphur. Top dressing should be completed before bulb initiation as delayed application results in thick neck or doubles. It is recommended to apply 3 tonnes of vermicompost/ha along with 50% of recommended dose of NPK ferlilizers to get better results. Application of micronutrients particularly copper and boron enhance bulb yield. It is recommended to apply *Azotobacter* @ 800g/ha or *Azospirillium* @800g/ha as seedling root dip. A suspension of bacterial inoculants is prepared in sufficient water. The seedlings are dipped in the bacterial suspension in shade for 2-3 hours in the afternoon hours and treated seedlings are transplanted immediately.

Irrigation

Onion being a shallow rooted crop requires light irrigations. The field should be irrigated immediately after transplanting. Soil moisture condition should be kept optimum. Frequent and light irrigation @ 2 acre-inch per irrigation at weekly interval is the best. Depending on rains, 8-10 irrigations are enough for kharif onion while late kharif or early rabi onion generally require 12-15 and Rabi onion generally requires 15-20 irrigations. Irrigation must be given at bulb formation stage. Irrigation should be stopped when tops mature and start falling for rabi onion and10 days before harvesting for kharif onion.

Interculture and weed control

Onion is a poor competitor of weeds hence, weeds must be kept down to get good yield. Onion is shallow rooted crops and planted closely and there is a chance of root pruning during weeding. Two light hoeings at the early stages of growth ensure good weed control. Chemical weed control gives better results. Basaline @ 1.0litre/ha or Stomp @ 3.35litre/ha should be applied immediately after transplanting and before first irrigation along wit.h one hand weeding. In kharif and early rabi onion neck fall generally does not occur particularly in eastern Indian condition due to higher moisture in the soil which delay the ceasation of top growth. In this condition, trampling of the top by rolling bamboo over the crop 6-7 days before harvesting cause ward off of moisture which help in stiffening the neck portion. Application of 10% salt solution with sticker 8-10 days before harvesting also cause neckfall and ensure proper maturity of the bulb.

Harvesting the crop

The rabi crop should be harvested one week after 50-70% neck fall stage. In kharif crop, harvesting is done soon after the colour of leaves changes to slightly yellow and tops starts drying and 6-7 days after trampling the top portion since tops do not fall. For green onion, the tender plants along with immature bulbs are harvested 75-80 days after transplanting. If soil is light the crop can be harvested by hand pulling or with hand implements or modified potato digger.

Post harvest handling and storage

After harvest the bulbs are cured in the field for big, small and multiplier onions. Purpose of curing or drying is to remove excess moisture from the outer skin and neck of the bulb which helps in minimizing shrinkage of bulb, developing skin colour and reducing the disease infection during storage. The rabi onions are cured in the field for 3-5 days after tops are cut leaving 2.0-2.2.5 cm above the bulb and then the bulbs are again cured in shade for 7-10 days to develop the skin and remove field heat.The kharif onions are generally cured for 3-4 days along with the tops. Cold storage temperature of 0°C with 65-70% relative humidity is ideal for 6-8 months long storage in dormant condition and reasonably free from decay.Onion bulbs can be stored at 24.0 to 30.0°C for 5-6 months without excessive loss in weight and sprouting. Dormancy of onion is longer both at lower (0°C) and higher (30°C) than at moderate temperature of 10-15°C.For low cost storage, a hut like structure is erected under the shade with thatched roof and sides covered by bamboo sticks with provision of good air circulation. Onion is stored in these sheds by spreading them on racks or tiers having 2-3 layers of bulbs. The stored onion should be inspected periodically and damaged, diseased and sprouted bulbs are removed immediately from the stock. Irradiation of bulbs with very low dose of 40-90 Gy gamma rays effectively controls sprouting of onion. Jute mesh bags or plastic netted bags are commonly used for packaging onion.

Yield

Bullb yield is 30-35t/ha for big sized common onion in rabi season; 25-30t/ha big sized common onion in kharif season; 16-20t/ha for small sized common onion in rabi season and 15-18t/ha for multiplier onion.

Garlic

Garlic, *Allium sativum* under the family Alliaceae (Amarylidaceae) having chromosome No. 2n = 2x =16 is the second most widely cultivated *Alliums* next to common onion. The five major garlic growing countries in the world are USA, Egypt, China, Korea and India and other important countries for garlic are Thailand, France, Spain and Yugoslavia. Garlic is used as a spice and condiment throughout India. The important garlic growing states in India are Madhya Pradesh, Gujarat, Odisha, Maharashtra and Uttar Pradesh.

Garlic of more than 40mm size is exported to Quatar, Saudi Arabia, UAE, Baharin, Kuwait, Bangladesh and Sri Lanka. Demand now from all over the world is for bigger cloved (40-60 cm diameter) with 10-15 cloves in each bulb and such big cloved garlic are the long day cultivars suitable for growing in the hill condition.

The crop

Garlic plants are closely related to and similar to onions and they have a similar, but stronger odour. They have a characteristic pungent, spicy flavour that mellows and sweetens considerably with cooking. The leaves of garlic plants are neither inflated like onion leaves nor tubular like those of bunching onions. Instead, they are flat, with a crease down the middle and are held erect in two opposite ranks. Most varieties stand about 30-60 cm tall at maturity. Garlic plant's underground bulbs, with the exception of the single clove types, is normally divided into numerous fleshy sections called cloves, closed in a common membranous coat (tunic), and are easily separable. Bulbs are divisible into 6-20 cloves.

Garlic is a particularly rich source of organosulfur compounds. The two main classes of organosulfur compounds found in whole garlic cloves are L-cysteine sulfoxides and γ-glutamyl-L-cysteine peptides. S-allyl-L-cysteine sulfoxide (alliin) accounts for approximately 80% of cysteine sulfoxides in garlic. When raw garlic cloves are crushed, chopped or chewed, an enzyme known as alliinase is released. Alliinase catalyzes the formation of sulfenic acids from L-cysteine sulfoxides. Sulfenic acids spontaneously react with each other to form unstable compounds called thiosulfinates known as allicin (half-life in crushed garlic at 23°C is 2.5 days). The formation of thiosulfinates is very rapid and has been found to be complete within 10 to 60 seconds of crushing garlic. Allicin breaks down *in vitro* to form a variety of fat-soluble organosulfur compounds, including diallyl trisulfide (DATS), diallyl disulfide (DADS), and diallyl sulfide (DAS).

Crushing garlic does not change its γ-glutamyl-L-cysteine peptide content. γ-Glutamyl-L-cysteine peptides include an array of water-soluble dipeptides, including γ-glutamyl-S-allyl-L-cysteine, γ-glutamyl methyl cysteine, and γ-glutamyl propyl cysteine. Water-soluble organosulfur compounds, such as S-allyl cysteine and SAMC are formed from γ-glutamyl-S-allyl-L-cysteine during long-term incubation of crushed garlic in aqueous solutions, as in the manufacture of aged garlic extracts.

The plants produce a leafless flower stem (a scape), but the flowers are sterile and produce bulbils (small cloves) rather than seeds. Garlic is generally propagated clonally from cloves and bulbils. Hundreds of cultivars are divided into two subspecies: 1) Hardneck garlic (*A sativum* ssp. *ophioscorodon*); and 2) Softneck garlic (*A. sativum* ssp. *sativum*). Hardneck garlic produces a flower stalk that coils like a snake, then straightens out and bears clusters of pea-sized bulblets or "bulbils" that are like miniature garlic bulbs.

Importance and use

Garlic is a common spice and condiment crop. It is a flavouring agent particularly for different non-vegetarian dishes. The flavour of garlic cloves is more powerful than that of the other bulbs crops. Immature garlic is sometimes pulled, rather like a scallion, and sold as "green garlic". The leaves and bulbils , milder in flavour than the bulbs are sometimes eaten. Green leaves are used in chutney and pickles. It is used in soups, sauces and for seasoning different foods. Dehydrated garlic is in great demand particularly for export.

Nutritional value

Garlic has higher nutritive value than other *Allium* bulbs crop. Garlic is the rich source of carbohydrates (29 g/100g fresh), protein (6.3g/100g fresh) and phosphorous (310mg/100g fresh). It is a rich source of thiamine (0.16 mg/100g fresh) and riboflavin (0.23 mg/100g fresh). It contains fair amount of calcium (30 mg/100g fresh). Ascorbic acid content is quite high in green garlic.

Medicinal value

Garlic has been used for culinary and medicinal purposes in many cultures for centuries. Garlic is a particularly rich source of organosulfur compounds, which are thought to be responsible for its flavour and aroma, as well as its potential health benefits. Organosulfur compounds derived from garlic have the potential to prevent and treat chronic diseases, such as cancer and cardiovascular disease. Observational epidemiological studies found that garlic consumption was associated with a lower risk of stomach cancer.

The highly biologically reactive organosulfur compounds have potential influence on both phase-I (carcinogen activation) and phase-II (carcinogen detoxification) enzymes. Garlic cloves contain different flavonoids, the plant secondary metabolites like, quercetin, kaempferol, myricetin, luteolin, etc. which are potential antioxidants. Allicin (diallyl thiosulfinate) which has hypocholesterolemic action reduces the cholesterol concentration in human blood. Garlic consumption also reduces serum lipids perhaps by decreasing their synthesis and/or increasing their excretion through the intestinal tract. Garlic oil also reduces blood sugar and this hypoglycemic effect may be due to increased glycogenesis in the liver and better utilization of glucose in the peripheral tissue. Garlic is described as hot stimulant, carminative and anti-rheumatic. Garlic oil is a powerful anticeptic mainly due to presence of allicin. It is used as vermifuge for expelling round worms and has long been recommended for cure of a number of ailments viz., wounds, ulcers, pneumonia, bronchitis, atopic dyspepsia and gastro-intestinal disorders. Inhalation of garlic oil or garlic juice has generally been recommended in cases of pulmonary tuberculosis, rheumatism, sterility, impotency, cough and red eyes. Garlic is a gastric stimulant and helps in the digestion of food and absorption of nutrients from it.

Origin and taxonomy

Garlic had its original home in Central Asia particularly the high mountain plateaus of Pamir and Tian Shan. Its wild ancestor, *Allium longicuspis* Regel is native to Central Asia which was known in Egypt as early as 3000 B.C. and also to the ancient Greeks and Romans. From the extended regions of Turkmenistan in the south to the north east-ward in Pamir and Tien Shan regions of Central Asia, it spread to Mediterranean region in Prehistoric times where its virtues are still cherished perhaps more than in any other region of the world. It was known in Egypt as early as 3000 B.C. and also to the ancient Greeks and Romans. It spread to the eastern Mediterranean region, Caucasus and North Africa which are considered to be the secondary centre of origin. It reached China and is grown in China and India since a long time. Single clove garlic (also called pearl or solo garlic) originated in the Yunnan province of China.

However, its introduction in Europe and the rest of Asia was much later. It was used in England as early as first half of the 16th century. The Spanish and Portuguese settlers introduced garlic in North and South America at the beginning of the 15th century. The Anglo-Saxon settlers in Australia and Central and Southern Africa started its cultivation in the beginning of the 19th century.

Garlic *Allium sativum* L. is classified under the class Monocotyledon and belongs to the section *Porrum* of the family Alliaceae. The classification of garlic is mainly based on bolting and non-bolting habit however, controversy remains on the name of the subspecies for bolting and non-bolting type. In one classification, the non-bolting species is called *Allium sativum* var. *sativum* and the two subspecies *Allium sativum* var. *ophioscorodon* and *Allium sativum* var. *scordoprasum* come under bolting type. In another classification the non-bolting subspecies is *Allium sativum* var. *vulgare* and the bolting subspecies is *Allium sativum* var. *sagitatum* and each subspecies is divided into different ecotypes. Three ecotypes namely, Continental, South Russian and East Mediterranean come under non-bolting subspecies and another three ecotypes viz., Asian, Caucasian and Seaside belong to the bolting subspecies.

Common garlic cultivars have a somatic chromosome number of 2n = 16 (with a karyotypic formula of six metacentric chromosomes, four sub-metacentric chromosomes, and six acrocentric chromosomes), although some garlic plants found in the Campania region of Italy were shown to be tetraploid (4n = 32).

Adaptation in India

Garlic from its original home in Central Asia reached India long ago. Under domestication, it has become exclusively vegetatively propagated crop, mostly through cloves and bulbils in some cases. During long period of garlic cultivation in India many cultivars have been developed mainly due to vegetative mutation and human selection. However, vegetative propagation has considerably restricted the variability needed for the development of large number of cultivars as seen in onion.

The classification of garlic is mainly based on bolting and non-bolting habit although most of the garlic types do not bolt naturally. In most of the garlic types grown in India bulbs are formed under short day condition and critical day length being 12.00 hour. Long day cultivars are suitable for high altitude condition. Bulbs are generally silvery-white in colour although pink coloured bulbs are also found. Bulbs of all the commercial varieties contain multiple cloves ranging from 20-40 of 3.5-5.0 cm size with creamy flesh colour. However, single cloved bulb is also found in some local cultivars. A number of local cultivars are grown in India in different names like, Fawari, Regalle Gaddi, Madrasi, Tabiti, Creole, Jamnagar, Katki, etc.

Clonally selected improved varieties

Varieties with silvery-white bulb: Agrifound White (G-41), Yamuna Safed (G-1), Yamuna Safed-2 (G-50), Yamuna Safed-3 (G-282), Yamuna Safed-4 (G-323), Agrifound Parvati, Raksha Sugandh, Fawari, Rajalle Gaddi, Madrasi, Tahiti, Jamnagar, Prinkindel, Sweta, HG.1, HG-6, Pusa Sel-10, LCC1, ARU-52, Almora, etc.

Varieties with pink bulb: Godavari

Varieties suitable for export: Agrifound Parvati, Yamuna Safed, Agrifound White, Jamnager Ladwa, Malek, etc.

Varieties suitable for hill: Agrifound Parvati

Cultural requirements

Garlic is basically a frost-hardy plant, requiring cool and moist period and relatively dry period during bulb maturity. Garlic generally grows in mild climate without extremes of high or low temperatures even though it can be grown under a wide range of climatic condition. Photoperiod and temperature both influence bulb formation and development. Bulb development is better under cool temperature condition. Exposure of dormant cloves or young plants around 13- 20°C depending on the varieties for 1-2 months hastens bulbing. Garlic is grown as a rabi crop in the plains of India. Both long day and short day cultivars are present in garlic. Almost all varieties grown in Indian plains are short day type varieties. Some long day type varieties are grown in hill condition. The short day garlic varieties grown in Indian plain require about 12.00 hours day length for proper bulb development. The long day garlic varieties grown in the hills require about 13.00 hours day length and lower temperature for proper development of bulbs. Garlic is grown on wide ranges of soil such as sandy loam, silt loam and heavy clay soils. Well drained loam or medium-loam soil rich in organic matter is ideal for garlic. Development of bulbs is not normal when garlic is grown in sandy or heavy clay soil. In heavy soil bulb development is restricted and crops mature late compared to light soils. Garlic is sensitive to high soil acidity and alkalinity and ideal soil pH ranges between 6.0 to 7.0. High alkaline and saline soils are not suitable for garlic cultivation.

Propagation

Garlic is vegetatively propagated by single cloves. Bulbils produced in some varieties are also used as planting material. Micropropagation through *in vitro* culture is employed to produce disease free clones.

Planting time

The time of planting differs from region to region and it is August-October in Madhya Pradesh, Karnataka, Maharashtra and Andhra Pradesh; September-November in Northern parts of India; October-November in Gujarat, West Bengal and Odisha; March-April and October-November in the hills.

Land Preparation

The field is ploughed to fine tilth by giving four to five ploughing with sufficient intervals between two ploughing. Planking should be done for leveling. The width of plots should be such that manual intercultural operations are possible. The normal width of a bed should be about 1.8m and length may vary accordingly to the level of land.

Planting of cloves

Cloves of 8-10 mm diameter are the best for planting. Bigger cloves from outer side of the bulb should be selected. Long and slender cloves present in the centre of the bulb should not be selected because such cloves give poorly developed plants. About 500 Kg cloves are required for planting in one hectare land. A population of 3, 33,000 plants generally remain in one hectare land. The field should be irrigated lightly just after transplanting or direct seeding.

Methods of planting

Dibbling : It is one of the commonly adopted planting method where fields are divided into small plots convenient for irrigation. Cloves are dibbled 5-7.5cm deep keeping their growing ends upward.

Furrow planting : It is also a common method of planting garlic where shallow furrows are opened with hand hoe. The cloves are dropped in the furrows and then are covered lightly with loose soil.

Broadcasting : Cloves are scattered evenly in well leveled field and then are covered by harrowing and field is divided into small plots for irrigation. This method is mostly followed in Gujarat.

Planting of bulbils : Bulbil is used as planting material for the cultivation of garlic in the Lahaul and Spiti districts of Himachal Pradesh. The bulbils are first sown in nursery and then transplanted after 45 days in main field at proper distance.

Spacing

The planting distance depends on varieties and different cultivation region of India. Wider spacing generally results in thicker necks of the bulbs. Spacing of 15cm x 10cm is given for big sized clove and 10cm x 8cm for small sized bul clove.

Nutrient management

Soil should be liberally manured and fertilized for getting good yield. Application of 15-20 tonnes of farmyard manure per hectare before planting the cloves is recommended. A general fertilizer dose of 125 kg N, 65kg P_2O_5 and 100kg K_2O per hectare is recommended for the improved varieties. Half N along with entire phosphate and potash fertilizers should be applied as basal and rest N should be top dressed one month after planting the cloves. Top dressing should be completed before bulb initiation as delayed application results in thick neck. Excessive nitrogen application also results in thick neck of the bulb and sprouting before harvest. Soil application of borax upto 10 kg/ha increase bulb size and yield. Foliar application of micronutrients particularly manganese (0.01% $MnSO_4$), boron (0.02% boric acid), copper (0.02% $CuSO_4$) and zinc (0.02% $ZnSO_4$) stimulate dry matter accumulation in the cloves.

Irrigation

Garlic being a shallow rooted crop requires light irrigations. The field should be irrigated immediately after planting the cloves. In general, garlic needs irrigation at an interval of 7-8 days during vegetative growth and of 10-15 days during bulb maturation. Frequent and light irrigation at weekly interval is the best to keep optimum soil moisture. Over watering should be avoided. Irrigation should be stopped when the crop reaches maturity at least 15 days before harvesting when the crops first begin to break over or become dry.

Interculture and weed control

Garlic is a poor competitor of weeds hence, weeds must be kept down to get good yield. Garlic is shallow rooted crop and planted closely and there is a chance of root pruning during hand weeding. First weeding should be done after one month of planting and second after two months of planting. Chemical weed control gives better results. Application of Pendimethalin @ 3.35 litre/ha after planting and before first irrigation along with one hand weeding at 45 days after planting is the best.

Harvesting the crop

The crop is ready for harvest when tops turn yellowish or brownish and show signs of drying up and bend over. It becomes ready for harvesting 4-5 months after planting depending on variety, climate and soil. It is generally harvested in March-April in the plains and July-September in the hills. Harvesting is done by pulling out the plants along with the tops. The bulbs are dried for a week or so under shade.

Post harvest handling and storage

After harvest, the bulbs should be cured in shade for about a week. Purpose of curing or drying is to remove excess moisture from the outer skin and neck of the bulb which helps in minimizing shrinkage of bulb and reducing the disease infection during storage. Curing also allow the bulbs to become compact and go into dormant stage. During curing the bulbs are covered along with the tops of each other to avoid damage from the sun. Curing in shade may be done on a floor having ventilation from bottom or on wire racks. After curing before the bulbs are put for storage or marketing, the thick necked, split, injured, diseased or bulbs with hollow cloves are sorted out. Thoroughly cured garlic bulbs keep fairly well in ordinary well ventilated rooms. For low cost storing, a hut like structure is erected under the shade with thatched roof and sides covered by bamboo sticks with provision of good air circulation. Garlic is stored in these sheds by spreading them on racks or tiers having 2-3 layers of bulbs. Garlic with dried leaves can be stored by hanging in well-ventilated room. Bulbs are generally stored for 6-8 months at relative humidity higher than 70% at any temperature however, after this period moulds are developed and root growth is started. Cold storage temperature of 0-2.2°C with 60-70% relative humidity is ideal for 6-8 months long storage in dormant condition and reasonably free from decay. Best cold storage condition is provided by packaging the bulbs in polyethylene bags, and stored at 0-1.6% C temperature with 32-35 % relative humidity. Garlic will keep longer if the tops remain attached. Peeled cloves may be stored in wine or vinegar in the refrigerator. Commercially, garlic is stored, in a dry, low-humidity environment under ambient condition. Irradiation of bulbs with very low dose of 60 Gy gamma radiation effectively controls sprouting and prolongs storage life of the bulbs.

Yield

Average bulb yield is 10-20 t/ha depending on the variety and climate. Recovery of clove in bulbs ranges from 86-96% depending on the variety.

Plant protection measures for onion and garlic

Disease

Damping off (*Pythium, Rhizoctonia, Phytophthora,* etc.)

It is very destructive fungal disease especially in nursery beds of kharif and early rabi crop of onion. Seed rotting and both pre-emergence and post-emergence damping off of seedlings cause thin or irregular stand of seedlings in the seed bed.

Control measures

- Treating the seeds with Thiram/Captan/Mancozeb @ 3.0g/kg of seeds.
- Drenching the nursery beds 7 days before sowing with Thiram/Captan or any copper fungicide @ 3g/litre of water.
- Covering the beds with transparent polythene sheets before sowing and left to open sun for at least 15 days, known as soil solarization.
- Spraying the young seedlings with 0.25% Mancozeb or Metalaxyl-Mancozeb (0.25%) or Cymoxanil-Mancozeb (0.25%) or Fosetyl-Al (0.25%) with a sticker at weekly intervals.
- Provision of proper drainage to the seed beds.
- Removal of the affected seedlings from the beds as soon as the symptoms are visible.
- Avoidance of flooding the beds to check spread of the disease.

Stemphylium blight (*Stemphylium vesicarium*)

It is a serious fungal disease of onion. This foliar disease is particularly serious in the seed crop. Infection occurs in the form of small yellow to pale orange spots or streaks on one side of the leaves or flower stalks causing fall over of seed stalks or leaves.

Control measures

- Spraying the crop with Mancozeb @ 2.5g/l or Chlorothalonil @ 2g/l or Difenconazole @ 0.5g/l or Propiconazole (0.1%) with a sticker 4-5 times starting from 15 days after transplanting.

Purple blotch (*Alternaria porri*)

It is destructive disease of both onion and garlic. The common symptoms of this important fungal disease occur on leaves and flower stalks as small, shunken, whitish flecks with purple coloured centre. Finally the leaves and flower stalks fall down from point of attack. The most favourable temperature for disease development is 28-30°C temperature with 80-90% relative humidity.The disease appears mostly in the kharif and seed crop of onion.

Control measures

- Atleast three summer-ploughings to reduce disease severity.
- Spraying the crop with Mancozeb @ 2.5g/l or Chlorothalonil @ 2g/l or Difenconazole@ 0.5g/l with a sticker 4-5 times starting from 15 days after transplanting.

Basal rot (*Fusarium oxysporum* f. sp. *cepae*)

This fungus attacks both onion and garlic. Common symptoms of this fungal disease are wilting and rapid dying-back of leaves from tip with the roots turns pinkish as the plants approach maturity. The bulbs become soft and a semi-watery decay is found advancing from base of the scales leaves. The infection may also be carried over to the storage where bulb rotting takes place in the same manner.

Control measures

- Adoption of long crop rotation (at least 5 years) excluding related crops like, garlic, leek, etc. and green manuring.
- Treating the seeds with Thirum @ 3g/kg of seeds
- Application of lime to raise the soil pH to 7.2
- Soil application of *Trichoderma viride* @ 3 kg/ha along with manure is effective.
- Adoption of mixed cropping with sugarbeet, maize or wheat in the infected field.

Downey mildew (*Peronospora destructor*)

This fungal disease is prevalent in cool and humid weather condition. Yellowish spot first appear on the upper surface of the leaves. Later, violate growth of the fungus appears on the infected leaves of onion and garlic and on flower stalks of onion which later becomes pale green-yellow and finally leaves or seed stalks collapse. In storage, bulbs of the infected plants become soft, spongy, shrivelled, amber in colour and of poor keeping quality.

Control measures

- Adoption of crop rotation with 4- year break in onion or garlic cultivation
- Control of weeds and keeping the field clean.
- Removal of the primary infected plants.
- Post harvest heating of mother bulb at 41°C to destroy the pathogen
- Spraying the young seedlings with 0.25% Mancozeb or Metalaxyl-Mancozeb (0.25%) or Cymoxanil-Mancozeb (0.25%) or Fosetyl-Al (0.25 %) with a sticker 4-5 times at an interval of 10 days.

White rot (*Sclerotium cepivorum*)

This soil borne fungus invades roots and leaf sheaths of onion and garlic resulting in killing of stem plate, roots and leaf sheath. Leaves become yellow and flaccid and finally whole plant is killed.

Control measures

- Treating the seeds with Thirum @ 3g/kg of seeds or Carbendazim @ 2.5-3 g/kg of seeds
- Incorporation of green manure is also effective.
- Soil application of *Trichoderma viride* @ 3 kg/ha along with manure is effective.
- Spraying the crop with 0.1% Carbendazim / Carbendazim (1g) + Mancozeb (2g) /l or Captan @ 2.5 g/l along with sticker checks the rapidity of the infection.

Black mould (*Aspergillus niger*)

This fungus cause the storage disease of both onion and garlic causing rotting of the bulb and development of black mould occur when the bulbs are stored in high temperature and high humidity condition.

Control measures

- Storage of bulbs in cool and well ventilated place.
- Spraying the bulbs with 0.2% Captan or Carbendazim 0.1% at the time of storage checks the disease.

Soft rot (*Erwinia carotovora pv. carotovora*)

This common bacterial disease of onion and garlic is caused in storage.Bulbs with green necks and which are not properly cured is attacked by these bacteria. The rot begins from the neck of the bulb. Cloves of the bulbs are water soaked and soft. Foul odour emits and bacterial oozes come out upon squeezing the neck of the bulb.

Control measures

- Avoidance of mechanical injury to the bulb during harvesting.
- Proper curing till the neck is tight and scale leaves are dry.
- Bulbs with green neck should not be kept in storage.
- Spraying the bulbs before storage with 0.5% each of Bavistin + Plantomycin is effective in keeping the bulb disease-free.

Insect pests

Thrips (*Thrips tabaci*)

It is the most injurious pest of onion. The nymphs and adults are slender, fragile and blackish or yellowish in colour. Both of them are found lacerating the leaf epidermis and lapping the exuding sap, producing silver white patches known as 'silver top'. Severe infestation resulted drying of leaves from tip to the base gradually.

Control measures

- Adoption of crop rotation.
- Destruction and burning of the weeds and plant debris on which adults hibernate.
- Application of insecticides like Acetamiprid @ 0.5g/l observing the infestation of thrips after transplanting.
- Spraying the crop with Malation @ 1ml/l or Endosulphan @ 1.5ml/l or Nuvacron @ 1ml/l of water 4 times at 15 days interval starting after transplanting.

Cut worm (*Agrotis ipsilon*)

It is a highly polyphagus insect and become active with the onset of cool weather. This nocturnal caterpillar remains in the soil crevices during day time and cut the seedlings of onion and garlic at ground level at night.

Control measures

- Clean cultivation and breaking the big soil clots.
- Mechanical destruction of hiding larvae at base of affected plant in the early infested crops is very effective to manage this pest.
- Dusting the soil around the plant with 4% Endosulphan @ 2kg a.i./ha.

Garlic mite (*Aceria tulipae*)

The mite attacks both onion and garlic and colonize along the mid rib vein of dorsal surface of leaf where they feed and multiply. All the active stages viz., adults, nymphs and eggs are found on the affected site. The leaf of the infected plant curled and twisted resulting stunted growth of the plant and yield is severely reduced.

Control measures

- Soaking of seed material into 0.05% Dicofol solution for ten minutes before planting is effective.
- Spraying of infected standing crop with wettable sulphur @ 2 gm/litre of water or 0.05% Dicofol is effcetive.

Onion fly (Hylemia antigua)

Adult appears like house fly. Maggots are small white and devoid of legs. Maggots enter the bulbs through roots and attack the tender portions. Infested plants turn yellowish-brown and finally dry up.

Control measures

- Crop rotation excluding onion and garlic.
- Application of granular insecticide like Phorate 10G @ 25kg/ha.
- Spraying the crop with Malathion 1ml/litre or Thiodan 1.5ml/litre of water.

Leguminous Vegetable Crops

- **Garden pea**
- **French bean**
- **Vegetable cowpea**
- **Hyacinth bean**
- **Cluster bean**

Garden pea

Pea, *Pisum sativum* var. *hortense* under the family Fabaceae (Leguminosae) having chromosome No.2n =14 is an important legume vegetable cultivated throughout the world. Green peas are all-time favourite vegetables in the world. As a cool season crop, it is extensively grown in temperate zone and winter season in the tropics and subtropics. Top five garden pea producing countries are China, India, USA, France and Egypt. In India, it is grown extensively in Uttar Pradesh, Madhya Pradesh, Himachal Pradesh, Punjab, Haryana, Rajasthan, Maharashtra, Bihar and Karnataka, contributing to 67% of the total production.

The crop

Garden pea is a self pollinated annual herb, bushy or climbing, glabrous, usually glaucous; stems weak, round, and slender, 30-150 cm long; leaves alternate, pinnate with 1-3 pairs of leaflets and a terminal branched tendril, leaflets ovate or elliptic, 1.5-6 cm long. The leaf type may be conventional, semi-leafless and leafless. Leaf size in most cases increases up to the first node bearing the first flower. Stipules are large, leaf-like and up to 10 cm long. The inflorescence of pea is a raceme arising from the axil of the leaf. Corolla is white or pink or purple. The flower has 5 fused sepals, 5 petals, 10 stamens (9 fused in a staminal tube and 1 stamen is free), and one carpel, which develops into a pod with multiple peas. Pea pods are botanically fruit, since they contain seeds and developed from the ovary.

Pods are swollen or compressed, short-stalked, straight or curved, 4-15 cm long, 1.5-2.5 cm wide, 2-10 seeded, 2-valved, dehiscent on both sutures. Colour of the pod is mostly commonly green, occasionally golden yellow, or infrequently purple. The name "pea" is also used to describe other edible seeds from the crops of the family Fabaceae such as pigeon pea (*Cajanus cajan*), cowpea (*Vigna unguiculata*), and grass pea (*Lathyrus* sp.). The node at which the first flower emerges is characteristic of a given variety.

In temperate regions the number of nodes at which the first flower emerges is reported to vary from 4 in the earliest to about 25 in late maturing types under field conditions. Flowers borne on the same peduncle produce pods that mature at different times, the youngest being at the tip. On a whole plant basis, flowering is sequential and upwards from node to node. Seeds are globose or angled, smooth or wrinkled, exalbuminous, whitish, gray, green or brownish. Wrinkled-seeded garden peas are sweeter than smooth seeded types. 100 seeds may weigh from 10 to 36 grams.

Importance and use

Peas are cultivated for the fresh green seeds, tender green pods, dried seeds and foliage. Green peas cooked as a vegetable and are marketed fresh, canned or frozen. In India, fresh peas are used in various dishes, some popular being *aloo matar* (curried potatoes with peas), *matar paneer* (paneer cheese with peas), *peas pulao* (a rice dish with peas) though they can be substituted with frozen peas as well. Ripe dried peas are used whole, split or made into flour. Green peas are the number one processed vegetable specifically frozen and dried. Green foliage of garden pea is also used as pot herb vegetable in parts of Asia (India, Myanmar in particular and Africa). Some cultivars are grown for their immature tender green pods, which are eaten cooked or raw. Pea is being used in a growing snack market. One snack item is prepared by soaking the peas overnight and frying them in palm oil or coating them with other food items such as rice flour before frying for the purpose of imparting different flavours. Another product is prepared by finely grinding the peas and extruding them under pressure to create different shapes. Cultivars such as Alaska, Super Alaska and Supergreen have long been the standard type of canning pea. Canning cultivars usually have a tough skin that holds its shape during canning. In some parts of the world including India, dried field peas are consumed split as dahl, roasted, parched or boiled.

Nutritional value

Peas are high in fibre, protein, vitamins (folate and vitamin C), minerals (iron, magnesium, phosphorus and zinc), and lutein (yellow carotenoid pigment that benefits vision). Dry weight is about one-quarter protein and one-quarter carbohydrates (mostly sugars). Fresh green peas contain per 100 g: 44 calories, 75.6% water, 6.2 g protein, 0.4 g fat, 16.9 g carbohydrate, 2.4 g crude fibre, 32 mg Ca, 102 mg P, 1.2 mg Fe, 6 mg Na, 24.8 μg K, 405 μg β -carotene equivalent, 0.28 mg thiamine, 0.11 mg riboflavin, 2.8 mg niacin, 65 μg folic acid and 27 mg ascorbic acid. Green peas are low in calories and contain no cholesterol.

Medicinal value

Peas contain phytosterols, especially β-sitosterol which help lower cholesterol levels inside the human body. It is also good source of vitamin K which has a role in the cure of Alzheimer's disease patients by limiting neuronal damage inside the brain. It also contains adequate amounts of anti-oxidant flavonoids and earotenoids such as β carotene, lutein and zea-xanthin which help to protect from lung and oral cavity cancers.

Origin and taxonomy

Peas appear to have been cultivated for nearly 7,000 years. The earliest archaeological finds of peas come from Neolithic Syria, Turkey and Jordan. In Egypt, evidence of peas dates from ca. 4800–4400 BC. The wild pea is restricted to the Mediterranean basin and the Near East.

Central Asia (Kazakhstan, Uzbekistan, Tazikistan, Turkmenistan, Kyrgyzstan, Afghanistan), Abyssinia (Ethiopia) and the Mediterranean basin are the primary centres of origin of pea with the near-east (Syria, Lebanon, Israel, Jordon) as the secondary centre of origin. In another view, *Pisum* is considered to have been originated in Ethiopia from where it spread during pre-historic times to Mediterranean region, Central Europe, the Near East, and subsequently to rest of the world. The wild progenitor of *Pisum sativum* is unknown.

Archaeological evidence shows that smooth seed form was cultivated in the Near East and Europe as early as 7000 BC. The earliest archaeological finds of peas date from the late neolithic era of current Greece, Syria, Turkey and Jordan. In Egypt, early finds date from *ca.* 4800–4400 BC in the Nile delta area, and from *ca.* 3800–3600 BC in Upper Egypt. The pea was also present in Georgia in the 5th millennium BC. Peas were present in Afghanistan *ca.* 2000 BC, in Harappa, Pakistan, and in northwest India in 2250–1750 BC. In the second half of the 2nd millennium BC, this pulse crop appears in the Ganges Basin and southern India.

Pea was taken to China in the first century. In early times, peas were grown mostly for their dry seeds. By the 17th and 18th centuries it had become popular to eat peas "green," or fresh, while they are immature and right after they are picked, especially in France and England. In England, the distinction between field peas and garden peas dates from the early 17th century. Green peas were introduced from Genoa to the court of Louis XIV of France in January 1660. Sugar pea was introduced to France from the market gardens of Holland and French called sugar peas *mange-tout*, for they were consumed pods and all.

The International Legume Database accepts three species, one with two subspecies:

- *Pisum abyssinicum* (syn. *P. sativum* subsp. *abyssinicum*)
- *Pisum fulvum*
- *Pisum sativum* - pea
  - *Pisum sativum* subsp. *elatius* (syn. *P. elatius*, *P. syriacum*)
  - *Pisum sativum* subsp. *sativum*
  - *Pisum sativum* subsp. *sativum* var. *hortense* (Garden pea)
  - *Pisum sativum* subsp. *sativum* var. *arvense* (Field pea)

  Other variations of *Pisum sativum* include:
  - *Pisum sativum* var. *saccharatum* is commonly known as the snow pea
  - *Pisum sativum* var. *macrocarpon* is known as the sugar snap pea or snap pea

Both snow pea and sugar snap pea are eaten whole before the pod reaches maturity. In snow pea, the pod is flat in shape, while in sugar/snap peas, the pod becomes cylindrical, but is eaten while still crisp, before the seeds inside develop.

The somatic chromosome number of pea is 2n = 2x = 14.

Adaptation in India

Archaeological evidences suggest the existence of pea in India before the advent of the Aryans. This pea must be round and non-sweet pulse type (*P. sativum var. arvense*). The sweet garden pea (*P. sativum* var. *hortense*) varieties developed in temperate regions of Europe might have been introduced in India during the British regime. Some of the hardy and round seeded garden pea varieties like Bonia, Kaip, Asauji, etc. may possibly be the early introductions of the Britishers.

Improved varieties

Early types: Asauji, Early Superb, Arkel, Harbhajan, Alaska, Hisar Harit, Early Badger, Meteor, Lincoln, Azad P-1, Jawahar Matar-3, Jawahar Matar-4, Jawahar Peas-54, Matar Ageta-6, Azad P-3, Pant Matar-2, Pant Sabji Matar-3, VL Ageti Matar-7, Kashi Nandini, Narendra Sabji Matar-6, Narendra Sabji Matar-4, Kashi Udai, Kashi Mukti, VRP-2.

Medium/Late types: Bonneville, Alderman, New Line Perfection, Arka Ajit, Jawahar Matar-1, Jawahar Matar-2, Jawahar Peas-83, NDVP-8, NDVP-

10, Vivek-6, Vl Matar-3, Pant Uphar, Punjab-88, Azad P-2, Jawahar Peas-83, Jawahar Peas-15, Vivek Matar-9, Narendra Sabji Matar-5, Kashi Shakti, Azad P-5, Swarna Mukti, Priya

Powdery mildew resistant variety: Jawahar Matar-5, Jawahar peas-83, Jawahar Matar-6, Jawahar Matar-15, Jawahar Matar-54

Edible podded type: Sylvia, Oregon Sweet Podded,JP-19, UN-53 (6)

Cultural requirements

Pea is basically a cool season crop. It requires a cool, relatively humid climate and are grown with temperatures from 7 to 30°C and production is concentrated between the Tropics of Cancer and 50° N. As a winter annual, pea tolerates frost to -2°C in the seedling stage, although top growth may be affected at -6°C. Winter hardy peas can withstand -10°C. The optimum temperature levels for the vegetative and reproductive periods of peas are 21 and 16°C, and 16 and 10°C (day and night), respectively. Temperatures above 27°C shorten the growing period and adversely affect pollination. A hot spell is more damaging to peas than a light frost. Hot dry weather reduces the number of pods per plant and ultimately total yield and degrades the quality also due to conversion of sugars into starch. Peas can be grown successfully during mid-summer and early fall in those areas having relatively low temperatures and a good rainfall, or where irrigation is available.

It prefers well drained loamy soil, however, it can be grown on light sandy soil to clay soil as well. Early crop can be taken in light soil but higher productivity is achieved in clay soil. Pea can not thrive on acidic soils. It is very sensitive to saline and alkaline conditions. The pH of 6.0–7.5 is ideal.

Sowing time

Pea varieties are generally classified into 3 groups as early, mid and late. Early crop usually suffer from *Fusarium* wilt and late varieties suffer from powdery mildew. It should be sown from first fortnight of October to mid November in Northern Indian plains. In the northern hills, the sowing can be done in September at the altitude of 3000-5000 feet. However, at the altitude of 9000 feet, the planting can be done in April where the pods are harvested in July as an off season vegetable pea which bring higher prices. The extra early variety may be sown during September and late maturing varieties can be sown up to December also. A second crop of pea may be taken in hills by sowing in March as summer crop or by sowing in May as autumn crop.

Land Preparation

The land should be prepared in well advance with at least 4-5 ploughing to a good tilth and to remove all stubbles, weeds, etc from the land. Well rotten organic manure preferably farm yard manure or compost @ 15-20 t/ha should be added during land preparation and the land is levelled properly.

Sowing of seeds

Viable, healthy, well-matured and pure seeds should be used for sowing. The seed rate for early variety is about 100-120 kg/ha whereas 80-90 kg/ha for mid and late varieties. The seeds must be treated before sowing to avoid losses due to fungal diseases. Mixture of Thiram + Bavistin (2g + 1g/kg seed) is recommended for seed treatment. The seeds should be mixed thoroughly with the required quantity of fungicide before sowing.

Pea being a leguminous crop, fixes atmospheric nitrogen through root nodules and the comparative nitrogen requirement is less. Soaking of seeds with Rhizobium bacteria (*Rhizobium leguminosarum*) in 10% molasses solution is effective for increased nodulation.

Pea seeds can be drilled by local implements or by tractor hauled seed drills. The seeds should be sown 2–3cm deep in the soil. Soaking seeds of wrinkled seeded type for overnight is essential. Sowing should be done in row 30 cm apart and plant to plant spacing may be maintained at 5-7.5 cm with seed placement at a depth of 3 cm.

Nutrient management

Pea being a leguminous crop, fixes atmospheric nitrogen through root nodules and the comparative nitrogen requirement is less. About 15-20 t/ha FYM and a general dose of 40 Kg N, 80 kg P_2O_5, 60 kg K_2O per hectare is recommended. Phosphatic fertilizers enhance yield and quality of pods. Pea is a potash loving plant and it should be preferably applied in the form of potassium sulphate. Prolonged application of DAP causes sulphur deficiency. Gypsum or pyrites may be recommended for the deficiency of sulphur.Application of small quantity of N (10kg/ha) as topdressing after periodical picking of pods improves yield potential.

Irrigation

Water requirement of pea is comparatively low. However, soil moisture deficit reduces growth and also affects nodulation. Pre-sowing irrigation and irrigation at flowering stage are highly beneficial. Normally 2–3 irrigations are needed for mid-season or late sown peas. If the soils are lighter and shallow, the number can be increased. Excess of moisture results in yellowing of the plant and reduces the yield. Furrow irrigation is normally adopted for irrigating peas but sprinkler system of irrigation is much better. Moisture stress at flowering and subsequent pod filling stage is most undesirable affecting the yield and quality of pods.

Interculture and weed control

Clean and shallow cultivation should be followed. Heavy weed growth is generally found in pea field right from sowing till harvest due to slow growth of pea during earlier stages and wide spacing given for hand picking of green pods. Usually 1–2 hand-weedings are sufficient. Weeds can also be controlled by mechanical means but deep harrowing damages the roots. Late interculture will also damage the crop due to trampling and mechanical breakage of tender and succulent stems and branches. The weeds can also be controlled by application of pre-sowing herbicide (Fluchloralin @ 0.75 a.i./ha) or pre-emergence herbicide (Pendimethalin @ 1.0 kg a.i./ha) application. Pea being a legume crop if planted year after year in the same field creates problem of wilt complex and other soil borne diseases. To check this damage, pea crop may be rotated with other vegetable crops. Besides, crop rotation also helps in maintaining balance of soil nutrients.

Harvesting the crop

Harvesting of green pods must be done at proper stage. Delay in harvesting cause deterioration of the pea quality. The green pod pickings may be done during the morning or evening. The pea plants are very tender with soft stem and therefore pickings should be done gently. A small jerk damages the vines thereby injuring the plants. Repeated picking should be avoided and only 2-3 pickings in early crop and 3-4 in mid-season/late crop should be done.

Post harvest handling and storage

Good quality pods are uniformly shiny green, fully turgid and free from defects and mechanical damage. Diseased, damaged, over-mature yellow and immature and under-sized pods are sorted out before packing for marketing. Green peas are perishable and cannot be stored for long period. However un-shelled peas can be stored longer than the shelled peas and are storable for about 2-3 weeks at 0°C and RH of 90-95%. Snow peas are stored at 0-2°C at 90-95% RH for 1-2 weeks. Green pea pods are packed for marketing in baskets, gunny bags or boxes of various types and sizes including corrugated fibre board boxes depending upon the availability and market requirements. The containers must be well filled to avoid bruising due to loose packing.

The tender peas must be put through the process of freezing shortly after being picked so that they do not spoil too soon. In order to freeze and preserve peas, they must first be picked, shelled and then blanched. The peas are boiled for a few minutes to inaitivate the enzymes that may shorten their shelf life. They are then cooled and removed from the water. The final step is the actual freezing to produce the final product. This step may vary considerably; some companies freeze peas by air blast freezing, where the peas are put through a tunnel at high speeds and frozen by cold air. Finally, the peas are packaged and shipped out for retail.

Yield

Pod yield is 2.5-4.0 t/ha in the early crop and 6.0-7.5t/ha in the mid-season/ late crop.

French bean

French bean, *Phaseolus vulgaris* under the family Fabaceae (Leguminosae) and having chromosome No.2n = 22 is an important leguminous vegetable crop grown worldwide for its tender pod and edible dry seeds. It is the most commonly consumed legume worldwide, and it is the most important legume produced for direct human consumption, with a commercial value exceeding that of all other legume crops combined. It is also known as common bean, kidney bean, dwarf bean, haricot bean, snap bean, string bean or garden bean. Brazil and India are the largest producers of dry beans, while China produces, by far, the largest quantity of green beans. Brazil, India, Indonesia, Turkey, Columbia, USA, Canada and Ethiopia are the other leading green bean producing countries. In India, French bean is grown mainly in Maharashtra, Jammu and Kashmir, Himachal Pradesh and Uttar Pradesh hills, Nilgiri (Tamil Nadu) and Palni (Kerala) hills, Chickmagalur (Karnataka), Chhotanagpur plateau regions of Jharkhand, Uttarakhand and Darjeeling hills of West Bengal.

The crop

French bean is a highly variable species that has a long history of cultivation. All wild members of the species have a climbing habit, but many cultivars are classified as "bush beans" or "pole beans", depending on their growth habit. Leaves are alternate, green or purple which are divided into three oval, smooth - edged leaflets, each 6–15 cm long and 3–11 cm wide. Bracts on the rachis of the inflorescences are persistent, and the size and shape of the bracteoles are distinguishing characteristics of bean cultivars

Flowers are generally purple, pink or white of about 1 cm long. The floral structure contributes to the high rate of self-pollination: anther dehiscence and stigma receptivity occur at the same time, before the flower is fully open, and the anthers and stigma are positioned near one another at the time of anther dehiscence and stigma receptivity

The flowers give way to pods 8–20 cm long and 1–1.5 cm wide. These may be green, yellow, or purple in colour, each containing 4–6 beans. The beans (seed) are smooth, plump, kidney-shaped, up to 1.5 cm long, range widely in colour, and are often mottled in two or more colours. Flowering is reduced once the seeds begin to form inside the pods.

The three commonly known types of green beans are: string or snap beans, which may be round or have a flat pod; stringless beans, which lack a tough, fibrous "string" running along the length of the pod; and runner beans, which belong to a separate species, *Phaseolus coccineus*.

As the name implies, snap beans break easily when the pod is bent, giving off a distinct audible snap sound. The pods of snap beans (green, yellow and purple in colour) are harvested when they are rapidly growing, fleshy, tender (not tough and stringy), bright in colour, and the seeds are small and underdeveloped (8 to 10 days after flowering). Raw or undercooked beans contain a toxic protein called phytohaemagglutinin which can be deactivated by cooking beans for ten minutes at boiling point (100 °C).

Importance and use

The main categories of common beans, on the basis of use, are snap beans (tender pods with reduced fibre which harvested before the seed development phase), dry beans (seeds harvested at complete maturity) and shell beans (seeds harvested at physiological maturity).

Most of the beans for processing (canned, frozen, and freeze-dried) are round podded, while fresh market cultivars are often flat or oval shaped. Yellow podded cultivars are also grown for culinary purposes. The immature seeds are boiled or steamed and used as a vegetable. The mature seeds are boiled, baked, pureed, ground into a powder or fermented into 'tempeh'. The powdered seed makes a protein-enriching additive to flour and can also be used in soups. The seeds can also be sprouted and used in salads or cooked. Young leaves - raw or cooked can be used as a potherb. The very young leaves are sometimes eaten as a salad, the older leaves are cooked. The straw can be use as fodder. *Phaseolus* species, is as a member of the legume family Fabaceae acquire the nitrogen they require through an association with rhizobia, a species of nitrogen-fixing bacteria.

Nutritional value

French bean is an especially valuable source of the amino acids lysine and tryptophan; the minerals iron, copper and zinc; and beneficial phytochemicals, antioxidants and flavonoids. Immature pods contains 91.4g moisture, 1.7g protein, 0.1g fat, 4.5g carbohydrates, 1.8g fibre, 0.5g minerals per 100g of edible green pods. Among the minerals, it contains 50 mg calcium, 28 mg phosphorus, 1.7 mg iron, 129 mg potassium, 37 mg sulphur, 4.3 mg sodium and 0.21 mg copper. Edible beans of 100 g also contain 221 IU vitamin-A, 0.08 mg thiamine, 0.06 mg riboflavin, 11.0 mg vitamin-C and 0.3 mg nicotinic acid.French bean is a good source of amino acids such as arginine, histidine, lysine, tryptophan, phenyl alanine, tyrosine, methionine, cysteine, threonine, leucine, isoleucine and valine.

Medicinal value

Beans are also said to be anti-diabetic and good for bladder, burns, cardiac carminative, depurative, diarrhoea, dropsy, dysentery, eczema, emollient, hiccups, itch, kidney resolvent, rheumatism, sciatica and tenesmus. The green pods are mildly diuretic. Dietary fibers are beneficial for the prevention of cholesterol and sugar levels after having a meal. French beans are also helpful in energizing the body as they are rich in iron, the nutrients found in hemoglobin which helps in giving energy to the body. The presence of copper in French beans, in turn, helps in the proper synthesis of hemoglobin. People suffering from severe migraine attacks can get benefit from eating French beans as they are a very good source of riboflavin, the source that helps in mitigating migraine attacks. French beans have many anti-inflammatory nutrients like β carotene and of course, Vitamin C. They also contribute in preventing fatal diseases like colon cancer. Dry seed flour is also used externally in the treatment of ulcers. When bruised and boiled with garlic they are used to cure intractable coughs. A homeopathic remedy is made from the entire fresh herb and is used in the treatment of rheumatism, arthritis and disorders of the urinary tract.

Origin and taxonomy

French bean is thought to have been originated in South and Central America. Southern Mexico and Central America are considered to be the primary centre of origin, while secondary centres lies in Peru-Bolivia-Ecuador region of American continent. It was actually domesticated in Mesoamerica first, and traveled south, probably along with squash and maize. Multi-locus sequence data have indicated that the domestication of common bean was initiated 8 000 years ago. Cultivation of French beans started by the American Indian tribes settled in Tehuacan Valley of Mexico and in Callejon de Huaylas, Peru. When Christopher Columbus returned from his second voyage to the New World in the year 1493 he brought French beans with him in the Mediterranean region. French beans were considered to be rare to find and expensive but soon became one of the commonly used beans in the 19th century. In France, French beans were introduced in the year 1597 by the Conquistadors. In Europe, the French bean spread rapidly in the 16th and 17th centuries and reached England by 1594. *Phaseolus aborigineus* is known as the progenitor of French bean (*Phaseolus vulgaris*) and the domestication of this species might have occurred from *Phaseolus aborigineus* in Brazil and northern America. During the process of domestication in common beans, several morphological changes like growth habit, testa colours, pod structure, etc. have occurred.

Chromosome number of French bean is 2n = 22

Adaptation in India

French bean has been introduced in India comparatively in recent times from Europe during the 17th century. Most of the local cultivars native to North-eastern region of India is pole types. Plant introduction plays very important role in breeding French bean in India. In fact, most of the popular varieties grown in India are exotic varieties.

Improved varieties

The French beans are classified into string and stringless beans based on the extent of fibre in pods. Wax-podded varieties are not common and mostly confined to home garden. The beans are grouped according to their growth habit *viz*., pole or climbing beans and bush beans, which are dwarf in nature. Important varieties grown in India are:

Bush type: Contender, Giant Stringless, Bountiful, Premier, Jampa, Masterpiece, Pusa Parvati, Arka Komal, Arka Subidha, Pant Anupama, VL Boni-1, YED-1, Phule Surekha, Kashi Param, Azad Rajmah-1, Phule Suyash, Swarna Priya, Swarna Lata.

Pole type: Kentucky Wonder, Pusa Himlata, TKD-1, KKL-1.

Resistant to angular leaf spot: Pant Anupama, Lakshmi, SVM-1.

Resistant to powdery mildew: Contender, Pusa Parvati.

Resistant to wilt: Jampa.

Resistant to mosaic disease: Pant Anupama, Contender, Pusa Parvati.

Cultural requirements

In plains, French beans are grown as a winter crop and in hills, they can be cultivated throughout the year except in winters as the crop is frost-sensitive. It cannot withstand drought as well as heavy rainfall. French bean is a cool weather crop but grow happily under mild warm season. It is sensitive both to frost and to very high temperatures. French bean cultivars are mostly day neutral types. The seeds do not germinate below 15°C. The optimum temperature required for better growth, pod set and crop maturity is in between 15.6° and 21.1°C temperature. During hot weather, blossoms and pods may drop. Similarly, heavy and continuous rains results in dropping of blossoms and pods.

French bean can be grown on a wide range of soil types but sandy loam soils are best suited. Good yields are seldom obtained from very heavy soils. The suitable soil reaction is between 5.5 and 6.0 pH.They cannot grow well in extreme acidic or alkaline soils. On sandy soils, French bean matures in a shorter time than on the heavier soils. High moisture content and high nitrogen delay maturity. It is very sensitive to high concentration of aluminium and manganese.

Sowing time

Timely sowing is an important aspect for harnessing more yield. Reduction in yield under late sown conditions may be attributed to poor development of yield components due to low temperature prevailing during reproductive stages of the crop. In the plains of India, French bean is sown twice a year, first sowing is done in September to October and the second in January to February. In hills, sowing is done from March to June. In South and Eastern India, it is cultivated from October to March. The seeds should never be soaked before sowing.

Land Preparation

Soil should be prepared well before sowing with at least 4-5 ploughings and all stubbles, weeds, etc are removed from the land. Well rotten organic manure preferably farm yard manure or compost @ 15-20 t/ha should be applied during land preparation and the land is levelled properly.

Sowing the seeds

Viable, healthy, well-matured and pure seeds should be used for sowing. About 30-40 kg/ha seed is enough for pole type and 80-90 kg/ha for bush type varieties. The seeds must be treated before sowing to avoid losses due to fungal diseases. Mixture of Thiram + Bavistin (2g + 1g/kg seed) is recommended for seed treatment. The seeds should be mixed thoroughly with the required quantity of fungicide before sowing. Seeds can be drilled by local implements or by tractor hauled seed drills. The seeds should be sown 2–3cm deep in the soil. The drilling or dibbling can be done in rows placed 90 cm apart for pole type and 45 cm apart for bush type. The spacing of 15-20 cm from plant-to-plant is ideal.

Nutrient management

French beans have the capacity of fixing the atmospheric nitrogen as nitrogen-fixing bacteria symbiotically lives in their roots. However, it has poor capacity to fix atmospheric nitrogen because of shy nodulating character. Hence, the nitrogen requirement of French bean is different from other pulse crops which vary with the genotypes. Soaking of seeds with Rhizobium bacteria (*Rhizobium phaseoli*) in 10% molasses solution is effective for increased nodulation. Being a shy nodulator, French bean crop readily responds to large doses of nitrogen. A general fertilizer dose of 80kg N, 60kg P and 50kg K/ha is recommended. Half of N and full doses of P and K should be placed in bands 7–8cm away from seed to avoid injury at the time of planting and the remaining N is top dressed at the time of flowering. French bean can absorb sulphur in great quantities and it is necessary to maintain the relation of nitrogen and sulphur in the plant to produce protein and application of sulphur between 10-20 kg/ha can mitigate sulphur deficiency. French bean also responds to micronutrients. The foliar application of B, Cu, Mo, Zn, Mn and Mg each of 0.1% is effective in enhancing quality and pod yield.

Irrigation

French bean is a shallow-rooted crop and is sensitive to both water excess and water stress conditions. Adequate moisture in the top soil layer at time of sowing is essential to ensure good germination of seeds. About 6–7 irrigations are needed during the growing season however, frequency of irrigation depends upon season, soil type and organic-matter content. The plants are susceptible to water stress at critical periods of growth—pre-blooming, flowering and pod-filling stages. Thus, the moisture level should be near to field capacity particularly during flowering and pod-formation stage. Deformed pods can result from water stress due to low moisture or excessive evapo-transpiration losses. Flooding in French bean causes anoxia leading to root rot. Therefore, moisture should be evenly distributed throughout the growth period. About 150–400 mm of water is usually sufficient for French bean.

Interculture and weed control

Very shallow cultivations should be done at early stage of crop growth. Deep cultivation will disturb plant growth due to root pruning because roots grow shallow near the surface. Cultivation during early stages will also help in keeping the weeds down. Once plants grow well, the large leaves form dense canopy which acts as a weed suppressor. The thick foliage cover completely shades the soil beneath and weeds, starved of sunlight, are unable to compete with. The weed-crop competition starts 2–3 weeks after planting. Inadequate weed management practices reduce the yield drastically. The narrow spacing reduces crop-weed competition and two weedings may be needed till the plant can smother the weeds.

Beans are susceptible to injury during cultural operations after plants begin to flower. Cultivation or working among plants, where foliage is wet should be avoided, because spores of anthracnose and angular leaf-spot diseases are easily spread under wet conditions. Bush beans do not require staking, while pole varieties are staked for obtaining high yield. The vine grows clockwise around the support. Two wooden or iron poles are used at the two ends of the rows and lines of the wires are stretched. Fine rope or jute string is fixed in zigzag fashion between the lines of the wires. Pole beans are also grown in mixture with corn or okra as the stalk is used for support of the vines. Application of GA_3 (50ppm) on 30 days old plants is effective in increasing plant height, number of leaves and pod yield. A pre-emergence application of Alachlor @ 2–2.5 kg a.i./ha is recommended for an effective control of weeds.In hills, French beans are intercropped with maize, amaranth and foxtail millets. The maize + French bean–wheat cropping sequence is commonly followed. In plains, French bean is intercropped with early potato. In mid-hills of Himachal Pradesh, pole beans are sometimes grown as relay after tomato crop to use the same stakes.

Harvesting the crop

French bean comes to harvest from 40 days onwards after germination depending on the variety. It takes about 10-14 days after flowering for the first picking of the pods for vegetable purpose. Harvesting depends upon the ways they are used. Green pods are usually harvested when they are fully grown but seeds start forming and remain tender. Quality of pod (uniformity in green colour), texture (tender, firm and non-fibrousness) and natural bean flavour are reduced when harvesting is delayed. Generally 2-3 pickings are done for bush beans and for pole type it is 3-5 pickings. The beans are generally harvested by hand. Mechanical pickers are also devised for harvesting of beans especially for processing type varieties. These employ 'Once over' destructive harvest, which strips leaves and remove pods from the plants. Shelled beans are harvested when their seeds achieve full size and become relatively firm. The seeds are separated from pods and empty pods are discarded because these become fibrous. Shelled beans have the characteristics of remaining firm.

Post harvest handling and storage

Green beans are highly perishable since their respiratory rate and moisture content, both is very high. Rapid cooling after harvesting is important to maintain quality. After harvesting, beans are washed and culled and diseased, inferior pods are discarded. Optimum storage and transit temperatures are 5°–7°C. Under these conditions storage life increases up to 20–25 days. At temperatures below 1°–2.5°C, chilling injury occurs 10–12 days after storage. In India, bulk of French bean is consumed as fresh while in western countries more than 80% crop is processed. Processors have strict requirements for specific varieties such as white seed coat colour and uniform crop maturity. The varieties with fleshy pods are widely used for commercial freezing.

Yield

Pod yield is 5.0-6.0 t/ha in the bush type varieties and 12.0-15.0 t/ha in pole type varieties. Dry seed yield is generally 1.5-2.0 t/ha. Productivity of fresh market beans varies from season to season and region to region and yields are generally higher in temperate than the tropical zones.

Vegetable cowpea

Cowpea, *Vigna unguiculata* under the family Fabaceae (Leguminasae) and having chromosome no. 2n = 2x =22 is in cultivation from very ancient times in the tropics of old world. Cowpea cultivars grown for the immature green pods as vegetable are variously known as asparagus bean, snake bean and yard long bean and when grown for dry or immature seeds, they are known as black-eye pea, kaffir pea, China pea and southern bean. Cowpea is grown extensively for dry seed in the African countries particularly, Nigeria, Niger, Burkina Faso, Ghana, Kenya, Uganda, Malawi, Tanzania, Togo and Senegal. The major cowpea producers in Latin America are Brazil, Venezuela, Peru, Panama, El Salvador and Haiti. Asian production includes that of yard long bean as vegetable and major vegetable cowpea producing countries are India, SriLanka, Bangladesh, Myanmar, China, Korea, Indonesia, Nepal, Pakistan, Phillipines, Thailand and Malayasia. In the USA, cowpea is grown both for immature pod and seed and dry seed. In India, cowpea is known since Vedic times. It is grown widely throughout the year for all forms – tender pod, dry seed, fodder, green manure and cover crops both as sole and mixed crop.

The crop

Cowpea is an annual herbaceous legume. Four subspecies of cowpea are recognised, of which three are cultivated. There is a high level of morphological diversity found within the species with large variations in the size, shape and structure of the plant. Vegetable cowpeas are mainly grown for its tender and immature pods.

Cowpeas can either be short and bushy (as short as 20 cm), prostrate or climbing annuals. Climbing vine may grow up to a height of 2 metres. It has a well developed root system and the tap root can penetrate to a depth of 2.4 m after eight weeks. All cowpeas are virtually glabrous in nature. The size and shape of the leaves varies greatly, making this an important feature for classifying and distinguishing cowpea varieties. Another distinguishing feature of the cowpeas are the long 20–50 cm peduncles which hold the flowers and pods. One peduncle can support 1-4 or more pods. Flower colour varies through different shades of purple, pink, yellow and white and blue. Seeds and pods of wild cowpeas are very small, while cultivated varieties can have pods between 10 and 110 cm long. A pod can contain 6–16 seeds that are usually cylindrical or kidney shaped depending on how crowded the seeds remain within the pod. Their texture and colour is very diverse. They can have a smooth or rough coat and be speckled, mottled or blotchy. Colours include white, cream, green, red, brown and black or various combinations.

Black-eyed pea, a common name used for the unguiculata cultivar group, describes the presence of a distinctive black spot at the hilum of the seed. Sesquipedalis in Latin means "foot and a half long", and this subspecies/cultivar group is characterised by unusually long pods, leading to the common names of yardlong bean, asparagus bean and Chinese long-bean.

Importance and use

The cowpea, whether utilized for green pods as vegetable or dry seed as pulse, forms an important component of farming systems from the arid to the humid tropics. In fact, it probably has the greatest potential among all food legumes in the semi-arid to sub tropical areas. Cowpea is well adapted to stress and has excellent nutritional qualities. It is a key dietary staple for the poorest sector of many developing countries of Africa and Latin America and greatly improves an otherwise bland and unbalanced diet based on sorghum and millets.

Cowpea is the crop of all round utilization, grown for tender pods and seeds, dry seeds as pulses, green leaves and even roots.The typical vegetable cowpea has higher monosaccharide: polysaccharides ratio.The immature pods and seeds as well as dry seeds besides being widely used as fresh vegetable and pulse are also frozen and canned. The seeds are usually cooked and made into stews and curries, or ground into flour or paste.

Some unusual utilizations of cowpea include baking powder for biscuits, vegetable milk and legume starch.It is tolerant to drought and most of the soil stress, thus can be grown over a wide range of environmental conditions. Effective cowpea – *Rhizobium* symbiosis can fix about 150 kg nitrogen per hectare.It is also grown for hay, silage, pasture for all types of live stock. It is also used as fodder and green manure crop.

Nutritional value

Cowpea pods are good source of protein, fibre, minerals, calcium and vitamins particularly vitamin A, B and C. It contains 8.0g carbohydrates, 4.3 g proteins and 0.6g fat, 2.0g fibre per 100g of edible portion.Amino acid profile particularly lycine, leucine and phenylalanine contents are relatively high in cowpea which greatly improves the protein quality of cereals.It provides high of 0.9g minerals per 100g of edible part. Tender fruits contain 80mg calcium, 74mg phosphorus and 2.5 mg iron per 100 g fresh.It provides 941 I.U of vitamin A per 100g edible portion. It is a good source of Vitamin B like, 0.09mg Riboflavin and 0.07mg Thiamine per100g fresh.It contains 13.0mgof vitamin C per100g fresh. The grain is a rich source of folic acid, an important vitamin that helps prevent neural tube defects in unborn babies. Antinutritional factors like trypsin inhibitor, chymotripsin, etc. present in dry seeds can easily be removed by boiling in fresh water before cooking.

Medicinal value

Cowpea is particularly important for the supply of vitamin B in the foods. Vitamin B_1 (thiamine) has received great attention for its role in protecting heart health. It can actively prevent heart failure and largely control the ventricles of the heart. Furthermore, various flavonoids are found in cowpeas that can reduce inflammation and promote more normal heart function. Cowpea contains high amount of dietary fibre which plays a particular role in the balance of cholesterol in the body, which can prevent heart attacks and strokes, as well as the build-up of plaque in the arteries. High amount of dietary fibre which gives protection against certain types of cancer particularly colon cancer, regulates transit and lowers blood cholesterol.

Cowpeas are being increasingly linked to lower levels of chronic illnesses like cancer. The antioxidants found in these legumes seek out and neutralize free radicals within the body, which cause chronic illness and cellular mutation. The Vitamin C and A alone that is found in cowpeas is enough to give the body a major detoxifying boost to protect the immune system. Significant levels of tryptophan in cowpeas, which can help the body relax and ease into better sleep patterns.

Origin and taxonomy

The cowpea is one of the most ancient human food sources and has probably been used as a crop plant since Neolithic times. All evidence points to its originating in Africa, although where the crop was first domesticated is uncertain. Although there is no archaeological evidence for early cowpea cultivation the centre of diversity of the cultivated cowpea is West Africa, leading an early consensus that this is the likely centre of origin and place of early domestication. New research using molecular markers has suggested that domestication may have instead occurred in East Africa and currently both theories carry equal weight.

Remains of charred cowpeas from rock shelters in Central Ghana have been dated to the second millennium BC. In 2300 BC the cowpea is believed to have made its way into South East Asia where secondary domestication events may have occurred. In India, cowpea is known since Vedic times. However, it is generally agreed that the cowpea is of African origin as conspecific wild forms are found in Africa but are absent in Asia. However, India is considered the modern centre of diversity of cowpea cultivars. From South East Asia, cowpeas traveled north to the Mediterranean, where they were used by the Greeks and Romans. The first written references to the cowpea were in 300 BC and they probably reached Central and North America during the slave trade through the 17th to early 19th centuries. The first written reference of the word 'cowpea' appeared in 1798 in the United States. The name was most likely acquired due to their use as a fodder crop for cows.

The earlier view was that the progenitor of cowpea was the wild var. *dekindtiana* of sub-species *dekindtiana* but later it was suggested that var. *mensensis* of sub-species *dekindtiana* was the most likely progenitor of cowpea.

In the African context, the role of cowpea is predominantly that of pulse and African use of pulse continues, but after the introduction of Unguiculata forms (sub-species *unguiculata*) to southeast Asia and India, two other subspecies were evolved: sub-species *sesquipedalis* (vegetable type) mainly in southeast Asia and sub-species *cylindrica* (fodder type) in India and southeast Asia under the predominant influence of human selection. The other sub-species *textilis* was probably selected from *unguiculata* forms in West Africa for fibre from its long peduncle.

Cowpea belongs to the sub-family Faboideae under the family Fabaceae (Leguminoceae) with in the order Fabales. Classification of cowpeas was a matter of debate as they were ranked as species, varieties or sub-species. In the general classification, cowpeas belong to the botanical species *Vigna unguiculata L.* with four cultivated sub-species, *unguiculata*, *sesquipedalis cylindrica* and *textilis* and two wild sub-species *dekindtiana* and *mensensis.* Some authorities however, do not consider the cultivated sub-species as distinct and lumped them under one sub-species *V. unguiculata,* sub-species *unguiculata* and differentiate them by intraspecific category "cultigroup". The sub-species *unguiculata, sesquipedalis, cylindrica* and *textilis* are renamed as cultigroup Unguiculata, Sesquipedalis, Biflora and Textilis. Similarly the wild sub-species *dekindtiana* and *mensensis* were put under single sub-species *dekindtiana* and distinguished them by varietal category.

It is diploid and its chromosome number is 2n = 22.

Adaptation in India

Cowpea is grown throughout India for its tender pods of different length and colour (green, dull green, whitish, purple) to be cooked as vegetable or for dry seeds used as pulses and foliage as fodder or green manure. In India, cowpea cultivars grown for tender pods as vegetable are either true Sequipedalis forms (sub-species *sesquipedalis*) originally introduced from southeast Asia or integrades (intermediate types) of Unguiculata (sub-species *unguiculata*) and true Sequipedalis forms, while the cultivars grown for dry seed are either true Unguiculata forms or integrades of Unguiculata and Biflora (sub-species *cylindrica*) forms and the cultivars grown for fodder are either ture Biflora (sub-species *cylindrica*) or integrades of Unguiculata and Biflora forms.

Improved varieties

Indeterminate, viny or semi-viny type: Pusa Barsati, Arka Garima, Yard Long Bean, Anaswara, Swarna Suphala

Determinate and bushy type: Pusa Phalguni, Pusa Dofasli, Pusa Komal, Asseem, Pusa Rituraj, Narendra Lobia-2, Bidhan Barbati-1, , Bidhan Barbati-2,Co-2, Arka Samrudhi, Cowpea 263, Arka Suman, Kashi Shyamal, Kashi Gouri, Kashi Kanchan, Bidhan Sadabhar.

Compact, bushy and dual type: Pusa Rituraj, Narendra Lobia-1, Asseem

Cultural requirements

Cultivars of the viny and vegetable type under *V. unguiculata,* sub-species *sesquipedalis* (cultigroup Sesquipedalis) are photosensitive in nature and basically short day crop.Integrades or intermediate types of the sub-species *unguiculata* (cultigroup Unguiculata) and *sesquipedalis* (cultigroup Sesquipedalis) mainly cultivated as vegetable cowpea show both photo sensitive and photo non-sensitive characters.Vegetable cowpea cultivars developed through combination breeding involving the cultivars of sub-species *unguiculata* (cultigroup Unguiculata) and *sesquipedalis* (cultigroup Sesquipedalis) generally show photo non-sensitive character.Cultivars under sub-species *unguiculata* (cultigroup Unguiculata) and sub-species *cylindrica* (cultigroup Biflora) generally show photo non-sensitive character.It grows best in long and warm growing season. It grows well under wide temperature range of 21%C-35%C. The best temperature for its seed germination is 18°C to 25°C. Better growth and development takes place at 21°C to 27°C. It is susceptible to low temperature and frost. It does not grow well at high temperature. Cowpea is a drought hardy crop and comes well under rain fed condition tolerating moderate dry spells. It can not withstand heavy rainfall and water logging condition. It can grow on almost all types of soil of average fertility with optimum pH range of 5.5 to 6.5.Well drained, fertile, sandy to sandy loam soils are ideal.

Land preparation

The field is prepared with 4-5 deep ploughing to a depth of 25 cm. All the weeds, stubbles are removed and the land is leveled by laddering.Manures should be mixed with the soil at the time of last ploughing.The plant is susceptible to water logging so land should be well drained. Cultivars of determinate and bushy types are generally grown on flat beds of convenient size. Cultivars of the indeterminate, viny types are generally sown in the mounds and trained along the stakes. Beds and mounds are prepared with the mixture of 15-20 tonnes of FYM per hectare and basal dose of fertilizers are mixed with the soil.

Sowing time

Cowpea can be grown in spring-summer, rainy seasons, rainy-autumn and early autumn season. In places having mild climate, where summer and winter are moderate, it can be grown round the year. However, sowing time is January-February for spring-summer season; May-June for summer-rainy season; July-August for rainy-autumn season and August-September for Early autumn season.

Seed sowing

Seeds are sown directly in the field. Seeds are sown by hand dibbling in the bed.The seeds can be sown by fertilizer cum seed drill or by dibbling in line which is operated by a tractor, bullock or manual labour.Sowing is also done in a well-prepared and leveled field by broadcasting method however, line sowing is the best. The seeds should be inoculated with *Rhizobium* culture before sowing for quick nodulation on the roots which fix atmospheric nitrogen. About 375 - 500 gram Rhizobium culture is sufficient for treating the seeds required for one hectare.Determinate compact types are sown in rows by drilling behind the plough.20-25 kg seed/ha is sufficient for bush type cultivars grown in spring-summer season; 12-15 kg seed /ha is sufficient for bush type cultivars grown in rainy season and 8-10 kg seeds/ha is sufficient for indeterminate, viny cultivars grown in rainy-autumn and early autumn season.Seeds require sufficient moisture for germination.It is better to sow the seeds in pre-irrigated moist soil than irrigate the field after sowing.

Spacing

Spacing is 75 x 60 cm for indeterminate, viny type cultivars as sole crop; 50-60 x 50-60 cm for bushy and vigorous type cultivars and 45-50 cm x 25-30 cm for bushy and semi-vigorous type cultivars.

Nutrient management

Although cowpea is a legume crop, it responds well to the application of fertilizers particularly phosphatic fertilizers. Phosphorus helps in better nodulation in the roots through *Rhizobium* symbiosis.Its requirement of nitrogen fertilizer is comparatively much less because of being the legume crop.Well rotten farm yard manure at the rate of 15-20 t/ha should be applied at the time of land preparation. A general fertilizer dose of 25 kg N, 75 kg P_2O_5 and 60 kg K_2O per hectare is recommended.Half N along with full P and K fertilizers and farm yard manure should be applied as basal, preferable in bands about 7-8 cm to the side and slightly deeper than the seeds and rest N should be top dressed 25-30 days after sowing during earthing up around the plant 6-7 cm away from the plant stem.

Irrigation

Cowpea is shallow rooted crop and requires comparatively less moisture for its growth and development. Pre-sowing light irrigation helps in better germination of seeds. Irrigation is given to the crop after germination if moisture is insufficient particularly during spring-summer season. It is a hardy crop and can tolerate low soil moisture. However, the crop must be irrigated during its most critical stages i.e. flowering and pod development stages. It is sensitive to heavy irrigation and water logging hence, light irrigation is most suitable for successful luxurious growth. Irrigation prior to the flowering helps in good pod setting and irrigation should be given after the pods have set. It is essential to maintain available soil moisture atleat above 50% during flowering and pod development. Irrigation regularly at 7-10 days interval during dry periods is recommended to get high yields particularly during spring-summer and early autumn season.

Interculture and weed control

Indeterminate, viny cultivars must be given support for better growth, development and pod yield. The crop needs to be trained along the stakes made up of bamboo or jute sticks upto about 1.0 m high above the ground. In mixed cropping, the pole type cultivars get support on the stalks of maize and sorghum after their earheads are harvested. At the initial stages of weed growth, shallow hand weeding and hoeing are sufficient to check the weed growth.At the later stages of crop growth, the weeds are kept under check due to the thick canopy of the crop.Earthing up about four weeks after sowing helps in root aeration as well as weed control.Spraying of growth regulator malic hydrazide at 50-200 ppm concentration just before flowering increase the yield of pod.Application of herbicides effectively controls weeds. Pre-sowing application of Fluchloralin @ 2litres/ha checks the weed growth for 20-25 days.

Harvesting the crop

The crop duration of the indeterminate, viny types is about 125-160 days and that of the indeterminate, bush types is 80-100 days after sowing depending on the varieties. The pods are harvested periodically by hand picking when they are fully grown and succulent and have not become more fibrous. Normally harvesting starts 60-65 days after sowing in bushy cultivar and 75-80 days after sowing in case of indeterminate, viny cultivars. About 5-9 pickings of tender pods in pole types and 2-3 pickings in bush types can be done depending on the variety. As harvest is delayed, total yield increase but the quality falls rapidly.

Post harvest handling and storage

All fibrous and damaged pods should be removed from the harvested lots. Quality of the tender pods deteriorates very quickly under ambient condition because of their very high respiration and transpiration rate. High transpiration rate makes the pods leathery rendering them unfit for marketing and consumption. The selected pods are washed in running water before sending to market. Fresh green pods can be kept for two weeks at O^{o}C and 90-95% relative humidity. For dry seed purpose, the pods are allowed to full maturity on plants and then the crop is harvested and threshed after proper drying.

Yield

Yield of cowpea varies according to region, cultivar and duration of the crop. Yield of tender pods is 12.0 to 18.0 t/ha for indeterminate, viny cultivars and 7.0 to 12.0 t/ha for bush type cultivars. Yield of dry seeds is 1.5-2.0 t/ha.

Hyacinth bean

Hyacinth bean, *Lablab purpureus* (L).Sweet. (Syn. *Dolichos lablab* Roxb.) under the family Fabaceae (Leguminaceae) and having Chromosome No. 2n = 2x = 20, 22 is one of the most ancient among the cultivated plants. It is also called field bean, Indian bean, Indian butter bean, lablab bean, dolichos bean, Egyptian bean, Autralian bean, bonavist bean, lobia bean, sem and waby salad bean. The name "*Lablab*" is an Arabic or Egyptian name describing the dull rattle of the seeds inside the dry-pod. It is a tropical crop and mainly grown in South Asia (India, Bangladesh), South-east Asia and Africa. It is commonly cultivated in West Bengal, Madhya Pradesh, Maharashtra, some ares of Gujarat, Tamil Nadu, Andhra Pradesh, Haryana and eastern Uttar Pardesh. In India it is grown as a monocrop, intercrop or mixed crop with ragi, sorghum, bajra, maize or castor. Indeterminate pole types as monocrop are grown over the bower. In mixed cropping, the pole type hyacinth bean plants get support on the stalks of ragi and sorghum after their earheads are harvested.

The crop

Hyacinth bean or dolichos bean is a short-lived perennial herb, frequently grown as an annual. Thick stem is usually twining and may reach to 1.5 to 18.0 m in length. However, bushy, semi-erect and prostrate forms exist. Probably no other legume shows such variation in form and habit. The tap-root is well developed with many laterals and well developed adventitious roots.

The trifoliate leaves are tinged purple each up to 15 cm long. They may be hairy on the undersides. The inflorescence is made up of racemes of many flowers. The white, pink or purple flowers give rise to 5 to 15 pods of several centimetres long and are generally flat and glossy. The fruit is a legume pod variable in shape, size, and colour. The pod colour may be green, light green, purple, white, green with purple sutures, etc. The pods generally contain 3-8 seeds, which may be white, cream, buff, reddish, brown or black, sometimes with a white hilum. Mature, dried seeds are reportedly toxic due to high levels of cyanogenic glucosides and should be boiled in two changes of water before eating to remove the toxins.

Two types of hyacinth bean in India and are sometimes considered as distinct species. *Lablab purpureus* var. *typicus* is a "vegetable type" hyacinth bean; perennial twining herb, cultivated mostly as an annual, distributed throughout the tropical and temperate regions of Asia, Africa and America. The pods are white, green, purple or purple-margined. Seeds are white, yellow, brownish, purple or black. *Lablab purpureus* var. *lignosus* is a "pulse type" hyacinth bean; semi-erect, bushy, perennial herb, cultivated as an annual. It shows little or no tendency to climb. Leaflets are innately trifoliate, smaller than those of var. *typicus*. Flowers are borne on a straight upright stalk, often a foot high on which they open in succession. Pods are oblong, flat and broad, firm-walled and fibrous, contain 4-6 seeds with their long axis at right angles to the suture. Seeds almost rounded white, brown or black. The plant emits a characteristic odour.

Chief characteristics of these two types are given below:

Lablab purpureus var. *typicus* (syn. *Dolichos lablab* var. *typicus*)

- A viny and indeterminate type with soft edible pods as the pod walls has less fibre.
- Perennial and photosensitive type
- Grown for tender pods as vegetable.
- The long axis of the seed is parallel to the suture.

Lablab purpureus var. *lignosus* (syn. *Dolichos lablab* var. *lignosus*)

- Generally bushy and determinate type with pods having high fibre content.
- Pods have characteristics aroma.
- Annual and photo non-sensitive type.
- Grown for fresh dry seeds as pulse.
- Seeds are at right angle to the suture.

Importance and use

It is an excellent pod vegetable and mostly grown for fresh whole pod. Young immature pods are cooked and eaten like green beans. They have a strong beany flavour and in some curry preparations it is mixed with other vegetables. The immature seeds can be boiled and eaten like any shelly bean. Dried seeds should be boiled in two changes of water before eating since they contain toxins cyanogenic glucosides. In Asia, the mature seeds are made into tofu and fermented for tempe. They are also used as bean sprouts. Young leaves are eaten raw in salads and older leaves are cooked like spinach. Flowers are eaten raw or steamed. The large starchy root tubers can be boiled and baked. It is an excellent nitrogen fixer and is sometimes grown as a cover crop or green manure crop or for livestock fodder. It is grown as ornamental plant in the USA.

Nutritional value

The pods are one of the rich sources of protein. It contains 6.7g carbohydrates, 3.8g protein and 0.7g fat per 100g of edible portion. It provides high of 0.9g minerals per 100g of edible part. Tender fruits contain 210mg calcium, 68mg phosphorus 55 mg sodium, 34mg magnesium and 1.7mg iron per 100 g fresh. It provides 312 I.U of vitamin A per 100g edible portion. It is a good source of Vitamin B like, 0.06mg Riboflavin, 0.1mg Thiamine and 0.7mg Nicotinic acid per100g fresh. It contains 9.0 mg vitamin C per100g fresh. Antinutritional factors like trypsin and chymotrypsin inhibitor etc. present in dry seeds can easily be removed by boiling in fresh water before cooking.

Medicinal value

Hyacinth bean is particularly important for the supply of vitamin B in the foods. It contains high amount of dietary fibre which gives protection against certain types of cancer particularly colon cancer, regulates transit and lowers blood cholesterol.

Origin and taxonomy

Hyacinth bean has been originated in India as wild forms of this bean are found in this country. From India, it was introduced to China, Western Asia, Egypt and other tropical countries of South and South-east Asia and Africa.

The hyacinth bean, *Lablab purpureus* (L.) Sweet belongs to the small genus *Lablab* under the family Fabaceae (Leguminaceae) and the order Fabales. Vegetable and pulse type hyacinth bean belong to two distinct botanical varieties under *Lablab purpureus* viz., *Lablab purpureus* var. *typicus* (soft podded vegetable type) and *Lablab purpureus* var. *lignosus* (fibrous podded pulse type).

The hyacinth bean has chromosome number 2n = 22.

Adaptation in India

Hyacinth bean has been originated in India and later it was introduced to China, Western Asia and Egypt. Various form and cultivars of both the cultivated types *viz.,* viny, soft and edible podded vegetable type (*Lablab purpureus* var. *typicus*) and fibrous podded pulse type (*Lablab purpureus* var. *lignosus*) are available in India. A number of improved varieties of particularly vegetable type (*Lablab purpureus* var. *typicus*) have been developed through purification and subsequent improvement over the indigenous cultivars. However, edible podded, bush type cultivars have also been developed in India through hybridization between the viny cultivars of *Lablab purpureus* var. *typicus* and bushy cultivars of *Lablab purpureus* var. *lignosus*.

Improved varieties

Viny, photo sensitive and vegetable type: Pusa Early Prolific, Deepaliwal, JDL-37, JDL-79, JDL-53, T-1, K-6802, HD-18, HD-60, CO-1, CO-2, CO-3, CO-4, CO-5, CO-10, Rajni, Dasarwal, KDB-403, KDB-405, Pusa Sem-2, Pusa Sem-3, Phule Gouri, Swarna Utkrist, BCDB-1

Compact, photo non-sensitive and vegetable type: Arka Joy, ArkaVijay, CO-6, CO-7,CO-8,CO-11, CO-13

Compact, photo non-sensitive and dual type: Konkan Bhusan,Vijay, Vani, Wal Konkan-1, Hebbal Avare-1, Hebbal Avare-3, Hebbal Avare-4,CO-11, CO-12, CO-13

Cultural requirements

Cultivars of the viny and vegetable type under *Lablab purpureus* var. *typicus* are photosensitive in nature and basically short day crop.Cultivars under *Lablab purpureus* var. *lignosus* and those developed through combination breeding involving the cultivars belonging to *Lablab purpureus* var. *typicus* and *Lablab purpureus* var. *lignosus* are photo non-sensitive in nature and can be cultivated through out the year in favourable temperature condition.It is a cool season and drought tolerant crops. The best temperature for its seed germination is 18°C to 27°C. Better growth and development takes place at 21°C to 27°C. Temperature range of 12°C to 18°C is considered most suitable for better fruit set.It is susceptible to frost.It does not grow well at very high temperature.It can grow on almost all types of soil of average fertility with optimum pH range of 6.5 to 7.8. Sandy loam, silt loam and clay loam soils are best for its cultivation. Very rich soil leads to more vegetative growth and less pod yield. It can not stand water-logging.

Land preparation

The field is prepared well with 4-5 deep ploughing to a depth of 25 cm. All the weeds and stubbles are removed and the land is levelled by laddering. Manures should be mixed with the soil at the time of last ploughing. The plant is susceptible to water logging so land should be well drained. As a monocrop, it is generally grown on flat or raised beds, about 1.0 to 1.5 m wide and seed are sown on both sides of the bed. Mounds in the bed method of planting are prepared with the mixture of 15-20 tonnes of FYM per hectare and basal dose of fertilizers are mixed with the soil.

Sowing time

Soft podded indeterminate cultivars belonging to *Lablab purpureus* var. *typicus* are photosensitive in nature and flowers under short day condition with the advent of cool temperature hence, are sown in August-September during late rainy to early autumn season. Determinate and compact cultivars are photo non-sensitive in nature and can be sown early during June-July.

Seed sowing

In general, seeds are sown directly by hand dibbling in the field. The seeds are inoculated with *Rhizobium* culture @ 2gm/kg of seeds before sowing for quick nodulation on the roots which fix atmospheric nitrogen. Determinate compact types are sown in rows by drilling behind the plough. For sowing in one hectare land, 20-30 kg seed for pole type cultivars and 35-40 kg seed for bush type cultivars is sufficient.Seeds require sufficient moisture for germination. It is better to sow the seeds in pre-irrigated moist soil than giving irrigation after sowing.

Spacing

Spacing is 1.5m x 75 cm; 1.5m x 1.0 m for pole type cultivars as sole crop; 1.0m x 75 cm for pole type cultivars as mixed crop and 75 cm x 75 cm for compact and bush type cultivars.

Nutrient management

Hyacinth bean responds well to manuring and application of particularly phosphatic fertilizers. Phosphorus helps in better nodulation in the roots through *Rhizobium* symbiosis. Its requirement of nitrogen fertilizer is comparatively much less because of being the legume crop.Well rotten farm yard manure at the rate of 15-20 t/ha should be applied at the time of land preparation. A general fertilizer dose of 20-30 kg N, 60 kg P_2O_5 and 60 kg K_2O per hectare is recommended. Half N along with full P and K fertilizers and farm yard manure should be given as basal and rest N should be top dressed 30 days after sowing by broadcasting around the plant 6-7 cm away from the plant base.

Irrigation

Pre-sowing light irrigation helps in better germination of seeds. The crop needs irrigation after germination if moisture is insufficient. It is a hardy crop and can tolerate low soil moisture.The crop must be irrigated during its most critical stages i.e. flowering and pod development stages. Moisture stress during flowering causes huge flower drop. Irrigation regularly at 7-10 days interval during dry periods is advocated to get high yield.

Interculture and weed control

Indeterminate, pole type cultivars grown on trellis gives better crop stand and pod yield mainly because of better use of sunlight by maximum number of leaves. The crop needs to be trained over low trellis of 1.5 m high above the ground. The determinate types are also given support by bamboo poles. In mixed cropping, the pole type cultivars get support on the stalks of ragi and sorghum after their earheads are harvested.At the initial stages of weed growth, shallow hand weeding and hoeing are sufficient to check the weed growth.At the later stages of crop growth, the weeds are kept under check due to the thick canopy of the crop.Application of herbicides effectively controls weeds. Pre-sowing application of Fluchloralin @ 2litres/ha checks the weed growth for 20-25days.

Harvesting the crop

The crop duration of the pole types is about 210-240 days and of the bush types is 90-160 days depending on the varieties. The pods are harvested periodically by hand picking when they are fully grown and succulent and have not become fibrous.Normally, harvesting starts two months after sowing in bush type cultivar and three months after sowing in case of pole type cultivars. In bush types generally harvesting is completed by 2-3 pickings of tender pods while in pole types about 9-12 pickings can be done.

Post harvest handling and storage

All fibrous and damaged pods are removed from the harvested lots. Quality of the tender pods deteriorates very quickly under ambient condition because of their very high respiration and transpiration rate.High transpiration rate makes the pods leathery rendering them unfit for marketing and consumption. The selected pods are washed in running water before sending to market. The best storage temperature is 4.5^{o} to 7.0^{o}C with 85-90% relative humidity.Fresh pods can be kept for about 15-20days in cold storage. Dried seeds can be kept for 2-3 years at room temperature under dry condition.

Yield

Yield of tender pods is 20.0 to 40.0 t/ha in pole type cultivars; 4.0 to 6.0 t/ha in bush type cultivars. Dry seed yield is generally 1.0 to 1.5 t/ha.

Cluster bean

Cluster bean, *Cyamopsis tetragonoloba* (L.) Taub.under the family Fabaceae (Leguminosae) and having chromosome No.2n = 14, a drought hardy and salt tolerant leguminous vegetable crop is grown for its young pods. About 80% of world production comes from India and Pakistan, but due to strong demand, the plant is being introduced into new areas. It has been growing in Myanmar, Sri Lanka, and Texas and Arizona states of the USA since long back. It is known popularly in India as Guar, Thupi, Urahi, Koth Avarai, Gavar, Gor Chikudu, Gorikaya, Kothavara, Guvar and Matki. In India, it occupies an important place in dry land agriculture of mainly arid and semi-arid regions of North-western states of Rajasthan, Haryana, Gujarat and parts of Punjab, Uttar Pradesh, Madhya Pradesh and near to the coastal areas of Kutch, Gujarat.

The crop

Cluster bean grows upright, reaching a maximum height of up to 2–3 m. It has a main single stem with basal branching along the stem. It has strong taproot system by which the plant can access soil moisture in low soil depths. Additionally, this crop develops root nodules with nitrogen-fixing soil bacteria Rhizobia in the surface part of its rooting system. The leaves and stems are mostly hairy, depending on the cultivar. The leaves have an elongated oval shape (5 to 10 cm length) and of alternate position. Clusters of flowers grow in the plant axil and are of white to bluish colour. The developing pods are rather flat and slim containing 5 to 12 small oval seeds of 5 mm length with test weight varying between 25 and 40 g. Usually, mature seeds are white or gray. The seeds have a very remarkable characteristic. Its kernel consists of a protein-rich embryo (43-46%) and a relatively large endosperm (34-40%), containing big amounts of the galactomannan which is a polysaccharide containing polymers of mannose and galactose in a ratio of 2:1 with many branches. Galactomannan polysaccharide is popularly known as guar gum or mucilage. Galactose exhibits a great hydrogen bonding activity showing a viscosifying effect in liquids. Three principle enzymes, *viz.* α-D-galactosidase, β-D-mannanase and β-D-mannosidase are responsible for the degradation of galactomannan in germinating seeds.

There are number of varieties in this crop, some of them are single stemmed, i.e., monopodial, and others are branched from the base. Single stemmed types are more suitable for green pod production.

Importance and use

Cluster bean is grown for vegetable, feed, fodder, green manure and gum production. The tender pods are generally used as vegetable for the preparation of curry and fried items. Leaves are used boiled or stir-fried as pot herbs. In some parts of India, green pods are dehydrated and stored for further use. Cluster bean is also grown as a forage and green manure crop for enhancing the soil fertility through fixation of atmospheric nitrogen (50-60 kg/ha) and also for incorporation of organic matters in the soil.

Dry seeds are widely being used for the extraction of gum, which is mainly used in about 25 major industries, like textiles, cosmetics, explosives, paper, mining, oil, food processing etc. apart from being used as adhesives on postage stamps, to impart smoothness and stability to bakery products and as a foam stabilizer in beer. The gum is also used as a stabilizer and thickener agent in food products *viz*., ice cream, bakery mixes and salad dressings. Guar gum is exported to different countries, like USA, Japan, UK, France, Italy, and Netherlands etc., and the USA by far is the major importer, while India and Pakistan are the major exporters.

Guar-meal after separation of gum is a potentially valuable source of protein for animal feed. The gum extraction factories utilize hardly about 40 percent of the total seed production and most of the remaining seeds are used as cattle feed by the farmers, however, growth and egg production of chick has been found adversely affected due to 10% saponin content in guar-meal, limiting its use to low levels in poultry feed.

Nutritional value

The young pods of cluster bean are as good as the French bean pods in food value because of their high nutrient constituents of 10.8 g carbohydrates, 3.2 g protein, 0.4 g fat, 130 mg calcium, 4.5 mg iron, 65.3 IU vitamin A, 0.09 mg thiamine, 0.03 mg riboflavin and 49.0 mg ascorbic acid per 100 g fresh bean.

Medicinal value

Pods and seed of cluster bean are used to cure inflammation, sprains, arthritis, as anti-oxidant, antibilious and laxatives whereas leaves are used in asthma and to cure night blindness. Importance of cluster bean as potential hypoglycaemic agents for the treatment of diabetes has recently been emphasized. In food, guar gum acts as bulk laxative and cholesterol controlling agent. Guar gum is one of the most widely stipulated food fibres that make the food more viscous in human digestive tract and slow down the absorption of glucose in intestine. It controls sugar metabolism in diabetes patients and in those having high serum lipids and high blood pressure. Boiled seeds are generally used for the treatment of plague, enlarged livers, head swellings and swellings on broken bones.

Origin and taxonomy

Origin of cluster bean is unknown, since it has never been found in the wild. Diverse opinions are advanced regarding origin of this crop. In one opinion, it was probably domesticated in dry regions of Africa and is assumed to have developed from the African species *Cyamopsis senegalensis*. In other opinion, India is considered the centre of origin of cluster bean. It was presumed to have further domesticated in India and Pakistan, where it has been cultivated for many centuries although, no wild form of this crop is found in India.

Chromosome number of cluster bean is 2n = 14

Adaptation in India

In one school of thought, India is considered the centre of origin of cluster bean. In other opinion, it was domesticated in dry regions of Africa and further domesticated in India and Pakistan. Whatever may be the presumption, cluster bean has been cultivated for many centuries in India and Pakistan which facilitated development of wide variability of indigenous strains.

Improved varieties

Single stemmed variety: Pusa Sadabahar, Pusa Naubahar, IC 11388, Sel Ches 13-7

Branched variety: Pusa Mausami, Sharad Bahar, P 28-1-1, IC 11704

Cultural requirements

Cluster bean grows well in semiarid areas because it is basically a drought tolerant crop, and can be grown with some returns even in the areas with a 100-150 mm annual rainfall and high temperature (up to 44°C) condition in arid zones. However, proper germination of seeds and root development takes place at a temperature between 25° and 30°C. It is very susceptible to frost. It grows well both during summer and *kharif* seasons. However, *kharif* crop gives higher yield because of enhanced flower bud production in moist conditions.

Cluster bean can be grown on almost all types of soil, but a well-drained sandy loam soil having pH 7.5-8.0 is favourable for this crop. It cannot withstand waterlogging conditions since root nodules are destroyed in heavy and poorly drained soils. It grows best in moderate alkaline conditions (pH 7-8) and is tolerant to salinity.

Sowing time

The crop is grown throughout the year in southern parts of India because of prevailing favourable climatic conditions. However, the crop sown in May-June gives better yield. In North Indian conditions, two crops are generally taken: one spring-summer crop by sowing seeds in February-March and another rainy season crop by sowing seeds in June-July. Sowing of seeds during first fortnight of July is the best. The crop for the fodder purpose can be sown from April to mid July under north Indian conditions.

Land Preparation

Soil should be prepared well before sowing with at least 4-5 ploughings and all stubbles, weeds, etc are removed from the land. Well rotten organic manure preferably farm yard manure or compost @ 15-20 t/ha should be applied during land preparation and the land is levelled properly.

Sowing the seeds

Seed rate varies with the viability percentage, spacing, and method of sowing. Generally a seed rate of 10-15 kg for vegetable and grain crop per hectare is sufficient. Both flat bed and ridge and furrow methods are generally followed for raising the crop. However, paired row system of sowing has been advantageous in some locations because of better weed control and microclimate in between two rows.

The crop grown as main crop or mixed with other crops, like cucurbits, cotton and sugarcane or as a border crop around main crop line can be sown by either broadcasting or dibbling method at a depth of 5-7 cm. Cluster bean is best fitted in intercropping system, particularly with pearl millet, because of difference in their growth and feeding habits. Being leguminous crop, it has the ability to fix atmospheric nitrogen through symbiotic association with *Rhizobium*. Seed inoculation with *Rhizobium* (Cowpea miscellany group) @ 10g per kilogram of seed along with 60 kg P and 80 kg K/ha significantly increased the seed yield, protein and gum content in the seeds. The seeds after inoculation should be kept under shade to keep the inoculums viable until they are sown in field.

Generally, the plant rows for vegetable purpose are spaced by 45 cm and 30 cm for late sown crops, and the plants are kept at a distance of 15-20 cm in rows, while narrow spacing within rows is maintained for seed crop.

Nutrient management

Fertilizer requirement depends on soil type, irrigation facilities, and weather conditions, and like most of the beans, it is also highly responsive to organic matter application. Application of 10-12 tones of farmyard manure before sowing the early crop under irrigated condition is highly beneficial. The bean crops in general show good response to phosphorus and potash and poor response to nitrogen application. Normally, full dose of NPK fertilizers containing 10-20 kg nitrogen and 50-70 kg each of phosphorus and potash per hectare are applied as basal 5-10 cm below the soil surface at the time of sowing. Phosphorus nutrition has been found vital for cluster bean, and significant correlations were revealed between phosphorus uptake by the plant, dry matter yield and phosphatase activity in the rhizosphere. Higher shoot and root weight, and N and P uptake could be recorded with *Glomus fasciculate* inoculation followed by application of phosphorus. Most of the studies under different soil conditions implicated the positive role of micronutrients like zinc through application of zinc sulfate @ 25 kg/ha along with recommended doses of fertilizers which significantly influence the number of nodules, nitrogenase activity of the nodules, and carbohydrate and protein contents of the leaves and shoot dry matter yield. Response of molybdenum is also found quite encouraging, and two sprays of 0.15% sodium molybdate at 15 and 30 days after seedling emergence give better yield.

Irrigation

Cluster bean is generally grown as a rain fed crop, however, single irrigation at 60 days after sowing increases the yield and harvest index of crop, while two to three irrigations at 10-12 days interval in spring-summer are sufficient to meet the evapo-transpiration requirement of the crop.

Interculture and weed control

Cluster bean seeds take a week to germinate, depending on soil moisture status in the field. To maintain desired plant population at appropriate spacing the seedlings are thinned 10-15 days after sowing when they become large enough to handle. Intense crop-weed competition for moisture, nutrients, sunlight and space under dry land condition cause as high as 70-90% yield reduction. It necessitates keeping the crop weed free at least for first 30 days after sowing by two to three hand weeding, which increase pods per plant, water use efficiency, and also yield.

Chemical weed control by pre-sowing application of different herbicides, like 2,4-D or disodium methane arsenate (DSMA) @ 2.0 kg a.i./ha against

Parthenium hysterophorus, Fluchloralin @ 2 litre/ha or Basalin @ 1 kg a.i./ha in upper 10 cm soil against annual grasses and broad leaved weeds, Trifluralin @ 1.5 kg/ha in upper 5 cm soil, and 2000 ppm Chlormequat are effective in controlling weeds in early stages, and thick crop canopy checks the weed growth at later stages of crop growth. Mulching with dry grass is useful to prevent soil moisture loss and to increase root growth, nodulation, shoot and plant growth and water use efficiency.

Harvesting the crop

Green pods, which mature at leaf axils simultaneously are harvested 40-45 days after sowing onwards, and generally continues up to 120 days, depending on the variety. Dry beans are harvested when most of the pods are fully ripe and turned yellow until the lower pods become dry enough for shattering, which generally take 90 to 100 days after sowing, depending upon the variety, soil type, rainfall and its distribution. After harvesting, the stalks are left for drying for 1-2 weeks, and then, trampled over by bullocks or tractor to separate the seeds from the pods.

Post harvest handling and storage

Tender green pods largely used as fresh vegetable do not have longer shelf life due to their high respiration rate associated with common deteriorative symptoms of shrivelling and chlorosis. Tender green pods can be kept for two days in usable condition with frequent sprinkling of water under ambient conditions but can be preserved as fresh for 15-20 days in cold storage at 0°C temperature and 85-90% relative humidity. Tender pods are also processed to make dehydrated products.

Yield

The yield of green pods varies from 30 to 80 q/ha and that of dry seeds from 6 to 10 q/ha, depending on the variety, soil type, season of cultivation and the management practices adopted during the course of its cultivation.

Plant Protection Measures of Peas and Beans

Physiological disorder

A number of temperature and nutrient-related disorders are found in French bean.

Small and mis-shaped pods: Flower initiation and development are greatly delayed under sub-optimal temperatures, especially below 10°C, where small and mis-shaped pods are produced because fertilization may not occur in this low temperature condition.

Blossoms drop: Blossom drop and ovule abortion are common problems at high temperature condition of about 35°C. Planting the crop at a suitable time and place mitigate this problem.

Cotyledon cracking: Transverse cotyledon cracking takes place when dry seeds of beans are sown in wet soils because seeds of the susceptible cultivars imbibe water rapidly. Use of resistant varieties with hard seed coat, optimum seed moisture content and planting of crop at suitable time are essential to avoid this disorder.

Necrosis of hypocotyl: Necrosis of hypocotyl is associated with low calcium content in seed after germination. Adequate Ca and Mg contents in the soil offset this problem.

Diseases

Powdery mildew (*Erysiphi polygoni* in pea, cowpea and French bean; *Laveillula taurica var macrospora* in hyacinth bean; *Laveillula taurica* in cluster bean)

This common fungal disease of peas, French bean, cowpea, cluster bean and hyacinth bean generally appears late in the season during dry summer. Faint, slightly discoloured specks are first formed on leaves and from these specks greyish-white powdery growth of the mycelium and spores spread over the leaves, stem and even fruits. This white, powdery fungal growth is primarily asexual spores called conidia. Older plants are affected first. Infected leaves usually wither and die. Severely infected plants are defoliated and weakened by premature drying up and death of infected leaves and fruits either do not set or remain very small.

Control measures

- Growing of tolerant varieties like, Jawahar Peas 54, Jawahar Peas 4, Azad Pea 2, Jawahar Pea 83, Jawahar Peas 15 of garden pea; GAUG 63 of cluster bean
- Spraying the crop with 0.5% Sulfex or other formulation of wettable sulphur as soon as disease appears.
- Spraying the crop with Bavistin or Benlate or Topsin M @ 1g/l of water 2 to 3 times at 15 days interval.
- Spraying the crop thrice with systemic fungicide like Bavistin or Tridemorph @ 0.1% concentration at weekly interval.

- Two sprays of 0.1% Bavistin followed by two sprays of 0.1% Karathane or Tridemorph effectively control the disease.

Alternaria leaf spot (*Alternaria cucumerina* pv. *cyamopsidis*)

This fungal disease of clueter bean is mainly prevalent in Haryana, Tamil Nadu, and parts of Rajasthan. The disease appears as dark brown round to irregular spots varying from 2 to 10 mm in diameter on leaf blades, which spread rapidly to form circular lesions under humid weather conditions. In severe cases, concentric rings of dark-brown conidiophores are developed and the leaflets become chlorotic, and usually drop off.

Control measures

- Spraying the crop with 0.2% Zineb twice at an interval of 15 days.
- Application of iprodione, which Iprodione, the hydantoin fungicide.
- Spraying the crop with 0.25% Mancozed or Metalaxyl at 7-8 days interval.
- Growing tolerant variety of cluster bean, like HG 182.

Myrothecium leaf spot (*Myrothecium roridum*)

This fungal disease of cluster bean is usually favoured by high temperature conditions, and appears as small to minute oil-soaked spots of 1-2 mm diameter, which later turn brown, coalesce and cover large area of leaf.

Control measures

- Spraying the standing crop with 0.2% Zineb five weeks after sowing.

Bean anthracnose (*Colletotrichum lindemuthianum* in French bean, cowpea, hyacinth bean; *Collectotrichum capsici* pv. *cyamopsicola* in cluster bean)

It is most serious seed borne fungal disease of French bean, cowpea, hyacinth bean and cluster bean. This disease becomes most severe in high rainfall areas of subtropical climatic condition. Tissues of the necrotic spots on leaves collapse, become thin and papery and produce pinkish slimy mass in high humidity condition. Similar lesions appear on the cotyledons and stems of the young seedlings. The most characteristic symptoms of the disease in four bean crops are mentioned.

French bean: Black, sunken, crater like cankers on the pods. The lesions remain isolated by yellow-orange margins and under high humidity, dull salmon coloured ooze is formed in the centre.

Cowpea: Dark brown, sunken lesions with raised, reddish or yellowish margins are found on pods. Under high humidity condition, dull salmon or pinkish coloured oozes or slimy mass is formed at the centre of the lesion.

Hyacinth bean: Most characteristics symptoms appear on the pods where dark brown, sunken lesions with raised, reddish or yellowish margins are found.

Cluster bean: This disease appears as brown to black spots on leaves, petioles, and stems during rainy season.

Control measures

- ❑ Use healthy and disease free seeds
- ❑ Clean cultivation and crop rotation.
- ❑ Treating the seeds with Carbendazim @ 2.5g/kg of seeds or Thirum @ 2.5g/kg of seeds.
- ❑ Spraying the crop with 0.1% Bavistin or Hexaconazole or 0.2% Chlorothalonil at 10 days interval.

Rust (*Uromyces pisi* in pea; *Uromyces fabae* in French bean)

It is a common fungal disease of pea and bean. Characteristic symptoms produced in pea and French bean are mentioned below.

Pea: Orange coloured pustules are developed on the foliage and all the above ground parts of the plant and the infected leaves usually wither and die.

French bean: Dark brown coloured rust pustules are developed on the foliage with chlorotic hallo and the infected leaves usually wither and die.

Control Measures

- ❑ Spraying the crop with Bavistin or Benlate or Topsin M or Tebuconazole @ 1g/l of water 2 to 3 times at 15 days interval.

Ashy stem blight/ dry stem blight (*Macrophomina phaseolina*)

This seed and soil borne fungal disease mostly appears in cowpea, hyacinth bean and cluster bean. This disease becomes severe in warm (28°-35°C) and humid climate.

Cowpea and hyacinth bean: In early stages, it causes damping off and killing of young seedlings where black cankerous lesions appear on the cotyledons. Collar portion of the plant becomes brown and vascular bundle of the roots also turns brown resulting in dry root rot.

Cluster bean: Reddish-brown discolouration on stem is caused initially, which becomes darkened with the formation of numerous small black sclerotia both inside and outside of tissues and on roots in advanced stage of attack, resulting production of very few pods.

Control measures

- Growing of tolerant varieties, like Kutch-8 and RGC 471 in cluster bean
- Use disease free and healthy seeds.
- Dry seed treatment with Carbendazim/Captan/Thirum @ 2-3g/kg of seed.
- Application of adequate organic matter in soil.
- Soil application of *Trichoderma viride* @ 3 kg/ ha along with manure is effective.
- Spraying the crop with 0.1% Carbendazim / Carbendazim (1g) + Mancozeb(2g) /l or Captan @ 2.5 g/ l along with sticker checks the rapidity of the infection.

White mold (*Sclerotinia sclerotiorum*)

It is an important fungal disease of French bean. Symptoms are usually first visible about one week after full bloom because blossoms generally are the first part of the plant to be colonized by the fungus. Leaves, stems, and pods in contact with the colonized blossoms then become infected, provided moisture is present. Initial symptoms on these tissues are pale colored, water-soaked lesions. These lesions enlarge and within a few days become covered with a white, cottony fungal growth. Leaves of severely diseased plants become yellow and eventually turn brown and fall off. As the disease progresses, the infected plants wilt and the leaf canopy opens. The fungus may eventually invade and kill all the above ground parts of the plants.

Control Measures

- Successful control with fungicides depends upon spraying at the proper time and completely covering the plants with the chemical, especially inside the plant canopy.
- Application of Carbendazim / Benomyl @1.5 g/lt or Carbendazim (1g) + Mancozeb (2g)/l at 10 days intervals for 2-3 times is very much effective.

Die back (*Colletotricum capsici*)

It is a fungal disease of cowpea. Twigs and branches dry up from tip to downwards due to the attack of this fungus. Small black, dot like structures appear on dried up portions. Infected pods shrivel.

Control measures

- Spraying the crop with 0.1% Bavistin or Hexaconazole or 0.2% Chlorothalonil at 10 days interval.

Downy mildew (*Peronospora parasitica*)

This fast growing fungus affects peas and beans. It may appear at any time of the season, but is most often a problem in humid areas, especially in cool wet weather because the spores need water to germinate and grow. It enters the plant through wounds and natural openings and first appears on older leaves as white, yellow or brownish spots on the upper surfaces and downy white, cottony patches that are formed on the lower surface of the leaflets. The upper surface turn yellow, brown and eventually dries up.

Control measures

- Spraying the crop with 0.25% Dithane M-45 or Metalaxyl-Mancozeb (0.25%) at 15 days interval during growth period.

Bacterial blight (*Xanthomonas campestris* pv. *phaseoli* in French bean; *Pseudomonas pisi* in pea; *Xanthomonas vignicola* in cowpea; *Xanthomonas axonopodis pv. phaseoli* in hyacinth bean; *Xanthomonas axonopodis* pv. *cyamopsidis* in cluster bean)

It is most serious soil and seed borne bacterial disease of peas and beans. The disease is most destructive where temperatures are moderate and abundant inoculum is available.

Pea: The affected plants develop watery, olive-green blisters on stems and leaf bases and water soaked oily spots appear on pods and leaves.

French bean: Leaf symptoms first appear as small, water-soaked spots on the lower leaf surface. These lesions rapidly become necrotic and are visible on both upper and lower leaf surfaces. Infection of expanding leaves may result in leaf distortion. A chlorotic zone of yellow tissue (halo) develops around the necrotic spots. In cases of severe infection, a generalized systemic chlorosis may occur. Symptoms also occur on pods and stems as water-soaked, red or brown lesions which may exhibit crusty bacterial ooze.

Cowpea: The primary damage is caused through high mortality of the seedlings, especially when the seeds used for sowing are from a severely diseased crop. Infected leaves show light yellow, irregular to circular spots with necrotic brown centre, later changing to straw colour. Dark green water-soaked spots of variable shape and size appear on pods which later becomes yellow and dry.

Hyacinth bean: Olive-green blisters appear on stems and leaf bases. Water-soaked spots also appear on leaf and pods. Infected leaves show irregular, sunken, yellow to brown necrotic spots with narrow yellowish halo. In severe infection, defoliation occurs. Dark green, water-soaked spots may appear on the pods which later become yellow brown and dry.

Cluster bean: This disease usually occurs in *kharif* season throughout the cluster bean growing areas of this country because high humidity, warm temperature (28-30°C) and spattering rains favour the spread of this disease. The disease appears as small, circular, and water-soaked spots on dorsal surface of the leaf, which coalesce to form bigger spots surrounded by chlorotic area on older leaves. Flaccidity of the affected portion due to invasion of vascular tissues turns the necrotic spots to brown and finally cause defoliation and occurrence of black longitudinal streak on stem.

Control measures

- Growing resistant varieties like, Pusa Komal in French bean; RGC 471, HG 75, HG 182 and GAUG 63 in cluster bean
- Use of disease free healthy seeds.
- Removal of plant debris before sowing
- Seed treatment with 0.02% Streptocycline solution for 3.0 hour before sowing.
- Spraying the standing crop with 0.01% Streptocycline + 0.2% Copper oxychloride.

Bean common mosaic virus (BCMV)

It is an important virus disease of beans worldwide. Typical symptoms in bean consist of green mosaic and downward cupping along the main vein of each leaflet. Green vein banding, blistering, and malformation are common in leaves of the same plant. Plants are reduced in size, and pods may be mottled and malformed.

Control measures

- BCMV is most effectively controlled by growing resistant cultivars and using virus-free seed.

Cowpea mosaic virus

Mosaic virus of cowpea is seed borne as well as vector transmitted (aphid). The affected leaves develop a typical mosaic of broad and raised dark-green patch along with chlorotic spots. The diseased plant remains stunted with reduced and malformed leaves.

Yellow mosaic of hyacinth bean

Yellow mosaic virus of hyacinth bean is transmitted by white fly while cowpea mosaic virus is seed borne and aphid transmitted. Characteristic mosaic pattern on the affected leaf occur due to the infection of these virus.

Control measure

- Use the seeds that have been harvested from mosaic free plants
- Removal and destruction of the affected plants as early as possible
- Spraying the crop with systemic insecticides like Rogor or Metasystox @ 2 ml / l of water to control the insect vector.

Insect pest

Stem fly (*Ophiomyia phaseoli*)

It is a polyphagous pest. The adult flies are metallic black with hyaline wings and the newly hatched maggots about a millimeter long are white but become yellow at later stages. The maggots mine the leaf, bore inside the petiole and tender stem tunneling inside which cause girdling effect and death of the plant. The adults puncture the leaf and the affected leaves become yellow and dry up.

Control measures

- Removal and destruction of all affected plants
- Spraying the crop with Nuvacron 1.5 ml/l of water twice at weekly interval from the time of complete germination of seeds
- Spraying the crop with 6% neem seed extract reduces the incidences.

Pod borers (*Adisura atkinsoni, Exelastis atomosa, Helicoverpa armigera* and spotted borer *Maruca testulalis* in French bean, cowpea and hyacinth bean and *Helicoverpa armigera* in pea and cluster bean)

In pea, pod borer is the most serious pest in some parts of the country. Among the different borer complex that attack cowpea, French bean and hyacinth bean, the spotted borer is found to be the most destructive under Bengal basin. The caterpillar after hatching bore into the developing pods and feed on the seeds therein. Severely infested crops are subjected to high economic yield loss leaving behind most of the affected fruits.

Control measures

- Removal and destruction of the infected pods at the initial stage of attack
- Spraying the crop with Nuvacron @ 1 ml/l, or Thiodan @ 1.4 ml/l or Malathion @ 2 ml/l or Sevin W. P. @ 4 g/l of water
- Application of a nuclear polyhedrosis virus (NVP) @ 250 LE/ha/week starting from first appearance of young leaves as biological control measure.
- Spraying the crop with 5% neem seed kernel extract reduces the incidence

Hairy caterpillers (*Ascotis imparata* and *Spilosoma obliqua*)

These insect pests generally attack French bean. The larvae cause characteristic skeletonization of the leaves during the early gregarious stage and later they completely denude the plants.

Control measures

- Removal and destruction of the leaves with gregarious larvae during the early stage.
- Spraying the crop with Nuvacron @ 1 ml/l, or Thiodan @ 1.4 ml/l or Malathion @ 2 ml/l or Sevin W. P. @ 4 g/l of water
- Application of a nuclear polyhedrosis virus (NVP) @ 250 LE/ha/week starting from first appearance of young leaves as biological control measure.
- Spraying the crop with 5% neem seed kernel extract reduces the incidence

Bihar hairy caterpillar (*Diacrisia obliqua*)

This insect pest generally attacks cluster bean. Larvae of this insect first feed on foliage, and under severe attack, the leaves are skeletonised and whole plant is destroyed.

Control measures

- Spraying the crop with 0.07% Endosulfan 35 EC (Thiodan) @ 1.4 ml/l or Malathion @ 2 ml/l.

Aphids (*Macrosipum pisi, Aphis craccivora, Myzus persicae*)

Aphis craccivora is the most damaging to cowpea, cluster bean and other leguminous crop. These gregarious and tiny insects suck cell sap from the tender aerial parts of the plant, which shows some characteristics symptoms, like curling of leaves, twisting of twigs and developing fruits, and sometimes shedding of flowers. They also transmit mosaic viruses.

Control measures

- Spraying the crop with 5% neem seed kernel extract along with 2% starch reduces the incidence.
- Spraying the crop with Rogor @ 1.5ml/l of water.
- Application of Carbofuran granules in the field (Furadon 3G @ 1.0 Kg a.i. / ha) before sowing of the seeds.

Jassid *(Empoasca fabae)*

This sucking insect pest generally attacks cluster bean. Both nymphs and adults of this insect suck cell sap usually from ventral surface of the leaf, and because of severe infestation, the affected leaves show typical hopper burn symptoms.

Control measures

- Spraying the crop with 5% neem seed kernel extract along with 2% starch reduces the incidence.
- Spraying the crop with Rogor @ 1.5ml/l of water.
- Application of Carbofuran granules in the field (Furadon 3G @ 1.0 Kg a.i./ ha) before sowing of the seeds.

Flower thrips (*Haplothrips verononiae; Frankliniella sulphurea)*

Flower thrips are very destructive to cowpea. The thrips cause severe bud and flower drops resulting poor yield. The adult and nymphs suck sap from the petals, calyx and developing ovary.

Control measures

- Spraying the crop with Acetamiprid @ 0.5 g/l at 10 days interval at the early stage.
- Application of neem seed kernel extract at later stage of the crop is useful.

Root knot nematode (*Meloidogyne* spp)

Cowpea, hyacinth bean and cluster bean are mostly infested by the nematodes. The nematodes attack the roots and produce galls there which cause stunting of the plants and severely reduce yield.

Control measures

- Ploughing the field in summer months.
- Application of oil cakes, more specifically neem cake in the field during land preparation
- Growing moderately resistant varieties of cluster bean like, JG-2, HGV-75 and GAUG-26
- Application of granular insecticides like Thimet 10G @ 25 Kg/ha or Furadon 3G @ 35kG/ha in soil during land preparation.

Leafy Vegetables

- **Vegetable amaranth**
- **Spinach beet or palak**

Vegetable amaranth

The genus *Amaranthus* is rather unique in having species which are used for grain, vegetable and ornamental purposes. Vegetable amaranth, *Amaranthus* spp. under the family Amaranthaceae (Chromosome No.: 2n = 2x = 32, 34) is the most common tropical leafy vegetable in Southeast Asian countries especially India, Malaysia, Indonesia, southern China and hot and humid regions of Africa. "Amaranth" derives from the Greek word "*amárantos*" meaning "flower". It fits well in a crop rotation because of its very short duration and large yield of edible matter per unit area. Four species of *Amaranthus* are documented as cultivated vegetables in eastern Asia: *Amaranthus cruentus*, *Amaranthus blitum, Amaranthus dubius*, and *Amaranthus tricolor*. Among the leafy types, *Amaranthus tricolor* L. occupies the predominant position in India with different morphological forms in colour and shape of leaves.

The crop

Amaranthus, collectively known as amaranth, is a cosmopolitan genus of annual or short-lived perennial plants. Approximately 60 species are recognized, with inflorescences and foliage ranging from purple, through red and green to gold. Members of this genus share many characteristics and uses with members of the closely related genus *Celosia*. *Amaranthus* shows a wide variety of morphological diversity among and even within certain species.

Members of this family have simple leaves that are opposite or alternate, with margins entire or coarsely toothed, and without stipules. Catkin-like cymes of densely packed flowers grow in summer or autumn. The flowers are solitary or aggregated in cymes, spikes or panicles and unisexual. The bracteate flowers are regular with 4 to 5 petals, often joined. There are 1 to 5 stamens. The hypogynous ovary has 3 to 5 jointed sepals.

Depending on the cultivar, photoperiod and cultural practices, flowering may start 4-8 weeks after sowing, making the plant less suitable for consumption. There are at least four times as many female flowers as male flowers. Pollination is effected by wind but the abundant pollen production, especially in the higher flowers, causes a high rate of self-pollination. Outcrossing varies from 0 to 40%. In *A. tricolor* and *A. cruentus,* the seeds mature after 3-4 months and then the plant dies. *A. dubius* will continue its generative stage for a much longer period and when cut regularly, the plant may become shrubby and perennial, but even at its mature stage, the leaves are succulent enough for consumption.

Cultivars of *Amaranthus tricolor* have striking yellow, red, and green foliage. *A. gangeticus*is is an annual flowering plant with deep purple flowers which can grow to 61– 91 cm tall. *Amaranthus blitum* is commonly called purple amaranth. *Amaranthus dubius* usually grows to a size of 80–120 cm. It has both green and red varieties, as well as some with mixed colours. The green variety is practically indistinguishable from *Amaranthus viridis*. *Amaranthus cruentus* is a tall annual herb topped with clusters of dark pink flowers which shares many morphological features of *Amaranthus hybridus*.

General characteristics of *Amaranthus tricolor*

- Erect annual, up to 1.5 m tall.
- Leaves elliptical to lanceolate or broad-ovate, dark green, light green or red.
- Clusters of flowers axillary, often globose, with a reduced terminal spike, but occasionally the terminal spike is well developed. Tepals 3.
- Fruit dehiscent, with a circumscissile (dehiscence extends all round) lid.
- Seeds black, relatively large; 1200-2900 seeds/g.
- Cultivated.

General characteristics of *Amaranthus dubius*

- Annual, sometimes biennial, up to 2 m tall, erect, strongly branching.
- Leaves ovate or rhomboid-ovate, shortly cuneate at base, dark green.
- Lower clusters of flowers axillary, upper clusters leafless and in lax panicled spikes. Tepals (3-5)
- Fruit dehiscent, with a circumscissile lid.
- Seeds black, very small; 3000-4800 seeds/g.
- Cultivated vegetable, sometimes escaped as weed.

General characteristics of *Amaranthus cruentus*

- Tall annual, up to 2.5 m. Leaves lanceolate, acute and often short-decurrent at base, greyish-green.
- Clusters of flowers in large axillary and terminal panicled spikes. Tepals 5.
- Fruit dehiscent, with a circumscissile lid.
- Seeds dark brown to black; 2500-3000 seeds/g. Seeds of grain types are light yellow.
- Cultivated as vegetable or grain.

General characteristics of *Amaranthus blitum* cv. group *oleraceus*

- Small annual, up to 75 cm tall.
- Leaves obovate or rhomboid- obovate, shortly cuneate at base, green or more or less purple.
- Lower clusters of flowers axillary, upper clusters leafless and in axillary and terminal panicled spikes. Tepals 3 (-5).
- Fruit indehiscent or finally bursting irregularly.
- Seeds small, dark.
- Weed, sometimes cultivated. It is a popular cultivated vegetable in India.

Importance and use

Leaf amaranthus types are generally used as pot herb vegetable. Young plants with tender leaves and stems are used as a stir-fry vegetable or in soups. It gives considerable biomass yield per unit area. It is easily cultivated both in kitchen garden and on commercial scale. The tiny seeds of grain amaranthus are popped or parched and milled for flour or gruel. It is also used to make a fermented product called "ogi". Amaranth being a C4 carbon fixation plant, which allows it to convert carbon dioxide into biomass at a more efficient rate than other plants.

Nutritional value

Amaranth leaves have a high content of essential micro- nutrients. They are an excellent source of β-carotene with 4-8 mg per 100 g edible portion, vitamin C 60-120 mg, iron 4-9 mg, Calcium 300-450 mg, potassium 341 mg, magnesium 247 mg, phosphorus 83 mg and 1g fibre. They are rich in fibre and folic acid and their protein content (20-38% based on dry matter) includes methionine and other sulphur-containing amino-acids. The general dry matter content is high (12-16%). The leaves and stems have nitrate and oxalate levels similar to spinach (*Spinacia oleracea*L.) and spinach beet (*Beta vulgaris* L.), but no adverse nutritional effects occur with a consumption of 100-200 g per day.

Medicinal value

Consumption of leaf amaranthus protects against cancers of digestive tract, stomach, colon and rectum because it contains considerable quantity of antioxidant compounds such as β carotene and ascorbic acid. It prevents vitamin A deficiency related night blindness and xeropthalmia in the childrens. Incidence of blindness in children due to poor availability of provitamin A in the diet can be reduced with a daily consumption of 50 to 100 g of amaranth leaves. Amaranthus leaves and tender stems contain considerably high dietary fibre which help in reducing the problem of constipation. Leaf amaranthus contains some anti-nutrient compounds like oxalates and nitrates which may cause kidney stone and methaemoglobin in blood, respectively if it is consumed in high quantity regularly.

Origin and taxonomy

The centers of diversity for *Amaranthus* are Central and South America, India, South East Asia and the secondary diversity is in West and East Africa.

Amaranthus tricolor, the predominant leafy amaranth species, originates from tropical Asia. Most probably it is native to India and from where it spread to the neighbouring countries. Its domestication took place in prehistoric times and the wild ancestor is not known.

Amaranthus dubius is native to South America and was introduced to Asia, Europe and Africa. It has much diversity in Central America, Indonesia and India.

Amaranthus blitum is native to the Mediterranean region and it is naturalized in other parts of the world, including much of eastern North America.

Amaranthus cruentus is probably native to tropical America.

All grain Amaranthus have originated in Central and South America particularly, Guatemala (*A.cruentus*), Argentina and Peru (*A. hypochondriacus*), high Andes of South America (*A. caudatus*) with Hisdustani subcontinent as an important secondary centre of diversity.

Amaranthus belongs to the genus *Amaranthus* and family Amaranthaceae. This family comprises of 65 genera and 850 species. The genus includes 50-60 species, both cultivated and wild, and used for grains and leaves. The genus *Amaranthus* is classified into two sections: *Amaranthus* and *Blitopsis*. All leaf amaranthus species come under the section *Blitopsis* and all grain type species under the section *Amaranthus* .The important leafy type cultivated *Amaranthus* species are: *A. tricolour*, *A tristis*, *A hypochondriacus*, *A polygonoid*, *A viridis*, *A lividus* (*A. blitum*) and *A caudatus*. The most popular grain species are *A. hypochondriacus, A. cruentus* and *A. caudatus* and the most important weed species are *A.spinosus, A.viridis* and *A.graecizans*.

The basic chromosome numbers in this genus are x = 16 and x = 17 and both numbers are found in sect. Amaranthus and sect. Blitopsis. The chromosomes in this genus are very small in size, hampering a detailed karyotype analysis. Chromosome numbers of different species of the genus Amaranthus are: *A. retroflexus* 2n=34, *A. caudatus* 2n=32, *A. hybridus* 2n=32, *A. spinosus* 2n =34, *A. cruentus* 2n=34, *A. hypochendriacus* 2n=32, *A. paniculdatus* 2n=32, *A. roxburghianus* 2n=34, *A. blitoides* 2n=32, *A. polygonoides* 2n=34, *A. albus* 2n=32, *A. viridis* 2n = 34, *A. lividus* 2n = 34 and *A. tricolor* 2n = 34.

Adaptation in India

Leaf amaranth belonging to *A. tricolor* is apprehended to be the native of India. Several cultivars of vegetable amaranth belonging to the different species viz., *A. tricolor*, *A. tristis*, *A. hypochondriacus*, *A. polygonoid*, *A. viridis*, *A. lividus* and *A. caudatus* are grown in India under different names. The cultivars vary considerably in plant height, size and colour of leaves, branching habit, flowering time and response to cutting. Interspecific and interval hybridization in nature have caused such wide variation. Mostly green and red types are cultivated and both these types are nutritionally similar. Improvement of *Amaranthus* in India has been done by selection of superior types from indigenous germplasm. The varieties may be grouped according to the species under which they belong.

Improved varieties

A. tricolor: CO-2, CO-5, Pusa Badi Chaulai, Pusa Kiran, Pusa Lal Chaulai, Arka Arunima, Arka Suguna, Arka Samraksha, Arka Varna, Utkal Mayuri, Amt-105, Amt-237, Annapurna

A lividus (*A. blitum*): Pusa Chhoti Chaulai, Pusa Kirti,

A. dubius: CO-1

A. tristis: CO-3

A hypochondriacus: CO-4

A polygonoid: Sirukeerai

Cultural requirements

Amaranthus is widely distributed in both temperate and tropical regions of the world. Amaranth grows from sea level to 2400 m altitude. *Amaranthus* species that grow under varying climatic condition differ in day length requirements and respond differently to change in photo and thermoperiodism. The different species may suit different altitudes. Generally it grows well within a temperature range of 22-30°C. A minimum temperature of 15 to17°C is required for seed germination. Amaranth is grown during both rainy and summer seasons, though irrigation is normally required for summer season crops since the rate of transpiration by the leaves is high. Frequent irrigation is required, related to the stage of growth of the crop and the moisture-retaining capacity of the soil. It can however tolerate periods of drought after the plant has become established. Amaranth grows well in loam or sandy-loam soils with good water-holding capacity, but it can grow on a wide range of soil types and soil moisture levels. Amaranth can tolerate a soil pH from 4.5 to 8.0 however, pH of 6.4 is ideal. *Amaranthus tricolor* is relatively salt-tolerant.

Land preparation

Amaranth requires thorough land preparation for good growth. The land should be prepared in well advance with at least 4-5 ploughings to a fine tilth. All stubbles and weeds should be removed. After three to four harrowing, the land is leveled to bring the field into a fine tilth. Well rotton organic manure (preferably farmyard manure) @ 25t/ha should be added at the time of last ploughing.

Sowing time

Amaranthus is warm season crop and it can be grown throughout the year in tropical and subtropical regions. In south and East Indian condition amaranthus is grown almost throughout the year. In north Indian condition, generally two sowings are practiced: February – March for summer crop and June – July for rainy season crop.

Sowing of seeds

Seeds are sown by broadcast method or by drilling in lines 20-30 cm apart at the rate of 0.5 to 1.0 g/m^2 of bed. Since amaranth seeds are very small, mixing seeds with sand at a ratio of 1 g seed to 100 g sand makes it easier to sow the seed and to obtain a uniform stand. Seeds after sowing should be covered lightly with a layer of compost immediately after broadcasting. When plants are to be grown in rows, furrows are to be made at 0.5 to1.0 cm depth and rows are spaced at 10 cm apart on the bed. Seeds are sown 5 cm apart within the row and cover with a layer of compost. In South India, there is also a practice of transplanting amaranthus either as a pure crop or along with the border of beds other vegetables. For transplanting, seedlings are grown in a seedbed, pulled and planted bare-root or they can also be grown in seedling trays, lifted with the root ball intact and planted. Seed rate is about 2kg/ha for direct sowing and 1kg/ha for transplanting crop. Seed treatment with IBA at lower concentration of 0.1-1.0 ppm increases germination.

Spacing

The plant spacing within the row is maintained at about 25-30 cm at the time of thinning. Because of its small seed size, shallow sowing about 1-1.5 cm deep is recommended. Sometime seedlings are transplanted during rainy season with a spacing of 45 cm from row to row and 20-25 cm from plant to plant.

Nutrient management

Amaranth is a low management crop and can grow in poor soils, but it will benefit from the application of fertilizer resulting in higher yield. Nitrogen required for the growth of a crop will vary depending on the N status of the soil and potential for mineralization. A general fertilizer dose of 90 kg N and 50 kg each of K_2O and P_2O_5/ha is recommended. Whole of $P_2O_5$and K_2O and one-third of N are applied before sowing or transplanting at the time of field preparation. The balance of N is applied 4 weeks after transplanting and next dressing after every cutting. Amaranthus is short duration crop and is good feeder of macronutrients. Application of N also increases the plant crude portion, carotene and chlorophyll contents in the leaves and stem. Five tonnes of vermicompost together with 50:50:50 N, P_2O_5, K_2O kg/ha give the highest vegetative yield as well as nutrient uptake, implying the synergistic effects of combined application of vermicompost and chemical fertilizers in amaranth production.

Irrigation

Although amaranth is relatively drought tolerant, yet insufficient water will reduce yield. Water should be applied especially just after sowing or transplanting to ensure a good stand. As a rule, the plants should be irrigated if wilting occurs at noon. In the absence of sufficient soil-moisture at the time of sowing, first irrigation should be applied immediately after sowing and further irrigation are to be given once a week with gentle flow of water. During summer, subsequent irrigations are given at 3-5 days intervals. In rainy season, irrigation is given as and when required.

Interculture and weed control

Weeds compete for light, water and nutrients, thereby resulting in reduced yield. Thorough land preparation is the first key to effective weed control. Amaranth is small-seeded crop and grows slow in the initial stages hence, weed control is essential early in the season. A seedbed free of weed seeds allows amaranth seedlings to get a head start on the weeds and establish a canopy that can shade out emerging weed seedlings. Shallow inter cultivation is practiced to remove the weeds from the field and generally 3-4 weeding are needed. Amaranthus is mostly sown directly to the field, so it should be thinned properly to get good crop stand. Generally 10-15 days after sowing, it is thinned to a spacing of about 20-25 cm between plants.

Harvesting the crop

In green types of *A. tricolor* the plants are pulled as a whole, washed and send to market as tender greens. Optimum stage of harvest in amaranthus is about 25-35 days after sowing as at this stage, performance of this types are better with leaf weight, stem weight, leaf length, leaf breath, stem diameter, plant height as well as nutrition and palatability. In *A. tristis*, a total of 10 clippings can be obtained with the first clipping starting on the 25 days and thereafter at weekly interval. In *A. blitum*, 6 clipping are made in 60 days duration. In case of intermediate and late flowering cultivars of *A. cruenta* at the start of the rains and continuously cutting back is more profitable method of harvesting than uprooting at the optimum commercial stage. The highest vegetable yield is obtained from black seeded cultivars and highest seed yield from the white seeded cultivars.

Post harvest handling and storage

Amaranthus does not stand storage for more than few hours under ordinary conditions. About 3-4 days storage life can be obtained when green leaves are stored in cool storage condition.

Plant Protection Measures

Diseases

Leaf spot (*Cercospora canescens*)

This disease is characterized by the presence of numerous small, brown, circular spots on the leaves. In the beginning, the spots are small, roundish with concentric rings but later these spots increase in size and some times they coalesce.

Control measures

- ❐ Spraying the crop with Blitox (0.3%) or Difenconazole (0.05%) or Chlrothalonil (0.2%) 2-3 times at 15 days interval.

White rust (*Albugo bliti*)

This disease is characterized by white, blister like circular or irregular pustules on the lower surface of the leaf and opposite each pustule on the upper surface, a yellow patch developes. Severe infection causes death of the leaves giving the field a blighted look.

Control measures

- Spraying the crop with 0.2-0.3% Mancozeb or 0.25% Metalaxyl-Mancozeb.

Insect pests

Leaf eating caterpillar (*Hymenia recurvalis*)

It is a serious pest of *Amaranthus* and many other vegetable crops. Larvae bind the leaves together by their webbing and eat away by scraping the leaf tissues. In severe cases the leaves dry up and the plant die.

Control measures

- Spraying the crop with neem seed karnel extract at regular interval as prophylactic measures as soon as the pest is observed.
- Sparying the crop with Malathion 0.05% or DDVP 0.05% is effective.

Red spider mite (*Tetranychus urticae*)

The mites attack all type of Amaranthus during summer months which colonize at the ventral surface of leaf and produce numerous whitish specks. In case of severe attack the leaf fall off and dry out.

Control measures

- Spraying the crop with botanical pesticides like neem or karanja products as prophylactic measures.
- Mass release of predatory mite, *Amblyseius longispinosus* is very effective against the red spider mite complex.
- Spraying the crop with Propergite @ 1.5 ml/litre or Dicofol @ 2.5 ml/litre is effective.

Yield

Green yield is generally 10-25 tonnes per hectare depending on the variety, number of cuttings, location, season and cultural practices. However, marketable yield per hectare increases as plant spacing is decreased. In low cost polyhouse, yield may be as high as 48-50 t/ha.

Spinach beet or palak

Palak or spinach beet or beet leaf, *Beta vulgaris var bengalensis* under the family Chenopodiaceae (Chromosome No.2n = 2x = 18) is one of the most common leafy vegetables of the tropical and subtropical regions. It is also known as beet leaf, leaf beet, desi palak, common palak, palang sag, palang and meetha palang. It is commonly grown in Uttar Pradesh, West Bengal, Punjab, Haryana, Delhi, Madhya Pradesh, Bihar, Maharashtra and Gujarat. Spinach is widely cultivated in different European countries and the USA.

The crop

Palak belongs to same species in which garden beet, sugar beet and Swiss chard are included. The plant is an annual herb; has large, leafy, thick-stemmed green looks like beets without swollen the root. Basal leaves few not forming a dense rosette, with long strong petiole, ovate to oblong-ovate, up to 20 cm long, apex obtuse, passing above into smaller narrower leaves; flowers bisexual, two or three in a cluster, on simple or branched terminal spike; perianth segments 5, united at base, narrowly oblong, incurved in fruit; stamens 5, filaments united at base; stigmas 2 or 3; utricle adnate to perianth at base.

Palak is a good substitute for spinach but both is not the same. Both palak or spinach beet and spinach belong to the same family but some prominent characteristics differentiate these two important leafy vegetables.

Difference between palak or spinach beet and spinach

Palak or spinach beet	Spinach
Botanical name: *Beta vulgaris* var. *bengalensis* Hort.	Botanical name: *Spinacea oleracea* L
Chromosome number 2n=18	Chromosome number 2n=12
Leaves are narrowed at base with entire margin and long petiole	Leaves are hastate with lobed margin
Bisexual sex form	Dioecious sex form
Produce hermaphrodite flowers	Produce either staminate or pistilate flowers
Perianth not hardened in fruit	Fruiting perianth enlarging, becoming hardened and often spiny
Tolerates high temperature and grow well in hot weather	Purely a cool season crop and cannot tolerate high temperature. In warm weather and long days it quickly tend to flower

Importance and use

Palak is generally used as pot herb vegetable. Young plants with tender leaves are used as leafy vegetable. It fits well in different crop rotations because of its adaptation to high temperature condition, short cropping duration and appreciable biomass yield per unit area. It is easily cultivated both in kitchen garden and on commercial scale. The fresh tender leaves of palak are delicious and can easily be cooked with other vegetables like, potato, onion, and brinjal or fried alone. Delicious vegetarian dishes are also prepared with cheese and potato.

Nutritional value

Palak is rich and cheap source of vitamin A, iron, essential amino acids, ascorbic acid etc. It is a very good source of fibre, vitamins A, C, E and thiamine (B1), riboflavin (B2), pyridoxine (B6) and vitamin K, besides minerals like, calcium, iron, magnesium, potassium and zinc. 100g palak leaves contain moisture (86.4 g), fat (0.8 g), fibre (0.7 g), protein (3.4 g), minerals (2.2 g), carbohydrates (6.5 g), phosphorus (30 mg), riboflavin (0.56 mg), calcium (380 mg), iron (16.2 mg), thiamine (0.26 mg), vitamin A (9770 IU), nicotinic acid (3.3 mg) and vitamin C (70 mg). Nutrient contents vary with the cultivars.

Medicinal value

Palak is the rich and cheap source of vitamin A as compared to carrot. It prevents vitamin A deficiency related night blindness and xeropthalmia in the childrens. Palak is mildly laxative besides other medicinal values. Palak provide enough roughage which is important for balanced diet. It is easily digestible and useful for the elderly persons. However, because of high oxalate content, peoples suffering from kidney stone problem are advised to restrict its consumption.

Origin and taxonomy

Palak has most probably been originated in the Indo-Chinese region, especially Bengal region of Indian sub-continent. It was known in China in 647 A.D. It was first recorded in India in the medical treatise, Carak Samhita, c. 600 B.C. The palak, *Beta vulgaris* var. *bengalensis* Hort. belongs to the family Chenopodiaceae. It is closely related to Swiss chard, beet root and sugar beet, all having the same chromosome number. The sea beet, *Beta vulgaris* ssp. *maritime* is the ancestor of beet root and palak.

Chromosome number is 2n=18.

Adaptation in India

Palak as a leafy vegetable is known from Charaka Samhita days, c. 600 B.C. which is the testimony of its long history of cultivation in India. In India, it was perhaps first cultivated in Bengal. Although, despite such a long history of cultivation the genetic variability observed in this species is not wide enough, commensurate with outcrossing breeding behaviour. However, all the improved varieties have been developed utilizing indigenous variability.

Improved varieties

Punjab Green, Punjab Selection, Pusa Jyoti, HS-23, All Green, Pusa Harit, Jobner Green, Pusa Bharati, Banerjee's Giant, Pusa Palak, Ooty-1, Pant Composite, Arka Anupama, Palak No.51-56.

Cultural requirements

It is a winter season crop in the plains of India. It can withstand frost better than other vegetables. It can also tolerate warm weather but relatively hot weather will result in quick bolting making the plant unfit for producing edible leaves. Palak thrives well in a comparatively cool climate and does best at a temperature range of 7-24°C. It is half-hardy and can withstand light frosts, although growth is retarded at low temperatures. Prolonged exposure to temperatures less than 5°C will induce seed production (bolting), usually in spring. During hot weather, leaves remain small and are of inferior quality. During hot weather leaves pass edible stage quickly.

It can be grown on a wide variety of soil types, provided they are well-drained, free of root knot nematodes, reasonably fertile and amply supplied with water. It does very well in sandy loam soil. It can tolerate slightly alkaline soil, but best quality and high yield can be obtained in the soil with pH 7.0. Palak is a salt tolerant vegetable and can be grown successfully in saline-sodic soils. Jobner Green gives excellent result in the soil with pH ranging between 7.0-10.5.

Land Preparation

The soil for this crop can be prepared by 3-4 ploughings followed by planking so that soil is well pulverized and levelled. After preparation of the soil thoroughly, beds of convenient size and irrigation channels are made.

Sowing time

Sowing time is January-February for early spring crop; June-July for rainy season crop and September-November for autumn-winter crop (main crop). In the hill condition, sowing is done in March-May.

Sowing of seeds

Seeds are sown by broadcast method or by drilling in lines 20-30 cm apart and 8-10 cm plant to plant (after thinning). In both the cases, seeds should be placed at a depth of about 4 cm. The seed germinates in about 8-10 days if proper soil moisture and temperature are available. Seed rate is 25-30 kg/ha for summer crop and 10-15 kg/ha for winter crop. Seeds should be soaked overnight to hasten the germination. Seeds should be sown to a depth of 3-4 cm.

Nutrient management

Palak is short duration crop and is good feeder of macronutrients. A basal dose of 25-30 tonnes of farm yard manure should be applied per hectare at the time of land preparation. A general fertilizer dose of 100 kg N and 60 kg each of K_2O and P_2O_5/ha is recommended. Whole of $P_2O_5$and K_2O and 1/2 of N are applied before sowing at the time of field preparation. The balance of N is applied 4 weeks after sowing for one time harvest by uprooting the plant. In case of periodical harvest by cutting, 1/3 of N is applied as basel and the rest is recommened as topdressing after every cutting. Application of 1.5 per cent foliar application of urea spray 15 days after germination and after each cutting up to third cutting give good leaf yield yield.

Irrigation

The plant has a moderately deep root system but like other leafy vegetable crops, it should not be allowed to suffer moisture stress. It thus requires fairly frequent irrigation to ensure that the soil does not dry out to less than 50 per cent available water. Soil moisture should never be limiting. In light soils, it is desirable to give irrigation soon after the sowing is complete, but in heavier soils the sowing should be done when enough soil moisture is available. In spring-summer season, irrigation is required at 6-7 days interval and in autumn-winter season, irrigation is required at 10-15 days interval. However, rainy season crop does not require much irrigation.

Interculture and weed control

Regular hoeing and weeding should be done to keep away the weeds from the field and to loosen the soil for proper aeration and generally 2-3 hoeing-cum-weeding are required. After every cutting, weeding and hoeing should be done.

Harvesting the crop

The crop is generally harvested either by uprooting the whole plant at tender stage or by cutting the tender leaves periodically. Its first flush of leaves become ready for picning 3-4 weeks after sowing. Subsequent cutting should be taken at an interval of 20-25 days depending upon variety and season. About 4-6 cuttings can be taken depending on the season and variety.

Post harvest handling and storage

Palak has a large leaf surface to weight ratio and has very high respiration rate. After harvest the leaves cannot be stored for longer period and hence, should be sent to the market immediately after harvesting. Before marketing, leaves are tied in small bundles. All the weeds, diseased and broken leaves should be discarded at the time of bunching. Washing of the leaves is not desirable because it causes rotting and yellowing of the leaves in the centre of the bundle. In the early morning the leaves are crisp and may break during harvesting. Under low temperature (0°C) and high relative-humidity (90-95%), leaves can be stored for 10-14 days.

Plant Protection measures

Diseases

Damping off (*Pythium sp.*)

Seedlings are attacked by the fungus. The affected seedlings look pale green and brownish lesions are found at the collar region at the advanced stage. Collar rotting occurs subsequently and the seedlings topple over the ground.

Control measures

- Seed treatment with Thiram or Captan @ 3 kg/ha of seed or with Bavistin 2 g/kg of seed.
- Soil drenching with 0.25% Mancozeb.
- Spraying the crop with Mancozeb (0.25%) or Metalaxyl-Mancozeb (0.25%).

Leaf Spot (*Cercospora beticola*)

Brown dead circular spots are found on the diseased leaves due to the attack of this fungus.

Control measures

- Spraying the crop with 0.2 % Chlorothalonil or 0.1% Propiconazole at 15 days interval.

Insect pests

Aphid (*Aphis* sp.)

Small, oval, soft bodied nymphs and adults are found in large number on the tender portions of the plant. They suck the cell sap from the tissue which causes curling and ultimately drying up of leaves. Black sooty moulds developed due to honeydew secretion by the aphid which hinder photosynthesis and devitalize the plant.

Control measures

- Spraying the crop with 0.05% Dimethoate (Roger 2 ml/l), 0.08% Malathian (Cythion 2 ml/l) at 10 days interval

Leaf eating caterpillar (*Laphygma exigua*)

Young caterpillars feed on the leaves by scraping gregariously. Older caterpillars feed voraciously the entire leaf. It causes economic defoliation ranging from 16-80%.

Control measures

- Spraying the crop with Malathian (Cythion 2 ml/l) at 10 days interval dissolved in 625 litres of water.

Yield

Leaf yield widely vary between 25-45 t/ha depending on the variety, number of cuttings, location, season and cultural practices.

Under-utilized Leafy Vegetables

Being rich in bio-diversity, there are several lesser known under-exploited and under-utilized plant species available in India which have tremendous potential to be used as leafy vegetables. Most of these vegetables are cheaper source of protective nutrients such as vitamins and minerals. Some of these vegetables have the potential to produce significantly high amount of food per unit area and are also rich in protein. These leafy vegetables are easy to cultivate, do not require high input and can thrive well on marginal and sub-marginal land and therefore, could be exploited for meeting the nut:.ent requirement of predominantly vegetarian population of the country. Moreover, these leafy vegetables have tremendous capacity for acclimatization and they not only contribute towards food but also act as genetic reservoir of several important traits. With drastic increase in country's population and fast depletion of natural resources there is an urgent need to explore new sources of vegetable and to diversify the vegetable cultivation to meet out the present day demand. Brief accounts of some under-utilized leafy vegetables are given below :

- **Basella or Indian spinach**
- **Water spinach**
- **Leafy greens of Brassica species**
- **Colocasia greens**
- **Bathua**
- **Portulaca**
- **Indian sorrel**

Basella or Indian spinach

Basella (Syn. Malabar spinach, Ceylon spinach, Indian spinach, Vine spinach, Chinese spinach, etc.), *Basella alba* and *B. rubra* under the family Basellaceae (2n = 44, 48*)* is a popular leafy vegetable grown in almost all parts of Tropical Asia and Africa. This leafy vegetable has a mild flavour and taste which is similar to spinach.

Use

In Indian cuisine, young leaves and soft stems are cooked after mixing with other vegetables and even fish. It is also used as stir fries with garlic and chili peppers, omelettes or just on its own as a steamed vegetable. However, it should not be cooked too long as the mucilage can give a slimy texture.

The young, juicy leaves make a great addition to salads and sandwiches. All leaves and shoots can be added to soups and stews. The succulent mucilage is a particularly rich source of soluble fibre and is especially useful as a thickener in soups and stews.

Nutritive and medicinal value

It is a highly nutritious leafy vegetable. The crop has high levels of vitamins A, B, and C, calcium, magnesium, phosphorous, manganese, zinc and decent amounts of Iron and copper. It has good levels of mucilage which is a valuable aid in detoxifying the body. Basella fruit is a rich source of betalains and has value-added potential for use in the development of food colourants and nutraceuticals. The major red pigment is gomphrenin I, a potent antioxidant and inflammatory inhibitor. Fresh leaves, particularly of *Basella rubra*, are rich sources of several vital carotenoid pigment antioxidants such as β carotene, lutein, zeaxanthin. Chinese sell the bright red juice in the form of a powder called "gintjoo".

Nutritional value per 100 g fresh edible vine as per USDA data base is given below :

Energy: 19 kcal	Calcium: 109 mg
Vitamin A equiv: 400 µg	Iron: 1.2 mg
Thiamine (B_1): 0.05 mg	Magnesium: 65 mg
Riboflavin (B_2): 0.155 mg	Manganese: 0.735 mg
Niacin (B_3): 0.5 mg	Phosphorus: 52 mg
Vitamin B_6 0.24 mg	Potassium: 510 mg
Folate (B_9): 140 µg	Zinc: 0.43 mg
Vitamin C: 102 mg	

Content of phenols, flavonoids, flavonols, pro-anthocynidins does not differ much between green (*Basella alba*) and purple (*Basella rubra*) types, however anthocyanin content is maximum in purple variety (1.65 mg/g dry weight) as compared to green variety (0.285 mg/g dry weight). Total carotenoid and β carotene showed high concentration of 2.48 and 0.229 mg/g fresh weight in green compared to 1.55 and 0.44 mg/g fresh weight in purple variety. Radical scavenging activity is high in purple variety compared to the green variety.

Various parts of the plant are used for treatment of the diseases as well as for different healing activities of human beings as well as animals across the globe especially in India and China. Its use has been discovered as asperient and rubefacient. Some of the compounds in basella are basellasaponins, kaempherol, betalin, etc. Regular consumption of basella in the diet helps prevent osteoporosis (weakness of bones), iron deficiency anaemia. Besides, it is believed to protect the body from cardiovascular diseases and cancers of colon. Because of the mucilaginous nature, the leaves and stem are used as poultice. The juice of leaves is prescribed against constipation especially for children and pregnant women. The leaf juice is a demulcent, diuretic, febrifuge and laxative. A paste of leaves is applied externally to treat boils. The flowers are used as an antidote to poison.

Like in spinach, basella too contains oxalic acid, which may crystallize as oxalate stones in the urinary tract in some people.

Origin and taxonomy

Basella is native to the Indian Subcontinent, Southeast Asia and New Guinea. It has been naturalized in the China, tropical Africa, Brazil, Belize, Colombia, the West Indies, Fiji and French Polynesia. There are two main species of this crop is *Basella alba*, which has green stems and thick fleshy leaves, and *Basella ruba* which has red stems. It is diploid (2n = 44, 48) and belongs to Basellaceae family.

Propagation

Basella is a short-day plant and flowering is induced during the short-day months of November to February. The plant develops red berries which should let to be dry on the vine. Seeds are collected in the following spring. It can also be propagated by tip cuttings of the vine.

Cultural requirements

It is a very warm season crop and is extremely heat tolerant. Optimum temperature for seed germination is 18-20°C and for its good growth is above 33°C. It is grown throughout the tropics as a perennial and in warmer temperate regions as an annual. It grows well under full sunlight in hot, humid climates and in areas lower than 500 metres (1,600 ft) above sea level. However, it produces larger juicier leaves if grown in partial shade.

Crop production

Basella grows well in hot and humid climate. It is generally grown by direct seeding when night temperature is above 16°C because it is highly susceptible to frost. Low temperature slows down growth rate and results in small leaves. It can also be grown by transplanting seedlings. For better germination, the seeds are soaked in water overnight the day before planting. In Northern and Eastern plains of India, seeds are sown from March to May while in Southern parts, it is sown in July and again in October-November.

A seed rate of about 13.0 to 16.0 kg is required for one hectare of land. When direct seeding is done, seeds are sown in rows in well prepared beds. Furrows of 1.0-1.5 cm depth are spaced at 10- 15 cm apart on bed. The seeds are sown 5 cm apart in rows. The seeds are covered with a layer of compost. The seedlings are thinned to stand 10-15 cm apart at two to three true leaf stage. The seedlings can also be raised in trays or seed beds. Seeds are then sown in furrows spaced at 5 cm apart and then covered with soil. Seed beds are then covered with insect proof nets to provide shade. Seedling trays are filled with different potting mixtures having good water holding capacity viz., peat moss, potting mixture, rice hulls, vermiculite and sand. Two or three seeds are sown per tray cell at 1.0 to 1.5 cm depth. Thinning is done to retain one seedling per tray cell at two to three true leaf stage. Later they are hardened by slowly exposing to direct sunlight prior to transplanting. Seedlings are ready for transplanting when they have five true leaves.

Recommended spacing varies with variety and harvest methods. For once over harvest, raised beds of 20-30 cm height at 90 cm width at top are taken at convenient length. Rows are spaced 10-15 cm apart with 15 cm between plants within rows. Irrigation should be given immediately after transplanting to establish good field stand. Stem cuttings of 20-25 cm length with three to four internodes are soaked in water overnight and planted at spacing of 20-30 cm between rows and 15-20 cm within a row at a depth of 5-10 cm. Two or three stem cuttings are planted per hole. Stem or root cuttings should be planted during monsoon month or in early summer.

Basal dressing of 20-30 tonnes of farm yard manure and 60:60:40 kg NPK/ha is recommended which should be applied before transplanting or sowing. Frequent irrigation promotes growth of plant and makes the plant succulent however, water stagnation should be avoided. Water stress induces early flowering and may lead to thin stems and small leaves. The crop requires irrigation at an interval of 5-6 days in summer and 7-8 days during cool season. When plant starts trailing, it should be trained on supports.

Plant protection measures

Main diseases recorded in this crop are damping off caused by *Pythium aphanidermatum*, leaf spot caused by *Acrothecium basellae, Fusarium moniliforme* and *Cercopsora* sp and mosaic due to unidentified virus. Seed treatment with Thiram/ Captan / Mancozeb @ 3.0g /kg of seeds and drenching the nursery beds 7 days before sowing with Thiram/ Captan or any copper fungicide @ 3g / litre of water are important preventive measures against the diseases. The crop is almost free from major insect attack.

Crops raised from seeds become ready for harvest with edible stems and leaves 8-10 weeks after sowing. Stem tips of 20-30 cm are harvested and repeated harvests of new growth stems can be made throughout the season. The plants raised from root or stem cuttings becomes ready for harvest in about 6 weeks after planting. Yield varies from 15-20 tonnes/ha depending on the growing condition. The crop can be stored for very brief period under ambient condition.

Water spinach

Water spinach, *Ipomoea aquatica* Forsk under the family Convolvulaceae (2n = 30), a semi-aquatic tropical plant, is found throughout the tropical and subtropical regions of the world. This plant is known in English as water spinach, swamp cabbage, swamp morning glory, water convolvulus, etc. It is known as "kangkong" in Southeast Asia, "phak bung" in Thai, "ong choy" in Cantonese, "rau muong" in Vietnamese, "kalmi shak" in Bengali and "hayoyo" in Ghana. It is a popular cultivated green vegetable in China, India, Malaysia, Africa, Brazil, the West Indies, and Central America. It not only grows wild but is also cultivated throughout Southeast Asia, and is one of the widely consumed vegetable in this region.

There are two common types of water spinach: one that grows on land and the other that grows in water. The two types bear different flowers and leaves. Land-grown water spinach has long, narrow leaves with pointed ends and bears white flowers. The succulent foliage and stem tips are light green in colour.

The water or marshy land grown type is a tender, trailing or floating perennial aquatic plant. Its stems are 2–3 metres or more long, rooting at the nodes, and they are hollow and can float. The leaves vary from typically sagittate (arrow head shaped) to lanceolate, 5–15 cm long and 2–8 cm broad. The flowers are trumpet-shaped, 3–5 cm in diameter and usually white in colour with a mauve centre, ovary superior, glabrous, 2-celled (sometimes 4) with 4 ovules; style simple; stigma capitate and simple with 2 or 3 rounded lobes; fruit a spherical or ovoid capsule 6 to 10 mm long, thin walled, surrounded by persistent sepals, dehiscing by 4 (rarely 6) valves, or irregularly; seeds 4 or less, greyish, densely pubescent or at times glabrous.

Use

This semi-aquatic tropical plant is grown as a vegetable for its tender shoots and leaves. Practically all parts of the young plant tissue of water spinach are edible although the shoot tips and younger leaves are preferred. Stir-fried water spinach is a popular vegetable dish in India and other countries of Southeast Asia. It is also cooked with other vegetables. It is also boiled, steamed, or pickled throughout Asia. In Singapore, Indonesia and Malaysia, the tender shoots along with the leaves are usually stir-fried with chili pepper, garlic, ginger, dried shrimp paste and other spices. Shoot tips and younger leaves can also be eaten raw but it makes a better salad if first cooked briefly. Coarse stems and leaves are often used for animal feeding.

Nutritive and medicinal value

Water spinach represents one of the richest sources of carotenoids and chlorophylls. The leaves contain adequate quantities of most of the essential amino acids in accordance with the WHO recommendation pattern for an ideal dietary protein. Consequently, when compared with conventional food crops such as soybeans or whole egg, it has potential for utilization as a food supplement. Nutritional value per 100 g fresh edible vine as per USDA data base is given below:

Energy: 79 kcal	Calcium: 77 mg
Vitamin A equiv: 315 μg	Iron: 1.67 mg
Thiamine (B_1): 0.03 mg	Magnesium: 71 mg
Riboflavin (B_2): 0.10 mg	Manganese: 0.16 mg
Niacin (B_3): 0.9 mg	Phosphorus: 39 mg
Vitamin B_6 0.096 mg	Potassium: 312 mg
Folate (B_9): 57 μg	Zinc: 0.18 mg
Vitamin C: 55 mg	

Extract of water spinach exhibits high antioxidant properties with hydrophilic-oxygen radical absorbance capacity (H-ORAC) and 2,2-diphenyl-1-picrylhydrazyl (DPPH) scavenging activity being 341.92 ± 1.32 and 37.67 ± 2.63 imol Trolox equivalent / gram of dry weight (TE/g dry weight), respectively. The total polyphenols content has been estimated to be 12.56 ± 0.08 mg gallic acid equivalent/gram of dry weight (mg GAE/g dry weight). According to various sources, water spinach has been used extensively as a medicinal plant in India: as a mild laxative, in the treatment of ringworm and as a poultice in febrile delirium treatment.

Indian traditional medicine system, Ayurveda has identified many medicinal properties of water spinach and it is effectively used against nosebleeds and high blood pressure. Most of the studies have focused on the inhibition of prostaglandin synthesis, effects on liver diseases, constipation and hypoglycaemic effects.

In Indonesia, people believe it has a calming effect and can be used as a sleeping tablet. The root of water spinach is believed to cure haemorrhoids. Because of its high iron content, doctors recommend it to patients suffering from anaemia.

However, if the leafy greens are harvested from contaminated areas, and eaten raw, water spinach which is hosts of snails (*Segmentina* spp.) may transmit *Fasciolopsis buski*, an intestinal fluke parasite of humans and pigs, causing fasciolopsiasis.

Origin and taxonomy

Water spinach (*Ipomoea aquatica*) is diploid (2n = 30) and belongs to Convolvulaceae family. It is native to China and has become naturalized in many parts of the world including other parts of Asia, Africa, South America, Central America and the West Indies. Europeans discovered this plant species in a variety of places in the Old World when they first arrived there.

Two main cultivar groups can be distinguished *viz*., aquatic type (*Ipomoea aquatica* var. *aquatica*) and upland type (*Ipomoea aquatica* var. *reptans*).

Propagation

The aquatic type is generally propagated by vine cuttings and upland type by both seed and vine cuttings. The aquatic type in particular is fast growing species with a capacity to increase in length approximately 10 cm per day under good growing conditions. It can reproduce both sexually by seeds and asexually by rooting at the nodes. To obtain seeds, harvesting of the plants is stopped to allow developing flowers to mature, from which seed bearing pods form. Flowers come in the vines in warm months and produce 175-245 seeds per plant during peak season.

Nodes of the existing stems readily root to establish new plants producing new plants when segmented. Once established, cutting may be made at any time.

Cultural requirements

Water spinach is usually confined to tropical and subtropical areas because it is intolerant of cold temperatures. It grows satisfactorily in mean temperature range of 24 and 30 °C with at least 760 mm of rainfall a year and does not grow well below 20°C. It grows well in moist soil or in still to flowing waters. These plants relish heat, humidity, water and nutrients. Optimum growth occurs in full sunlight condition. Marshy lands and waterlogged soils are ideal for growth of this crop. Low temperature, shade and salinity are limiting factors for growth of water spinach. It grows poorly in cold weather but can tolerate light frost that affects only the outer leaves.

Crop production

The upland type can be grown in beds provided there is plenty of moisture. Fresh, mature seeds display primary dormancy within 15 days after harvest. Natural germination occurs following an after ripening period and scarification of the seed coat. It can be grown in a range of soil types from sandy soils to heavy textured clay loams, but friable well-drained soils high in organic matter are preferred. The ideal pH range is 5.5-7.0. However, it is not tolerant of brackish or salt water.

For the upland types, moist raised beds are usually prepared. Seedling or stem cuttings are transplanted into these beds. Seeds can also be directly sown at relatively close spacing 10 to 15 cm in order to maximize yield. The crop is responsive to supplemental fertilization and application of NPK fertilizer @ 20:30:20 kg /ha along with organic manure is the best. Under dry land conditions, it grows as an erect herb. Without standing water, the plant roots at every node and becomes woody and inedible.

Roots are produced at stem nodes that come in contact with water or moist soil. New plants can root within a week. Once roots are established, the plant grows as a trailing vine. Along waterways, the stems spread out over the water surface, forming a dense, tangled network that can obstruct water flow and access to it. Stems that have grown out over water have round, hollow stems and petiolate, basally lobed leaves.

The up land type water spinach is ready for harvesting 30 days after planting and is harvested several times as re-growth of shoots readily occurs. Growth is usually rapid with harvest beginning at 4 to 5 weeks. The yield is 10 to 15 tonnes/ha.

Under good conditions, aquatic type water spinach can produce 19 to 20 tonnes/ha in 9 months duration.

Leafy greens of *Brassica* species

A variety of *Brassica* species are widely consumed as leafy greens. The predominant species is *Brassica rapa* L. (syn. *Brassica campestris* L.) with several botanical varieties. Chinese cabbage comes under *Brassica rapa* var. *pekinensis*, which includes subtypes that form heads, known as "Won bok" and "Napa cabbage" as well as loose headed types. "Mibuna" and "Mizuna" greens are *B. rapa* var. *nipposinica* which form fairly large clumps with many stems bearing narrow leaves. "Pak choy" (*Brassica rapa* var. *chinensis*) includes types which are called "Spoon cabbage", "Choy sam" (Singapore), "Pechay" (Phillipines), and "Taisai" and "Shirona" (Japan), as well as others. The preferred type of "Pak choy" has dark green leaves and long, white, somewhat wide petioles. Some other varieties have shorter, slightly green petioles. Some varieties are grown for the flowering stems rather than the leaves.

The flowering Brassicas include "Choy sum" (*B. rapa* var. *parachinensis*), "Purple flowering pak choy" (*B. rapa* var. *purpurea*), and Chinese broccoli or Kailaan or Chinese kale (*B. oleracea* var. *alboglabra*). Chinese kale is among the 10 most important market garden vegetables in some Southeast Asian countries, such as Thailand and China. Young flowering shoots and small leaves are consumed as green either raw as salad or cooked vegetable.

Leaf mustards (*B. juncea*) are consumed in great quantities in China and other Asian countries.It is also called brown, mustard and Indian mustard. Its forms are variable, with leaves that are smooth or hairy, entire or divided, and petioles that are either narrow or wide. Compared to "Pak choy", leaves of "Leaf mustards" are lighter green while its petioles are green and shorter.

Use

In India, young tender leaves of *Brassica* greens are consumed as cooked leafy vegetable. The *Brassica* greens are often cooked with ham or salt pork, and may be used in soups and stews. Boiling these vegetables may reduce the content of the beneficial phytonutrient compounds but steaming, microwaving, and stir frying do not appear to do so.

It can also be used in salads or mixed with other salad greens. Older leaves with stems may be eaten fresh, canned or frozen, for potherbs, and to a limited extent in salads. In the flowering Brassicas like Chinese kale, young flowering shoots and small leaves are consumed as green either raw as salad or cooked vegetable.

Nutritive and medicinal value

All leafy greens of *Brassica* contain a variety of other essential vitamins and minerals particularly Vitamin A and C, and iron and a good source of fibre. Per 100 g fresh leaf of Indian mustard is reported to contain 24 calories, 2.4 g protein, 0.4 g fat, 4.3 g total carbohydrate, 1.0 g fibre, 160 mg Ca, 48 mg P, 2.7 mg Fe, 24 mg Na, 297 mg K, 1825 μg β carotene equivalent, 0.06 mg thiamine, 0.14 mg riboflavin, 0.8 mg niacin, and 73 mg ascorbic acid. All leafy greens of *Brassica* are also low in calories, fat, and sodium.

Brassica vegetables are known to for sulfur-containing phytochemicals viz., glucosinolates, 3,3'-diindolylmethane, sulforaphane, etc with potent anticancer properties. *Brassica* vegetables are rich in indole-3-carbinol, a chemical which boosts DNA repair in cells *in vitro* and appears to block the growth of cancer cells *in vitro*. The phytochemical, 3,3'-diindolylmethane is a potent modulator of the innate immune response system with potent antiviral, antibacterial and anticancer activity. Indian mustard is a folk remedy for arthritis, foot ache, lumbago, and rheumatism. In the traditional medicinal systems of India, the plant has been reported to have anti-diabetic potential.

These *Brassica* vegetables also contain goitrogens, some of which suppress thyroid function. Goitrogens can induce hypothyroidism and goiter in the absence of normal iodine intake.

Origin and taxonomy

Brassica rapa L. (synonymous with *B. campestris* L., $2n = 20$) has been originated from the highlands near the Mediterranean sea rather than from the Mediterranean coastal areas. Headed Chinese cabbage (*B. rapa* var. *pekinensis*) has its centre of diversity in northern China and has some relationship to the oilseed type that is grown there. *B. rapa* var. *chinensis* (Pak choy) is a leaf vegetable which differentiated from oilseed rape types of middle China. *B. rapa* var. *parachinensis* (Choy sum) is a derivative of the var. *chinensis*, and var. *campestris* is the most primitive leaf vegetable.

Primary centre of origin of *Brassica juncea* ($2n = 36$) is Central Asia (north-west India), with secondary centres in Central and Western China, Eastern India, Burma, and through Iran to Near East.

Brassica napus ($2n = 38$) is an amphidiploid species derived from interspecific crosses between *B. oleracea* ($n = 9$) and *B. rapa* ($n = 10$).Wild forms of *B. napus* have been reported to occur on the beaches of Gothland, Sweden, the Netherlands and Britain.

Propagation

All the leafy greens of *Brassica* are seed propagated vegetable crops. All these oriental greens are slower to bolt (flower) when temperatures are cool. They generally flower under long day when temperature starts rising up.

Cultural requirements

All leafy greens of *Brassica* species are cool season crop which are ideally suited to the regions with mild summer and moderately cool spring and autumn.

With suitable climatie conditions as prevail in the hills of India and careful scheduling, it is even possible to harvest three crops of *Brassica* greens in one growing season.

Crop production

Oriental leafy green plants grow best in well drained, moderately acidic to neutral soil (pH 6.0 to 6.8 is optimum) with a good level of soil organic matter. In the plains, seeds are sown in very early spring for spring crop and in the autumn for winter crop. Seeds are sown 30–45 cm apart and plants are then thinned to about 15 cm as they become crowded in the row. Control of weeds at the early stages by 1-2 inter-cultivations is essential. Seed rate of 4–6 kg is sufficient for one hectare. All these *Brassica* greens require fertile sandy loamy soil for satisfactory crop. They are well responsive to fertilizer and application of 50–75 kg N, 100 kg phosphate and 50 kg potash per hectare is generally recommended. A steady supply of water is important for good plant growth and quality. Insufficient moisture may result in tip burn, slow growth, and less tender leaves.

Oriental *Brassica* leafy vegetable plants are harvested as they near maturity, about 45–60 days after sowing depending on variety and weather conditions. On an average 12 tonnes of leaf is produced per hectare. Greens are cooled to near 0°C immediately after cutting and kept at or near that temperature during transportation and marketing. Humidity is kept at 90–95% by use of ice over the load or in the packages.

Colocasia greens

Plants under the genus *Colocasia* is thought to be one of the oldest cultivated plants in the world, being cultivated in Asia for more than 6000 years. *Colocasia* includes six species of herbaceous perennials from tropical Asia with a corm on or just below the ground surface. They are native to swamps and other moist areas and can be used in large aquatic containers in the garden or in the ground in moist soil. The arrow-shaped (sagittate shape), sometimes rounded leaves are large and mostly green, sometimes with prominent veins. The leaves may be large to very large, 20–150 cm long and the "elephant's ear plant" gets its name from the leaves, which are shaped like a large ear or shield. The petioles are thick and succulent and often purplish. The leaf attachment is referred to as "peltate", which means that the petiole attaches near the centre of the leaf. The cultivated plants rarely bloom.

Depending upon cultivars, the runners, leaves or petioles are used as vegetable. The leaves and petioles of *Colocasia esculenta* var. *esculenta*, *Colocasia esculenta* var. *antiquorum* is eaten as vegetable to increase the intake of greens as a source of micronutrients. However they may cause stomach upset if eaten raw and the sap may irritate skin. In the beginning, the primary concern about the raw plants was microscopic needle like raphides of calcium oxalate monohydrate, a crystalline chemical that may annoy the mucus membranes. However, it's now thought that the calcium oxalate simply behaves as a carrier for various other harmful toxins that enter in the body from the injury to the mucus membranes brought on by the toxic irritant.

Use

Leaves and petioles of *Colocasia* species are used as cooked vegetable in different parts of the India. The leaves are also used to make pickles. "Badi", a very common food item is made by mixing the stems and leaves of the *Colocasia* plant with lentil and drying it. In Manipur, the leaves are used to make one of the types of a Manipuri ethnic cuisine, locally known as "Utti". In different states of India viz., Bihar, West Bengal, Karnataka, etc. leaves and petioles are used to make curry. In Maharastra and Gujarat the leaves as used to make a fried snack which is prepared by applying besan (gram flour) on the leaves and then frying them. In Kerala, the leaves are used to make "chembila curry". In Maharashtra, the leaves are used to make a sweet sour curry with peanuts and cashew nuts. In Nagaland, the leaves are dried, powdered, kneaded into dough and baked into biscuits that are burnt and then dissolved in boiling water before being added into meat dishes to create thick, flavourful dry gravy.

The crop produces runner (stolon) which is preferred as a vegetable and is very popular in Assam, Bihar, West Bengal and adjoining Bangladesh.

Nutritive and medicinal value

Colocasia leaves and petioles eaten as a vegetable are excellent sources of provitamin A carotenoids, calcium, fibre, and vitamins C and B_2 (riboflavin), and they also contain vitamin B_1 (thiamine).

Nutritional value per 100 g fresh greens of *Colocasia* as per Nutrition data is given below:

Energy: 11.8 kcal	Calcium: 30 mg
Vitamin A equiv: 1351 IU	Iron: 0.6 mg
Thiamine (B_1): 0.1 mg	Magnesium: 12.6 mg
Riboflavin (B_2): 0.1 mg	Manganese: 0.2 mg
Niacin (B_3): 0.4 mg	Phosphorus: 16.8 mg
Vitamin B_6 0.0 mg	Potassium: 181 mg
Folate (B_9): 35.3 μg	Zinc: 0.1 mg
Vitamin C: 14.6 mg	

It has been found that the nutrient content of the prepared recipes of the greens of *Colocasia esculenta* var. *antiquorum* especially with reference to dietary fibre, β carotene, calcium and iron contents are higher than the spinach. No significant difference was observed in the sensory attributes (appearance, taste, flavour and overall acceptability) of *Colocasia* recipes with that of the spinach.

Colocasia greens are good source of dietary fibre and potassium which help to maintain the blood pressure. The leaves of *Colocasia esculenta* show anti-inflammatory activity.

Colocasia greens also contain different anti-nutrient factors. Raw leaves of *C. esculenta* contain very high concentration of oxalate (13.23 mg/g), moderate concentrations of saponins (9.94 mg/g), tannins (7.38 mg/g), alkaloids (6.62 mg/g), lectins (4.63 mg/g) and phytates (3.41 mg/g), low concentration of trypsin inhibitors (2.04 mg/g) and very low concentration of cyanogenic glycosides (0.97 mg/g). Hence, *Colocasia* greens should be consumed after adequate boiling to help decrease the quantity of oxalates.

Origin and taxonomy

Colocasia is thought to have originated in the Indo-Malayan region, perhaps in eastern India and Bangladesh, and spread eastward into Southeast Asia, eastern Asia, and the Pacific islands; westward to Egypt and the eastern Mediterranean; and then southward and westward from there into East Africa and West Africa, whence it spread to the Caribbean and Americas.

The edible aroids belong to the genus *Colocasia*, within the sub-family Colocasioideae of the monocotyledonous family Araceae. Because of a long history of vegetative propagation, there is considerable confusion in the taxonomy of the genus *Colocasia*.

Cultivated taro is classified as *Colocasia esculenta* however, the species is considered to be polymorphic and there are at least two botanical varieties: *Colocasia esculenta* (L.) Schott var. *esculenta* and *Colocasia esculenta* (L.) Schott var. *antiquorum* (Schott) Hubbard & Rehder.

Chromosome numbers reported for *Colocasia* include 2n = 22, 26, 28, 38, and 42. The disparity in numbers may be due to the fact that the chromosomes are liable to unpredictable behaviour during cell divisions. The most commonly reported results are 2n = 28 or 42.

Propagation

Colocasia species is an herbaceous plant which grows to a height of 1-2m. The plant consists of a central corm (lying just below the soil surface) from which leaves grow upwards, roots grown downwards, while cormels, daughter corms and runners (stolons) grow laterally. The root system is fibrous and lies mainly in the top one meter of soil. The plant is generally propagated vegetatively by corms, cormels, daughter corm and runners.

Cultural requirements

Colocasia requires an average daily temperature above 21ºC for normal production. It cannot tolerate frosty conditions. Partly because of its temperature sensitivity, taro is essentially a lowland crop. At the same time, partly because of their large transpiring surfaces, the plants have a high requirement for moisture for their production. Normally, rainfall or irrigation of 1,500-2,000 mm is required for optimum yields. Taro thrives best under very wet or flooded conditions. Dry conditions result in reduced yields of both top and corm. Yields at high altitudes tend to be poor.

Crop production

Although *Colocasia* is almost evergreen perennial in tropical climates, it is grown as an annual. Slightly acidic, moist soil, rich in organic matter is best suited for this crop. It does best in partial shade, but tolerates full sun if it gets plenty of water. It grows reasonably well in average soil provided it is moisture retentive. The plants should not be left to go dry for too long which causes wilting of leaves.

The side-growing cormels weighing about 20-25 g are the good planting material. About 800 kg/ha of cormel is required. Cormels are generally planted in April-May in irrigated condition and in June-July under rain-fed condition at the spacing of 30×100 cm. Application of 25 tonnes of FYM, 20 kg N, 30 kg P and 60 kg K/ha as basal and 20 kg N, 30 kg P and 60 kg K/ha on 45 days after planting has been found satisfactory.

Colocasia can be grown in paddy fields where water is abundant or in upland situations where water is supplied by rainfall or supplemental irrigation. *Colocasia* is one of the few crops (along with rice and lotus) that can be grown under flooded conditions. This is due to air spaces in the petiole, which permit underwater gaseous exchange with the atmosphere. It does best in soil of pH 5.5-6.5. It is able to form beneficial associations with vesicular-arbuscular mycorrhizae, which therefore, facilitate nutrient absorption. Some cultivars can tolerate salinity also.

Soft rot, bacterial blight, corm and root rots, and dasheen mosaic virus are common diseases of *Colocasia*, while aphids, whiteflies, and spider mites also occur.

The runners are periodically harvested through long growing span but the plants are uprooted at the early stages generally 3-4 months after planting to harvest the leaves and petioles.

Bathua

Bathua or pigweed or lamb's quarters (*Chenopodium album* L) though cultivated in some regions as a nutritious leafy vegetable, the plant is elsewhere considered as a weed. It is cultivated and consumed in many parts of India as a food crop. It is equally widely distributed in both the northern and southern hemispheres, occurring in Asia, North America and Europe.

The plant is an erect, branched (occasionally unbranched) annual herb, green, more or less coated with white mealy pubescence. It tends to grow upright at first, reaching heights of 10–150 cm (rarely to 3 m), but typically becomes recumbent after flowering (due to the weight of the foliage and seeds) unless supported by other plants. Stems are yellowish to green, green-striated and sometimes reddish or green with red spots at leaf axils. Lower and medium leaves petiolate, blade usually 2-6 cm, variously trullate, rhombic-ovate to lanceolate, clearly longer than broad, base narrowly to broadly cuneate, margins irregularly serrate to entire, often somewhat 3-lobed, teeth mostly acute, often unequal in size; uppermost leaves lanceolate, usually entire. The leaves are dull green with a pale pink centre with a waxy coating and wavy margin. Inflorescence is a variable spiciform or cymosely branched panicle, mostly terminal, 10–40 cm long. Flowers are perfect, small, sessile, green; calyx of 5 sepals that are more or less keeled and nearly covering the mature fruit; petals 1; stamens 5, pistil 1, with 2 or 3 styles, ovary single-celled, attached at right angles to the flower axis. Fruit is an achene (seed covered by the thin papery pericarp). Seed nearly circular in outline, oval in cross section, side convex, glossy, black, mean size 1.5 mm x 1.4 mm in diameter, weight 1.2 mg. The seeds exhibit considerable polymorphy; some are smooth, some striate, and others possess a raised reticulum. Testa colour also varies significantly and can be black and shiny, brown or brownish-green. All of these variations in colour and form may be found in the seeds of a single plant.

Use

The leaves and tender, young shoots are consumed as leafy vegetable and cooked like any other leafy green. Leaves are also mixed with curd and wheat flour to prepare some tasty preparations like, bathua raita and bathua paratha. Dried leaves are ground to make powder that can be added to regular flour. Bathua seeds are edible which are very nutritious and contain whole set of essential amino acids.

Nutritive and medicinal value

Bathua leaves are very good source of high quality protein, antioxidants and vitamins, particularly vitamin C and vitamin A. The leaves are good source of potassium, iron, calcium, and zinc. 100 g fresh leaves contain 2.9g carbohydrate, 3.7 g protein, 0.4 g fat, 0.8 g fibre, 1.74 mg β carotene, 37 mg vitamin C, 0.3mg riboflavin, 0.1mg thiamine, 265mg sodium, 288 mg potassium, 258 mg calcium, 80 mg phosphorus, 23 mg magnesium and 0.3mg zinc.

In Ayurveda, it is very useful in treating kidney stone and also reduces formation of stone. It is also used for inner and external swellings, jaundice, irregular period, curing infections after delivery, anemia and for blood purification. Bathua is good for heart, liver, spleen and gall bladder. It also improves haemoglobin level in blood.

Origin and taxonomy

Its native range is obscure due to extensive cultivation, but includes most of Europe, from where Linnaeus described the species in 1753. It is equally widely distributed in both the northern and southern hemispheres, occurring in Asia, North America, Europe, India, South Africa, Australia and South America. It is present throughout North America. In tropical regions it is mostly found at higher altitudes. It is domesticated in the Himalayan region where it is grown as a grain crop and it is cultivated as a traditional leafy vegetable in India. There is archaeological evidence to suggest it was cultivated as a pseudocereal in Europe in prehistory.

Chenopodium album L is somewhat poorly circumscribed taxonomically, and genotypes cultivated in the Himalayas that are assigned to *C. album* bear little similarity to the weedy form of *C. album*. However, *C. album* and related hexaploid species have a single flavonoid profile that supports the recognition of a single species. Analysis of DNA content suggests that this species is an alloploid derivative of a cross between unknown diploid and tetraploid species. *Chenopodium album* L belongs to the family Chenopodiaceae and having chromosome number is 2n = 54 (hexaploid).

Propagation

The crop is propagated through seeds which are slightly smaller than mustard seeds. Average seed production varies between 3000 and 20,000 seeds/plant. Seeds remain viable for extended periods in the soil.

Cultural requirements

The plant is tolerant of a wide range of cultural conditions, climates, soil types, fertility and pH. It occurs from sea level to altitudes of 3600 m, and from latitudes 70°N to more than 50°S. It can grow in wide temperature range of 5–30°C. It is frost tolerant. It seems to grow most vigorously in temperate and subtemperate regions however, it is also a potentially serious weed in almost all winter-sown crops of the tropics and subtropics.

Crop production

Two improved varieties developed and released in India are Pusa Bathua No.1 and Ootty 1. Seeds can be directly sown in field or seedling can be raised in nursery and transplanted at spacing of 30 x 15cm. Brown seeds germinate rapidly but black seeds persist in soil for longer time. Immature seeds are not capable of germination. It grows best in fertile sandy loam soils. Crop growth increases with the increase in Mg content in soil. Pants grow vigorously at wider spacing.

After about 35-40 days, either seedling can be uprooted or clipped near ground level and used as leafy green vegetable. The plants generally grow to a height of about 40 cm and yields about 30.0 t/ha in in 50 - 60 days in plains.

The plant is vulnerable to leaf miners, making it a useful trap crop as a companion plant. Growing near other plants, it attracts leaf miners which might otherwise have attacked the crop to be protected. It is a host plant for the beet leafhopper, an insect which transmits curly top virus to beet crops.

Portulaca

Portulaca or common purslane (*Portulaca oleracea* L.), an ancient, cosmopolitan species is an annual succulent herb.

Purslane is mostly an annual, but it may be perennial in the tropics. Stems are glabrous, fleshy, purplish-red to green, arising from a taproot, often prostrate, forming mats. Purslane has a taproot with fibrous secondary roots and is able to tolerate poor, compacted soils and drought.The leaves (also fleshy) are alternate, sub-alternate or opposite, obovate to spatulate with an obtuse or truncate-emarginate apex. The leaves may range from 40 mm x 15 mm up to 60 mm x 25 mm in fertile soils. Apical whorls have 2-5 leaves, usually 4. Axillary hairs are missing, inconspicuous or barely visible. The sour taste is due to oxalic and malic acid, Also present are two types of betalain alkaloid pigments, the reddish betacyanins (visible in the coloration of the stems) and the yellow betaxanthins (noticeable in the flowers and in the slight yellowish cast of the leaves). The yellow flowers have five regular parts and are up to 6 mm wide.

There are 5 yellow petals ranging from 3 to 10 mm long by 2 to 8 mm wide with 6-15 stamens. The style branches are 3-6. Depending upon rainfall, the flowers appear at any time during the year. The flowers open singly at the center of the leaf cluster for only a few hours on sunny mornings. Flowers donot open on cloudy days or days when the temperature is below 21°C. The flowers are self-fertile and do not exhibit apomixis. Pollens are very sticky, a characteristic that is not present in windborne pollen. Flowers are in a group at the end of the stem. The 2 sepals are fused at the base of the ovary and may form a wing-like carina 3-4 mm long that can cover the fruit. Seeds are formed in a tiny pod, which opens when the seeds are mature. Tthe capsule ranges from 4 to 9 mm, opening at or just below the middle.

Capsule production and overall plant growth increase with day length. Seeds are black when mature, but may be red or brown when immature. The seeds are 0.6-1 mm long, usually with granulate to flat-stellate surfaces. However, other patterns, with raised stellate and tuberculate surfaces can occur. The species is self-compatible.

C4 metabolism allows *P. oleracea* to optimize photosynthesis in conditions of high heat and bright sunlight while enduring periods of limited water availability.

Use

Leaves and young shoots are used as cooked green leafy vegetable. It has a slightly sour and salty taste and is eaten throughout much of Europe, the Middle East, Asia and Mexico. It may be used fresh as a salad, stir-fried or cooked as spinach is. Because of its mucilaginous quality it also is suitable for soups and stews. In Pakistan, it is cooked as in stews along with lentils, similarly to spinach, or in a mixed green stew.

It is highly tolerant of both chloride- and sulphate-dominated salinities hence, appears to be an excellent candidate for inclusion in saline drainage water reuse systems. It is also a source of a gum with emulsification properties that can be used in the food industry.

Nutritive and medicinal value

It is a rich source of potassium (494mg/100g) followed by magnesium (68mg/100g) and calcium (65mg/100g) and possesses the potential to be used as vegetable source of omega-3 fatty acid (alpha-linolenic acid in particular). It is very good source of alpha-linolenic acid (ALA) and gamma-linolenic acid (4mg/g fresh weight). It contains the highest amount of alpha-tocopherol, Vitamin E (22.2mg per 100g of fresh) and fairly high ascorbic acid (26.6mg per 100g of fresh) content. It also contains other vitamins (mainly vitamin A, and B,) and iron. Both reddish betacyanins and yellow betaxanthins pigment types are potent antioxidants.

Origin and taxonomy

It has an extensive distribution, assumed to be mostly anthropogenic, throughout the Old World extending from North Africa and Southern Europe through the Middle East and the Indian Subcontinent to Australasia. In historic contexts, seeds have been retrieved from a protogeometric layer in Kastanas, as well as from the Samian Heraion dating to seventh century BC. It was highly esteemed in ancient Egypt and cultivated in Europe as far back as the middle ages.

The origin has not been universally agreed upon. Different reports list it as native to South America, North Africa and western Asia and Europe, India, the Sahara, and even Australia. It belongs to the family Portulacaceae. Cytologically, *Portulaca oleracea* is characterized bý a sequence of polyploids, from a base number of x=9. There are diploid races (2n=18) in Africa, Central and North America; tetraploid races (2n=36) in India, Central and North America and hexaploid races (2n=54) in India, Africa, Europe, North America and Hawaii. There are additional reports of 2n=45 in India and 2n=52 in Japan.

Propagation

This crop is propagated by seeds. Purslane reproduces primarily from seed. Over 6000 seeds can be produced after the first flush of flowers (5-6 weeks of growth). One plant can produce between 100,000 and 242,000 seeds over an entire season. Generally, purslane grows rapidly, producing flowers, fruits and seeds within 6 weeks of germination. The plant can also be propagated vegetatively through development of adventitious roots from the base of cut shoots.

Cultural requirements

It grows from sea level to 2600 m and is most common in the temperate and subtropical regions, although it extends into the tropics and higher latitudes. Common latitudes are between 45°N and 40°S, with extension to 58°N in North America and 54°N in Europe. It has a wide tolerance of photoperiod, light intensity, temperature, moisture and soil type.

Crop production

The crop can be grown in any type of soil. The seeds are sown directly in the field from April to September in hills and March to June in plains. Light is required for germination, but the temperature requirement is variable. Seeds can germinate at wide temperature range of 10-40°C. Regular irrigation is required for raising good crop. It matures quickly and the plant becomes ready for first harvest in about a month after sowing. The leaves do not remain fresh for any length of time. So, for continuous supply of leaves seeds are sown at 15 days interval.

Indian sorrel

Indian sorrel, *Oxalis corniculata* under the family Oxalidaceae (2n = 48) is a subtropical plant. It is also known as creeping wood sorrel or the sour docks. It is a very popular perennial herb, delicate appearing, low growing and herbaceous plant. The plant is a procumbent herb, stems rooting and having pubescent with apprised hair. Leaves are tri-foliate with three heart shaped leaflet, a beautiful bright green above, but of a purplish hue on their under surface. The long slender leafstalks are often reddish towards the base. The leaflets are usually folded somewhat along their middle, and are of a peculiarly sensitive nature. Only in shade are they fully extended: if the direct rays of the sun fall on them they sink at once upon the stem, forming a kind of three sided pyramid, their under surfaces thus shielding one another and preventing too much evaporation from their pores. At night and in bad weather, the leaflets fold in half along the midrib, and the three are placed nearly side by side to 'sleep,' a security against storm and excessive dews. The 5-petaled flowers appear on tall stems above the foliage and may be white, pink or red, depending on the species. They will grow to 6"-12" in height. It is distributed throughout the warmer parts of India.

Use

Neither the flowers nor any part of the plant has any odour, but the leaves have a pleasantly acid taste due to the presence of considerable quantities of binoxalate of potash. This, combined with their delicacy, has caused them to be eaten as a spring salad from time immemorial. Sorrel is the best, not only in virtue, but also in pleasantness of taste. It has been found that incorporation of Indian sorrel leaves in the prepared products increases the nutrient density or nutritional qualities.

Nutritive and medicinal value

It is particularly rich in iron (365 ppm) and fibre (8.7 %) contents. It is well known for its medicinal value and as a good appetizer. In the traditional system of medicine, the whole plant is employed for treating anaemia, wounds, piles and skin eruptions. It is also used in the treatment of influenza, fever, urinary tract infection, diarrhoea and poisonous snake bites. The plant is a good source of vitamin C and is used as an anti- scorbutic in the treatment of scurvy.

As the characteristic of this genus, Indian sorrel contains oxalic acid giving the leaves and flowers a sour taste which can make them refreshing to chew. However, in very large amounts, oxalic acid may be considered slightly toxic, interfering with proper digestion and kidney function.

Origin and taxonomy

Place of origin of *Oxalis corniculata* (2n = 48) is unknown, but it is considered an Old World plant. India is considered as its place of origin.

Propagation

The plants are herbaceous and grow from bulbs or tubers in little clumps.

Cultural requirements

Indian sorrel prefer cool to average (14-18°C) night time temperatures but not warmer than 25°C during the day time. It also prefer bright indirect light however, they can tolerate some direct sun. This plant does not like swampy or soggy soil.

Crop production

Indian sorrels are generally grown from bulbs. In hot summer regions they will need protection from the afternoon sun. They should only be planted in very well drained soil. It grow best when they are grown in bright light but not direct sun, relatively cool temperatures and plenty of fresh air. The soil should be kept evenly moist when the plants are actively growing, but it prefers to be kept on the dry side during winter while they are dormant.

It is not a heavy feeder. The plants can be feed every two weeks in the spring and summer with a balanced liquid fertilizer diluted in half. All spent flowers by reaching into the foliage should be carefully pinched off to encourage good vegetative growth. Leaves are harvested periodically.

Okra

Okra, *Abelmoschus esculentus* under the family Malvaceae (Chromosome No.: 2n = 126, 130, 132, 144) is an important vegetable in the tropical and sub-tropical parts of the world. Top 10 okra producing countries are India, Nigeria, Sudan, Iraq, Pakiatan, Cameroon, Ghana, Egypt, Benin and Malaysia. Other prominent okra producing countries are Turkey, Iran, Yugoslavia, Bangladesh, Afghanistan, Japan, Brazil, Ethiopia, Cyrpus, Thailand, Myanmar and the Southern United States. In India, it is commonly grown in Maharashtra, Gujarat, Karnataka, Tamil Nadu, Haryana, Punjab, Uttar Pradesh, Bihar, West Bengal and Odisha. Okra is known by many local names in different parts of the world viz., ladies finger in England, gumbo in the USA, bhindi in India, etc.

The crop

Okra is generally a tall, upright annual plant with thick, stiff upright stem. Plant is characterized by indeterminate growth. Stem is robust, variable in height and branching depending on the variety. Plant height varies widely between 50 and 400 cm depending on the genotype however, most of the varieties generally reach 180- 200 cm height.

Leaves are alternate and can appear very different at different stages of the plant growth. Very young plants hve rounded leaves with serrated edges, they look more like small saucers. As they mature, they become larger many lobed sometimes as few as three lobes other usually palmately five lobed. Leaves still have a surface large area and slightly lighter veins, often slightly yellow and serrated edges. As the plant become taller, the leaves become more lobed with much less leaf area. In all the stages, the leaves are covered with fine bristles or spines that make them coarse to touch. Flowering is continuous and produced individually at a leaf node near the top of the stem. A flower bud appears in the axil of each leaf above 6th to 8th leaf depending upon the cultivar. The crown of the stem at this time bears 3-4 underdeveloped flowers but later on during the period of profuse flowering of the plant there may be as many as 10 undeveloped flowers on a single crown. As the stem elongates, the lower most flower buds open into flowers. There may be a period of 2, 3 or more days between the time of development of each flower but never does more than one flower appear on a single stem. Flower is axillary and solitary, 3.75-8.75 cm diameter, with five yellow to pale yellow petals arranged in a cone, often with a red or purple spot at the base of each petal. The flower holds a large upright style with deep burgundy coloured stigma. Flowers open only once in the morning and close after pollination on the same day. On the following morning the corolla withers. Okra has perfect flowers (male and female reproductive parts in the same flower) and is predominantly self-pollinating. The fruit is a capsule and grows quickly after flowering. The greatest increase in fruit length and diameter occurs during 4th to 6th day after pollination. It is at this stage that fruit is most often plucked. The fruit colour may be green, light green and even purple containing numerous white seeds. As the fruits ripe, seed colour changes to brownish-black.

Importance and use

Okra is quite popular in India because of easy cultivation, dependable yield and adaptability to varying moisture conditions. Okra is mainly used for its tender green fruits as vegetable in many countries. In Iran, Egypt, Lebanon, Israel, Jordan, Iraq, Greece, Turkey and other parts of the eastern Mediterranean, okra is widely used in a thick stew made with vegetables and meat. In Indian cooking, it is often fried or added to gravy-based preparations. Tender green fruits are used as fried vegetable or in curry and sambar. It is also used in soups. Sun dried bhindi are mainly used in Africa and India for year round consumption. Frozen and sterilized fruits are also important market product mainly in USA. The roots and stems are useful for clearing sugarcane juice in preparation of jaggary. Ripe seeds are roasted, ground and used as a substitute for coffee in Turkey. Crude fibre from dry fruits and stems are used in the paper industry and making cardboard. Dry seeds contain 13-22% edible oil and refined oil may be used as a substitute of common edible oil particularly cotton seed oil. Oil from seeds is used in perfume industry. The seed cake is used as animal feed. In Far East countries like, Papua New Guinea, Fiji and Solomon Islands the wild species, *A manihot* is used as leafy vegetable.

In India, okra has tremendous export potential as fresh vegetable which accounts for 30% exchange earnings from export of vegetables other than onion. It is generally exported to our neighbouring countries and Middle-East countries like, Bahrin, Quatar, Kuwait, Saudi Arabia, Muscat, Tehran and Abu Dhabi and South-East Asia particularly Singapore, Mauritious, Malayasia, Sri Lanka and Bangladesh. The major export oriented okra producing areas in India are Nasik, Ozar, Saikheda, Dindori, Kolhar, Naraingaon and Sholapur in Maharashtra.

Nutritional value

It is the rich source of iodine, an essential micro-nutrient. It is a good source of minerals particularly calcium (66mg/100g fresh) and phosphorus (56mg/100g fresh). The fruits also contain good amount of sulfur and sodium. Tender fruits contain Vitamins A (88 IU), thiamine (0.07 mg), riboflavin (0.10 mg) and vitamin C (13 mg) per 100 g fresh. Tender fruits contain 1.9% protein,

6.4% carbohydrate and 1.2% fibre. The mucilage on hydrolysis give polysaccharides composed of galacturonic and glucuronic acids and minor contents of galactose, rhamnose, glucose and arabinose. Dry seeds contain 20-24% protein.

Medicinal value

Okra has long been favoured as a food for the health-conscious. It contains potassium, vitamin B, vitamin C, folic acid, and calcium. It is low in calories and has high dietary fibre content (1.2%) which includes a large spectrum of molecules *viz*., cellulose, non-starch polysaccharides and lignin which gives protection against certain types of cancer particularly colon cancer, regulates transit and lowers blood cholesterol. The mucilaginous properties of okra are useful in treating digestive disorder. Okra has been suggested to help manage blood sugar in cases of type 1, type 2, and gestational diabetes. It is said to be very useful against genito-urinary disorders, spermatorrhoea and chronic dysentery. Its use in regular diet is reported to increase sperm count in men. Its medicinal value has also been reported in curing ulcers and relief from haemorrhoids. Root juice can be used to treat cuts, boils and wounds.

Origin and Taxonomy

Okra has been an important crop from time unknown in Afghanistan, India, Iran, Pakistan, Turkey and Yugoslavia. Hence, cultivars of these countries have certain adaptation characteristics making them different from those of other countries.

Abelmoschus esculentus is found all around the world from Mediterranean to equatorial areas. Cultivated and wild species clearly show overlapping in Southeast Asia, which is considered as the centre of diversity. There are two hypotheses concerning the geographical origin of *A. esculentus*. Some authors argue that one putative ancestor (*A. tuberculatus*) is native to Uttar Pradesh in northern India, suggesting that the species originated from this geographic area. Belief of Indian origin of okra gets strong support from the mentioning of 'Tindisha' and 'Gandhamula' designating okra in ancient Sanskrit literature. Others, on the basis of ancient cultivation in East Africa and the presence of the other putative ancestor (*A. ficulneus*), suggest that the area of domestication is north Egypt or Ethiopia, but no definitive proof is available today. For *A. caillei*, only found in West Africa, it is difficult to suggest an origin outside.

Okra was introduced to America during the early post-Columbian period. Okra was earlier botanically named as *Hibiscus esculentus*. Since in okra, calyx, corolla and staminal column are fused together at the base and fall together after anthesis, it has been renamed as *Abelmoschus esculentus*, distinguishing it from *Hibiscus* in which calyx is persistent. There are 34 *Abelmoschus* species, including 30 species in the old world and 4 in the new world. The wild species particularly *Abelmoschus manihot* and *Abelmoschus tetraphyllus* have been extensively used in breeding okra resistant to yellow vein mosaic virus.

There are significant variations in the chromosome numbers and ploidy levels of different sepceis in the genus *Abelmoschus*. The lowest number reported is 2n=56 for *A. angulosus*, whereas the highest chromosome number reported are close to 200 for *A. manihot var. caillei*. Most of the cytological reports suggest that okra is a natural amphididiploid of *A. tuberculatus* with 2n = 58 and an unknown species with 2n = 72. Chromosome numbers of *Abelmoschus esculentus* as 2n=72, 108, 120, 132, 144 are the indication of a series of polyploids of x =12.

Adaptation in India

There is not much evidence available as to when and how the cultivated okra was introduced into India. Scientists believed it to have originated from India owing to mention of 'Tindisha' and 'Gandhamula' designating okra in ancient Sanskrit literature and availability of one putative ancestor (*A. tuberculatus*) in northern India. Significant native diversity of okra and its allied species in India points towords its secondary diversity here. However, much variability in okra cultivars is not found in India. Eight *Abelmoschus* species occur in India. Out of these, *A. esculentus* is the only known cultivated species. *A. moschatus* occur as wild species and is also cultivated for its aromatic seeds, while the rest six are truly wild types. The wild species occupy diverse habitats. The species *A. ficulneus* and *A. tuberculatus* is spread over the semi-arid areas in north and northwestern India; *A. crinitus* and *A. manihot* (*tetraphyllus* and *pungens* types) in tarai range and lower Himalayas; *A. manihot* (*tetraphyllus* types), *A. angulosus*, and *A. moschatus* in western and eastern ghats; and *A. crinitus* and *A. manihot* (mostly *pungens* types) in the northeastern region depicts their broad range of distribution in different phyto-geographical regions of the country.

Open pollinated improved varieties

Improved varieties: Pusa Makhmali, Pusa Sawani, Pusa A-4, Hissar Unnat, Azad Kranti, Arka Anamika, Arka Abhay, Punjab Padmini, CO-1, MDU-1, Punjab-7, Punjab-8, Parbhani Kranti, Gujarat Bhindi-1, Harbhajan Bhindi, Varsha Uphar, Selection 2-2, VRO-6, MDU-1, P-13, Azad Bhindi-1, Gujarat Okra-2, Utkal Gaurav, Phule Kirti, Kashi Saatdhari, Hisar Naveen, Susthira, Kashi Mohini, Kashi Mangali

Varieties resistant to YVMV : Arka Anamika, Arka Abhay, Pusa A-4, VRO-6, Azad Krishna, Kashi Mohini, Kashi Mangali

Varieties suitable for export: Pusa Sawani, Parbhani Kranti, Punjab-7, Varsha Uphar.

Cultural requirements

Okra is basically a hot weather crop and thrives well during hot and humid climatic condition. It is grown in tropical and subtropical climate in the plains and in summer months in the hill. It is susceptible to drought and low night temperature and frost. It can be successfully grown under the temperature ranging between 25-30°C. The plant grows shorter under high temperature and low humidity condition. Day temperature more than 42°C causes flower drop. Optimum temperature range for seed germination is 25-35°C and seeds do not germinate below 17°C. Adequate sun shine is very important for growth and yield of the crop. It can grow well in wide range of soil, sandy to clay soil if those are well manured with enough organic matter and good drainage facility. Loose, friable, well manured loam soils are the best for the crop. Water logging is harmful for the crop. Ideal pH range is 6.0 to 6.8 for most of the cultivars.

Land preparation

The field should be prepared with 2-3 ploughing upto a depth of 20-25 cm to make into a fine tilth before seed sowing. All the weeds and stubbles are removed and the land is levelled by laddering. The plant has well developed root system and is a good feeder. Soil should be made rich in organic matter content by adding 20-25 tonnes of FYM per hectare at the time of land preparation. Summer ploughing during hot summer helps to a great extent in controlling weed and soil borne micro-organisms. The field should be divided into several plots with irrigation channels depending on irrigation source and slope of the land.

Sowing time

Okra grows successfully in tropical and subtropical climate hence, can be grown year round in the southern states due to prevailing mild winter there. However, the crop is grown in two main seasons: spring-summer and rainy season. Sowing is done during late January – early February for Spring-summer crop in eastern region; February-March for spring-summer crop in north and western region; May-July for rainy season crop in most areas; mid August for rainy-autumn crop and November for winter crop in southern states.

Seed sowing

Okra gives little success on transplanting and thus seeds are sown directly in the field. Seeds are sown by hand dibbling or behind the plough or by seed drill. Broadcasting is not recommended due to high seed rate as well as inconvenience in cultural operations and harvesting. 18–22 kg seed/ha is sufficient for spring-summer crop and 8-10 kg/ha for the rainy season crop. The seeds should be soaked overnight in water for better germination. Soaking the seeds in 0.2% Baviatin over night protect the seedlings from wilt disease. Soaking the seeds with 400 ppm GA or 20 ppm NAA or 20 ppm IAA enhances germination. Seeds are sown on ridges in spring-summer season on the side facing sun and on flat beds in rainy season. It is better to sow the seeds in pre-irrigated moist soil than irrigate the field after sowing. Pre-sowing soil application of 20-22 kg Furadon 3G helps in protecting the plants from root knot nematodes and other pests during initial 4-5 weeks. On an average, 50-70 thousand plants can be accommodated in one hectare land.

Spacing

For spring-summer crop, spacing should be 45 cm x 30 cm for branching type and 30 cm x 15 cm for non-branching type; for rainy season crop spacing is 60 cm x 45 cm and 60 cm x 30 cm (50000 plants/ha) for branching type and 45 x 30 cm (66000 plants/ha) for non-branching type. For harvesting of small fruits for export, 3 row group of 20cm x 60 cm between groups is given.

Nutrient management

The plant needs to produce 5 tonnes of dry matter to produce 10 tonnes of green fruit equivalent to 1 tonnes of dry matter in fruits. A general fertilizer dose of 125 kg N, 75 kg P_2O_5 and 60kg K_2O per hectare is recommended for open pollinated improved varieties.One-third N along with other fertilizers should be given as basal and rest N should be top dressed in two split doses at 30 days after sowing and at flowering. For the ratoon crop the same nutrient schedule is followed and basal dressing should be done immediately after pruning followed by irrigation and plant protection measures. A general fertilizer dose of 200 kg N, 100 kg P_2O_5 and 100kg K_2O per hectare is recommended for the hybrids. Half N along with other fertilizers should be given as basal and rest N fertilizer should be top dressed in two split doses at 30 days interval after sowing. In spring-summer, application of N in splits after every 3-4 pickings increase crop duration and yield. Foliar spray of 0.3% zinc sulfate and 0.02% ammonium molybdate in the standing crop enhance fruit yield. Soil application of calcium is beneficial for the seed crop. Vermicompost @ 4-6t/ha can be applied as broadcasting during final land preparation.

Irrigation

Seed should be sown when the soil has sufficient moisture. Irrigation should be given at the initiation of first true leaf during spring-summer and at its expansion during rainy season. Subsequest irrigation needs to be applied at 5-6 day's intervals during summer and as and when required for the kharif crop depending on the rain. However, the crop must be irrigated during its most critical stages i.e. flowering and fruit setting stages and water stress in these stages considerably reduce yield. It is to be kept under consideration that over irrigation by flooding may cause wilting of the plant. Drip irrigation system increases the yield as well as saves 70-80% irrigation water.

Interculture and weed control

Proper weed management is urgently needed to check yield loss. At least two weedings are required till the crop canopy covers the soil surface. Application of herbicides reduces the number of weeding to zero in summer and one during rainy season. Application of Fluchloralin (Basalin 48 EC) @ 1.5 Kg ai/ha as pre-sowng soil incorporation and Pendimethalin (Stomp 30EC) @ 0.75 kg ai/ha as post-sowing and pre emergence soil surface spray is recommended. Surface application of herbicides is effective for 4-5 weeks. Okra generally does not require staking, training and pruning. Spraying the standing crop with 100-500 ppm ethrel reduce vegetative growth and weaken apical dominance. For ratoon crop from quick branching varieties, pruning the spring-summer crop 20-25 cm above the ground after summer fruiting with the onset of rains followed by manuring the crop and plant protection cover ensures good establishment of the crop.

Harvesting the crop

First harvesting is generally done 40-50 days after sowing depending on the variety. The tender fruits are harvested by bending the pedicel with a jerk or with the help of a knife at every alternate day. Use of cotton cloth gloves to protect fingers and harvesting in the morning when hairs are soft is the best for nearby market. For distant markets, the fruits should be harvested in the late evening and transported the produce in the cool condition at night.

Post harvest management and storage

The fruits for export purpose should be green, tender, 6-8 cm long, 4-5 ridged, slender and roundish with pointed tips. For export the pre-cooled fruits in airconditioned room are packed in suitable size perforated paper cartoons (usually 5.0 kg size) and transported in refrigerated vans. In air-tight containers, the fruit may turn pale during transit due to heat generated through respiration of the fruits. Okra fruit should not be washed before storing as it makes the fruits slimy. The fruits after harvesting are kept in plastic bag which should be directly placed in the refrigerated hold. In this method, fruits can be stored at 7-9^{0}C temperature and 80-90% relative humidity for 7-10 days. Fruits can be frozen as whole pods. It can also be dried for longer storage and use but it should not be bleached before drying.

Post harvest treatment of fruits with cycocel (100 ppm) enhances self-life. Post-harvest treatment of fruits with 250 ppm ascorbic acid retains chlorophyll of the fruit and keeps them in good condition for 8-9 days in ambient condition.

Plant Protection measures

Diseases

Yellow vein mosaic disease (Yellow vein mosaic virus)

This severe single-stranded DNA virus is transmitted by the insect vector white fly (*Bemicia tabaci*). This destructive disease infects at all stages of the crop. In the leaf, homogeneous interwoven network of yellow veins enclosing islands of green tissues within are developed. In extreme cases, the entire leaf turn complete yellow or cream coloured. The infected plants are stunted and bear very few, yellow coloured fruits.

Control measures

- Growing of resistant or tolerant varieties like, Arka Anamika, Arka Abhay, Pusa A-4, VRO-6, Azad Krishna, Kashi Mohini, Kashi Mangali, etc.
- Adjustment of the sowing time according to white fly population. Early sowing of crop during January-February and autumn sowing crop to some extent reduces the intensity of disease.
- Removal of diseased plants as soon as they are noticed in the field.
- Application of Carbofuran (Furadon 3G @ 20-22 kg /ha) to the soil at the time of sowing.
- Control of whitefly by 4-5 foliar sprays of Dimethoate (Rogor 2ml/l) or Imidachlorpid (3.5 ml/ 10 litre) or Thiomethoxam (3.5 ml/ 10 litre).

Powdery mildew (*Erysiphe cichoracearum*)

This fungal disease appears under prolonged dry conditions. White floury or powdery pustules appear on the lower surface of leaf extending to upper surface causing yellowing and death of the leaves.

Control measures

- Spraying the crop with Bavistin or Benlate @ 1g/l of water 3 to 4 times at 15 days interval.
- Two sprays of 0.1% Bavistin followed by two sprays of 0.1% Karathane or Tridemorph effectively control the disease.

Insect pests

Jassids/ leaf hoper (*Amrasca biguttula biguttula*)

The adults are wedge shaped, about 2.0 mm long, pale green with a black dot on posterior portion of each fore wing. Hoppers suck the sap from the lower surface of leaves causing cupping and later drying and falling of leaves. Early sown crops are highly susceptible to jassid attack and found to be very active during February to June with a peak population attained in the month of May.

Control measures

- Application of granular insecticides like Carbofuran (Furadon 3 G @ 20-22kg /ha) at the time of sowing.
- Spraying the crop with Imidaclorpride @ 0.3ml/l at the early flowering stage and if required, use of neem based insecticide for effective and healthy management of the pest.
- Spraying the crop with 0.05% Phosphamidom (Dimecron 1.5 ml/l) or 0.05% Endosulphan (Thiodan 1.5ml/l) at 10 days interval.

Red spider mite (*Tetranychus* spp.)

Four spider mite species, *Tetranychus urticae* Koch, *T. macferlanei* Baker & Pritchard, *T. ludeni* Zacher and *T. neocaledonicus* Andre are commonly found to attack vegetable crops. They are tiny, microscopic and morphologically alike without visual difference among these four species. However, *T. urticae* is found everywhere to attack the crops. Okra is severely infested by these mite species and found to feed from the ventral surface of the leaf. All the active stages of the mite suck sap resulting whitish appearance of the crop. Severely infested crops are covered by their heavy webbing. Summer crops are worst affected.

Control measures

- Application of balanced fertilizer and proper irrigation keeps the plants healthy and reduce mite attack.

- Spraying the crops with neem based acaricides during first occurrence of mite in the crop.
- Heavy infestation requires acaricidal application (Dicofol @ 2.5 ml/l; Propergite @ 1.5 ml/l or Diafenthiuron @ 1ml/l) to save the crop.

Shoot and fruit borer (*Earius vitella*; *E. insulana*)

The caterpillars are brownish-white with a number of black and brown spots on their body. They bore into the stem, flower buds and fruits which results in drying up of shoots and dropping of flower buds and developing fruits.

Control measures

- Destruction of infested plant parts along with the larvae inside and maintain sanitation in the field.
- Avoidance of growing of alternate host plants like, *Malva parviflora*, *Althaea officinalis*, cotton, etc, in nearby field.
- Spraying the crop with Spinosad @ 0.25 ml/l at 15 days intervals.

Yield

Fruit yield is 7-8 t/ha in open pollinated varieties grown in spring-summer; 11-12 t/ha in open pollinated varieties grown in rainy season and 15-18 t/ha in the hybrids grown in rainy season.

Tropical Tuber Crops

- **Elephant foot yam**
- **Sweet potato**

Elephant foot yam

Among the tropical tuber crops, elephant foot yam, *Amorphophallus paeoniifolius* (Dennst) Nicholson under the family Araceae (Chromosome No. 2n=26,28) has become a popular aroid vegetable due to high productivity in a short growing season and high net returns. It contains vitamins, minerals, energy and also has medicinal and therapeutic values. Elephant foot yam, is very much popular in Philippines, India, Malaysia, Indonesia, China, Sri Lanka and many other Southeast Asian countries. In India, this crop is traditionally cultivated in West Bengal, Andhra Pradesh, Tamil Nadu, Bihar, Gujarat, Kerala and Jharkhand.

The crop

Elephant foot yam is a tropical tuber crop and perennial herb. It is grown as annual plant for its round corm but can be perpetuated perennially through underground corm. The stems can be 1 to 2 meters tall. Leaves (usually one or two on a stem) are about 50 cm long and consist of several oval leaflets.

A corm is a short, vertical, swollen underground stem that serves as a storage organ which consists of one or more internodes with at least one growing point, generally with protective leaves modified into skins or tunics. The tunic of a corm is formed from dead petiole sheaths, remnants of leaves produced in previous years. On the top of the corm, one or a few buds grow into shoots that produce normal leaves and flowers. Corms are sometimes confused with true bulbs. Corms are modified stems that are internally structured with solid tissues, which distinguish them from bulbs, which are mostly made up of layered fleshy scales that are modified leaves. As a result, when a corm is cut in half it is solid, but when a true bulb is cut in half it is made up of layers.

Corms of elephant foot yam are round; usually attain 2-6 Kg in one season depending on the size of planting material, duration of growth and nutrition. Corm weight increases many fold depending on the number of seasons that the crop was grown before harvest. Along with the vegetative growth in the rainy season, the plant may release some side-corms however, propensity of releasing cormels depends of the variety.

Elephant foot yam has two morpho-types: Rough type of petioles (*Amorphophallus paeoniifolius* var. *sylvestris*) and Smooth types of petioles(*Amorphophallus paeoniifolius var. hortensis*). Both these types show numerous leaf morphological variations. The rough type is associated with acrid corms. The smooth type exhibits acridity at the immature stage but it is substantially reduced with the maturity of the crop. Improved varieties of smooth petioled types are non-acrid, and endowed with excellent cooking qualities. However, the morpho-types do not correspond to different genetic groups. Therefore, breeding to enhance palatability to eliminate tuber acidity and reduce oxalic content becomes the main goal of the plant breeders.

The acridity in the corm is believed to be due to the combination of needle shaped crystals (raphides) of calcium oxalate and a chemical irritant, the nature of which has not been resolved till date, although it is reported to be either a diglucoside of 3,4-Dihydroxy benzaldehyde or an unidentified proteinase or hormone or sapotoxin.

Cultivars of elephant foot yam are classified according to corm shape, corm surface shape and colour, corm flesh pigmentation, petiole size, petiole pigmentation, leaf structure, inflorescence shape, size and colour, yield and quality characteristics.

The inflorescence emerges in the fourth year after planting from the seed or cormel, and biannual flower bearing is common after the first flowering. The flower bud emerges from the corm as a purple shoot, and later blooms as a purple inflorescence. The most spectacular and largest inflorescences (2.5m.) are produced by *A. titanium* Becc. which rarely flowers out of its original environment in Indonesia. The pistillate (female) and staminate (male) flowers are borne on the same plant and are crowded in cylindrical masses as an inflorescence. The top part is responsible for secreting mucus that gives off putrid, rotten-meat odour that attract pollinating beetles, the middle part of the inflorescence contains staminate and the base of the inflorescence contains pistillate flowers. The stigmas of the female flowers become receptive on the first day of the bloom, when the foul smell will draw pollinating insects inside, and the inflorescence will close, trapping them for a night to allow the pollen deposited on the insect to be transferred to the stigmas. Later in the second day, the female flower will no longer be receptive of pollens, the male flowers will start to bloom, and the inflorescence will open again. This allows the pollen to be deposited on the emerging insects to be pollinated on different flowers, while preventing the pollens from the same inflorescence to fertilize itself, preventing inbreeding. The reproductive system exhibits a dichogamous barrier to prevent selfing.

After 24-36 hours of the first bloom of the inflorescence, the female flowers will start developing into bright red berries and other parts of the inflorescence will start wilting away. The berries are red when ripe and are not quite round, being subglobose or ovoid. Mature berries pertinently drop around mother plants; nevertheless, long-distance dispersal by birds has been reported.

Importance and use

The corms are commonly used as a vegetable after cooking and in preparation of indigenous ayurvedic medicines. The corms are the cheapest source of carbohydrates mainly starch and fibres, vitamins and minerals. It plays an importantl role in food security and is the important staple or subsidiary food for a large group of population. Petioles of unopened leaves can also be used as vegetable. The corm can be also used for making various value added products like flour, dried cubes, pickles, medicines etc.

Nutritional value

Among the edible aroids, elephant foot yam is reported to have the lowest starch (15-16%) and highest protein content (2.5-3.5%) on fresh weight basis . The corm is also rich in fibre (around 1.45%) and potassium (620 mg/100g) and low in sodium (11 mg/100g). The high potassium-sodium ratio makes the corm a good source of food capable of maintaining a good potassium-sodium balance in the body and thereby protect against osteoporosis and heart diseases. In addition, glucose, galactose, rhamnose and xylose are also present as free sugars.

Corms have considerable potential to be an alternative food source, but it has not been realized its full potential because of the palatability problems associated with the content of certain anti-nutritional factors like oxalate, trypsin inhibitors and acridity. The soluble oxalate content of smooth corm bearing non-irritant cultivars is found to be lower and safe from the view point of accumulation of urinary oxalate leading to kidney stones.

Medicinal value

Mature corm is used locally as medicine and disinfectants in many Asian countries. Corms are stomachic, carminative, used in piles, dysentery and having blood purifier property. Fresh corms have stimulant and expectorant properties. Konjac glucomannan is a water-soluble dietary fibre extracted from tubers of *Amorphophallus konjac*, a native of China. The fresh tuber is reported to contain around 8-12% glucomannan. Purified powder is reported to contain about 90-97% glucomannan and is used for treatment of constipation, obesity, high cholesterol and Type 2 diabetes in China. It is externally applied to relieve pain of rheumatic swellings.

Origin and taxonomy

Elephant foot yam is found across Australasian and African countries. It has been used as food in Southeast Asia since the prehistoric era.Elephant foot yam is believed to have originated in Asia and Asia-Pacific region probably in India or Sri Lanka. It belongs to the family Araceae and sub-families Lasioideae. The genus *Amorphophallus* includes some 100 species from Africa, India, and Malaysia to Australia. The species that are utilized are *A. paeoniifolius* (Dennst.) Nicolson), *A. konjac,* C. Koch (syn. *A rivieri* Durien var. *konjac* Engler*), A. oncophyllus* Prain, *A. variabilis* Blume and A. *konjac* is cultivated and utilized in Japan and warmer parts of China. *A. oncophyllus* and *A. variabilis* are utilized in Indonesia. The most important species is *A. paeoniifolius* (Dennst.) Nicolson (Syn. *A. campanulatus*).

The somatic chromosome number is 2n=2x=26,28. The induced tetraploids do not show any superiority over normal diploids.

Adaptation in India

The Old World genus *Amorphophallus* has long been dispersed with the migration of the sea-faring people of Indo-Malaysia-Pacific region and believed to predate the culture of rice. The cultivation of *A. paeoniifolius* is rather restricted to Malaysia, and the drier parts of India and Sri Lanka. It is, however, reported to grow wild in Southeast Asia extending to Java, the Philippines and the Pacific. Initially, local irritant types are primarily grown in homestead areas in the tropical parts of the country. After the introduction of smooth corm and non-irritant types from east Godavari region of Andhra Pradesh during early 80's , the crop is gaining tremendous popularity in the Gangetic plains of eastern India. Now the crop is regarded as the most important cash crop in this region.

Improved varieties

Smooth corm, non-irritant type : Gajendra, Sree Padma, Bidhan Kusum, Sree Athira, NDA-9.

Cultural requirements

Elephant foot yam grows well in hot climate at a temperature of 25-35°C. Humid climate helps in the initial stages of the crop growth whereas, dry climate facilitates corm bulking. Well-distributed rainfall of 1000-1500 mm is helpful in good crop growth and corm yield. Well-drained, fertile, sandy loam soil is ideal for elephant foot yam cultivation. The crop can be grown in other soil types including laterite soils also by raising the crop in pits filled with well-decomposed farmyard manure and good sandy loam soil. Water stagnation at any stage adversely affects crop growth, crom bulking and quality.

Planting time

Planting time of the corm is February-April as irrigated crop in the Gangetic plains and June-July as rainfed crop.

Land Preparation

The land should be prepared well in advance with repeated ploughing (at least 4-5 ploughing) to a fine tilth. All stubbles, weeds, etc should be removed from the land. Well rotten organic manure preferably farm yard manure or compost @ 20-25 t/ha need to be applied during land preparation and the land is levelled properly.

Production of seed corm

Huge quantity of seed corms is required to plant per unit area and availability of quality seed material is the major constraints in elephant foot yam cultivation. In intensive cropping zone, the farmers are not keen to keep the crop in their field up to maturity and go for early winter vegetables, instead. In this situation, the crisis of seed corm during planting is of common occurrence particularly in West Bengal. The vast mono-cropped areas of eastern part of India can be exploited through seed corm production programme by which they can meet the requirement of quality seed materials for commercial growing areas at reduced rate.

Planting the seed corm

Whole corm size of 500-750 g is recommended for commercial cultivation. However, due to non- availability of the recommended size, farmers usually cut the corm into small pieces from big corm. The big size corm of 2-3 kg is cut into 4-6 pieces vertically retaining part of the apical bud. Dipping of the cut corms in thick cow-dung slurry (1:1) mixed with Captan (0.3%) reduces the corm rot in the field and also increases sprouting. After removing the corms from cow-dung slurry, it should be dried in shade for 6-8 hours before planting. Small and whole corms are preferred over cut pices for commercial cultivation. If nursery facility is available, cut pieces after treatment can be planted on raised beds during the first week of June for sprouting. The cut corms should be closely kept in lines 20 cm apart on nursery beds. The bed should be lightly covered with soil and paddy straw and light irrigation should be regularly provided. The corms sprout in about 3 weeks. The sprouted corms should be transferred to the main field with the onset of monsoon. For obtaining a good commercial crop, the planting should be done in pits (50 cm x 50 cm x 30 cm) filled with Farm Yard Manure, good soil and fine sand in the ratio of 1:1:1. The distance of the pits is decided depending on the size of planting material being used.

For seed materials ranging from 500-600g, a distance of 75 cm x 75 cm; for more than one kg, 90 cm x 90 cm spacing and for 500g and below, 60 cm x 60 cm spacing is recommended. The cut corms are planted 4-6inch below the soil surface with the apical portion upward and the pits should be covered with soil followed by light earthing up after planting. Paddy straw is usually used as mulch material.

Nutrient management

Nutrient requirement of elephant foot yam is rather high. Well decomposed farmyard manure @ 20-25 t/ha should be given before 15 days of final land preparation. If the planting is done in pits, the FYM mixed with soil should be filled in pits only. The fertilizer dose should be decided depending on the soil type and nutrient status. A general fertilizer dose of 150:100:150 kg/ha of nitrogen, phosphorus and potash in the Gangetic plains of eastern India has been found to be optimum. Full dose of phosphorus and one-third quantity of nitrogen and potash is applied at the time of planting. The remaining dose of nitrogen and potash is applied in two split doses at 60 and 90 days after planting, immediately after weeding and followed by earthing-up. *Azotobactor* may be applied either as corm dipping (2% solution) or soil application (9.0 kg/ha) at 35 days after planting which increases the corm yield.

Irrigation

If the planting is done during February-April, frequent irrigation should be provided immediately after planting to initiate sprouting of corms. Mulching with paddy straw after planting the corms helps in conservation of moisture, early sprouting, checking early weed growth and reduces the number of irrigation. Depending on the soil moisture availability, irrigation should be given at regular intervals till the arrival of monsoon. Care should be taken to prevent water stagnation at every stage of crop growth. Atleast 8-10 irrigations are required throughout the growth period.

Interculture and weed control

After 30-35 days of planting, when the sprouts start emerging, weeding is done followed by first earthing-up. Second earthing up should be followed after first top dressing of N and K at 60 days of planting. Light earthing up if required, may be done after second top dressing at 90 days of planting. To keep the crop healthy, plant protection measures should be strictly followed. During the initial period of 2 months after planting, crops like leafy vegetables, cowpea, etc. can be taken up as inter-crop for additional benefits. Crop rotation of elephant foot yam (mid-March to mid-November) - wheat/ mustard/ Bengal gram/peas (mid-November to Mid-March) or Elephant foot yam (mid-June to mid-December)-

okra/colocasia/maize/mustard (January to mid-June) has been found suitable. Inter-cropping of elephant foot yam is very successful in coconut, banana and other plantations for getting additional income.

Harvesting the crop

Yellowing of the leaves and drying up of the plants indicate that the crop is ready for harvest. Corms are harvested by digging using crowbars or spades after 6-7 month. While harvesting, care should be taken to avoid injury to the corms.

Post harvest handling and storage

After harvesting, soil and roots should be removed from the corms. Damaged corms should immediately be disposed-off for local sale as vegetable. The corms should be graded and dried in shade for 4-5 days before marketing. Use of baskets made of bamboo/palm leaves has been found to be ideal. The corms can be kept in layers sandwiched with paddy straw or dry banana leaves. The corms are always in great demand due to their acceptability as a popular vegetable and usefulness as major component of several indigenous medicinal preparations. The corms can be also used for making various value added products like flour, dried cubes, pickles, etc.

Plant Protection measures

Diseases

Collar Rot (*Sclerotium rolfsii*)

The collar region of the plant is attacked by the fungus and water soaked lesions appear on them. The leaves become yellow. The stem shrinks and collapses due to rotting at the collar region. A thick white spreading mat of mycelia with globular sclerotia can be seen on the affected tissue. Heavy rains, high humidity, heavy soils, poor drainage are some of the pre-disposing factors.

Control measures

- Since the pathogen is soil-borne, cultural practices such as crop rotation, removal of plant debris and improvement of drainage system will minimize the severity of the disease.
- Corm should be treated with Carbendazim (0.2%) before planting.
- Application of *Trichoderma viridae* @10 kg/ha as soil treatment is effective.
- Drenching of the collar region with Carbendazim (1g) + Mancozeb (2g) or Thiram (2g) + Carbendazim (1g) / litre of water twice at monthly interval commencing from the first appearance of the symptom effectively control the disease.

Mosaic (*Dasheen mosaic virus*)

The characteristic symptom of this virus disease is mosaic mottling on the leaves. The mottled plants produce small corms. The virus is spread through aphids.

Control measures

- Corms should be collected from heathy, disease free crops for planting.
- Rouguing of infected plants will help in minimizing the secondary spread.
- Spraying of Dimethoate (Rogor) @ 1.5 ml/l is effective against the vectors.

Yield

The yield of elephant foot yam depends on the size of planting material used and planting time. The ratio between quantity of planting material used and corm yield is generally 1:9. For commercial cultivation, if 75 cm x 75 cm spacing is provided and approximately 500g size planting material is used, corm yield may go up to 50-60 t/ha.

Sweet Potato

Sweet potato, *Ipomoea batatas* under the family Convolvulaceae (Chromosome No. 2n = 6x = 90) is an important tuber crop grown mostly in tropical and sub-tropical regions of Asia, Latin America, the Pacific islands, Papu New Guinea and Africa. Top 10 sweet potato producing countries are China, Nigeria, Indonesia, Uganda, Tanzania, Vietnam, Ethiopia, USA, India and Rwanda. "Ipomoea" was derived from two Greek words "*ipos*" meaning "bind weed" and "*homoios*" meaning "resembling". The species name "batatas" was originally the Taino name for sweet potato. In India it is largely cultivated in Uttar Pradesh, Bihar, West Bengal, Orissa, Madhya Pradesh, Tamil Nadu, Kerala, Karnataka and Andhra Pradesh. It is commonly known as "sakarkand" in India.

The crop

Sweet potato is an herbaceous and perennial plant. However, it is grown as an annual plant by vegetative propagation using either storage roots or stem cuttings. Its growth habit is predominantly prostrate with a vine system that expands rapidly and horizontally on the ground. The types of growth habit of sweet potatoes are erect, semi erect, spreading and very spreading. Stem is cylindrical and its length, like that of the internodes, depends on the growth habit of the cultivar and of the availability of water in the soil. The erect cultivars are approximately 1 m long, while the very spreading ones can reach more than 5 m long. Some cultivars have stems with twining characteristics. The internode length can vary from short to very long, and, according to stem diameter, can be thin or very thick.Depending on the cultivar, the stem colour varies from green to totally pigmented with anthocyanins (red-purple color). The hairiness in the apical shoots, and in some cultivars also in the stems, varies from glabrous (without hairs) to very pubescent. Leaves are simple and spirally arranged alternately on the stem in a pattern known as 2/5 phyllotaxy (5 leaves spirally arranged in 2 circles around the stem for any two leaves be located in the same vertical plane on the stem). Depending on the cultivar, the edge of the lamina can be entire, toothed or lobed. The base of the lamina generally has two lobes that can be almost straight or rounded. The shape of the general outline of sweet potato leaves can be rounded, reniform (kidney-shaped), cordate (heart-shaped), triangular, hastate (trilobular and spear-shaped with the two basal lobes divergent), lobed and almost divided. Lobed leaves differ in the degree of the cut, ranging from superficial to deeply lobed. The number of lobes generally range from 3 to 7 and can be easily determined by counting the veins that go from the junction of the petiole up to the edge of the leaf lamina. However, toothed leaves have minute lobes called teeth which could number from 1 to more than 9. Some cultivars show some variation in leaf shape on the same plant. The leaf colour can be green-yellowish, green or can have purple pigmentation in part or the entire leaf blade. Some cultivars show purple young leaves and green mature leaves. The leaf size and the degree of hairiness vary according to the cultivar and environmental conditions. The hairs are glandular and generally are more numerous in the lower surface of the leaf. The leaf veins are palmate and their colour, which is very useful to differentiate cultivars, can be green to particularly or totally pigmented with anthocyanins. The length of the petiole ranges from very short to very long. Petioles can be green or with purple pigmentation at the junction with the lamina and/or with the stem or throughout the petiole. On both sides of the insertion with the lamina there are two small nectarines.

The root system consists of fibrous roots that absorb nutrients and water, and anchor the plant, and storage roots that are lateral roots, which store photosynthetic products.The root system in plants obtained by vegetative propagation starts with adventitious roots that develop into primary fibrous roots, which are branched into lateral roots. As the plant matures, roots of pencil thickness that have some lignification are produced. Other roots that have no lignification are fleshy and thicken a lot, are called storage roots. Plants grown from true seed form a typical root with a central axle with lateral branches. Later on, the central axle functions as a storage root. A tuberous root or storage root is a modified lateral root, enlarged to function as a storage organ. The enlarged area of the root-tuber or storage root can be produced at the end or middle of a root or involve the entire root. It is thus different in origin but similar in function and appearance to a stem tuber. Root tubers are perennating organs, thickened roots that store nutrients over periods when the plant cannot actively grow, thus permitting survival from one year to the next. The massive enlargement of secondary roots have the internal and external cell and tissue structures of a normal root. They produce adventitious roots and stems which again produce adventitious roots. In root-tubers, there are no nodes and internodes or reduced leaves. Root tubers have one end called the proximal end, which is the end that is attached to the old plant; this end has crown tissue that produces buds which grow into new stems and foliage. The other end of the root tuber is called the distal end, and it normally produces unmodified roots. In stem tubers the order is reversed, with the distal end producing stems.A transverse section of the storage roots shows the protective periderm or skin, the cortex or cortical parenchyma that, depending on the cultivar, varies from very thin to very thick.In the cambium ring, the latex vessels are found. The amount of the latex formed depends on the maturity of the storage root, the cultivar, and the soil moisture during the growing period. The latex drops are produced when the storage roots are cut and they darken very quickly due to the oxidation. The storage root skin colour can be whitish, cream, yellow, orange, brown-orange, pink, red-purple and very dark purple. The intensity of the colour depends on the environmental conditions where the plant is grown. The flesh colour can be white, cream, yellow or orange. However, some cultivars show red-purple pigmentation in the flesh in very few scattered spots, pigmented rings or, in some cases, throughout the entire flesh of the root.

Sweet potato cultivars differ in their ability to produce flower. Under normal conditions in the field, some cultivars do not flower, others produce very few flowers, and some others flower profusely. The inflorescence is generally a cyme in which the peduncle is divided in two axillary peduncles; each one is further divided in two after the flower is produced (biparous cyme). In general, buds of the first, second and third order are developed. However, single flowers are also formed. The flower buds are joined to the pedundle through a very short stalk called pedicel. The colour of the flower bud pedicel and peduncle varies from green to completely purple pigmented. The flower is bisexual. Besides the calyx and corolla, they contain the stamens and the pistil. The calyx consists of 5 sepals, 2 outer and 3 inner, that stay attached to the floral axle after the petals dry up and fall.The corolla consists of 5 petals that are fused forming a funnel, generally with lilac or pale purple limb and with reddish to purple throat (the inside of the tube). Some cultivars produce white flowers.The androecium consists of five stamens with filaments that are covered with glandular hairs and that are partly fused to the corolla. The length of the filaments is variable in relation to the position of the stigma. The anthers are whitish, yellow or pink, with a longitudinal dehiscence. The pollen grains are spherical with the surface covered with very small glandular hairs.The gynoecium consists of a pistil with a superior ovary, two carpels and two locules that contain one or two ovules. The style is relatively short and ends in a broad stigma that is divided into two lobes that are covered with glandular hairs. At the base of the ovary there are basal yellow glands that contain insect-attracting nectar. The stigma is receptive early in the morning and the pollination is mainly by bees.

The fruit is a capsule, more or less spherical with a terminal tip, and can be pubescent or glabrous. The capsule turns brown when mature.Each capsule contains from one to four seeds that are slightly flattened on one side and convex on the other. Seed shape can be irregular, slightly angular or rounded; the colour ranges from brown to black; and the size is approximately 3 mm. The embryo and endosperm are protected by a thick, very hard and impermeable testa. Seed germination is difficult and requires scarification by mechanical abrasion or chemical treatment. Sweet potato seeds do not have a dormancy period but maintain their viability for many years.

Importance and use

Its large, starchy, sweet-tasting, tuberous roots are consumed as vegetable after boiling, baking or frying. Being a major source of carbohydrate it is often used as a substitute for cereals. Yellow and orange fleshed cultivars are rich in β carotene, the precursor of vitamin A. It can also be eaten fresh, canned, frozen or dehydrated and use in variety of food products, like pie fillings, purees, candied pieces, soufflés, baby foods, etc. Peeled tubers are sliced and dried in the sun to produce chips. Dehydrated tubers are also ground into flour and often mixed with the cereal flour. Sweet potato flour is used as a supplement to cereal flour in the preparation of bakery products, puddings and milk jelly. Sweet potato is a good raw material for the fermentation production of industrial alcohol, lactic acid, acetone, butanol, vinegar, etc. The tender tops and leaves are used as vegetables in Africa, Indonesia and Philippines.

It is a source of starch for use in textile, paper, cosmetics and food industries, and for preparation of glucose and adhesive. The sweet potato starch is suitable for sizing paper and textiles and for the use in laundries. Tubers are also used as high carbohydrate feeding stuff for cattle, pigs and in poultry. Feeding raw tubers to cattle and swine can pose problems due to presence of protease inhibitors and when tubers are infested due to phytoalexines. The vines also serve as good source of fodder for livestock.

Nutritional value

Besides simple starches, raw tubers are rich in complex carbohydrates, dietary fibre and β carotene (a provitamin A carotenoid), while having moderate contents of other micronutrients, including vitamin B_5, vitamin B_6 and manganese. When cooked by baking, small variable changes in micronutrient density occur to include a higher content of vitamin C . Sucrose and a few reducing sugars are present in the tuber. General quality of sweet potato is mostly dependent on its sugar content (mostly sucrose) which varies from 3-6%. Sugar content increases during storage and cooking. β-carotene content, the precursor of vitamin A ranges between 0.5 to 20.0 mg per 100 g fresh. Yellow and orange fleshed cultivars are rich in β–carotene. The tubers contain 1.5 g protein and 24.0 mg ascorbic acid (vitamin C) per 100 g fresh. The tubers are also rich in minerals particularly potassium (370mg/100gm), phosphorus (49.0mg/100g) and calcium 946.0 mg/100g) contents. Sweet potato leaves contains about 3.2% protein on fresh weight basis and also rich in β carotene, vitamin C and some minerals (Ca, P and Fe).

Medicinal value

Varieties with orange or yellow fleshed tubers are excellent source of β carotene, which apart from being precursor of Vitamin A, is ascribed with anti-carcinogenic properties. The tuber contains very high dietary fibres that give protection against constipation, colon cancer, heart diseases and diabetics. It is rich in potassium with a very high K: Na ratio and can be recommended as a part of low salt diet.

Origin and taxonomy

The origin and domestication of sweet potato is thought to be in either Central America or South America. In Central America, sweet potatoes were domesticated at least 5,000 years ago. In South America, Peruvian sweet potato remnants dating as far back as 8000 BC have been found. It was introduced in Europe, probably in Spain at least 60 years before Irish potato. The introduction and spread in Africa and Asia closely follows the establishment of trading settlement by Spanish and Portuguese. In India, it has also been introduced through Portuguese and British trade and colonization. It spread into Polynesia by Spanish expedition, which started from Peru in the 16th century.

Of the approximately 50 genera and more than 1,000 species of Convolvulaceae, *Ipomoea batatas* is the only crop plant of major importance. It is the only tuber bearing species although a few species are known to have thickened roots. Perhaps, *Ipomoea trifida* is the closest relative of sweet potato although sweet potato is not found in wild state. Sweet potato, *Ipomoea batatus* belongs to the section Batatas of the family Convolvulaceae. It is a natural hexaploid ($2n = 2x = 90$) with the basic chromosome number is $x = 15$ however, it is usually free from any deleterious effect of high autoploidy. The hexaploid sweet potato probably derived by amphidiploidy from a tetraploid ($2n=60$) and a diploid ($2n=30$) to produce a triploid ($2n=45$), followed by subsequent doubling of chromosome to produce hexaploid ($2n=90$).*I. trifida* was the direct ancestor of sweet potato and its diploid predecessors were thought to be *I. leucantha*.

Adaptation in India

Though a tuber crop similar to sweet potato was cited in the literatures of the "Vedic" period, yet generally it is considered to have been introduced in India by Portuguese in early 16th century. A number of cultivars have been developed by clonal selection from these germplasm resources of India. Wide variation exists among the cultivars for different morphological characters like shape, size, colour of skin and flesh of tuber, flower colour, flesh texture, depth and period of rooting, etc.

However, huge exotic collections mainly from Japan, Taiwan, USA, Nigeria, Sudan, Peru, Puerto-Rico, USA, Nairobi and Argentina enriched the germplasm resources of India. A number of cultivars have been developed by clonal selection from these germplasm resources of India.

Improved varieties

White/cream skinned varieties: Pusa Safed, Rajendra Sakarkand-5, Sree Nandini, CO-1, CO-2, Sree Varun, Sree Gautam.

Brown skinned varieties: Cross-4, Samrat, V-35, Kalmegh, Rajendra Sakarkand-35, Rajendra Sakarkand-43, Sree Kanaka

Red or pink skinned varieties: H-41,H-635, H-620, CO-3,Pusa Lal, Pusa Sunheri, H-42, Varsha, Sree Vardhini, Sree Bhadra, Sree Rethna, Konkan Ashwini, Kiran, Bhuban, Gouri, Shankar, Rajendra Sakarkand-47, VL Sakarkand-6, Sree Arun, Kalinga, Sree Kishan, Sree Sourin, Bidhan Jagannath

Orange fleshed varieties: Kamala Sundari, Gouri, CO-3, Kiran, Sree Vardhini, Sree Rethna

Light yellow fleshed varieties: VL Sakarkand-6

White/creamy fleshed varieties: Bhuban, CO-1, CO-2, H-41, H-42, Rajendra Sakarkand-5, Rajendra Sakarkand-43, Rajendra Sakarkand-47, Sree Bhadra, Samrat, Sree Nandini, Varsha, Sree Kishan, Sree Sourin and Sree Gautam

Resistant trap crop for nematode: Sree Bhadra

Resistant to SPFMV: Sree Vardhini

Cultural requirements

Sweet potato is grown in tropical and sub-tropical climate and also in mild temperate regions. It is widely grown from sea level up to the elevation of 2500 m however, its performance is better in the places of low elevation. It is susceptible to frost and its growth and tuber development are adversely affected at temperature below 20°C. Warm sunny days and cool nights are favourable for better tuber development. Night temperature of 15-25°C promotes tuber formation and growth, which prevail in the rabi season in the plains. Day temperature between 25°C and 30°C is ideal for tuber formation and development. Good rains during the period of early growth followed by dry sunny weather during the period of tuber bulking and maturity is ideal. It can tolerate drought moderately however, tuber yield reduces if draught prevails at the initial stage of growing period. The tuberization is almost absent at 75% shade and severely affected at 55% and not much affected at 25 % shade. It is grown in a wide variety of soils. It grows best in well drained and fertile loam, sandy-loam or clay-loam soil. The soil should be rich in organic matter, friable and well-drained with a pH range between 5.6 and 6.6. Heavy clay soil checks the development of the tubers and causes development of long cylindrical pencil-like tubers. Highly fertile soil favours luxuriant vine growth which eventually reduces tuber development. As a rainfed crop, it requires at least 500 mm rainfall during the growing season. If grown under high rainfall areas, produce vigorous vine growth and poor yield. A well distributed annual rainfall of 750-1000 mm is the best. It does not withstand heavy rainfall and water logging condition. It can somewhat tolerate slightly acidic soil but can not tolerate alkaline and saline soils.

Land Preparation

Well prepared soil is the prerequisite for growing good sweet potato crop. The field should be ploughed 3-4 times to a depth of 25-35 cm and clods are broken. The field should be properly leveled to provide good drainage. Well rotten organic manure preferably farm yard manure or compost @ 10-15 t/ha should be added to the soil before preparation of ridges.

Propagation and planting material

Sweet potato is propagated from vine cutting taken from freshly harvested vines or from medium sized weevil free sprouted tubers planted in the nursery. The vine cuttings from the freshly harvested crop are also planted in the nursery to get the cuttings for planting in the main field. Using vines obtained from freshly harvested crop should be avoided as it significantly increases weevil infestation in tubers. Vines should be produced in nursery to get healthy vigorous plants. The cuttings are selected from apical and middle portion in order to get higher sprouting and better yield. Vine cuttings for planting in the main field are about 20-35 cm long with 3-4 nodes.

Production of planting material in the nursery

Use of vine cuttings as planting materials over the years leads to reduced yield and more incidence of weevil (*Cylas fomicarius*) infestation. Therefore, healthy vine cuttings are prepared through tuberlets planted in primary and secondary nurseries.

Primary nursery

Raised nursery beds of 15-20cm height with fine tilth and good drainage facility should be prepared about 3 months prior to planting in main field. About 100 kg of medium sized weevil free tuberlets (100-125g each) planted in 100 square meter primary nursery bed can provide sufficient vine cuttings for planting one hectare of land. The tubers are soaked in Monocrotophos or Fenitrothion 0.05% for 10 minutes. The tubers are planted at a spacing of 20cm on ridges formed 60cm apart. Proper irrigation should be given on every alternate day for 10 days and thrice a week thereafter. Application of urea @ 1.5kg/100 square meter 15 days after planting ensure quick growth. For replanting in the secondary nursery, 45 days old vines are clipped to a length of 20-25 cm.

Secondary nursery

Vines obtained from primary nursery are planted in the second nursery. Secondary nursery area of 500 square metres is needed to multiply the vines obtained from primary nursery for 1 hectare land. The beds are prepared to a fine tilth with the addition of 500 kg FYM. The 20-25 cm long vines are planted from first nursery at a spacing of 20 cm on ridges spaced at 60 cm. 5 kg urea need to be top dressed in two split doses, 15 and 30 days after planting. The nursery need to be irrigated whenever required. Vines cutting for tuber production in the main field can be taken after 45 days.

Planting time

Sweet potato is grown throughout the country as rainfed during kharif and irrigated crop during rabi season. However, major sweetpotato area comes under early rabi and rabi seasons because of prevailing warm and sunny days, cool night and moderate rainfall which is favourable for getting high yield. Planting is done in September-October for autumn-winter (early rabi) crop; October-December for winter (rabi) crop; June-July for kharif crop; November for cultivation in the diara land and April-May for growing in the hill regions.

Planting of vines

Cuttings of 20-25 cm long from apical part of the vine having at least 3-5 nodes are ideal for tuber production. The vine cuttings with intact leaves are kept in bundles under shade for 2-3 days prior to planting to generate more roots and better establishment. There are several method of planting viz. mound-method, ridge and furrow method, bed method and flat method. Ridge and furrow method is most common one and it is adopted across the slope in sloppy lands for controlling soil erosion. Mound method is practiced in areas experiencing problems of drainage. The vine cuttings should be planted in furrows made with the help of tyne at an angle of $45^0$and burry 2-3 nodes horizontally in the soil. The vine cuttings are planted to a depth of 5-10 cm with only middle portion of vines buried in the soil keeping both ends exposed. In mound method of planting, 5 cuttings /mound are planted at a spacing of 80 cm x 75 cm.

Spacing

Spacing of 60 x 20 cm is mostly adopted which accommodate plant population of 83,000 plants/ha.

Nutrient management

Nitrogen is the primary limiting nutrient in sweet potato production. Potassium helps in better tuberization and yield. The quantity of NPK fertilizers to be applied varies with the soil, climate and location. Application of 10-15 tonnes of FYM or compost at the time of field preparation; along with a general fertilizer dose of 75 Kg N, 50 Kg P_2O_5 and 75 Kg K_2O per hectare is recommended. Full dose of P and half doses of K and N is mixed well at the time of planting and rest half of N and K is top dressed on the sides of the ridges 30 days after planting followed by earthing up. Application of 30 kg N and 2 kg *Azospirillium* per ha as vine dipping and 10 kg *Azospirillum* as soil application along with full dose of P and K is recommended.

Irrigation

Moist bed is required for 4-5 days after planting for proper sprouting and establishment of vines. If sufficient moisture is not present, daily irrigation is required for first 3 days followed by irrigation on alternate days for one week. Irrigation should be given at 15 days interval in rabi season ensuring at least 5-8 irrigations during crop period supplying 112 to 150 cm of water. The summer crop requires irrigation at more frequent intervals. Irrigation should be given in the most critical stages of plant growth i.e. tuber initiation, early bulking and late bulking to get higher number of tubers/plants. The crop can tolerate considerable periods of drought however, water shortage 50-60 days after planting when the storage roots started bulking considerably reduce yield.

Interculture and weed control

It has been found that spraying the crop with the growth retardant Cycocel @ 500-1000 ppm concentration twice, 30 and 45 days after planting increase tuber yield. Weeding at early growth stages between 7 and 14 days and 30 and 45 days after planting is essential. Second earthing up should be done at 60 days after planting. It is quick growing crop and it covers the soil quickly and suppresses most of the weed. For chemical control, Alachlor @ 3.5-6.5 kg a.i. /ha is recommended. The vines should be turned over after a month of planting to prevent production of adventitious roots at the nodes that are in contact with the soil which helps in better tuber development.

Crop rotation

In eastern zone of India, Jute and upland rice is followed by sweet potato. In Northern zone, sweet potato is rotated with pea + barley. In Bihar and Andhra Pradesh, maize-cowpea-sweet potato rotation is practiced. In Bihar, two-tier sweet potato planting with shallow and deep-bulking varieties in alternative row is found successful.

Harvesting the crop

Sweet potato generally takes about 90 to 140 days after planting depending on the variety, season and location. Harvesting should start at 90 days after planting (DAP) for early varieties, 110 DAP for medium duration varieties and 120 DAP onwards for late maturing varieties. The yield can be increased if the crop remains in the field for longer period but the tubers become fibrous, less palatable and damaged by weevil. Harvesting of tubers should start when older leaves turn brown and the adjacent soil cracks near the base of the plant. Harvest maturity of the tubers can be judged by cutting the tuber. In mature tuber, the latex dries up quickly without blackening while in immature tuber, the cut surface turns dark greenish colour. The vines are first cut and then the tubers are dug up without causing injury.

Post harvest management and storage

Sweet potato is subjected to several forms of post-harvest losses due to physio-biochemical processes in storage, physical damage, sprouting, microbial decay and weevil infestation. The harvested tubers are cured by spreading in sun at about 29-40°C with 80-90% relative humidity for 5-7days which results in the healing of wounds. For dehydration, cut tuber pieces are spread out in the sun to dry. The healthy tubers that are free from insect damage and mechanical injury should be stored in well ventilated room. For low cost storage, the tubers are piled in heaps and covered with a layer of straw. The heaps are then plastered with a mixture of clay soils and cowdung under the shade. Tubers remain fresh for few months. In another method, 50cm deep pits are dug, the tubers are stored in the pits and covered it with dried banana leaves and soil and thatch is made over pit. Keeping the tubers in earthen pots reeled with fine net or cloth also prevents the entry of insect pests and can be stored up to 3 months. The weevil free tubers can also be stored in alternate rows of dry sand up to 60 days at ambient condition. Ideal storage temperature is 13-16°C at 85-90% relative humidity and expected storage life is 4-6 months.

Sweet potato is processed into chips or flour in commercial scale, which can be stored for year round consumption for use such as in bread and cakes, or processing into fermented and dried products. Different machines are also available used for preparation of starch and noodles.

Plant protection measures

Diseases

Cercospora leaf spot (*Cercospora batate, C. ipomoeae*)

This fungus incites leaf spot disease which is characterized by the appearance of yellowish brown spots which gradually turn deep brown, with circular ovoid or irregular margin ultimately covering major portion of the leaf. Hot, wet and humid weather is favourable for this disease.

Control measures

- Maintainence of field sanitation to reduce its incidence
- Spraying the crop with 0.25% Dithane M-45 or Zineb at 15 days interval commencing from first appearance of the disease.

Alternaria leaf spot (*Alternaria spp.*)

This fungal leaf spot disease appear as brown lesion mostly on old and mature leaves with concentric rings and well defined margins. In later stage infected tissue may crack and fall off. Humid weather is favourable for this disease.

Control measures

- Spraying the crop with 0.3% Dithane M-45 at 15 days interval

Collar rot (*Sclerotium rolfsii*)

It is an important fungal disease causing severe damage in the nursery. At first small, oval, straw or brown coloured lesions appear at the collar region which gradually enlarge and encircle the stem base. The tissue becomes necrotic resulting in rotting of the sprout and plant easily breaks off.

Control measures

- Selection of the land where collar rot was not a problem for at least 3 years.
- Removal of crop debris and maintenance of field sanitation.
- Drenching the soil around the plants with 50ppm Vitavax.

Sweet potato scab (*Sphacaeloma batatas*)

This fungal disease is most severe in places where frequent fog, rain or dew occurs. Small brown lesions appear first on the veins and leaves, later become corky in texture and veins shrink causing curls on leaves. A scab like structure forms on the stem as the lesions coalesce hence, the name.

Control measures

- Adoption of crop rotation
- Use disease free planting material
- Spraying the crop with 0.25% Mancozeb at 10 days interval.

Black rot (*Rhizoctonia solani*)

This fungal disease occurs both in the field and storage. When disease occurs in the field it affects the sprouts and veins which are referred to as 'Black Shank' or 'Black root'. Small, slightly sunken black spots appear on tubers, finally covering the entire tuber. Affected tubers have an unpleasant taste and colour when cooked.

Control measures

- Adoption of crop rotation.
- Selection of disease free planting material.
- Avoidance of storing infected tubers with healthy ones.
- Curing the tubers at 30-35°C temperature and 85-90% relative humidity for 5-10 days.
- Treating the planting material with 0.2 % Thiobendazole or Benomyl.

Sweet potato feathery mottle virus (SPFMV)

Sweetpotato has been reported to be attacked by more than 12 different viruses belonging to seven different taxonomic groups. SPFMV is found almost wherever the crop is grown. Atleast three different strains have been reported which include Ring spot, Russet crack, and probably Internal Cork strains.

The most characteristic leaf symptom are the appearance of ringspots or chlorotic spots, irregular chlorotic patterns (feathering) along the main ribs which may or may not have purple pigmented borders. Mostly the foliar symptoms are transitional and the infected plants may not show any external sumptoms depending on the growth stage, genotype and environment. The infected roots exhibit external or internal symptoms. The external symptoms are the formation of annular necrotic lesions which gridle the fleshy fibrous roots caused by SPFMV (Russet crack strain). The tubers give a poor appearance and losses the market value. The internal symptom on tuber is caused by SPFMV-IC (Internal cork strain) which is characterized by the formation of corky brown hard tissues within the flesh. Infected tubers do not cook well and give a bitter taste. The virus is transmitted through aphids in a non persistent manner. Under experimental conditions, the virus could be transmitted through grafting (*Ipomoea setosa*), and through sap. *Chenopodium amaranticolor* and *C. quinoa* are local lesion hosts and *Nicotiana benthamiana* is a good propagative host.

Control measures

- Collection of planting materials from diseased fields should be avoided.
- Protection of the healthy plant material plots with systemic insecticides to prevent the spread through vectors.
- Plants showing virus disease symptoms amongst disease free population must be identified and destroyed.
- Weeds belonging to convolvulaceae from the vicinity of sweet potato fields should be eradicated.
- Virus free planting material could be obtained by meristem tip culture.

Insect pests

Sweet potato weevil (*Cylas formicarius*)

It is the most serious problem of sweet potato. Adult beetle is black with reddish-brown pro-thorax appendages. Eggs are white, oval in shape and covered with faecal matter. Adults and grubs cause damage to crop both in the field and in the storage. Adult weevils feed on all parts while the attack of grubs is restricted to vines and tubers. Feeding holes on the leaf, tuber and swelling of collar region of vines are some symptoms of its infestation.

Contrl measures

- Destruction of the plant residues and alternate hosts (other *Ipomoea* species) from the field.

- Use pest free planting material.
- Disinfection of the material by dipping in 0.05%Monocrotophos.
- Adoption of the cultural practices like, earthing up at 60 days after planting and frequent irrigation.
- Installation of pheromone traps @ 1trap/100 square metre to kill the male weevil.
- Collection of weevils from traps and periodical change of the detergent.
- Spraying the crop with fenthion or Fenitrothion @ 0.05% or Methyl demeton @ 0.05% at monthly interval.
- Avoidance of leaving the crop for long time in the field and harvest early to avoid infestation.

Yield

Tuber yield is 20-25 t/ha in irrigated condition; 10-12t/ha in rainfed condition. Apart from tuber yield, 10-12t/ha of vine is also available as green fodder or planting materials.

Perennial Vegetable Crops

- **Drumstick**
- **Agathi**
- **Chekkurmanis**
- **Curry leaf**

Drumstick

Drumstick, *Moringa oleifera* Lam. (syn: *Moringa pterygosperma* Gaertner) under the family Moringaceae and having chromosome number 2n = 28 is a multipurpose tree vegetable crop. Different parts of the plant are used for a variety of purposes. It is also known as horse- radish tree, ben oil tree and moringa. It is a fast-growing and drought-resistant tree. In India, it is grown all over for its highly nutritious pods, leaves and flowers. India is the largest producer of drumstick and among the states, Andhra Pradesh leads in both area and production followed by Karnataka, Tamil Nadu and Kerala. Apart from India, it is now cultivated throughout the Middle East, North Eastern and South Western Africa, Sri Lanka and in almost the whole tropical belt. It was introduced in Eastern Africa from India at the beginning of 20th century. In Nicaragua, the crop was introduced in the 1920s as an ornamental plant and as a live fence.

The crop

Drumstick is a fast growing, perennial tree which can reach a maximum height of 7-12 m and a diameter of 20-40 cm at 1m height. The main root is thick. The stem is normally straight but occasionally is poorly formed. The tree grows with a short, straight stem that reaches a height of 1.5-2 m before it begins to flower. The extended branches grow in a disorganized manner and the canopy is umbrella shaped. The alternate, twice or thrice pinnate leaves grow mostly at the branch tips. Leaves are 20-70 cm long, grayish-downy when young, long petiole with 8-10 pairs of pinnae each bearing two pairs of opposite, elliptic or obovate leaflets and one at the apex, all 1-2 cm long; with glands at the bases of the petioles and pinnae .

The flowers, which are pleasantly fragrant, and 2.5 cm wide are produced profusely in axillary, drooping panicles 10 to 25 cm long. They are white or cream colored and yellow-dotted at the base. The five reflexed sepals are linear-lanceolate. The five petals are slender-spatulate. They surround the five stamens and five staminodes and are reflexed except for the lowest. The zygomorphic flowers show a forenoon (6.00 h-12.00 h) pattern of anthesis after which pollen dehiscence takes place (7.00 h-13.00 h).

The pollen grains of drumstick are spherical with smooth exine, and had 3 germpores. The average diameters of fertile and sterile pollen grains were 43.5 and 33 μm, respectively. The average pollen production per anther during the rainy and summer seasons was 7 250 and 7 500, respectively. Pollen viability on acetocarmine staining and germination reached 97 and 88%, respectively. The best medium for pollen viability test through *in vitro* pollen germination is 10% sucrose supplemented with 200 μg/ml H_3BO_3, where 94±2% germination generates pollen tube with an average length of 1692 μm. Under ambient conditions, the pollen grains lose their viability within three days however, these can be stored for 7 days under refrigerated condition.

Natural populations yield an average of 10.28% fruit and successful pollination of the flowers requires a large number of insect (Orders Thysanoptera, Hymenoptera, Lepidoptera and Coleoptera) visitations and among these, *Xylocopa* is most common. Flowers favour cross pollination due to a delayed stigma receptivity. Biochemical studies of stigmas reveal that some extra proteins and esterase enzymes contribute towards its receptivity.

The fruits or pods are pendulous, triangular, tapering at both ends, 30 to 120 cm long and 1.8 cm wide, and contain from 20 to 35 seeds embedded in the pith. The pods split lengthwise into three parts when mature. The seeds are dark brown with three papery wings, round with a brownish semi-permeable seed hull. The hull itself has three white wings that run from top to bottom at 120-degree intervals. Each tree can produce between 15,000 and 25,000 seeds/year. The average weight per seed is 0.3 g and the kernel to hull ratio is 75:25.

Different bioactive compounds have been identified and isolated from different parts of the plant viz., anthomine and pterygospermine from root which is the condensation products of two benzyliso-thiocyanate molecules with one benzoquinone molecule; moringinine and spirochine (Sulphorated amino bases) from root bark; hydroxymellein, vanillin, octacosanoic acid, beta-sitosterol and beta-sitostenone from stem extract; benzylisothiocyanate derivative (a glycoside of mustard oil) and 4(4-acetyl-L-L-rhamnosyloxy)-benzoisothiocyanate from seed.

Importance and use

Its young seed pods and leaves are used as vegetables. The young green pods are very tasty and can be boiled and eaten like green beans and other vegetables. In India, the fruits are used to prepare a variety of delicious dishes like, curries, sambar, kormas, dals etc. Scraped drumstick pulp can be made into tasty dish called "drumstick vartha". It is also preserved by canning and exported worldwide.

The young leaves are edible and are commonly cooked and eaten like pot herb or used to make soups and salads. In the Philippines, the leaves are widely eaten. Bunches of leaves are available in many markets, priced below many other leaf vegetables. The leaves are most often added to a broth to make a simple and highly nutritious soup. The leaves are also sometimes used as a characteristic ingredient in "Tinola" — a traditional chicken dish, composed of chicken in a broth, drumstick leaves and either green papaya or another secondary vegetable.

The flowers can be eaten after being lightly blanched or raw as a tasty addition to salads. The resin from the trunk of the tree is also useful for thickening sauces.

Seeds should be eaten green before they change colour to yellow. Dry seeds can be ground to a powder and used for seasoning sauces. The roots from young plants can also be dried and ground for use as a hot seasoning base with a flavour similar to that of horseradish. This is why the drumstick tree has been given the name "Horseradish Tree". A tasty hot sauce from the roots can also be prepared by cooking them in vinegar.

The leaves and the residue obtained after the recovery of oil and coagulants can be good sources of proteins for animal feeds. Productivity of fresh material per unit area of drumstick is high compared with other forage crops.

Nutritional value

Fruit and leaves of drumstick are rich sources of proteins, vitamins and minerals β-carotene, amino acids and various phenolics and have low levels of anti-nutrients. Young leaves are an exceptionally good source of β-carotene, vitamins B, and C, minerals (particularly iron), and the sulphur-containing amino acids viz., methionine and cystine. Drumstick also provides a rich and rare combination of zeatin, quercetin, β-sitosterol, caffeoylquinic acid and kaempferol.

Medicinal value

Drumstick is very much valued for different medicinal properties. Various parts of this plant such as the leaves, roots, seed, bark, fruit, flowers and immature pods act as cardiac and circulatory stimulants, possess antitumor, antipyretic, antiepileptic, anti-inflammatory, antiulcer, antispasmodic, diuretic, antihypertensive, cholesterol lowering, antioxidant, antidiabetic, hepatoprotective, antibacterial and antifungal activities. These plant parts are being employed for the treatment of different ailments in the indigenous system of medicine, particularly in South Asia.

Drumstick is a vital component in Ayurveda, Siddha, Unani medicine systems of India. Fresh root which is acrid and vesicant is administered in cases of in intermittent fever. The root is applied externally as a poultice in cases of inflammation, as a valuable rubefacient in palsy and dropsy, and for bites from rabid animals. An infusion of the roots is recommended for asthma, and is useful in ascites caused by diseases of the liver and spleen. Root and root bark and stem bark are used as an abortifacient (cause abortion). The paste of the root bark is used orally for urinary calculi. The bark and leaf induce sweating used in anorexia and external ulcers. Leaves are galactogogue (substance that promotes lactation in humans and other animals), refrigerant, laxative and improve digestion. The tender leaves reduce phlegm and are administered internally for scurvy and catarrhal conditions. Flowers, irritant in action, are used to heal inflammation of tendons and abscesses. The unripe pods act as a preventive against intestinal worms. The fruit is sweet and pungent in taste, an appetizer preventing eye disorders and increasing semen both qualitatively and quantitatively.

Origin and taxonomy

Drumstick is native to the southern foothills of the Himalayas in north-western India, and widely cultivated in tropical and subtropical areas. There is evidence that cultivation of drumstick tree in India dates back many thousands of years.

Thirteen different species of *Moringa* (*M. drouhardii, M. hildebrantii, M. ovalifolia, M. stenopetala, M. concanenesis, M. oleifera, M. peregrina, M. arborea, M. borziana, M. longituba, M. pygmaea, M. rivae,* and *M. ruspoliana*) belongs to the Family Moringaceae. Out of the 13 species, *Moringa oleifera* (syn: *Moringa pterygosperma* Gaertner) is the economically most valuable species. Taxonomic position of the family Moringaceae is not yet clear and although, it has some features similar to those of Brassicaceae and Capparidaceae, the seed structure does not agree with either of the above families.

Moringa oleifera is a diploid species with 28 chromosomes (2n = 28).

Adaptation in India

Moringa oleifera Lam. (syn: *Moringa pterygosperma* Gaertner) is a native of India, occurring wild in the sub-Himalayan regions of North India and now grown world-wide in the tropics and sub-tropics.

Open pollinated improved varieties

Perennial Drumstick

A number of perennial cultivars of drumstick are cultivated in different parts of India giving economic yield (400-600 fruits per plant per year) for many years.

- **Jaffna (Yazpanam Moringa):** An introduction from Jaffna, Sreelanka which bears long pods of 60-90 cm length and with soft flesh.
- **Chem Moringa :** An introduction from Jaffna, SriLanka which flowers and bears fruits throughout the year.
- **Chavakkacherri Moringa :** An introduction from Chavakkacherri near Jaffna which bears long pods (90-120 cm).
- **Pal Moringa :** An introduction from Jaffna type with thick pulp and better taste.
- **Kodikal Moringa :** An annual type grown in Thruchirappalli district of Tamilnadu as a standard for training betel vine which bears thick fleshed short pods (20-25 cm).
- **Konkan Ruchira :** It is a selection from Vasai Local which bears dark green and medium long fruits.
- **Coimbatore 1 :** A high yielding variety with medium fruit length of (45-60 cm) and weight (60-70 g).
- **Coimbatore 2 :** A short fruited (25-35 cm) variety.

Annual Drumstick/ Seed Moringa

Annual drumstick show precocious bearing habit without any definite fruiting peaks. The tree becomes exhausted and appears to senesce at the end of one year giving it the name "annual". It is seed propagated and spreading fast in India due to their fast growth, high yield and adaptability to varied soil and climatic conditions.

- **PKM 1:** It is a dwarf tree growing to height of 4-6 m and is suitable for ratoon crop. Pods are 60-70 cm long with 6.3 cm girth weighing 120g, very pulpy with 70% edible portion. The average yield is 50-54 tonnes/ha (220-250 fruit per tree).
- **PKM 2 :** This variety is a hybrid derivative of the cross between MP 31x MP 28. It is a dwarf tree growing to height of 4.8 m in a period of six months and suitable for ratoon crop. Pods are 125 cm long with 8.3 cm girth weighting 280g. The average yield is 80-90 tonnes/ha (230-240 fruit per tree).

- **KM 1:** It is a selection from an annual seed moringa. Plant is short statured bearing short (32- 37cm) and thick (5.5- 6.0 cm) pods weighing 65- 82 g. The average yield is 30-35 tonnes/ha (226- 328 fruit per tree).

Cultural requirements

Drumstick is generally classified as a tropical plant however, is adapted to a wide range of agro-climatic conditions. It is most commonly cultivated in tropical or subtropical semi-arid regions. It grows best at temperatures between 25 and 35°C, but will tolerate up to 48°C in the shade and can survive a light frost although, growth is slowed down by temperatures below 10°C. It is drought-tolerant and grows well in areas receiving annual rainfall amounts that range from 250 to 1500 mm. It can also tolerate high rainfall in excess of 2000 mm, but most cultivars are sensitive to flooding and waterlogged conditions. Altitudes below 600 m are best for successful cultivation, but it can also grow in altitudes up to 1200 m in the tropics.

It can be grown in a variety of soil types and conditions from well-drained sandy loamy soils to heavier clay loam soils; however, it prefers a well-drained sandy loam or loam soil. It will not survive under prolonged flooding and poor drainage condition. Drumstick can tolerate a wide range of soil pH from acidic to alkaline reaction (pH 5-9).

Land preparation

The plant requires thorough land preparation and a well-prepared seedbed. In newly opened land with thick vegetation, the field should be cleared by mowing or cutting tall shrubs, grasses, etc. This should be followed by ploughing and harrowing to a maximum depth of 30 cm. It can be planted on ridges or in furrows. In areas with high rainfall, planting on ridges or raised beds is better than planting on furrows formed between ridges to reduce the risk of water logging. Bed width for annual moringa can vary from 60 to 200 cm depending on plant spacing and density. The beds may be covered with plastic mulch to prevent weed growth and holes are dug at 15-20 cm for transplanting the seedlings of annual moringa.

Propagation methods

Drumstick is planted either by direct seeding, transplanting of seedlings or using long stem cuttings. Perennial drumstick is propagated by stem cuttings and annual moringa is generally propagated by seed. Another method known as bioculture technique developed in the Philippines which utilizes small stem cuttings rooted in rooting medium with vermicompost.

Vegetative propagation: In perennial types, limb cuttings 50 cm -1.5m in length with a diameter of 14-16 cm are planted *in situ* during rainy season. Elite trees are cut down, leaving a stump with a 90cm head from which 2 to 3 branches are allowed to grow. From these shoots, cuttings of 100 cm long and 4 to 5 cm in diameter are selected and used as planting material. Cuttings are dried in the shade for three days before planting in the nursery or in the field.

Shield budding is successful, and the budded trees begin to bear fruit in 6 months. In Tamil Nadu condition, September and December is the best season for budding.

Seed propagation: Seeds do not show any dormancy periods and can be sown as soon as they are extracted. At present, seed propagated annual drumstick occupies about 70 per cent of the total area under drumstick cultivation in southern India.

Planting of stem cutting of perennial drumstick

The limb cuttings are planted in pits of 60 x 60 x 60 cm at a spacing of 4 x 4 m, during the months of June to August. The monsoon rains during the period facilitate easy rooting and further growth. While planting, one-third of the cutting should be kept inside the pit.

Seedling production of annual drumstick

Seedlings of annual drumstick can be grown in divided trays, individual pots, plastic bags, or seedbeds. A 50-cell tray with 3–4 cm deep and wide cells is suitable. The tray is filled with a potting mix that has good water-holding capacity and good drainage. Peat moss, commercial potting soil or a potting mixture prepared from soil, compost or rice hulls and vermiculite or sand can be used. The mix is sterilized by autoclaving or baking at 150°C for 2 hours. The seedlings are grown under shade or in a screen house with 50% shade. Two or three seeds are sown per cell. One week after germination, thinning should be done keeping the strongest seedling. Seedlings are transplanted one month after sowing or when they reach 20–30 cm in hight.

The seeds of annual drumstick can also be directly dibbled in the pit to ensure accelerated and faster growth of the seedlings. The best suited season for sowing the seeds is September under South Indian conditions. The time of sowing has to be strictly adhered to because the flowering phase should not coincide with monsoon seasons, which results in heavy flower shedding.

A plant spacing of 2.5 x 2.5 m between rows and plants should be adopted, giving a plant population of 1600 plants/ha. Pits of 45 x 45 x 45 cm in size are dug and the seeds are sown in the centre of the pit. The seed germinates 10 to 12 days after sowing. The seed requirement per hectare is 625g. Treatment of drumstick seeds with *Azospirillum* cultures at the rate of 100 g per 625 g of seeds before sowing result in early germination, and increased seedling vigour, growth and yield.

Nutrient management

Drumstick as a deep rooted tree grows well in most soils without additions of fertilizer. Once the tree is established, the extensive and deep root system of drumstick is efficient in mining nutrients from the soil. However, in poor soils, low in nutrient content, fertilizer application is essential for optimum growth and yield. Drumstick responds well to organic fertilizer application and application of farmyard manure at the rate of 1-2 kg/tree significantly increases crop performance in terms of growth and yield components. Manure or compost should be applied before the rainy season when the trees are about to start an intense vegetative growth and the second application may be done before the dry season for maintaining vegetative growth.

For optimum growth and yield, fertilizers @ 100-100-100 kg/ha of N, P_2O_5 and K_2O should be applied every year. Trenches are dug around the base of the plant (10-20 cm from the base) and commercial NPK fertilizer is applied at 300 g per tree per year in two equal split doses.

Irrigation

Drumstick is drought tolerant and does not require frequent irrigation. Seeds germinate and establish well if sown in moist soil and this is common during the wet season which is the recommended planting time. For optimal growth, irrigation should be given regularly during the first 2-3 months after sowing or transplanting. Once the trees are established, regular irrigation scheduling is not necessary particularly in humid tropics where rainfall is uniformly distributed throughout the year. However, during the dry season, irrigation is necessary to keep the trees producing leaves. For commercial cultivation, drip irrigation can be adopted with daily application rates of 12 to 16 liters of water per tree during the dry season and half this rate during the wet season.

Interculture and weed management

Weeds compete with drumstick for soil nutrients, water and light and if they are not timely controlled, the competition becomes very critical, especially at seedling and plant establishment stage. Weeds can be managed by manual weeding using hand hoes which will also loosen the soil for good aeration. An alternative to mechanical or manual weed control is mulching. Organic mulch can be applied with dry grass or weeds, rice straw, compost or dry leaves. Black polyethylene mulch is commonly used and is effective against weeds. Mulch also conserves soil moisture and reduces evapo-transpiration thus, minimizing the need for irrigation. Organic or polyethylene mulch can be applied around the base of each young tree and a weed-free planting area can be maintained by regularly cultivating between beds and rows.

After care and pruning

Drumstick is often pollarded to encourage lateral bud and new shoot growth. Pinching the terminal bud on the central leader stem is necessary when it attains a height of 75cm about two months after sowing. This will promote the growth of many lateral branches and reduce the height of the tree. In addition, pinching 60 days after sowing reduces the damage due to heavy wind and makes harvesting much easier.

Ratooning

Annual drumstick tree, after the harvest, is cut down to a height of one metre above ground level for ratooning. These ratoon plants develop new shoots and start bearing 4-5 months after ratooning. During each ratooning operation, the plants are supplied with the recommended dose of N, P and K nutrients along with 20-35 kg of FYM per plant.

Perennial types should also be pollard back to a height of 1.5 m from ground level during October November, followed by manuring with organic matter @ 25kg/plant and the recommended dose of fertilizers.

Harvesting the crop

Perennial types raised by cuttings take nearly a year to bear fruit. The pods are harvested mainly between March and June. A second crop is normally harvested from September to October. The pods should be harvested as early as possible when they reach maturity,i.e. when they turn brown and dry. A single tree can yield 50-70 kg of pods in one year under favourable conditions.

Annual drumstick types are seasonal in terms of fruit- bearing and the crop sown during September comes to harvest within six months. Fruit of sufficient length and girth are harvested before they develop fibre. The harvest period extends for 2-3 months and each tree bears 250-400 fruit depending on the type.

Harvesting for leaf production can be done by two methods. In the first method, the leaves are picked by the rachis one by one, while in the second method the trees are pruned at a definite height from the ground level (25-75 cm).

Post harvest handling and storage

Fruits should be packed either in polythene bags or Corrugated Fibre Board boxes with coir waste as filling material for reducing the physiological loss in weight. Shelf life of drumstick pods can be increased to approximately three to four times in controlled atmosphere storage at refrigerated conditions up to 40 days at 14°C and 4% O_2 and 5% CO_2 concentration in storage. Drumstick leaves are processed mainly for making powder, teas, coffee, juice and energy drinks. Leaf powder is used as ingredient of several food products including bread, noodle, biscuits, milk, crackers, potato chips and other snack food. The powder is also the main ingredient in food supplements (vitamins) in the form of "Moringa capsules" which are now widely available in the health food sector.

Matured and de-hulled Moringa seed contain approximately 20-40% oil depending on the extraction method and the genotype.

Different parts of the tree are processed into various products and important among them are listed below :

1. Moringa leaf powder
2. Moringa tea powder
3. Moringa soup powder
4. Moringa juice powder
5. Moringa pickle
6. Moringa oil
7. Moringa seed kernel
8. Moringa cake powder
9. Moringa capsule

Plant protection measures

Diseases

Fruit rot (*Fusarium semitectum, Drechslera haraiensis*)

Fruit rot caused by the fungus *Fusarium semitectum* is characterised by elliptic spots with distinct dark brown margin and light brown centre on the developing fruits. Symptoms appear as small water soaked spots of 1.0 mm diameter which later enlarge and coalesce with adjacent lesions. Light pinkish conidial mass of the fungus become visible on the affected areas. The spots are deep seated and causes fruit rot. In advanced stages, the pods dry up prematurely leaving uneven raised spots over them.

Fruit rot caused by another fungus *Drechslera haraiensis* is characterized by elliptical or elongated sunken spots with reddish brown raised margins on green pods. Diseased pods are shrunken to thinner dimensions at their stigmatic ends.

Control measures

- Use disease free and healthy seeds.
- Dry seed treatment with Carbendazim/Captan/Thirum @ 2-3g/kg of seed.
- Application of adequate organic matter in soil.
- Soil application of *Trichoderma viride* @ 3 kg/ ha along with bio-manure is effective.
- Spraying the crop with Carbendazim (1g) + Mancozeb(2g)/l along with sticker checks the rapidity of the infection.

Brown leaf spot (*Cercospora* spp)

The symptoms appear as scattered brown spots appear on the leaves and then spread to cover them entirely. Coalescing of spots leads to irregular and blighted appearance of the leaves. The leaves turn yellow and fall off prematurely. Conidia are disseminated by wind and rain splashes to the leaf surface where they germinate and cause infection. The fungus sporulate abundantly at 20 to 30ºC and disease development is favoured by high humidity and temperature.

Control measures

- Spraying the crop with 0.2 % Chlorothalonil or 0.1% Propiconazole or 0.25% Mancozeb at 15 days interval.

Insect pest

Pod fly/Fruit fly (*Gitonia distigmata*)

The adult fly is small, yellowish with red eyes. It lays eggs in grooves between the ridges of fruits. The newly hatched out maggot enters into the fruits and starts feeding on the internal content of fruits and causes gummosis leads to drying of fruits from tip. In addition to drying, it also causes splitting of fruits from tip. The maggots pupate in the soil.

Control measures

- Collection and destruction of all the fallen and damaged fruits
- Raking up the soil under the trees or ploughing the infested field to destroy the pupae
- Use attractants like citronella oil, eucalyptus oil, vinegar (Acetic acid), dextrose or lactic acid
- Setting up of fish meal trap to attract and kill the adult fly
- Application of Fenthion 80 EC 0.04 % during the vegetative and flowering stage.
- Application of Nimbecidine 0.03 % at 150 ppm during 50 % fruit set and 35 days after
- Soil application of Neem seed kernel extract (NSKE) @ 2 lit / tree at 50 % fruit set

Stem borer (*Batocera rubus Linn.*)

Adults are medium sized beetles and yellowish brown with white spots on elytra. Grubs are stout, about 10 cm long, yellowish with well-defined segmentation. Eggs are laid singly in cracks or crevices in the bark of the tree. On hatching, grubs make zig-zag burrow beneath the bark, feed on internal tissues, reach sapwood and cause death of the affected branch or stem. Pupation takes place within these tunnels.

Control measures

- Cleaning of all webbed material and excreta
- Plugging the holes with cotton wool soaked in fumigants like chloroform, formalin or petrol and sealing it with mud.

Caterpillars (*Eupterote mollifera, Noorda blitealis, Tetragonia siva, Metanastria hyrtaca*)

These insect pests are serious defoliators. They also infest unopened flower buds, causing them to drop.

Control measures

- Mechanical destruction of the gregarious caterpillars at the initial stage of attack
- Spraying the crop with Nuvacron @ 1 ml/l, or Thiodan @ 1.4 ml/l or Malathion @ 2 ml/l or Sevin W.P. @ 4 g/l of water
- Spraying the crop with 5% neem seed kernel extract reduces the incidence

Yield

Fruit yield is generally low (80-90fruit / year) in the first two years of fruit-bearing which gradually increases to 250-300 fruits/tree/ year in the fourth years onwards.

Agathi

Agathi (*Sesbania grandiflora* Pers) is a perennial, tropical, quick growing and soft wooded tree belonging to family Fabaceae. It is also called agati, scarlet wistaria tree, vegetable hummingbird, West Indian pea, etc. It is cultivated in many parts of India and Sri Lanka. There are two forms of agathi, one with red flowers and the other with white flowers; the white flowered agathi is suitable for kitchen garden. However, it is not grown on large scale for vegetable purpose. It is often cultivated on the low dikes between rice fields or in association with Guinea grass.

Agathi is a erect, fast-growing perennial (4-5 m in just 6 months), sparsely branched and soft wooded, legume tree that reaches 10-15 m in height and a diameter up to 12 cm. Its lifespan is about 20 years. Its roots are heavily nodulated and some floating roots may develop in waterlogged conditions. The trunk is straight with few branches. Leaves are paripinnately compound up to 15-25 cm long with 20-50 leaflets in pairs, dimensions 12-44 x 5-15 mm, oblong to elliptical in shape and opposite arrangement. Single leaflet is 2-4 cm long and 10-15 mm in breadth, linear, oblong, mucronate, stipulus lanceolate or setaceous deciduous. The bark is light gray, corky and deeply furrowed and the wood is soft and white. The flower clusters hanging at the leaf base. Flowers are oblong, 7.5-10.0 cm long with short axillary racemes, curved about 3 cm wide before opening. 2–4 large flowers with showy white or red petals are borne in the racemes. The calyx is campanulate and shallowly 2-lipped. Five petals are differentiated into a standard, wing and keel petals. The standard petal is usually upright, the wing petals spread out on either side of the flower, and the keel is boat-shaped and in this species is curved down and away from the flower. The fruit looks like flat, long and thin green beans. Pod is sub-cylindrical, straight or slightly curved up to 30-45 cm long and 5-8 mm wide, straw-colored or reddish-brown, glabrous and indehiscent, and hang vertically, containing 15-50 dark brown seeds. The seed is 3-4.5 mm x 2 mm x 2 mm, sub-cylindrical or bean like, elliptical, olive-green or red brown, 6-8 in a pod and 55-80 seeds weigh one gram.

Use

Leaves are used as green pot herb and the flowers are also eaten as vegetable for making fried preparations. The flowers are eaten as a vegetable in South Asia and Southeast Asia, including India, Laos, Thailand, Java in Indonesia, Vietnam, Maldives , Sri Lanka, and the Ilocos region of the Philippines. The young pods are also eaten.

Agathi has several environmental benefits. As a fast-growing, N-fixing legume, it is used for the reforestation of eroded areas and to improve soil fertility. It is often planted to make fence lines or as shade tree, windbreak and support for other crops like, blackpepper, betelvine, vanilla, etc. In Tamil Nadu, it is grown around banana plantation as a wind-break and around coconut seedlings as a shade plant. The bark yields good fibre and gum.

Nutritive and medicinal value

Leaves are rich in protein (8.4 g), vitamin A (9000 IU), vitamin C (169 mg), fibre (2.2 g) and minerals like calcium (1.13 g), phosphorus (80 mg) and iron (3.9 mg) per 100 g edible part. Leaves and flowers have medicinal properties.

The leaf extract may inhibit the formation of advanced glycation (the bonding of a sugar molecule to a protein or lipid molecule without enzymatic regulation) end-products. The leaf extract contains linolenic acid and aspartic acid, which have been found to be the major compounds responsible for the anti-glycation potential of the leaf extract. The leaves have the potential to be considered as a candidate for preparing the new treatment of diabetes mellitus.

Origin and taxonomy

It is a native of Malaysia. The plant is now widespread in most humid tropical regions of the world. Agathi (*Sesbania grandiflora* Pers) belongs to the family Fabaceae and having chromosome number 2n = 24.

Propagation

Agathi is propagated through seeds. However, stem cutting treated with different concentrations of IBA, NAA and Boric acid and their combinations induce rooting.

Cultural requirements

The tree thrives under full exposure to sunshine and is extremely frost sensitive. Optimal growth conditions are 22-30°C mean annual temperatures, 2000-4000 mm annual rainfall, at an altitude from sea level up to 800-1000 m. It does not thrive in temperatures below 10°C. It is adapted to a wide range of rainfall zones and soil types. It can be grown on heavy clay, alkaline and saline soils, as well as poorly drained soils and poorly fertilized soils. During waterlogging and floods, it develops floating adventitious roots and protective spongy tissue. Agati withstands acidic soils, 6- to 7-month drought periods and can survive with 800 mm annual rainfall. It is intolerant of high winds that can break stems and branches.

Crop production

It grows best in black cotton soils and comes up quickly when the surface soil is loose and uneven. It is resistant to drought. The seeds are first sown in the nursery. The seedlings are transplanted in the pits when they are 30-45 cm in height at a spacing of 90-100cm either way. Two months after transplanting small quantity of ammonium sulphate is applied to the seedlings.

In early stages of growth, it suffers severe attack by seedling blight caused by the fungus *Colletotrichum capsici*. The disease can be effectively controlled by spraying the crop with 0.4% Blitox.

Leaves are produced throughout the year and tender leaves are harvested periodically. First harvest can be taken 3-4 months after planting. A tree yields 4.5 - 9.1 kg of leaves / year. The plants come to flower by September - December and to fruiting during summer.

Chekkurmanis

Chekkurmanis (*Sauropus androgynous* Merr.) is a perennial shrub grown in some tropical regions as a leafy vegetable crop. It is known as Star gooseberry, Sweet leaf bush, Phak waan baan in Thailand, Cekurmanis in Malaysia, Katuk in Indonesia, Binahian in Philippines and Dom nghob in Cambodia. The plant grows wild in the evergreen forests of the Western Ghats and in the southern parts of Kerala. This plant is widely cultivated in Southern India, Indonesia, Singapore, Bangladesh, Guangdong, Guangxi, Hainan, and Yunnan.

It is an erect shrub with multiple upright stems that can reach up to 5.0m in height. It is flaccid and has cylindrical or angled branches. The leaves are dark green, measuring 2.0–7.5cm and oval or obtuse in shape. The plant has monoecious sex form bearing separate male and female flowers on the same plant. Flowers are without petals, small but prominently produced in leaf axils. The male flowers are disk-shaped, entirely or nearly so and have 6 creamy sepals enclosing 3 reddish stamens. Female flowers have 6 red sepals surrounding a single creamy-coloured pistil. Fruits are nearly globular, up to 1.5cm in diameter and whitish, explosive, 1.2 x 1.7 cm with persistent red sepals and many black seeds.Leaves contain an opium alkaloid called papaverine to the extent of 580mg per 100 g, which is excessive considering only 200mg of papaverine per day is required as an antispasmodic drug.

Use

It is one of the popular leafy vegetables in South and Southeast Asia and is notable for high yields and palatability. The leaves can be cooked like other greens. Leaves are used for the preparation of soup in Java. The leaves are used to give a light green colour to pastry and to fermented rice in the Dutch East Indies. The shoot tips are sold as tropical asparagus. In Vietnam, the locals cook it with crab meat, minced pork or dried shrimp to make soup. In Malaysia, it is commonly stir-fried with egg or dried anchovies. Flowers and small purplish fruits are also eaten. In Indonesia, the leaves are used to make infusion, believed to improve the flow of breast milk for breast-feeding mothers.

The plant is also useful for growing as a hedge around home gardens. In Java, it is often planted in live fences and in midst of garden beds to provide light shade for other vegetables in beds. Besides other uses, leaves are used as cattle and poultry feed in certain parts of the country. In some other places, the plants are grown as a soil binder to prevent soil erosion.

Nutritive and medicinal value

Chekkurmanis is also called as multi vitamin and multi mineral packed leafy vegetable. It has unique position in the list of leaf vegetables because of its high nutritive value and multi-purpose uses. Its leaves are very rich in protein, minerals and vitamins A, B and C. The leaves contain approximately 7.4g protein per 100g of fresh leaves whilst, for comparison, spinach has 2.0g, mint 4.8g, and cabbage about 1.8g. It also has high level of β carotene (pro-vitamin A) especially in freshly picked leaves, as well as high levels of vitamins B and C and minerals. Leaves are rich in carbohydrate (11.6 g), fat (3.2 g),vitamin A (9510 IU), thiamine (0.48 mg),riboflavin (0.32 mg), vitamin C (247 mg), fibre (1.4 g) and minerals like calcium (570 mg g), phosphorus (200 mg),iron(28 mg) per 100g edible part. The more the leaves mature, the higher the nutrient content of the leaves. Leaves contain different phytochemicals viz., sterols, resins, tannins, saponins, alkaloids, flavonoids, terpenoids, glycosides, phenols, catechol, cardiac glycosides, and acidic compounds.

The leaves have several medicinal properties. The juice of leaves pounded with roots of pomegranate (*Punica granatum*) and leaves of jasmine (*Jasminum sambac*) is used against eye troubles. A decoction of its roots is often recommended against fever in rural areas. Pounded roots and leaves are used as poultice for ulcers in the nose.

Origin and taxonomy

Chekkurmanis (*Sauropus androgynous* Merr.) belongs to the family Euphorbiaceae. It is native of India and Burma. It is found in Sikkim Himalayas, Khasi, Abor and Arka Hills at 1200 m mean sea level.

Propagation

It is generally propagated by stem cuttings which root easily. Both hardwood and semihard-wood cuttings of 20-30 cm length after trimming the leaves are used as planting materials. The crop can also be propagated through fresh seeds. Seed longevity is poor and it remains viable for only a few months.

Cultural requirements

A warm humid climate with good rainfall is the best suited for the luxuriant, succulent growth of leaves and twigs. The crop comes up well in mild humid locations also. Growth slows during the cold months and crops become unpalatable. The plant grows well in all types of soil. It can tolerate shade to some extent. It is fairly drought tolerant.

Crop production

The land is ploughed and leveled well. Well rotten FYM is added @ 2 kg/ sq.m. Rooted cuttings can be produced in nursery beds also or in pots and then transplanted in the shallow furrows spaced 30 cm and at a distance of 10-15 cm in 2-3 rows. The cuttings are planted at least 15 days earlier to the onset of monsoon during April-May. Subsequently frequent irrigations are given until root initiation takes place. Even though it withstands the hot dry weather for a long period, watering of plants in such condition is desirable for getting constant growth of new leaves. After the onset of monsoon, it does not require much irrigation. The cuttings come up very well and would be ready for harvest within 3 - 4 months of planting. Tip of the plant is nipped off which enables the plants in putting forth new branches. The tender shoots and leaves can be harvested intermittently for several subsequent years.

Curry leaf

Curry leaf (Murraya *koenigii* (L.) Sprengal) is a spicy leaf vegetable crop. Curry leaf grows throughout India including Andaman and Nicobar Islands upto an altitude of 1500m mean sea level. It is a backyard crop in many south Indian homesteads. In India, on a commercial scale, it is cultivated in Tamil Nadu, Andhra Pradesh, and Karnataka of India. It is also available in other part of Asian regions of Guangdong, Shainan, Bhutan, Laos, Nepal, Pakistan, Sri Lanka, Thailand and Vietnam.

The plant is semi deciduous, aromatic small spreading shrub or small tree with strong woody but slender stem which is dark green to brownish in colour. The tree grows up to 4.0–8.7 m in height and 15 to 40 cm in diameter.

Curry leaves are aromatic in nature having characteristic aroma. Leaves are shiny and smooth with paler undersides. Leaves are pinnate, exstipulate, having reticulate venation and having ovate-lanceolate with an oblique base, with 11-21 leaflets which are 2.0-3.5 cm long 1.00-2.0 cm wide. The leaflets are short stalked, alternate, gland dotted and having 0.5 cm long petiole. Leaf margins are irregularly serrate. Compounds found in curry tree leaves, stems, bark and seeds contain cinnamaldehyde and numerous carbazole alkaloids, including mahanimbine and girinimbine. The leaves are slightly pungent, bitter and feebly acidic in taste. The peculiar aroma is due to sulphur containing essential oil. The most important chemical constituents responsible for its intense characteristic aroma are P-gurjunene, P-caryophyllene, P-elemene and O-phellandrene. Curry leaf loses flavour when cooked. To retain fresh flavour, the leaves should not be removed from branches until ready for use. Leaves yield volatile oil and crystalline glucoside- 'Koenigiin'.

Inflorescence is a terminal cyme. Each cluster bears approximately 60 to 90 flowers which are bisexual and self-pollinated. Flowers are small, white, fragrant and ebracteate; calyx deeply five cleft, pubescent. A glucoside 'Murrayin' is obtained from flowers. Petals five, free, whitish, glabrous and with dotted glands. Stamens are 10 in number and small of about 4 mm, dorsifixed, arranged into circles, with long superior gynoecium of 5 to 6 mm size. Fruits are borne in close clusters, small ovoid or sub-globose, glandular, black with shiny appearance, thin pericarp enclosing one or two seeds having spinach green colour. In general, neither the pulp nor seed is used for culinary purposes.

Use

Leaves are used in culinary preparation to enhance flavour and taste of different food preparations of Indian cuisine like, curry, sambar, vada, rasam, kadhi, etc. Curry leaves ground with mature coconut kernel and spices form an excellent preserve. Spice powder is prepared for export. Oil is used as fixative for heavy type of soap and perfume which is extracted from leaves by solvent extraction method.

Nutritive and medicinal value

Leaves are very rich source of iron and vitamin.A. Leaves contain 6.1g protein, 1.0g fat, 18.7g carbohydrate, 83 mg calcium, 57 mg phosphorus, 0.93 mg iron, 7.56 mg β carotene and 4 mg vitamin C per 100 g fresh.

Curry leaves have a versatile role to play in traditional medicine. Green leaves are eaten raw for cure of dysentery, diarrhoea and for checking vomiting. Leaves and roots are also used traditionally as antihelmintic, analgesic, curing piles, inflammation, itching and are useful in leucoderma and blood disorders. External applications of the leaves have been beneficial in bruises, eruption and to treat bites of poisonous animals.

Origin and taxonomy

Curry leaf plant has been originated from east and south part of India, Pakistan, Sri Lanka and China. It grows wild in the Himalayas, Deccan plateau and in Western Ghats. Curry leaf (Murraya *koenigii* (L.) Sprengal) belongs to the family Rutaceae and having chromosome number is 2n=18.

Propagation

The plant is propagated by seeds as well as one year old root suckers. Seeds must be ripe and fresh. Seeds from dried or shriveled fruits are not viable. The whole fruit can be planted, but it is best to remove the pulp before planting in potting mix that is kept moist but not wet. Stem cuttings can also be used for propagation.

Cultural requirements

Curry leaf plant grow luxuriantly throughout the spring, summer and in rainy season in every part of the tropical region up to the height of 1500 to 1655 m from sea level. The leaves drop off during its' resting period in the winter months. It is a high temperature tolerant crop. It tolerates maximum temperature ranging 36-37°C but when temperature falls below 16°C, the vegetative buds become dormant. The plant prefers full sun and dry, well-drained soil for adequate growth.

Crop production

Some improved and highly aromatic varieties *viz*., DWD 1 (Suwasini), DWD 2 (Senkaampu), etc have been developed in India through clonal selection and selection from open pollinated seedling progeny. Generally seedlings are raised in the nursey and transplanted in main field. Well ripe fruits are harvested from high yielding mother trees, seeds extracted and sown in nursery beds or poly bags within 3-4 days. Each fruit contains 2-3 seeds. Nursery beds of size 1 x 1 m with 30cm height are prepared and adequate farm yard manure is incorporated. Seeds are sown at a spacing of 10cm and germination takes place with in three weeks.

The main field is prepared through ploughing to fine tilth and adding well rotten farmyard manure @ 20t/ha. Pits of $30cm^3$ size are dug at 1.2 to1.5m spacing and one year old seedlings are planted at the centre of each pit. Pits are irrigated immediately after planting. Second irrigation is given on third day and thereafter at weekly intervals. The interspace is kept free of weeds by periodical hoeing and in the first year, one intercrop of a pulse crop can be taken. A fertilizer mixture comprising of 150g N, 25g $P_2O_5$and 50g K_2O per plant along with 25 kg FYM is recommended annually for the crop. After each harvest, farm yard manure is applied @ 20kg/plant and mixed with soil. Plants should be pruned at the age of 8-10 months after planting at a height of 30cm from ground level retaining 5-6 branches which helps to maintain bushy structure of the plant.

In early stages of growth, it suffers severe attack of leaf spot and basal stem rot caused by the fungus, *Sclerotinia rolfsi*. These diseases can effectively be controlled by spraying the crop with 0.1% Carbendazim. Several insect pests viz., Citrus butterfly, Psyllid bug and Scale insects cause severe defoliation. These insect pests can be checked by spraying the crop with 0.05% Dimethoate (Roger 2 ml/l), 0.08% Malathian (Cythion 2 ml/l) at 10 days interval.

Harvesting of leaves starts 10-12 months after planting. At the end of first year, 250-400 kg leaves can be harvested from one hectare of land. In the second and third year, harvesting can be done once in 4 months and total yield per year is approximately 5400 kg/ha. In fourth year, harvesting is done once in 3 months and 2500 kg leaves can be obtained in each harvest. From fifth year onwards harvesting is done at three months interval and average yield will be 20t/ha of green leaves. The leaves retain flavour even after drying and hence, these are marketed both in fresh and dried form.

Chapter 14

Vegetable Based Cropping System

Cropping system is defined as the cropping pattern followed on a farm and its interactions with farm resources, other farm enterprises and production technology. The yearly sequence and spatial arrangement of crops and fallow on a given area is termed as cropping pattern.

India has over 17% of world's population living on 2.4% of the world's geographical area. Per capita agricultural land has reduced by 67% from 0.48 ha in 1951 to 0.16 ha in 2008 due to explosive increase in population. In this situation, it is very difficult to get horizontal expansion in vegetable cultivation. Hence, total production per unit area can be increased by increasing the productivity per unit area through (a) growing improved high yielding varieties and hybrids, (b) adequate and scientific crop management practices and (c) increasing cropping intensities following different cropping patterns.

Multiple Cropping

It is a cropping system representing the philosophy of maximum crop production per unit area of land within a calendar year or other relevant time span with minimum soil deterioration. In its simplest form, multiple cropping is a one-year cropping system in which two or more crops are grown in succession within a year. If all the crops are same it will resemble monoculture.

Basic principles of multiple cropping

- Utilization of crop varieties and hybrids for specific climatic requirements or duration.
- Soil management.
- Method of tillage.
- Crop residue management.
- Judicious use of fertilizer and other chemicals.
- Provision of adequate irrigation water (excess or lack of water should not be a limiting factor in any month of the calendar year).

- Compatibility of the crop combination with capital, enterprise preference and farmers' skill.

Some of the consequences of multiple cropping are outlined briefly which amply justifies that there is no danger or any permanent deterioration of soil productivity if this cropping system is followed with proper management practices.

- In multiple cropping systems, plant nutrients present in soil are utilized exhaustively due to intensive cropping, particularly with vegetable crops. Any loss in soil productivity due to such depletion is temporary and can be corrected by balanced application of organic manure and inorganic fertilizers.
- In multiple cropping, much organic residue and crop roots are added to the soil enriching soil organic matter obviously under adequate organic matter management practices.
- Soil microbial population is in a dynamic equilibrium with one another in a given environment. Any change with one another in this equilibrium due to intensive cropping is merely temporary. On the other hand, enhanced crop residues in various stages of decomposition can sustain diversified flora and fauna.

Multiple cropping in vegetable crops

Vegetable cultivation is a capital intensive venture. Hence, effective land utilization through intensive cropping is urgently needed. In market garden, where vegetable crops are cultivated on high-priced land, proper multiple cropping is mandatory. Various forms of multiple cropping in vegetable crops are crop rotation, succession cropping, intercropping and relay cropping. Multiple cropping in vegetable crops has several merits like (i) generation of more income from the farm through increased production per unit area in the year, (ii) generation of more employment through intensive cropping throughout the calendar year, (iii) maintenance of soil fertility and health through balanced nutrition and scientifically based crop alterations, (iv) optimum use of natural resources, particularly nutrients and water, (v) rational use of capital and time and *(vi)* integrated management of pest, diseases and weeds.

Crop rotation: It is a cropping system of growing different crops in a regular sequence on the same land repeatedly for the period of two, three or more years. Here the cycle of cropping sequence takes more than one year to complete.

Succession cropping: The system of growing two or more crops in succession on the same land within a year. This cropping system is generally followed in most market gardens where the aim is to keep the high-priced land occupied with the cash crops for most part of the year.

Relay cropping: It is also the system of growing different crops on the same land within a year but in this system, the succeeding crop is sown / planted before the preceding crop is ready for harvest. So, the growing span of two crops overlaps for a short period.

Intercropping: It is a cropping system where two or more crops are grown simultaneously in alternate rows or otherwise in the same land showing significant amount of intercrop competition. The crops may or may not be sown/ planted and harvested at one time. It is mainly aimed at increasing the yield of the companion crop without reducing the yield of the main crop.

If these cropping sequences are practised in the farm in an ordered manner for two, three or more years, the cropping system may be termed as crop rotation.

Examples of succession cropping followed in West Bengal

1. Vegetable cowpea (June to Oct) — mid-season cauliflower (Oct to Jan) — onion (Jan to May).
2. Pumpkin (Sept to Feb)—taro (Feb to June) — Vegetable cowpea (June to Sept).
3. Potato (Nov to Feb) — Bitter gourd (Feb to June) — okra (July to Nov).
4. Early radish (Aug to Oct) — French bean (Nov to Feb) — bitter gourd (Feb to June) — amaranth (June to Aug).
5. Cabbage (Aug to Nov)—French bean (Nov to Feb) — ridge gourd (Feb to June) — cowpea/ rice bean for green manure (June to July).
6. Early cauliflower (July to Oct)—Tomato (Oct to March) — amaranth (March to June).
7. Early tomato (August to Dec)—onion (Dec to May) — green manure crop (June to July).
8. Chilli (June to Nov) — pea (Nov to March) — amaranth (March to June).
9. Sweet pepper (Oct to Feb) — taro (Feb to July) — vegetable cowpea (July to Nov).
10. Chilli (Oct to April) — okra (April to Aug) — palak (Aug to Oct).
11. Late tomato (late Dec to April) — okra (April to August) — hyacinth bean (Aug to Dec).
12. Radish (Oct to Dec) — watermelon (Dec to April) — chilli (April to Oct).
13. Elephant-foot yam (March to Oct) — pumpkin (Oct to Feb).
14. Okra (Feb to July) — brinjal (July to Feb).

Note: To make such multiple cropping successful, cucurbit seeds need to be raised in polyethylene packets. When these are to be started during winter, the seedlings must be raised under polyethylene shed to avoid cold injury.

Examples of relay cropping followed in West Bengal

1. Potato — pumpkin (potato is harvested in March and pumpkin seeds are sown in last week of January. After harvesting of potato, pumpkin is continued up to August).
2. Early cauliflower — pumpkin (cauliflower seedlings are transplanted in August and harvested in November. Pumpkin seeds are sown in September in the standing crop of cauliflower and continued up to March).
3. Brinjal — ridge gourd/bitter gourd/watermelon (brinjal is grown during July to March. Seedlings of the cucurbits grown in polyethylene packets are to be planted in December and continued up to May).
4. Bitter gourd — okra (bitter gourd is grown during October to March. Pre-germinated seeds of okra are to be sown in February and continued up to June).
5. Bottle gourd /summer squash — bush type vegetable cowpea (the cucurbits are grown during January to May following ridge and furrow method. Cowpea seeds are sown in the beds between two furrows in April and continued up to July).
6. Cabbage — watermelon/cucumber (cabbage is grown during late October to first week of February, cucurbits is started with the seedlings in polyethylene packets from December and continued up to April).

Some examples of intercropping

1. Cabbage + radish (cabbage is the main crop. The companion crop radish is of short duration and harvested early).
2. Tomato + radish + lettuce (tomato is the main crop grown at 75-90 cm spacing. Of the companion crops, radish is harvested 50 days and lettuce 80-85 days after sowing. So, enough space is left after harvesting the companion crops to continue tomato successfully till 150 days after planting. In this intercropping, tomato must be staked and trained).
3. Cucumber + cabbage/cauliflower (cucumber is the main crop and is sown in October with a view to capitalize the advantage of early harvest in spring. Cabbage/cauliflowers are grown in the interspaces successfully for the initial 80-day period).
4. Carrot + peas (carrot, the main crop is very slow growing for initial 45 days. Early pea varieties, like Arkel can successfully be grown in between for the period of 85 days. At the time of last harvesting of peas, the plants are to be uprooted and earthing up in carrot be done simultaneously. Carrot is continued for 130-135 days after sowing).

5. Tomato + palak (palak can be grown successfully in between tomato for 75 to 80 days duration. Harvesting of palak by uprooting the whole plant starts 35 days after sowing).
6. Pointed gourd + palak/radish/early cauliflower (vine or root cuttings of pointed gourd, the main crop is planted in October at the side of the raised bed of 2.5 to 3 m width. Initial growth of the regenerated vine is slow. Short-duration crops (75 to 80 days) can be grown profitably as companion crop in the beds. Pointed gourd is continued up to September.)
7. Cassava + onion/cowpea (cassava is the main crop of long duration (200 days). Onion can be grown successfully in between the rows of cassava).
8. Swamp taro + garlic (it is widely practiced intercropping in the low lying areas of North Bengal. Regenerated plants of swamp taro from the runner are planted at 75 cm spacing in both the ways in October in depressed swampy land. With the advent of dry spell in winter, excessive moisture from the soil dries out leaving it fit for planting garlic in December. Garlic is harvested in March and swamp taro is continued up to last week of August with harvesting from June onwards when the land generally remains in waterlogged condition.)
9. Elephant-foot yam + okra or elephant-foot yam + sweet potato/cucurbits/ cowpea (elephant-foot yam is planted in April with wide spacing of 75 to 90 cm in both ways and continued up to January. Successful intercrops can be raised up to September.

Different vegetable crops can be grown as intercrop in the early stages of orchard development (first 2-3 years). Afterwards, shade provided by the orchard trees hinders successful vegetable cultivation. However, some hardy crops like elephant-foot yam, taro, sweet potato, cowpea can be grown as intercrop for some more periods. Similarly, vegetable crops like cowpea, cluster bean, brinjal, tomato, sweet pepper, onion can be grown as intercrop with long-duration field crop like sugarcane.

Principles for arranging the cropping sequence

The economic effect of crop rotation, succession cropping and relay cropping depends on the set of crops. However, there should be definite planning in arranging the cropping sequence based on the following principles.

1. Repetition of the crops having common diseases and pests should be avoided. For example, Solaneceous crops should be avoided in the rotation when there is problem of *Fusarium* and bacterial wilt in tomato and brinjal. Many serious diseases and pests can be controlled by scientific crop rotation. For example, club root of cabbage and other crucifers caused by the fungus

Plasmodiophora brassicae can effectively be controlled by keeping the land free from cruciferous crop for a period of at least three years. Similarly, root-knot nematodes (*Meloidogyne* spp., *Heterodora* spp.) can be controlled by including non-host crops like marigold, corn, wheat, rice, *etc*. in the cropping sequence. Rice-sweet potato-cowpea rotation has been recommended for the control of sweet potato weevil.

2. The relation of individual crops to their predecessors must be taken into consideration. Generally, vegetable crops of different families grow well after majority of the crops. But growing cucurbitaceous vegetable crops after Solanaceous fruit vegetable crops is not profitable, in general. Similarly, cabbage has marked depressing effect on the yield of onion that follows. On the other hand, a number of vegetable crops, like onion, spinach, tomato, etc. are good forerunners for the other crops when the compatibility conditions are fulfilled.
3. The deep-rooted crops should be followed by shallow-rooted ones. Deep-rooted vegetable crops, like pumpkin, tomato, peas and beans, carrot, etc. able to use nutrients extracted from deep layers of the soil, and these crops should be alternated with the crops having weak and shallow root systems like, onion, leafy greens, lettuce, cucumber, etc.
4. Heavy feeder crops like, cabbage, cauliflower, potato, brinjal, etc., should be followed by low-feeding crops like okra, cucumber, pumpkin, *etc*.
5. Specific nutritional requirements of various crops should be taken into consideration. Cabbage, cauliflower, tomato, leafy greens, etc. extract large amount of nitrogen from the soil and these crops should be followed by leguminous vegetable crops like cowpea, French bean, hyacinth bean, etc. Symbiotic nitrogen fixation by the *Rhizobium* bacteria that grows in the root nodules of the leguminous vegetable crops provides benefit to the following crops.
6. The crops which efficiently utilize the organic manure residues should be grown after the crop which do not utilize fully. For that matter, immediately after the addition of organic matter to the soil, cucumber, pumpkin, summer squash, cabbage, leek, etc. should be grown and the root crops which can utilize the organic residues fully should be followed.
7. Leguminous vegetable crops should be included in the cropping sequence which not only upgrade the protein status of the farm produce but also enhance soil fertility. However, it is to be kept in mind in this context that legumes increase the susceptibility of the following crops to moisture stress because legumes are large users of water and normally leave the soil dry after harvest.
8. Green manuring crops should be accommodated in the rotation in order to increase the organic matter status in the soil.

9. Crop rotation without applying fertilizers does not maintain high productivity levels. A combination of crop rotation and judicious fertilizer application produce the highest yields and maintain adequate soil organic matter levels. Where previous cropping has depleted nutrients, addition of adequate fertilizers can reverse the trend of soil organic matter and nutrient depletion.

Chapter 15

Post-harvest Handling and Storage of Vegetable Crops

Quality Components of the Produce

Various components of quality are used to evaluate the products in relation to specific grades and standards and responses to various environmental factors and post-harvest treatments. The relative importance of each factor depends upon the commodity and its intended use, fresh or processed. The main quality components of vegetables are mentioned below:

Appearance

Many defects influence the quality appearance of vegetable crops.

- **Morphological defects**: These are sprouting of potatoes, onions and garlic, rooting of onions, seed germination inside tomato, sweet pepper and beans, floret opening in broccoli, etc.
- **Physical defects**: These include shrivelling and wilting of all the products, internal drying of some fruits, mechanical damage such punctures, cuts and scratches, spitting and crushing, skin abrasions, deformation and bruising.
- **Temperature related disorders**: These include sunburn, chilling and sunscald disorders
- **Nutrient deficiency physiological disorders**: These include blossom end rot of tomato, sweet pepper and watermelon, black heart of celery, etc.
- **Pathological defects**: These defects due to decay incited by fungi and bacteria.

Size

For most products consumers have a definite preference as to the desirable size, which is the most widely used quality parameter. This preference is generally expressed as weight, diameter, circumference, length or width of the product. Where produce is graded according to size, it is a normal practice to package those of similar size together. The uniformity of size allows the produce to be placed into a container in a regular packing array, with the result firstly to have a more efficient use of packing space so that either more produce can be placed in the container or the size of the container is reduced.

Shape

While shape differs greatly between commodities, it is one of the major recognition factors used by the consumers who will place a lower value on a commodity, which lacks the expected characteristic shape.

Colour

Many produce show distinctive colour changes during maturation which have been correlated by the consumers with the development of the other desirable quality attribute so that the correct colour of the skin is often the basis for a decision to purchase the commodity displayed on the shelves of the shops or supermarkets.

Conditions

It is a quality attribute usually referring to freshness and stage of senescence or ripeness of a commodity. The factors which detract from the desirability of a commodity include:

- Wilting of leafy vegetables.
- Shrivelling of vegetables due to water loss.
- Skin blemishes such as bruises, scratch marks and cuts.
- Surface contamination by soil, birds or insects excretions, plant secretions such as latex straining.
- Residues from chemicals applied during the growing season.

Texture

Texture, important for eating and cooking quality is the feeling a food gives in the mouth. It is a combination of sensation derived from the lips, tongue and walls of the mouth, teeth and even the ears.

Flavour

Flavour is the sensory impression of food or other substance and is determined primarily by the chemical senses of taste and smell. Flavour comprises two factors: taste provoked by sugar and acidity of the produce and aroma influenced by volatile organic compounds. Evaluating flavour quality involves perception of taste and aroma.

Nutritive value

Fresh vegetables play an important role in human nutrition, especially as a source of vitamins, minerals and dietary fibres. Physical damage may reduce the nutritive value of a commodity.

Safety

Safety factors include the presence of toxicants such as the greening of potato, chemical residues and mycotoxin produced by fungi.

Factors influencing quality

Many pre and post-harvest factors influence the composition and quality of fresh produce which may include:

- Genetic factors viz., cultivar selection and rootstocks for grafts.
- Pre-harvest environmental factors like, climatic conditions, cultural conditions and time and method of harvesting.
- Harvesting which include maturity, ripeness and physiological age of the produce.
- Post-harvest treatments which include handling methods, storage time between harvesting and consumption, etc.
- Interaction among the different factors.

Objective of post-harvest handling

Both quantitative and qualitative losses occur at all stages in the post-harvest handling system of the distribution chain of perishables (from harvesting, through handling, packing, storage and transportation to final delivery of the fresh produce to the consumer). The objectives of post-harvest handling is, therefore, the creation of an understanding of all the operations concerned from harvesting to distribution so as to enable people to apply the proper technology in each step and in such a way to minimize losses and maintain quality as high as possible during the distribution chain. Special and careful attention to the following steps of the post-harvest chain is required to maintain the quality of the produce:

- Market demand for the produce
- Market requirements and buyers
- Knowledge of the fresh produce
- Cultivation practices
- Factors affecting post-harvest deterioration
- Harvesting and field handling
- Packing in the field
- Handling and packing in the packing house
- Common storage and refrigeration
- Transportation
- Sale to agents, traders or consumers
- Market handling
- Shelf-life of the produce

Advantages due to minimization of post-harvest losses

- It helps in extending the period of marketing, preventing stagnation of the produce and a crash in prices during peak harvest time.
- It helps in regulated supply of high quality produce to the consumers, which ensures higher remuneration to the producer and longer availability of nutritious and healthy produce to the consumer at reasonable price.
- The growers will naturally be inclined to produce more because they can reap profit for the produce which would have gone to waste otherwise.
- Reduction of post-harvest losses reduces the dependence on imports and saves a substantial amount of foreign exchange.
- Wastage of energy for production and marketing of the produce is eliminated.
- It increases the food supply and provides better nutrition to the people with the same amount of non-renewable resources such as, land, water, energy and labour.
- It will reduce colossal waste, the problem of garbage disposal and consequent pollution.

Nature and Causes of Loss of the Harvested Vegetables

Fresh vegetables are perishable in nature because they are composed of living tissues. A number of physiological and biochemical processes take place in the freshly harvested vegetables which contribute to different changes within the product which subsequently result in the deterioration of the quality, condition and appearance of the vegetables. Again, all vegetables are covered with countless bacterial and fungal spores and some of which can cause decay of the fresh produce. Hence, nature and causes of losses of the fresh vegetables may be categorized as: (i) losses due to physiological processes, (ii) losses due to enzymic changes and (iii) losses due to microbial invasion.

Losses due to physiological processes

Different physiological processes like transpiration, respiration, senescence, sprouting and seed germination which are continued in the fresh vegetables are the main causes of this type of loss.

Transpiration

Majority of the vegetables contain more than 80 per cent water and some of them like, tomato, cabbage, spinach, palak, asparagus, French bean, lettuce, cucumber, etc. contain as high as 90-95 per cent water. Vegetables lose water mainly by transpiration through stomata (e.g., leafy greens). Other paths of water loss are stem scars (e.g., tomato), lenticels (e.g., potato) and cracks that are resulted from mechanical injury. When the harvested produce loses even 5.0 percent of its fresh weight, it begins to wilt which make the tissue tough or mushy, less crispy and palatable and eventually render the produce unmarketable. This happens to several leafy greens and root crops within few hours after harvest. Tissue structure of the vegetables and surface area to volume influence the rate of water loss. Cuticular waxes as found in the fruits of tomato, sweet pepper and leaves of cabbage play a pivotal role in limiting transpirational water loss across the primary plant surface. For this reason, water loss is least in tomato, moderate in cabbage and sweet pepper, high in summer squash and very high in French bean, leafy greens, etc. Water loss is also associated with degradation of pigments and vitamins. Hydrolysis of protein (Protein is hydrolyzed or broken down into its component amino acids) also occurs under wilted condition. To extend the usable life of produce, its rate of water loss must be as low as possible.

Respiration

Respiration is the process by which plants take in oxygen and give out carbon dioxide. On the basis of their respiration rate and ethylene production patterns during maturation and ripening, fruits can be classified in two groups: climacteric fruits (they exhibit a large increase in carbon dioxide and ethylene production rates coincident with their ripening) and non-climacteric fruits (which exhibit no changes in their generally low carbon dioxide and ethylene production rates during ripening).

In general, there arises a sudden spurt in the rate of respiration in the vegetables immediately after harvest followed by a gradual decline with the advent of senescence. Enzymatic oxidation of sugars and loss of other compounds such as, protein, lipids and organic acids are also channelized through respiration. Respiratory rate of vegetables is closely related to their storage life. Higher the rate of respiration, shorter will be the shelf-life of the produce and *vice versa.* Respiration in the harvested vegetables also depends on temperature of the storage atmosphere.

Respiration results in the reduction of both reducing sugar (sugar that bear a free aldehyde (- CHO) or ketonic (-CO) group e.g. glucose, maltose, lactose, melibiose, cellobiose, etc.) and non-reducing sugars (sugars that do not have such group e.g., sucrose, trehalose, etc.), and even protein content. Respiratory activity is evaluated by the evolution of CO_2 per kg of the produce per hour. Carbon dioxide output may be less than 40 mg/kg/hour in the vegetables with low respiratory activity, while it may be more than 120 mg/kg/hour in the vegetables with very high respiratory activity. Vegetables which respire less after harvest deteriorate less rapidly and can be stored for relatively long period if other factors remain in favour. In accordance with the respiration rate by the freshly harvested edible plant parts, the vegetable crops can be classified as follows:

Classification of vegetables as per respiratory activity of the produce

Respiratory activity of the produce	Vegetable crop
Very low	Potato, onion
Low	Cabbage, sweet potato, turnip, garlic, elephant foot yam, taro, cucumber
Moderate	Chilli, carrot, tomato, beet, sweet pepper, lettuce
High	Hyacinth bean, French bean, peas, lettuce, lima bean, radish, cauliflower
Very high	Green onion, Broccoli, spinach, okra, leafy greens, sweet corn , parsley, muskmelon, watermelon

Ethylene production

Ethylene is a colourless gas that is naturally produced by plants which functions as a plant growth regulator. It triggers specific events during a plant's natural course of growth and development, such as ripening. Through this action, it induces changes in certain plant organs, such as textural changes, colour changes and tissue degradation. Some of these changes may be desirable qualities associated with ripening. In other cases, it can bring damage or premature decay. It is active even at small traces.

In addition to being naturally produced by plants, ethylene is produced by a variety of other sources which include internal combustion engines, cigarette smoke, and natural gas leaks. Even low concentrations of ethylene throughout the postharvest life of a commodity can affect quality, so care must be taken to minimize exposure from both natural sources (i.e. climacteric fruit being stored with non-climacteric ones) or to artificial sources (engine exhaust, heaters, etc). All ethylene-producing sources should be considered when optimizing postharvest storage conditions as inadvertent exposure to ethylene can contribute to loss of quality in some fruits and vegetables.

Classification of vegetable crops according to ethylene production rates

Class	Range at 20°C ($\mu L\ C_2H_4$/kg/hr)	Vegetable
Very low	less than 0.1	Artichoke, asparagus, cauliflower, leafy vegetables, root vegetables, potato
low	0.1-1.0	Cucumber, brinjal, pumpkins, watermelon, Cassava, okra, olive, pepper (sweet pepper and chilli)
Moderate	1.0-10.0	Tomato, muskmelon
High	10.0-100.0	Cantaloupe

Impacts of ethylene on post harvest quality of fruits and vegetables

- Russet spotting of lettuce (dark brown spotting on the mid-ribs of lettuce leaves).
- Yellowing or loss of green colour (for example, in cucumber, broccoli, kale, spinach).
- Increased toughness in turnip and asparagus spear
- Increase in sprouting in potato.
- Yellowing and abscission (dropping) of leaves in Brassica vegetables.
- Softening, pitting and development of off-flavor in peppers, summer squash, and watermelon.
- Browning and discoloration in brinjal pulp and seed.
- Discoloration and off-flavor in sweet potato.
- Increased ripening and softening of mature green tomato.
- Development of bitter taste in carrot and parsnip.

Ripening

The process of ripening is the sequential changes in sensory factors, colour, texture and taste of the fruit which marks the completion of development and commencements of senescence and may be termed as a form of programmed organ death. The most widely studied aspects of ripening include ethylene biosynthesis, signal transduction and softening. Ripening process is complex and involves the sequential, orderly participation of a number of cell wall components, including structural polysaccharides and enzymic and non-enzymic proteins. Increased solubility of pectic polysaccharides due to pectin depolymerization is one of the universal features of ripening fleshy fruits which happens largely due to pectinolytic enzymes, including polygalacturonases and pectin methyl esterases.

Ripening makes those vegetables unfit for marketing which are generally consumed at their immature stage. The vegetables which are consumed at their ripe stage are spoiled due to over-ripening when depolymerised pectin is finally converted to pectic acid and alcohol.

Senescence

It is the last stage of development of the tissue when anabolic biochemical processes give way to catabolic processes leading to death of the tissue. Increased membrane permeability, depolymerisation, respiration, phosphorylation and shift in protein synthesis lead to deteriorative processes and eventually senescence. During senescence, the most obvious change takes place is the degradation of chlorophyll and development of carotenoid pigments. Excessive senescence of the vegetable tissues causes serious loss of the produce. Senescence process invariably evolves ethylene leading to autocatalysis of several physio-biochemical reactions including increased respiration and ripening. Ethylene induces changes in permeability of the mitochondrial membrane, thus facilitating the increased ATP movement and thereby initiating several reactions including respiration.

Sprouting

This process of active resumption of accelerated growth causes deterioration of potato, onion, garlic in storage. Sprouting is increased with the increase in temperature and relative humidity in the storage atmosphere.

Seed Germination

Seed germination in some vegetables, like chow-chow (*Sechium educe*), cowpea, French bean, etc. sometimes takes place in the harvested fruits itself which considerably lowers the product quality.

Losses due to enzymic changes

Enzymes which are endogenous to plant tissues can have undesirable or desirable consequences. The major factors useful in controlling enzyme activity are: temperature, water activity, pH, chemicals which can inhibit enzyme action, alteration of substrates, alteration of products and pre-processing control.

Carbohydrate conversion: Conversion of sugar to starch is very much detrimental to most of the vegetable crops. Starch is a carbohydrate consisting of a large number of glucose units joined by glycosidic bonds. Plants produce starch by first converting glucose 1-phosphate to ADP-glucose using the enzyme glucose-1-phosphate adenylyl transferase. This step requires energy in the form of ATP. The enzyme starch synthase then adds the ADP-glucose *via* a 1,4-alpha glycosidic bond to a growing chain of glucose residues, liberating ADP and creating amylose.

Conversion of starch to sugar in storage is harmful for potato. At low temperature some of the carbohydrate splitting enzymes may get activated and play some important role for the sweetening of potato. The activities of invertase, amylase, β-galactosidase and cellulase have been found to be increased from harvesting to cold stored potatoes.

Oxidation of phenolic substances: Oxidation of phenolic substances in plant tissues in fruits and vegetables (chlorogenic acid, caffeic acid, coumaric acid, cinnamic acid derivatives in brinjal; tyrosine, caffeic acid, chlorogenic acid derivatives in lettuce; chlorogenic acid, caffeic acid, caffeylamide in sweet potato, etc.) by the enzyme polyphenoloxidase leads to browning of plant tissues. In the presence of oxygen from air, the enzyme catalyzes the first steps in the biochemical conversion of phenolics to produce quinones, which undergo further polymerization to yield dark, insoluble polymers called melanins.

Demethylation of pectic substances: Post-harvest demethylation of pectic substances in plant tissues leading to softening of plant tissues during ripening.

Losses due to chemical changes

Sensory quality: The two major chemical changes which occur during the processing and storage of foods and lead to a deterioration in sensory quality are lipid oxidation and non-enzymatic browning. Chemical reactions are also responsible for changes in the colour and flavour of foods during processing and storage.

Lipid oxidation: The rate and course of lipid oxidation reaction is influenced by light, local oxygen concentration, high temperature, the presence of catalysts (generally transition metals such as iron and copper) and water activity. Control of these factors can significantly reduce the extent of lipid oxidation in foods.

Non-enzymic browning: It is one of the major causes of deterioration which occurs during storage of dried and concentrated foods. These non-enzymatic reactions or Maillard reaction take place between nitrogenous compounds and sugar, nitrogenous compounds and organic acids, sugar and organic acids or organic acids among themselves which may independently cause browning of the product.

Losses due to colour changes

Chlorophylls: Almost any type of food processing or storage causes some deterioration of the chlorophyll pigments. Phenophytinisation (with consequent formation of a dull olive-brown phenophytin) is the major change and this reaction is accelerated by heat and is acid catalysed. Other reactions are also possible. For example, dehydrated products such as green peas and beans packed in clear glass containers undergo photo-oxidation and desirable colour is lost.

Anthocyanins: Anthocyanins are water soluble pigments responsible for colours from red to blue and their rate of destruction is pH dependent, being greater at higher pH values. Complex of anthocyanin is also found in nature. As for example, commelinin, a blue pigment from the blue flowers of *Commelina communis* L.is a complex composed of a reddish-purple anthocyanin, a pale yellow flavone and magnesium. Blue pigment of cornflower, protocyanin is a complex pigment comprising anthocyanins, flavone glycosides and metals (Fe^{3+}, Mg^{2+}, $2Ca^{2+}$). Some anthocyanins in the products form complexes with metals of the packages such as, Al, Fe, Cu and Sn. These complexes generally result in a change in the colour of the pigment (for example, red sour cherries react with tin to form a purple complex) and render it undesirable. Since metal packaging materials such as cans could be sources of these metals, they are usually coated with special organic linings to avoid these undesirable reactions.

Carotenoids: The carotenoids are a group of mainly lipid soluble compounds responsible for yellow, orange and orange-red colours of the products. The main cause of carotenoid degradation in foods is oxidation. The mechanism of oxidation in processed foods is complex and depends on many factors. The pigments may auto-oxidised by reaction with atmospheric oxygen at the rates dependent on light, heat and the presence of pro- and antioxidants.

Losses due to flavour changes

Fat auto-oxidation: In fruit and vegetables, enzymically generated compounds derived from long-chain fatty acids play an extremely important role in the formation of characteristic flavours. In addition, these types of reactions can lead to significant off-flavours. Fat content in the vegetables though very low yet is important from quality and storage point of view. Enzyme-induced oxidative breakdown of unsaturated fatty acids occurs extensively in plant tissues. Rancidity due to fat auto-oxidation imparts undesirable flavour to the stored product.

The permeability of packaging materials is of importance in retaining desirable volatile components within packages or in permitting undesirable components to permeate through the package from the ambient atmosphere.

Losses due to microbial invasion

Decay is an important cause of spoilage of the harvested vegetables and in most of the cases, moulds (fungi) and bacteria causes the decay. Soft rot causing bacteria causes great harm to the vegetables. Tender and succulent nature of the vegetables makes them vulnerable to the micro-organisms. Again, majority of vegetables are grown near the soil line which makes easy reach for the microbes to the vegetables. Besides, rough and careless handling, cuts and bruises make easy entry of the decay causing organisms. The most common fungi causing rots in vegetable are *Alternaria, Diplopia, Phomopsis, Fusarium, Rhizopus, Botrytis, Aspergillus, Penicillium,* etc. and most common bacteria causing rots are *Erwinia, Pseudomonas* and *Ceratocystis.* High temperature and relative humidity favour the development of these post-harvest decay causing micro-organisms. Some of the important storage diseases are: neck rot of onion caused by *Botrytis allii,* black mould of onion by *Aspergillus niger,* dry rot of garlic by *Fusarium oxysporium* and *A. solani,* soft rot of cabbage by *Erwinia carotovora,* storage rot of brinjal by *Botrytis cineria,* etc.

Factors responsible for post-harvest losses

1. Inadequate knowledge of maturity standards and harvesting of vegetables.
2. Improper handling of the produce at harvest time.
3. Lack of pre-cooling facilities immediately after harvest to remove field heat.
4. Improper curing which leads to huge loss of potato and onion.
5. Non-availability of adequate multipurpose cold storage facilities.
6. Lack of post-harvest treatment with fungicides.
7. Inadequate use of packaging materials.
8. Lack of quick, efficient and cost-effective transport systems.

9. Non-utilization of packing materials like SO_2 releasing pads and ethylene absorbents.
10. Seasonal stagnation of the produce.
11. Limited utilization (hardly 2.0%) of the produce for processing.

Control of post-harvest losses

There are certain techniques related to cultural operations, harvesting, post-harvest handling and storage of the produce which can control post-harvest losses of vegetables considerably.

Pre-harvest operations

Post-harvest decay need to be controlled before harvest, because many post-harvest diseases begin while the crop is still in the field. Some harvested commodities carry latent infection that may not be detected at harvest. Injuring the commodity from harvesting to marketing process needs to be avoided because bruises, wounds and other mechanical injuries serve as ports of entry for micro organisms. Soil adherent to the commodity need to be cleaned immediately after harvest, because these may be the carriers of micro-organisms from the field.

Proper pre-harvest cultural operations sometimes help in prolonging the sheif-life of vegetables. Some of the pre-harvest operations are given below:

- Avoidance of heavy application of nitrogenous fertilizers and at the same time, ensuring essential supply of potassic fertilizers to improve the keeping quality of vegetables because potassium activates or stimulates enzyme activity that are involved in carbohydrate and protein metabolism and is essential for cell organization and structure of cell walls.
- Irrigation should be discontinued 15 days before harvesting potato, onion and garlic for obtaining tubers/bulbs with maximum storage life as it checks sprouting.
- Pre-harvest spray of fungicides like, 0.25% Captan or Ziram 15 days before harvest reduces storage loss in onion.
- Three pre-harvest sprays of 0.2% Captan or Ziram at an interval of 10 days are effective in controlling post-harvest diseases of tomato. In tomato, foliar sprays of chelated iron before flowering lower the decay percentage and weight loss in storage.
- Blanching in cauliflower (tying of leaves to protect the curd from scorching sunlight) ensures good quality and comparatively long storability of the curd.

- Mulching also somewhat prolongs the shelf-life of vegetables because mulching materials obstruct reflection of soil heat to the produce and thereby keep them less hot which checks deterioration.

Harvesting Techniques

Development of vegetables starts with the formation of the edible part such as, the setting of fruits, the emergence of the seedling, the swelling of a root, tuber or bulb, the elongation of the stalk or petiole, etc. Maturity in vegetables reaches when the edible portion has attained some desirable conditions. Harvest is a single deliberate operation to separate the edible plant part from the growth medium. Time of harvest is a complicated consideration as great differences occur in the rate of development and maturation of different vegetables. Harvesting should always be done at proper stage of maturity because it not only determines the quality of the product but also prolongs its shelf-life. Tomato harvested at breaker stage ensures good quality, better shelf-life and proper development of colour. Harvesting should be done with care to avoid damage, bruising and cracking of the product which checks easy entry of the decay causing micro-organisms. Harvesting during hot period of the day should be avoided as immense field heat causes wilting and shrivelling of the produce. Evening harvested tomato has longest shelf-life, uniform ripening and better quality. Harvesting during or immediately after rain should not be done because most favourable conditions for multiplication of the decay causing micro-organisms is provided at that time. The ripe fruits should be separated from the unripe fruits either in the containers or storage rooms, because ripe fruits trigger ripening of the unripe fruits due to evolution of ethylene.

Handling of the produce

Washing, trimming and grading

After harvesting, the produce should be thoroughly washed with clean water to reduce the load of micro-organisms and to remove field heat from the produce. Washing of vegetables should preferably be done by sprayers. The produce should be kept dry after washing because free moisture on the surface of the produce enhances the growth of spores of micro-organisms.

Rotten, diseased, insect damaged and discoloured leaves of the vegetables should be trimmed off and at the same time damaged vegetables should be discarded. The harvested vegetables should be graded in accordance with shape, size, colour, maturity, freedom from disease and pest attack, *etc.* which helps the storage process convenient and prevents the healthy produce from being contaminated by decay causing micro-organisms.

Temperature management practices

Temperature management is the most important tool to extend shelf-life of freshly harvested produce. Temperature management begins with a rapid removal of the field heat from freshly harvested produce before packaging or transportation.

Pre-cooling reduces weight loss and shrinkage, causes delay in ripening and retards ethylene production which restricts enzymatic and respiratory activity and also reduce refrigeration load while storing at low temperature. Prompt cooling to required temperature also inhibits growth of decay causing micro-organisms. Pre-cooling is particularly important when the vegetables are transported to long distance. Various methods of pre-cooling are hydro-cooling, air-cooling, room cooling, forced air-cooling, top icing and vacuum cooling.

Pre-cooling with ice-cold water to 10-12°C with some fungicides increases shelf-life by reducing respiration and ripening rate and also ensures good surface colour. In India, pre-cooling of vegetables is generally done by immersing the produce in water. Water is an excellent medium to which heat from the vegetables is transferred. This hydro-cooling will be more effective if ice-cold water is used. Most modern method of pre-cooling is vacuum cooling. In the developed countries, pre-cooling is mostly done by keeping the produce in refrigerated trucks and by placing ice in the packages.

Control of relative humidity

Proper relative humidity should be 95-98% for vegetables (except onions and pumpkins at 70-75%) and 95-100% for some root vegetables. Appropriate relative humidity is important to control the i) water losses, ii) decay development, iii) incidence of some physiological disorders and iv) uniformity of ripening. Relative humidity can be controlled by the following methods:

- Addition of moisture to air by humidifiers.
- Regulation of air movement in relation to produce.
- Maintaining coil temperature to 1°C difference to air temperature.
- Wetting the floor in the storage room.
- Addition of crushed ice.

Supplemental post-harvest treatment

The following treatments (curing, waxing, sorting for defect elimination, hot water treatment, treatment with post-harvest fungicides or bactericides, use of sprout inhibitors, special post harvest chemical treatment, fumigation for insect control and films wrapping) may be applied to enhance the shelf-life of different vegetable produce.

1. Curing

It is a post-harvest treatment of tuber and bulb crops. In this treatment, the produces are exposed to relatively high temperature and relative humidity at farm level which trigger the synthesis of suberin in the outer tissue of potato, sweet potato and other tuber crops and drying of scale leaves and compaction of neck in onion and garlic.

Suberin is an essential plant biopolymer (waxy material) that is synthesized after wounding to protect the tissue from moisture loss and to reduce its susceptibility to bacterial and fungal attack. Curing stimulates suberization, wound healing and reduces respiration. A molecular rationale for suberin's protective function and development has proven elusive, in part because the polymer occurs as an integral unit together with cell-wall carbohydrates and does not dissolve in either aqueous or organic solvents.

The protective layer due to formation of special cell wall thickening may be of two types: primary and secondary. i.e., periderm. The primary protective layer is formed as a result of the lignification and suberization of parenchyma cells in this region or of cells arising from them by irregular cell division. Different cell wall materials like, cellulose, microfibrils, wound gums, etc. are deposited in the cell wall. These tough coverings act as effective barrier against microbial infection and water loss by transpiration. Proper curing can considerably reduce spoilage of potato, onion and other bulb and tuber vegetables like garlic, sweet potato, cassava, elephant-foot yam, yam, etc in storage. Curing of onion and potato in perforated plastic crates for 15 days at room temperature is most effective for reduction of losses due to rotting in storage.

2. Waxing

Application of fungicides and sometimes growth substances and other chemicals along with edible wax is quite helpful in prolonging the shelf-life of vegetables. Waxing is particularly helpful for tender fruit vegetables like okra, ridge gourd, pointed gourd, bitter gourd, snake gourd, brinjal, etc. Fruits of bitter gourd, snake gourd and watermelon dipped in solution of wax emulsion (6-12%) + sodium phenyl phenoate for 30-60 seconds prolong the shelf-life up to 3, 14 and 8 days. respectively at room temperature. In okra, fruit dipping in wax emulsion (12%) + cycocel 100 ppm is effective in minimizing the physiological losses in weight and in retention of chlorophyll up to 6 and 12 days at 32°C and 10°C, respectively. Coating of muskmelon with wax emulsion and thiabendazole (a systemic benzimidazole fungicide) increases storage life up to 7 days.

3. Hot water treatment

Hot Water rinsing and brushing is commercially applied to several fruits and vegetables in order to reduce decay development and maintain fruit quality after subsequent prolonged storage and shelf-life. This technology cleans and disinfects the freshly harvested produce at a relatively high temperature of 45-62°C, with the produce passing over revolving brushes for a very short time of 15-25 seconds. Hot water rinsing while brushing (HWRB) bell peppers immediately after harvest at 55° C for 12 second significantly reduces decay incidence, while maintaining fruit quality, compared with untreated fruit.

4. Post-harvest treatment of chemicals

Plant protection chemicals

Control of decay causing micro-organisms is of prime importance in storage. Pre-harvest sprayings along with post-harvest application of fungicides and antibiotics significantly prevent the decay causing fungi and bacteria. Fungicides, like Benlate, Thiabendazole, Benomyl, Dithane M-45, Captan, etc. have been found very effective against many decay causing pathogens. Application of lime and formalin is sometimes found effective. Dusting of lime @ 12 kg/ha at the cut surface of leaf blade during harvesting effectively reduces black mould of onion. Application of 0.3% formalin dip is effective against dry rot of garlic. Benomyl is effective in preventing growth of most damaging fungi in tomato. Bacterial soft rot of packed spinach and lettuce can be checked by dipping in 1000 ppm streptomycin. Application of vitamin K5 (synthetic types of vitamin K) at 50-500 ppm concentration markedly inhibits the activity of a number of micro-organisms and prolongs shelf-life of vegetables.

Antitranspirants

Both transpiration and photosynthesis show close dependence upon leaf resistance. Since CO_2 entering the sub-stomatal cavity has to traverse the same pathway as the escaping H_2O molecules during transpiration, there arise corresponding resistance for CO_2 and H_2O diffusion. Photosynthesis is more likely limited by the diffusive resistance, the internal process than mediated by stomatal capacity. The use of antitranspirants as an aid to increased water conservation and consequently enhanced water use efficiency is based on the anticipation that chemically induced increased diffusive resistance cause transpiration to decline more than photosynthesis. Application of antitranspirants (Mobileaf, Kaoline, etc.) is particularly beneficial to green leafy vegetables.

Growth substances

Deteriorative physio-biochemical processes that go on inside the harvested vegetables are mediated by indigenous growth hormones, principally auxin, gibberellins and cytokinins. Exogenous application of growth substances can alter different metabolic process and some of these alterations are beneficial to extending the shelf-life of vegetables. Yellowing due to degradation of chlorophyll is one of the main factors that limit the shelf-life of cauliflower, cabbage, broccoli and Brussels sprout. Senescence which causes yellowing can effectively be retarded by pre or post-harvest application of BA (N6-benzyladenine), a synthetic cytokin compound at lower concentrations (5-15 pm). Cytokinin controls development of chloroplast through stimulating the conversion of pro-plastids into chloroplasts in grana. Auxinic growth substances like 2,4-D, 2,4,5-T have profound effect on delaying leaf abscission of cauliflower when applied before or after harvest. Ripening of tomato can be delayed and degradation of chlorophyll in leafy greens can be checked by post-harvest application of gibberellins. Growth inhibitors like, CIPC (chloro-iso-propyl carbamate) and MENA (methyl ester of naphthalene acetic acid) are effective in controlling sprouting of tuber and bulb crops. However, application of both pre- and post-harvest application of maleic hydrazide has now been banned due its carcinogenic effect. Potatoes treated with 1% aqueous emulsion of CIPC and kept in cold storage do not have any sprouting for 5 months. Synthesis of solanine in potato is inhibited by Alar (N, N-dimethyl aminosuccinamic acid). Cycocel (2-chloroethyl-trimethyl ammonium chloride), another growth retardant can prolong the shelf-life of lettuce, asparagus, broccoli, etc. However, effective concentration and right stage of application is very important for obtaining maximum benefit.

In broccoli, post-harvest dip in 20 ppm BA for 15 seconds and held at 25°C under polyethylene cover prolongs the storage life by retaining green colour. In Brussels sprouts, spraying with GA 100 ppm extends the shelf-life to 9-10 days at 20°C. In head lettuce, application of 8 ppm kinetin for 30 seconds and packed in perforated polyethylene bags and stored at 6°C and 90% relative humidity retard senescence. In tomato, application of cycocel 100 ppm decreases fruit decay and weight loss. Post-harvest treatment with BA at 5-20 ppm increases the shelf life of radish. In okra, fruit dipping in the solution of 100 pm GA for 10 minutes and packed in polyethylene bags extend the shelf-life up to 9 days. In pointed gourd, fruit dipping in 100 ppm GA solution extends the shelf-life up to 7 days in ambient condition. Applicatibn of BA at 50 ppm concentration is effective in reducing decay loss of ridge gourd stored at 0.5°C temperature.

Other chemicals

In cucumber, application of borax (4%) reduces fruit decay when stored at room temperature. In tomato, potassium permanganate ($KMnO_4$) increases the length of storage by absorbing ethylene from around the stored fruits. In broccoli, dipping in sodium benzoate solution at 0.01 M for 3 minutes and then storing at 20°C effectively reduce the rate of ethylene production, respiration and yellowing. In okra, fruit dipping in the solution of ascorbic acid 250 ppm and then packed in polyethylene packages prolong the shelf-life up to 12 days at 32°C. Post-harvest application of calcium nitrate or chloride in the form of dip or pressure infiltration extends the shelf-life of tomato and other vegetables mainly by reducing weight loss and checking CO_2 and C_2H_4 production. In pointed gourd, fruit dipping in 250 ppm sodium benzoate solution extends the shelf-life up to 7 days in ambient condition.

5. Wrapping in film roll

In this method, fruits and vegetables are wrapped in heat-shrinkable plastic film. The main advantages of film wrapping of fruits and vegetables are i) reduction of weight loss, ii) minimization of fruit deformation, iii) reduced chilling injury and iv) reduced decay by preventing secondary infection of fruits packed in the same box.

6. Reduction of ethylene exposure during storage

- Transportation or storing of green leafy vegetables in containers holding ripening fruit (apple, pear, mango, tomato, banana, *etc.*) should not be done.
- If possible, it is better to use electric powered equipment in storage areas instead of gas powered equipment.
- Overripe or rotting fruit from storage loads which produce higher amounts of ethylene should be removed from the storage immediately.
- Ethylene sensitive products should not be stored with the products that produce high levels of ethylene.
- Ventilation rate of the storage area should be enhanced for efficient ward off of the evolved ethylene from the storage.
- Use of ethylene scrubbers or ethylene absorbants in storage areas to remove ethylene in the air. Several chemical formulations like brominated activated carbon, celite with $KMnO_4$, $KMnO_4$ on vermiculite, charcoal with palladium chloride have been proved to be effective as ethylene scrubbers or absorbants. Application of $KMnO_4$ 1000 ppm in paper lining can effectively be used as cushioning material for packaging tomato which acts as ethylene absorbant.

Packaging

It is an important consideration in prolonging the shelf-life of vegetables because of the protection provided by the packages from external injuries and from being shrivelled. Adequate packaging coupled with effective marketing can reduce the post-harvest losses considerably. The two main functions of packaging are: to assemble the produce in convenient units for handling and to protect the produce during transportation and marketing operations. The common packages are wooden boxes, wire-bound boxes, bamboo baskets, arhar (*Cajanus cajan*) sticks baskets, corrugated fibre board boxes, plastic crates and jute bags. Cheap materials like, polyethylene films, paperboard boxes lined with polyethylene can effectively be used in packaging. Wrapping of vegetables with newspaper and tissue paper may be useful because it reduces physiological loss in weight. Plastic crates are better in case of bulk transportation from the point of view of less bruising loss. Polyethylene packaging reduces decay and softening of the produce but the thickness and permeability of the film to CO_2, O_2 and water vapour need to be standardized for each vegetable. Handling of the produce during packaging should be prompt to minimize deterioration.

Transportation

Most of the post-harvest losses of vegetables occur during transportation and distribution. It has been estimated that spoilage of the vegetables takes place to the tune of one-third between farm and road head and another one-third between road head and consumer. In our country, transportation network for perishable commodities like vegetables is very weakly developed. Net work for refrigerated transport of vegetables is yet to be developed. Spoilage of the produce during transportation occurs mainly because of decay owing to uncontrolled temperature and relative humidity condition in the carriages. This deterioration in transit is enhanced due to inadequate as well as loose packaging. In this situation, quick, efficient and at the same time, cost effective transport network should be provided in the potential vegetable growing areas. Rail transport is 8-10 times more efficient than road transport with respect to the use of energy for the movement of the same tonnage. But unfortunately, long-distance transport of vegetables is mostly done by road because rail transport has not yet proved to be dependable for vegetables in our country. It is necessary to modify the long-distant transport carriages by introducing refrigeration, more ventilation and by improving loading and unloading systems.

Marketing

Interests of the producers as well as consumers are poorly served due to present faulty marketing systems for vegetables in our country. Moreover, huge quantity of wastes is also dumped due to poor marketing. The losses at both wholesale and retail sale levels vary from commodity to commodity. Very often, certain vegetables get stagnated in particular areas resulting in surfeit owing to lack in proper distribution system and climatic factors. Thus a substantial quantity of the produce is wasted. This huge wastage can be minimized by good distribution and effective marketing system for vegetables.

Storage

Storage is one of the most important aspects of the post-harvest handling of vegetables. Storage extends the period of availability of fresh vegetables by arresting the metabolic breakdown and decay which is achieved by controlling the temperature, relative humidity, concentration of certain gases in the atmosphere, chemical treatment and by irradiation. Forms of storage may range from primitive on-farm storage to highly sophisticated controlled environment storage.

On-farm storage

The time lag between harvest and marketing sometimes is quite high owing to poor transport network. In this situation, on-farm storage of some hardy vegetables like cabbage, beet, carrot, turnip, etc. for very short period sometimes checks post-harvest losses to some extent. The crops are generally placed in piles on the ground and covered with straw at first and then with soil. Sometimes, on-farm storage of these vegetables is done in trench or pit.

Home storage

Limited quantity of vegetables which require relatively high temperature and low relative humidity for storage like, ripe pumpkin, ripe winter squash, sweet potato, yam, well cured potato, onion and garlic, etc. can be kept in the shady and cool corner of the house for quite a long time.

Low cost storage without refrigeration

Ventilated storage structure for onion and garlic

Onion and garlic are the vegetable crops of great economic significance in our country. They are the major export earners and have great demand in domestic market. A large quantity of onion and garlic is lost due to inadequate post-harvest handling. The recommended ideal storage conditions for onion and garlic are 0°C temperature with 70 to 75 per cent relative humidity. However, under Indian condition, onion and garlic are stored in hut-like ventilated storage structure kept in open where considerable storage loss due to drying, rotting and sprouting occurs. A considerable reduction of spoilage of onion and garlic can be achieved by modifying the existing storage structure with the introduction of top and pile ventilation which is accomplished by perforated bamboo pipes and by keeping the storage structure under shed. Storage of onion in perforated plastic crates stacked in a ventilated room or shed reduces the storage loss to a considerable extent.

Zero energy cool chamber

This low cost cool chamber has been developed at the Division of Horticulture, IARI, New Delhi. Based on the principle of evaporative cooling, this chamber achieves its objective by maintaining low temperature and high humidity as compared to the ambient conditions outside. The cool chamber is provided with a wet porous bed, the passage of air which causes cooling and humidification by evaporation of water. Vegetables are kept in plastic crates inside the cool chamber to avoid crushing and to prevent direct contact of the vegetables with wet bricks. Due to low temperature and high humidity inside the chamber, the shelf-life of vegetables is increased considerably. The floor of the storage space is made of a single layer of bricks and the side walls of double layer of bricks with approximately 7.5 cm space between the bricks. The empty space between the two brick walls is filled with clean riverbed sand while the top of the storage space is covered with khaskhas mat or gunny cloth fitted in a bamboo frame. After construction, the bricks, sand and top cover is made saturated with water. Later, sprinkling of water daily, once in the morning and again in the evening is sufficient to keep required temperature and relative humidity inside. However, care should be taken in sprinkling water so that sand may not flow out of the wall and stored vegetables may not get direct contact with water. Water supply may be regulated by a drip system consisting of plastic pipes and microtubes connected directly to a water source which is controlled by a stopper. It is possible to bring down the temperature inside the cool chamber by 8-10°C with relative humidity within 90-95 per cent. This cool chamber is very effective in summer months particularly in the areas where low humidity prevails like, Rajasthan, Haryana, Punjab, western Uttar Pradesh and Delhi due to rapid evaporation of water causing quick cooling inside.

The cool chamber should be constructed under a shed to keep away direct sunlight, but good air flow should be ensured. The inside portion of the cool chamber should be clean and free from rotten plant materials. The cool chamber should be disinfected periodically by sprinkling chlorine water, bleaching powder or by sulphur fumigation.

Cold storage

Cold storage or refrigerated hold has nowadays become a part of the moderns living. Marketability of perishable produce like vegetables is closely linked with the development of cold storage. Low temperature storage is the best method of slowing down metabolic processes and decay of vegetables and thereby most effectively extends the shelf-life of the produce. It is always desirable to retard the physio-biochemical processes that go on inside the harvested vegetables with a view to preserve the quality and nutritive value of the produce. Low temperature has profound effect on the rate of respiration of the harvested vegetable and on the rate of multiplication of the decay-causing micro-organisms. Higher the temperature, the more is the rate of respiration and at the same time multiplication of the micro-organisms. Low temperature also reduces the moisture loss from the produce because moisture holding capacity of the air at low temperature is much less than it is at warm temperatures. Besides low temperature, relative humidity of the storage chamber is another important factor. Low relative humidity of the storage atmosphere leads to rapid moisture loss from the produce while higher relative humidity favours microbial growth. Cold storage facilities should be well constructed and adequately equipped and should have the following features.

- Good construction, and insulation and vapour barrier
- Strong floor
- Adequate doors for loading and unloading
- Effective distribution of refrigerated air
- Properly located controls
- Enough refrigerated coil surface
- Capacity adequate to the expected needs
- Appropriate stacking of the produce

In the cold storage, both temperature and relative humidity are considered simultaneously and in fact, the requirements are specific for each vegetable. For example, asparagus requires 0°C and 95% relative humidity whereas pumpkin requires 10-12°C and 70--75% relative humidity for storage. Improper low temperature leads to freezing injury of the produce whereas improper high temperature leads to shrivelling of the produce and at the same time favours the growth of decay-causing micro-organisms. Each vegetable has its own specific requirements of temperature and relative humidity hence, they cannot be put together in the cold storage. In the multi-commodity cold storage, temperature and relative humidity can be adjusted and maintained in accordance with the requirements of each or a group of vegetables.

Recommended storage temperature and relative humidity for different vegetables

Vegetables	Temperature (0°C)	Relative humidity (%)	Storage life (weeks)
Brinjal	10.0-11.1	92	2-3
Tomato (unripe)	8.9-10.0	85-90	4-5
Tomato (ripe)	7.2	90	1.0
Chilli (green)	7.2	85-90	3-5
Chilli (ripe)	5.6-7.2	90-95	2.0
Potato	3.0-4.4	85	34.0
Cabbage (early)	0.0-1.7	92-95	4-6
Cabbage (late)	0.0-1.7	92-95	12.0
Cauliflower (late)	0.0-1.7	85-95	7.0
Bitter gourd	0.6-1.7	85-90	4.0
Cucumber	10.0-11.7	92	2.0
Muskmelon	1.7-3.3	85-90	1.5
Pumpkin	10.0-12.0	70-75	24-36
Winter squash	12.8-15.6	70-75	24-36
Watermelon	7.2-15.6	80-90	2.0
Beet (topped)	0.0-1.7	90-95	8-14
Beet (bunched)	0.0	90	1.5
Carrot (topped)	0.0	95	20-24
Radish (topped)	0.0	88-92	3-5
Turnip	0.0	90-95	8-16
Hyacinth bean	0.6-1.7	90	3.0
Lima bean	4.4-7.2	90-95	1.5-2.0
Pea	0.0	88-92	2-3
Garlic	0.0	65	28-36
Onion (bulb)	0.0	70-75	20-24
Onion (leaves)	0.0	90-95	2.0
Taro	11.1-12.8	85-90	21
Cassava	0.0-1.7	85	23
Sweet potato	10.0-12.8	80-90	13-20
Yam	26.7	66-70	3-5
Ginger	7.2-10.0	75	16-24
Okra	8.9	90	2.0
Sweetcorn	0.6-1.7	90-95	1.0
Celery	0.0-0.6	92-95	8.0
Lettuce(head type)	0.0	90-95	3.0
Lettuce (leaf type)	0.0	95	1.0
Coriander leaves	0.0-1.7	90	5.0
Asparagus	0.0	95	3-4

Uttar Pradesh, Maharashtra, West Bengal, Punjab and Gujarat account for more than 60% cold storage capacity followed by Andhra Pradesh, Haryana and Madhya Pradesh.

Low temperature injury

All fresh produce is subject to damage when exposed to extremes of temperature. Commodities vary considerably in their temperature tolerance. Their levels of tolerance to low temperatures are of great importance where cool storage is concerned:

- **Freezing injury**: All produce is subject to freezing at temperatures between 0 and -2°C. Frozen produce has a water-soaked or glossy appearance. Although a few commodities are tolerant of slight freezing. It is better to avoid such temperatures because subsequent storage life will be short. Produce which has recovered from freezing is highly susceptible to decay.
- **Chilling injury**: Some types of fresh produce are susceptible to injury at low but non-freezing temperatures. Such crops are mostly of tropical or subtropical origin, but a few temperate crops may also be affected.

Susceptibility of fruits and vegetables to chilling injury at low but non-freezing temperatures

Commodity	Approximate lowest safe temperature °C	Chilling injury symtoms
Brinjal	7	Surface scald, *Alternaria* rot
Avocado	5-13	Grey discoloration of flesh
Banana (green/ripe)	12-14	Dull, gray-brown skin colour
Bean (green)	7	Pitting, russeting
Cucumber	7	Pitting, water-soaked spots, decay
Grapefruit	10	Brown scald, piking, watery break down
Lemon	13-15	Pitting, membrane stain, red blotch
Lime	7-10	Pitting
Mango	10-13	Grey skin scald, uneven ripening
Honeydew melon	7-10	Pitting, failure to ripen, decay
Watermelon	5	Pitting, biker flavour
Okra	7	Discoloration, water-soaked areas, piking
Orange	7	Pitting, brown stain, watery break down
Papaya	7	Pitting, failure to ripen, off-flavour, decay
Pineapple	7-10	Dull green colour, poor flavour

Potato	4	Internal discoloration, sweetening
Pumpkin	10	Decay
Sweet pepper	7	Pitting, *Alternaria* rot
Sweet potato	13	Internal discoloration, pitting, decay
Tomato: Mature green	13	Water-soaked, softening, decay
Tomato: Ripe	7-10	Poor colour, abnormal ripening, *Alternaria* rot

Controlled Atmosphere (C.A.) Storage

Controlled atmosphere storage means the cold storage with the changed composition different gases than what is normally found in the air of the atmosphere. Natural air composition is 79% of nitrogen, 21% of oxygen and traces of carbon dioxide. In C.A storage condition under perfectly sealed room, composition of air surrounding the commodity is changed by reducing oxygen and elevation of carbon dioxide level. The use of controlled atmosphere can be considered only as a supplement to the proper temperature and humidity conditions for storage. This storage condition markedly retards ripening, respiration and senescence and thereby prolongs shelf-life of vegetables. This storage proves beneficial for vegetables that deteriorate rapidly or those that complete ripening after harvest and when low temperature is unfavourable for a particular vegetable.

Hypobaric storage

It refers to storage of vegetables under low atmospheric pressure where the products are kept in vacuum, tight and refrigerated chamber. Low-pressure inside the storage chamber reduces oxygen tension which decreases respiration and facilitates the removal of ethylene and volatile compounds produced by the vegetables resulting in retardation of ripening and senescence.

Irradiation

Irradiation has been proved to inhibit microbial growth, delay ripening and extend the shelf life of fruits and vegetables. Gamma irradiation can inhibit sprouting in potato and onion and kill pests in grain. The microbiological safety of food can be improved and its shelf life prolonged without substantially changing its nutritional, chemical and physical properties using irradiation. Elimination of pests from the product can also be achieved, thus reducing food losses and the use of chemical fumigants and additives. Food irradiation up to an overall dose of 10 Gy has been considered as a safe and effective technology.

Processing

Processing of vegetables into different durable products, particularly during surplus production and seasonal super-abundance can reduce colossal post-harvest losses effectively. Vegetables are, in general, processed to manufacture different products like canned product (potato, pea, beans, okra, cauliflower, carrot, sweet potato, drumstick); dehydrated product (potato, pea, onion), juices (tomato, watermelon), jams (tomato, watermelon, muskmelon), jelly (watermelon), pickles (chilli, carrot, cucumber, cabbage, pumpkin, cauliflower, turnip, beet), sauerkraut (cabbage), preserve and candy (wax gourd, carrot, beet), frozen product (French bean, sweet pepper, potato) and different tomato products (puree, paste, sauce, ketchup).

Vegetable processing has immense potential as an agro-based industry. Processing units based on modern technology should be established in the vegetable growing belts so that the products can be manufactured in a mass scale for both domestic and export markets. Unfortunately, less than one per cent of the vegetables produced are processed in India as against 40-60% in Brazil, USA, South Africa, Israel and Australia. Canned and frozen vegetables are not popular in our domestic markets due to high cost of the products and availability of different kinds of fresh vegetables throughout the year. In our domestic markets, tomato products, like sauce, ketchup and dehydrated potato and pea are most common. However, good demands for processed vegetable products exist in the defence establishments and big hotels in our country. There is continued demand for frozen vegetables and tomato products in the foreign countries like Germany, Denmark, UK, New Zealand, etc. With the establishment of cold chain, the present problem of distribution and export of frozen vegetables can be overcome. Tomato paste is a potential item for export. Potential of exporting canned French bean, dried onion, cucumber pickle and sauerkraut (lactic acid fermented product of cabbage) to the European countries is high.

The vegetable processing industries, in India, have not been able to create a mass consumer base on account of the expensive nature of the products that are produced through the use of high energy and high cost technique. Therefore, it is necessary to introduce low cost techniques like solar drying, pickling and lactic acid fermentation using chemical preservatives so that the products come within the reach of the average consumer.

Chapter 16

Marketing and Export of Vegetables

Marketing is one of the most important factors in determining the success of any vegetable farming enterprise. Marketing includes all the operations and decisions taken by the producers. These decisions range from determining the most marketable crops for production to deciding how to best deliver quality produce to the consumers at a profit. However, contrary to popular belief, marketing does not begin after a crop is produced. Instead, marketing alternatives need to be considered even before the start of production.

However, vegetable marketing in India has not yet been organised so far. The long route from the grower to the retailer and ultimately to the consumer involves a chain of middlemen, transport contractors and wholesale merchants. Very often, there arises stagnation of the produce in the peak season or in the year of good production. The large number of intermediaries in the vegetable marketing channel grab maximum share of the consumer's price and as a result, the producers' profit becomes marginal. Vegetable cultivation is a capital intensive venture, and huge production alone cannot fetch profit for the growers. Organised marketing, storage and transport facilities are the key factors for making vegetable farming profitable which ultimatily boost vegetable cultivation.

Important Steps in Handling the Produce During Marketing

Quality of vegetables is the most important criterion for achieving success in vegetable marketing. Quality of the product can be improved considerably by improved agro-techniques, timely harvesting, grading, packaging and other handling methods. Physical handling of the product which is very important from their marketing point of view is discussed below.

Harvesting

Harvesting should all the time be done at proper stage of maturity because it not only determines the quality of the product but also prolongs its shelf-life. Generally, no hard and fast rules are followed with regard to the time of harvesting.

Maturity index

Fruits and vegetables are considered to be commercially mature when at the stage of physiological development that consumers consider to be the most desirable. The factors for determining the harvesting of vegetables according to consumer's purpose, type of commodity, etc can be judged by visual means (colour, size, shape), physical means (firmness, softness), chemical analysis (sugar content, acid content), computation (heat unit and bloom to harvest period), physiological method(respiration). These are indications by which the maturity is judged.

- **Visual indices:** It is most convenient index. Certain signals on the plant or on the fruit can be used as pointers. e.g., colour development in tomato, yellowish-brown ground spot in watermelon, etc.
- **Seed development**: It can also be used as an index of fruit maturity of peas and beans.
- **Calendar date**: Calendar date for harvest along with visual indices is a reliable guide to commercial maturity of tuber and bulb crops.
- **Heat units**: Harvest date can be predicted with the help of heat unit. For each cultivar, the heat requirement for fruit growth and development can be calculated in terms of degree days. Maturity at higher temperature is faster as the heat requirement is met earlier. This heat unit helps in planning, planting, harvesting and factory programmes for crops such as corn, peas, lettuce and tomato for processing.

The stage, at which a commodity reaches commercial maturity varies greatly. Many leafy vegetables are commercially mature at an early stage of plant development while fruits are often ready for harvest at a fully developed stage. Harvesting crops at the proper maturity allows the handlers to begin their work with the best possible quality produce. Produce harvested too early may lack flavour and may not ripen properly, while produce harvested too late may be fibrous or overripe and have a shorter shelf-life.

Maturity indices for different vegetable crops

Crop	Maturity indices
Potato and onion	Tops beginning to dry out and topple down while the leaves are still green
Garlic	When the leaves start turning yellow and show the sign of drying up
Yam and ginger	Large enough (Tough and fibrous if over-matured)
Sweet potato	When surface cut of the tuber dries clearly it indicates its maturity while in immature stage, the surface cut turns black.
Radish and carrot	Large enough and crispy (over-mature if pithy).
Green onion	Leaves at their broadest and longest size.
French bean, hyacinth bean	Immature tender pods soon after they reach edible maturity.
Garden pea	When colour of the pod changes from dark to light green and is well filled but not hard.
Cowpea, yard-long bean, snap bean, winged bean	Well-filled immature pods that snap readily.
Lima bean and pigeon pea	Well-filled pods that are beginning to lose their greenness.
Okra	Immature tender stage when ridges are not prominent and fruit surface gives a velvety touch.
Snake gourd, bottle gourd	Desirable size reached and thumb nail can still penetrate flesh readily (over-mature if thumbnail cannot penetrate flesh readily).
Bitter gourd, slicing cucumber	Desirable size reached but still tender (over mature if colour dulls and seeds are tough).
Pumpkin and winter squash	Fruits are generally harvested at fully ripe stage when the rind of the fruits has sufficiently been hardened.
Summer squash	Fruits should be harvested at tender stage when the fruits have grown hardly one-third of the size.
Muskmelon	When fruits are easily separated from vine with a slight twist leaving clean scar.
Honeydew melon	Change in fruit color from a slight greenish white to cream; aroma noticeable and sugar level reaches around 10 per cent.
Cantaloupe melon	Abscission layer developed at maturity provides good index of harvest.
Watermelon	When dull sound gives out on thumping the fruit or when the ground spot turns yellowish-brown coupled with drying of the tendrils.

Brinjal	Fruits should be harvested at tender stage when the fruit skin remains glossy.
Tomato	Green fruit surface colour turning pink.
Sweet pepper	Deep green color turning dull green or specific colour (red, yellow, violet, orange, etc.).
Sweet corn	Exudes milky sap when thumb nail penetrates kernel.
Baby corn	Small cobs are harvested before being pollinated and fertilized.
Cauliflower	Curd compact (over mature if curd segments elongates and become loose).
Broccoli	Bud cluster compact (over mature if loose).
Lettuce	Big enough before flowering.
Cabbage	Head compact (over mature if head cracks).
Celery	Big enough before it becomes pithy.

The crops like hyacinth bean, French bean, cowpea, okra etc. should be harvested soon after they reach edible maturity otherwise their quality will get deteriorated badly. In these crops, succulence and at the same time, quality of the fruits is diminished due to fibre deposition resulting toughening of tissues. In okra and cowpea, fibre deposition in the fruits takes place rapidly 8 and 10 days after anthesis. In all these crops, frequent periodical harvesting is necessary. In ridge gourd, sponge gourd, bottle gourd, bitter gourd, cucumber, brinjal, etc.over maturity though gives bigger fruit yet it lowers the quality drastically. If bottle gourd is not harvested at tender stage, the flesh becomes coarse and dry and the seeds become hard. Flesh of ridge and sponge gourd will be spongy if not harvested at green tender stage. Watermelon will be insipid if harvested at immature stage and if allowed to over ripen, the fruit loses its sweetness and soon develops an off flavour. In watermelon, the principal quality criteria are flesh crispness, good flesh colour and sweetness which are adequately manifested at full ripe stage. If cantaloupe is harvested before full development of abscission layer, flavour will not be developed. Maturity of some varieties of muskmelon may also be judged by skin toughness, flesh texture and sugar: acid ratio. If muskmelon is left too long in field, storage life is lost. Potato, yam, elephant-foot yam and taro should be harvested at full maturity stage when the above-ground portion starts yellowing normally. Radish should be harvested at proper stage so that they may not become fibrous, spongy and hollow inside. If turnip is harvested late, the root becomes hard and fibrous. Harvesting of tomato by keeping very small (8-10 mm) pedicel along with calyx enhances the shelf-life of the fruits.

Climacteric fruits like, tomato, muskmelon, etc. when are to be sent to the distant markets, should be harvested long before they fully ripe. Otherwise, the fruits will become overripe and ultimately spoiled in transit. For distant markets, tomato should be harvested at turning or pink stage and muskmelon, before softening of the fruit.

In all the vegetables, harvesting should be done with care to avoid injury and bruising as far as possible which considerably reduces the secondary infection by decay-causing micro-organisms. Harvesting during hot period of the day should be avoided as immense field heat may cause wilting and shrivelling of the produce. Promptness in removing the vegetables from the field is important especially in very hot or wet weather.

Harvesting

The goals of harvesting are to gather a commodity from the field at the proper stage of maturity with a minimum of damage and loss, as rapidly as possible and at a minimum cost. This is achieved through hand-harvesting in most vegetable crops and mechanical harvesting in some crops.

Hand harvesting : Hand harvesting has a number of advantages over machine harvest. People can accurately determine product quality, allowing accurate selection of mature product. This is particularly important for crops that have a wide range of maturity and need to be harvested several times during the season. Properly trained workers can pick and handle the product with a minimum of damage. The main problem with hand harvesting is labour management. In spite of these problems, quality is so important to marketing fresh- market commodities successfully that hand harvesting remains the dominant method of harvest of most vegetables.

Mechanical harvesting : Vegetables that are grown below ground (radish, potato, garlic, carrot, beet and others) are always harvested only once and the soil can be used to cushion the product from mechanical injury. Mechanical harvesting is currently practiced for these crops. A number of products destined for processing such as tomato, bean, pea and some leafy green vegetables are machine harvested because harvest damage does not significantly affect the quality of processed product. These crops have also been amenable to new production techniques and breeding that allow the crop to be better suited to mechanical harvest. The cultivar of the particular crop must be grown to accept mechanical harvest. The main demerits of mechanical harvesting are that machines are rarely capable of selective harvest and the harvesting machines are quite expensive. At the same time, mechanical harvesting will not be feasible until the crop or production techniques can be modified to allow one time harvest.

Post-harvest handling and preparation of the product for the market

Being living organs, vegetables continue to respire even after harvesting when they have a limited source of food reserves. In addition to degradation of respiratory substrates, a number of changes in taste, colour, flavour, texture and appearance take place in the harvested commodities which make them unacceptable for consumption by the consumers if these are not handled properly. Post harvest technology starts immediately after the harvest of fruits and vegetables and the whole process is categorized as "Handling of fresh produce". Post harvest Technology of fresh fruits and vegetables combines the biological and environmental factors in the process of value addition of a commodity. Many vegetables require some special preparation before they are offered for sale in attractive manner.

Precooling

Pre-cooling (prompt cooling after harvest) refers to rapid removal of field heat from the freshly harvested vegetables before packaging or transportation which delays ripening and shrinkage of the produce. Precooling is important for most of the fruits and vegetables because they may deteriorate as much in 1 hour at 32°C. In addition to removal of field heat from commodities, precooling also reduces bruise damage from vibration during transit. Cooling requirement for a crop vary with the air temperature during harvesting, stage of maturity and nature of crop. In our country, it is done mostly by immersing the harvested vegetables in water to remove the field heat from the produce which retard ripening and senescence process (hydrocooling). However, there are many methods of precooling viz, cold air (room cooling, forced air cooling), cold water (hydrocooling), direct contact with ice (contact icing), evaporation of water from the produce (evaporative cooling, vacuum cooling) and combination of vacuum and hydrocooling (hydrovac cooling). Some chemicals (nutrients/growth regulators/ fungicides) can also be mixed with the water used in hydrocooling to prolong the shelf life by improving nutrient status of crop and preventing the spread of post harvest diseases.

Washing, Cleaning, Trimming, Tying in bunches

Before fresh fruits and vegetables are marketed various levels of cleaning are necessary which typically involves the removal of soil, adhering debris, insects and spray residues. Chlorine in fresh water is often used as disinfectant to wash the commodity. Radish, carrot, beet, turnip, palak, celery, lettuce, asparagus, spinach, pointed gourd, etc. should be thoroughly washed to remove the soil adhered to them which not only gives a lustrous appearance but also prevents them from wilting. Vegetables, like tomato, brinjal, cucumber, sweet pepper, muskmelon, watermelon, pumpkin, etc. are often wiped with moist or dry cloth to clean the fruit to bring brightness in them.

Many vegetables need trimming, cutting and removal of unsightly leaves or other vegetative parts before marketing. Removal of decayed and diseased leaves of palak, spinach, lettuce, amaranth and other leafy vegetables, stripping off of older leaves in radish and non-wrapper leaves in cabbage not only improves the appearance of the product but also checks the development of secondary infection in transit. In general, carrot, beet, turnip and knolkhol are marketed after removing the tops or leaves. In cauliflower, the leaves should be trimmed above the curd and inedible portion like stem be discarded before sending to the market.

Some vegetables like, green onion, radish, carrot, palak, coriander, asparagus, etc. are often tied in bunches to make their retail sale convenient. It also gives a neat appearance to the produce.

Sorting, Grading and Sizing

Grading, a marketing function which facilitates the movement of the produce means sorting of the unlike lots of the produce into different lots according to quality specifications like, size, shape, colour, stage of maturity, freedom from disease and pests, etc. Each lot has substantially the same characteristics so far as quality is concerned. Standardization means making the quality specifications of the grades uniform among buyers and sellers over space and time. The characteristiecs *i.e.* quality specifications are the bases of which products are standardized and are called "Grade Standards". Grading makes orderly marketing and ensures higher price for the well-graded products. Buyers often pay less when some inferior quality specimens are found in the lot of a good quality product, meaning that price of the product is often governed by these inferior specimens. Well-graded products of inferior quality are often sold better than poorly or ungraded products of superior quality. Moreover, selling of well-graded products is considered to be an honest dealing and it also simplifies the whole marketing process.

Sorting is done by hand to remove the produce which are unsuitable to market or storage due to damage by insects, diseases or mechanical injuries. The remainder product is separated into two or more grades on the basis of the cosmetic quality of the produce. After sorting and grading, sizing is done either by hand or machine. Machine sizers work on two basic principles: weight and diameter. Sizing is done mostly on the basis of fruit shape and size (tomato, gherkin, etc.).

Developed countries have comprehensive sets of regulations for fruits, vegetables and root crops marketed in the countries themselves or for export. Many developing countries have framed the same regulations to export their produce to developed ones. They also often specify the type and size of the containers (packages) that can be used, labeling requirements, recommended storage and transport conditions and permitted post-harvest treatments. In most of the countries the regulations are mandatory and in a few of them they are merely guidelines.

The regulations usually categorize produce into three or four classes, with the lowest class being considered only just acceptable for marketing. The appearance of the produce in each class is decided on the basis of shape and colour, the type and extent of blemishes that can be present and physical characteristic specific to that commodity.

These pre- packaging operations would result in large savings in packages and also in transportation. In modern packaging, these operations are done in the "Packaging station" near the major vegetable producing areas.

Curing

Curing is an effective operation to reduce the water loss during storage from hardy vegetables viz, potato, onion, garlic, sweet potato and other tropical tuber vegetables. The curing methods employed for tuber crops are entirely different than that from the bulbous crops (onion and garlic). The curing of tuber crops develops periderms over cut, broken or skinned surfaces for wound restoration. It helps in the healing of harvest injuries, reduces loss of water and prevents the infection by decay causing pathogens. Onion and garlic are cured to dry the necks and outer scales. For the curing of onion and garlic, the bulbs are left in the field after harvesting under shade for a few days until the green tops, outer skins and roots are fully dried.

Waxing

Quality retention is a major consideration in modem fresh fruit and vegetable marketing system. Waxes are esters of higher fatty acid with monohydric alcohols and hydrocarbons and some free fatty acids however, coating applied to the surface of fruit is commonly called waxes whether or not any component is actually a wax. Waxing generally reduces the respiration and transpiration rates, but other chemicals such as fungicides, growth regulators, preservative can also be incorporated specially for reducing microbial spoilage, sprout inhibition, etc. However, it should be remembered that waxing does not improve the quality of any inferior product but it can be a beneficial adjunct to good handling. The advantages of wax application are i) Improved appearances of fruit, ii) Reduced moisture losses and retards wilting and shrivelling during storage of fruits, iii) Less spoilage specially due to chilling injury and browning, iv) Creates diffusion barrier as a result of which it reduces the availability of O_2 to the tissues thereby reducing respiration rate, v) Protects fruits from micro-biological infection, vi) Considered a cost effective substitute in the reduction of spoilage when refrigerated storage is unaffordable and vii) Wax coating are used as carriers for sprout inhibitors, growth regulators and preservatives. The principal disadvantage of wax coating is the development of off-flavour if not applied properly. Adverse flavour changes have been attributed to inhibition of O_2 and CO_2 exchange thus, resulting in anaerobic respiration and elevated ethanol and acetaldehyde contents. Paraffin wax, Carnauba wax, Bee wax, Shellac, Wood resins and Polyethylene waxes are used commercially.

Some vegetables like, tomato, brinjal, sweet pepper, cucumber, muskmelon, carrot, etc. are generally waxed with a water emulsion by dipping or spraying to retard the moisture loss from the product and at the same time, to improve their lustre.

Packaging

Proper or scientific packaging of fresh fruits and vegetables reduces the wastage of commodities by protecting them from mechanical damage, pilferage, dirt, moisture loss and other undesirable physiological changes and pathological deterioration during the course of storage, transportation and subsequent marketing. For providing uniform quality to packed produce, the commodity should be carefully supervised and sorted prior to packaging. Packaging cannot improve the quality but it certainly helps in maintaining it as it protects the produce against the hazards of its journey.

Packaging is an important consideration in the process of marketing. Proper packaging makes transportation and retail sale of the vegetables more convenient. It also increases the shelf-life of the products because of the protection provided by the packages from external injuries and from being shrivelled. Adequate packaging coupled with effective marketing can reduce the post-harvest losses of the vegetables. Packages also provide a measure of the contents which is also very important from the marketing point of view.

In our country, the concept of packaging is primitive where bamboo baskets, arhar stick baskets, gunny bags, and in some cases, wooden boxes (for sweet pepper, tomato) are used. Indiscriminate loading directly on to trucks without proper packaging is a common feature which accounts for huge spoilage. Most of the time, vegetables like pumpkin, watermelon, muskmelon, etc.

are transported in unpacked condition. It is again mentioned that most of the post-harvest losses in vegetables occur during transportation and distribution, and this deterioration in transit is aggravated due to poor as well as loose packaging.

Nowadays, alternative packaging materials which are recyclable are receiving attention. Corrugated fibreboard (CFB) boxes, paperboard boxes lined with polyethylene, polyethylene film, plastic crates and moulded paper trays are suitable alternatives. Recently, processing industries and some co-operative societies started using plastic crates for transportation of vegetables from the field as well as for storage. Wider adoption of modern packaging is limited due to (a) high cost of the packaging materials, (b) involvement of additional labour charges for packaging and pre-packaging operations, (c) shortcoming in the system of collection of the used packaging materials and (d) lack of information about the specification of different packaging materials for different vegetables.

Storage

A number of storage techniques (ground storage, ambient storage, refrigerated storage, air cooled storage, zero energy storage, modified atmospheric storage, hypobaric storage and controlled atmosphere storage) are being used for vegetables depending upon the nature of the commodity and the storage period intended.

Transportation

Timely and speedy delivery of vegetables with minimum spoilage and at reasonable cost is an essential feature of effective vegetable marketing. Effective vegetable production can be increased without putting more areas under the crop if potential areas can be brought under good transportation network. Moreover, vegetables which are grown in one area with the advantage of climatic conditions can be catered to the distant markets where those vegetable crops are not grown at that time due to climatic unsuitability. Delay in transport coupled with improper packaging is accounted to be the causes of huge post-harvest spoilage of vegetables in transit. Road transportation is costly compared to rail transport. However, rail transport which is 8-10 times more efficient than road transport has not proved to be dependable for the vegetables in our country. The concept of refrigerated transport for the vegetables is yet to be developed. Effective refrigerated transport of perishables in India has amply been executed in fish and milk collection and distribution system. So, it is not impossible for vegetables. Necessary steps to strengthen the cold chain and at the same time to provide quick, efficient and cost effective transport network at least in the potential vegetable growing areas should be taken immediately.

Marketing system

Marketing of vegetables faces a number of constraints due to their highly perishable nature, seasonal market arrivals and bulky nature. Assembling and subsequent marketing of the produce is further lopsided due to lack of proper storage facilities and quick transport systems. Very often the producers are forced to dispose of their produce at a very nominal price due to seasonal gluts. Another major defect in vegetable marketing is the involvement of several intermediaries who dominate the trade and reap huge profit. Consequently, producers' margin in the consumers' price become very low. In vegetable marketing, the following four channels are predominant.

1. Producer → Commission agent → Wholesale trader → Retailer → Consumer.
2. Producer → Wholesale trader → Retailer→ Consumer.
3. Producer → Commission agent → Wholesale trader → Consumer.
4. Producer → Retailer → Consumer.

Majority of the produce are marketed through channel 1 and 2. Market channel 4 only operates where the producing area is situated near markets or cities (the concept of market garden of vegetables). Vegetables are generally marketed in the big hotels or other establishments through channel 3. Organized effort of marketing through co-operatives is inadequate in our country. It is urgently felt that more horticultural producers' co-operative marketing societies should be established at village and district levels to control the activity of the intermediaries and to regulate the vegetable marketing. At present, National Agricultural Co-operative Marketing Federation (NAFED) and several State co-operative marketing societies have taken up procurement and marketing of onion, potato, ginger and other vegetables. National Horticulture Board has started catering information regarding the prevailing prices of vegetables at various wholesale markets on daily basis.

The existing systems need to be streamlined and monitored and the facility of the co-operative societies should be extended to the grass-root level. Moreover, closer co-ordination among the Agriculture Marketing Board, National Horticulture Board and State Department of Agriculture/ Horticulture should be ensured to formulate an action plan for marketing of vegetables.

Establishment of marketing cooperatives have been encouraged to provide marketing facilities to small farmers. The anticipated advantages are to increase the bargaining strength of farmers, removal of intermediaries and direct interaction with consumers. Facilities should be extended for availing of credit and cheaper transport, storage facilities, grading and processing of agricultural produce to fetch better prices.

Export trade of vegetables

India's diverse climate ensures availability of all varieties of fresh vegetables. It ranks second in vegetables production in the world, after China. As per National Horticulture Database published by National Horticulture Board, during 2014-15 India produced 86.602 million tonnes of fruits and 169.478 million tonnes of vegetables. The area under cultivation of fruits stood at 6.110 million hectares while vegetables were cultivated at 9.542 million hectares. The vast production base offers tremendous opportunities for export. During 2016-17, India exported fruits and vegetables worth Rs. 10,369.96 crores/1,552.26 million US dollar which comprised of fruits worth Rs. 4,448.08 crores/667.51 million US dollar and vegetables worth Rs. 5,921.88 crores/884.75 million US dollar.

Major export items

The prime vegetable which is exported from India is onion. Apart from fresh onion, other important traditional vegetables which are exported in fresh condition are okra, bitter gourd, chilli, potato and garlic. Different fresh traditional vegetables, like bottle gourd, tinda, taro, radish, carrot, peas, cauliflower, cabbage, ginger, tomato, pumpkin, brinjal, pointed gourd, drumstick, elephant-foot yam, sponge gourd,; non-traditional vegetables like, asparagus, celery, bell pepper, broccoli, sweet corn and baby corn, French bean and lima bean; organically grown vegetables and hybrid seeds are also being exported from India in good quantities.

Export standards for different vegetables

Commodity	Parameters for export
Big onion	4-6 cm in diameter, light to dark red in colour, round shaped, strong pungency for Gulf and South East Asian markets; 3-4 cm in diameter, light red and round shape in Bangladesh; Yellow/brown colour, 7-8 cm in diameter, globose or round shaped for European and Japanese markets
Small onion	Dark red, 2-3 cm in diameter and round shape
Potato	White, oval, 4.5-6.0 cm in size, Bangladesh demands red types and Iran and Iraq demands yellow fleshed types.
Okra	3-5 inch length, green, tender; packing 5 Kg.
Tondali/coccinia	1-2 inch , green, tender; packing 5 Kg.
Bottle gourd	12 inch length, greenish, tender, straight; packing 5 Kg.
Bitter gourd	Green, 20-25 cm long, spindle shaped.
Gherkin	Green small sixed having 160-300 fruits/kg in premium grade.

French bean	Straight 10-12 cm long, round, green pods in bush beans, flat beans 12-13 cm long.
Peas	4-5 inch in length, green, tender, straight; packing 5 Kg.
Cluster bean	4-5 inch in length, not over matured; packing 5 Kg.
Suran/yams	Cleaned, weighing around 5-10 Kg.
Green chilli	3-4 inch in length, green; packing 5 Kg.
Drum sticks	24 inch in length, straight, thick; packing 5 Kg.
Mode of transport	By air or by sea.

The processed vegetables which are exported include frozen mixed vegetables, dehydrated onions, flakes and powder, dehydrated garlic, dehydrated garlic powder, dried potato, preserved cucumber and gherkin, canned vegetables and other processed vegetables.

Export oriented production

Big onion for export is mainly grown in Nasik, Pune and Satara districts of Maharashtra, Periyar and Coimbatore districts in Tamil Nadu, Badaun in Uttar Pradesh, Patna in Bihar and Rajkot in Gujarat. Small onion is grown in Kolar and Bengaluru in Karnataka, Cuddapah in Andhra Pradesh, Madurai, Coimbatore and Salem districts of Tamil Nadu for export. Similarly okra for export is grown in Nasik, Ozar, Dindhori, Kolhar and Sholapur in Maharashtra. Chillies for export are produced in Pen, Alibaugh and Chaul in Raigad district, Dindhori and Niphad and Igatpur taluk in Nasik district of Maharashtra. For export purpose, watermelon is grown in Panvel; bottle gourd and bitter gourd in Nasik and Pune districts, French bean and cluster bean in Dindhori in Nasik district of Maharashtra. Gherkin, baby corn and tomato are grown for export in Maharashtra, Andhra Pradesh and Karnataka states. Potato for export is grown in Jalandhar and Ludhiana in Punjab, Kurukshetra and Karnal in Haryana and Indore in Madhya Pradesh. Recently gherkin has become an important vegetable for export which is mainly processed in areas like Tumkur, Hasan, Kolar and Dharwad in Karnataka.

Major export destinations

The major destinations for Indian fruits and vegetables are United Arab Emirates (UAE), Bangladesh, Malaysia, Netherland, Sri Lanka, Nepal, UK, Saudi Arabia, Pakistan and Qatar. Fresh onion is mainly exported to Bangladesh, United Arab Emirates, Singapore, Malaysia, Saudi Arabia, Sri Lanka, and Mauritius. Non-traditional vegetables like asparagus, celery, sweet pepper, sweet corn, lima bean and organically grown vegetables have good export potential in both fresh and processed forms to Europe and South Asia. Edible podded garden pea has good demand in European countries.

There is continued demand for frozen vegetables and tomato products in the foreign countries. Tomato paste and ketchup are the potential items of export particularly to Middle East, UK, Africa, Australia and Italy. Canned vegetables are mainly exported to Germany, Denmark, UK and New Zealand and dehydrated vegetables to Germany, UK and New Zealand. There is good demand for canned peas, beans and pickling cucumber (gherkin) in Europe. Among dehydrated vegetables, peas, onion and garlic are important and among frozen vegetables, prospects for peas, cauliflower, French bean, carrot and okra are bright. Chilli oleoresin is another important export item to USA, Europe and Japan.

Though India's share in the global market is still nearly 1% only, there is increasing acceptance of horticulture produce from the country. This has occurred due to concurrent developments in the areas of cold chain infrastructure and quality assurance measures. Apart from large investment pumped in by the private sector, public sector has also taken initiatives and with APEDA's assistance several centres for perishable cargoes and integrated post-harvest handling facilities have been set up in the country. Capacity building initiatives at the farmers, processors and exporters' levels has also contributed towards this effort.

Main constraints of export

- Non-availability of air cargo space as well as high freight rates are the two major constraints in increasing exports.
- High air freight from India to Gulf countries and UK compared to from Kenya, Jordan, Lebanon, etc.
- Inadequate and inappropriate storage and distribution infrastructure.
- Since vegetables are the main food items in India, domestic prices and demand are high which hinder regular export.
- Domestic production base has not been geared to meet export demand revealing lack of consistency in supply and quality.
- Lack of exportable varieties in many crops.
- Lack of post-harvest treatment and quality control facilities.
- Lack of pack house from the main producing areas to the port.
- Lack of technical support for the agro-industrial sector.

References

Bautista, O.K. and Mabesa, R.C. (Eds). 1977. *Vegetable production*. University of the Philippines, Los Banos. Philippines.

Bernard, O.L. (Ed.). 1979. *Hawk's Physiological Chemistry*, Tata McGraw Hill.

Bhat, K.L. 2007.*Minor vegetables, Untapped Potential*, .Kalyani Publishers, Ludhiana.

Bose, T. K. and Som, M.G. 1986. *Vegetable Crops in India* , Naya Prokash, Calcutta.

Brady, N.C. 1980. *The Nature and Properties of Soil*. MacMillan Publishing House, INC., USA.

Brown, H.D. and Hutchison, C.S. 1949. *Vegetable Science*, J.B. Lippincott Co., New York.

Butani, D.K. and Jotwani, M.G. 1984. *Insects in Vegetables*, Periodical Expert Book Agency, Delhi.

Byrne, R. and McAndrews, J. H. 1975. Pre-Columbian puslane (*Portulaca oleracea* L.) in the New World. *Nature*, **253**(5494): 726–727.

Celine, V. A. 2012. Drumstick. In: *Hand book of Vegetables* (Eds. Peter, K.V. and Hazra, P.), Studium Press LLC, Houston, Texas, USA, pp. 231-256.

Chatterjee, B.N., Maiti, S. and Mandal, B.K.1989. *Cropping Systems- Theory and Practice*, Oxford & IBH.

Chattopadhyay. A. and Hazra, P. 2008. Cluster bean. In: *Scientific Cultivation of Vegetables* (ed. Rana, M. K), Kalyani Publishers, Ludhiana, pp. 48-61.

Dixon, G.R. 2007. *Vegetable Brassicas and related Crucifers*. CABI, *Wallingford*, UK, 327p.

Donahue, R.L., Miller, R.W. and Shickluna, J.C. 1990. *Soils: An Introduction to Soils and Plant Growth*, Prentice-Hall of India.

Edmond, J.B., Musser, A.M. and and Andrews, F.S. 1964. *Fundamentals of Horticulture*, The Blakiston Co., INC., Toronto.

Epstein, E. and Bloom, A.J. 2004. *Mineral Nutrition of Plants: Principles and Perspectives*. Sinauer Associates, Sunderland, Massachusetts, USA.

Erickson, D.L., Smith, B.D., Clarke, A.C., Sandweiss, D.H. and Turos,S N. 2005. An Asian origin for a 10,000-year-old domesticated plant in the Americas. *Proceedings of the National Academy of Sciences,* **102**:18315–18320.

Fallik, E. 2011. Hot Water Treatments of Fruits and Vegetables for Postharvest Storage. In: *Horticultural Reviews* (Ed. Janick, J.), Wiley-Blackwell, **38**: 191 – 212.

FAO, Food and Agriculture Organization 1998. *Evaluating the potential contribution of organic agriculture to sustainability goals. Environment and Natural Resources Service.* Sustainable Development Department. FAO's technical contribution to IFOAM's Scientific Conference, Argentina, November 1998. (http://www.fao.org/DOCREP/003/AC116E/AC116E00.HTM .

FiBL (Forschungsinstitut fuer Biologischen Landbau) and IFOAM (International Federation of Organic Agriculture Movements). 2012. *Survey, Organic Agriculture Worldwide: Current Statistics*. Helga Willer, Research Institute of Organic Agriculture (FiBL), Frick, Switzerland. BioFach Congress (http://www.fibl.org).

Fritsch, R.M. and Friesen, N. 2002. Evolution, Domestication, and Taxonomy". In: *Allium Crop Science: Recent Advances* (Eds. Rabinowitch, H.D and Currah. L.). CABI, Wallingford, UK. pp. 9–10.

Gangopadhyay, S. 1984. *Advances in Vegetable Disease* , Associated Pub., New Delhi.

Ghosh, S.P., Ramanujam, T., Jos, J.S., Moorthy, S.N. and Nair, R.G. 1988. *Tuber Crops*, Oxford & IBH.

Grubben, G. J. H., and Denton, O. A. 2004. *Plant Resources of Tropical Africa 2. Vegetables*. PROTA Foundation, Wageningen; Backhuys, Leiden; CTA, Wageningen.

Gupta, U.S. (Ed.). 1978. *Crop Physiology* Oxford & IBH, New Delhi.

Hansen, E.V., Israelsen, O.W. and Stringham, G.E. 1979. *Irrigation Principles and Practices*, John Wiley & Sons, New York.

Hazra, P. and Dutta, A. K. 2011. Vegetable Breeding for Quality Traits. In: *Breeding and Protection of Vegetables* (ed. Rana, M.K.). New India Publishing Agency, New Delhi.

Hazra, P. and Som, M. G. 2015. *Vegetable Science* (Second revised edition), Kalyani Publishers, Ludhiana, 598 p.

Hazra, P. and Som, M. G. 2016. *Vegetable Seed Production and Hybrid Technology* (Second revised edition), Kalyani Publishers, Ludhiana, 459 p.

Hazra, P., Banerjee, M.K. and Chattopadhyay, A. 2012. *Varieties of Vegetable Crops in India (Second edition)*, Kalyani Publishers, Ludhiana, 199 p.

Hazra, P., Chattopadhyay, A., Karmakar, K. and Dutta, S. 2011. *Modern Technology for Vegetable Production*, New India Publishing Agency, New Delhi, 413p.

Hazra, P., Ghosh, S.K., Malty, T.K., Pandit, M.K. and Som, M.G. 2010. *Glossary of Horticulture* (3rd Edn.), Kalyani Publishers, New Delhi, 304p.

Heuze, V., Tran, G., Boval, M. and Lebas, F. 2016. *Agati (Sesbania grandiflora). Feedipedia, a programme by INRA, CIRAD, AFZ and FAO*. https://feedipedia.org/node/254.

Hill, J. B., Henry, W. P. and Alvin, R. G. Jr. 1992. *Botany*. Tata McGraw-Hill, New Delhi.

IFOAM. 1998. *Basic Standards for Organic Production and Processing*., Tholey, Theley, Germany.

Indira, P. and Peter, K.V. 1988. *Unexploited Tropical Vegetables*, Publication Unit, Directorate of Extension, Kerala Agricultural University.

Jacob, A. and Unexhull, H.V. 1960. *Fertilizer Use,* Varlags Gesellschaft fur Ackerban Mbh, Hannover.

Jobling, J. 2000. Postharvest ethylene: A critical factor in quality management [Online]. Sydney, Australia. Sydney Postharvest Laboratory PO Box 52 North Ryde, NSW, Australia 2113 (http://postharvest.com.au/EthylenePDF.PDF).

Kader, A.A. 1983. Post-harvest quality maintenance of fruits and vegetables in developing countries. In: Post-Harvest physiology and crop preservation (Ed. Lieberman, M.), Plenum Publishing Corporation. pp.455-469.

Kemmler, G. and H. Hobt, H. 1985. *Potassium- A Product of nature*, Kali-und Salz Ag, FR Germany.

Konokov, P.F. and Kiran, V.1988. *Vegetable Growing in Homegardens of Tropical and Subtropical Areas*, Mir Publisher, Moscow.

Lampkin, N. 2002. *Organic Farming*, Old Pond Publishing, Ipswich, England.

Leopold, A.C. and Kriedemann, P. E. 1981. *Plant Growth and Development*, Tata McGraw-Hill, New Delhi.

Loy, J.B. 2012. Breeding squash and pumpkins, In: Genetics, genomics and breeding of cucurbits (Eds. *Wang*, Y.*H*., Behera, T.K. and Kole, C.), CRC Press, Boca Raton, FL,USA. pp 93–139.

Lutz, J.M. and Hardenburg, R.E. 1966. *The commercial storage of fruits, vegetables and florist and nursery stork* , Agricultural Handbook No. 66, USDA, Washington.

Mandák, B., Trávnícek, P., Paštová, L., Korínková, D., 2012. Is hybridization involved in the evolution of the *Chenopodium album* aggregate? An analysis based on chromosome counts and genome size estimation., Flora (Jena), 207(7):530-540.

Maynard, D.N. and Hochmuth, G.J. 1980. *Knott's Handbook for Vegetable Growers*, John Wiley & Sons, New York.

Nanda, K.K. and Kochhar, V.K. 1985. *Vegetative Propagation of Plants* , Kalyani Publishers, New Delhi.

Nath, Prem. 1976. *Vegetables for the Tropical Region* , ICAR, New Delhi.

Palada, M. C. and Crossman, S.M. A. 1999. Evaluation of Tropical leaf vegetables in the Virgin Islands. In: *Perspectives on new crops and new uses* (ed. Janick, J.), ASHS Press, Alexandria, VA. USA, pp. 388-393.

Pantastico, E.B, Pantastico, E.B.(eds.). 1975. *Post-harvest physiology, handling and utilization of tropical and subtropical fruits and vegetables* AVI Publishing Co., Westport, Connecticut, USA, 560p.

Partap, T., Joshi, B.D. and Galwey, N.W. 1998. *Chenopods. Chenopodium spp. Promoting the conservation and use of underutilized and neglected crops*, International Plant Genetics Research Institute, 67 p.

Peter, K.V. and Hazra, P. (Eds.). 2012. *Hand book of Vegetables Volume I.* Studium Press LLC P.O. Box 722200, Houston, Texas 77072, USA.

Peter, K.V. and Hazra, P. (Eds.). 2015. *Hand book of Vegetables Volume II.* Studium Press LLC P.O. Box 722200, Houston, Texas 77072, USA.

Peter, K.V. and Hazra, P. (Eds.). 2015. *Hand book of Vegetables Volume III.* Studium Press LLC P.O. Box 722200, Houston, Texas 77072, USA.

Peter, K.V. (ed) (2007, 2008, 2009, 2010). *Under-utilized and Under-exploited Horticultural Crops.* Vol. I to V, New India Publishing Agency, New Delhi.

Sadhu, M.K. 1989. *Plant Propatation*, Wiley Eastern Ltd., Calcutta.

Schieberle, P., Ofner, S. and Grosch, W. 1990. Evaluation of Potent Odorants in Cucumbers (*Cucumis sativus*) and Muskmelons (*Cucumis melo*) by Aroma Extract Dilution Analysis. *Journal of Food Science*, **55**: 193–195.

Shanmugavelu, K.G. 1993. *Production Technology of Vegetable Crops*, Oxford & IBH, New Delhi.

Simmonds, N. W. (Ed.). 1979. *Evolution of Crop Plants*, Longman, London.

Sitte, P., Ziegler, H., Ehrendorfer, F. and Bresinsky, A. 1999. *Lehrbuch der Botanik*, Aufl. Spektrum Akademischer Verlag, Heidelberg, 1007 p.

Suslow, T.V. 2000. Post-harvest handling for organic crops. Organic Vegetable Production in California Series. Pub. 7254. University of California Davis, USA. (http://ucanr.org/freepubs/docs/7254.pdf).

Tandon, S. 1989. *Soil Fertility and Fertilizer Use*, IFFCO.

Thamburaj, S. and Singh, N. 2001. *Textbook of Vegetables, Tubercrops and Spices* , I.C.A.R. New Delhi.

Thompson, H.C. and Kelley, W.C. 1979. *Vegetable Crops*, Tata McGraw Hill.

Walters, T.W. and Decker Walters, D.S. 1989. Systematic reevaluation of *Benincasa hispida* (Cucurbitaceae). *Economic Botany*, 43: 274-278.

Widjaja, E.A. and Lester, R.N. 1987. Morphological, anatomical and chemical analysis of *Amorphophallus paeoniifolius* and related taxa. *Reinwardtia*, **10**: 271–280

Wien, H.C.(Ed.). 1997. *Physiology of Vegetable Crops*, CABI, *Wallingford*, UK.

Wilson, L. G., Boyette, M. D. and Estes, E. A. 1999. Postharvest handling and cooling of fresh fruit, vegetables, and flowers for small farms. North Carolina Cooperative Extension Service, USA. (http://www.ces.ncsu.edu/depts/hort/hil/hil-800.html).

Zohary, D. and Hopf, M. 2000. *Domestication of plants in the Old World*, Oxford University Press, Oxford (third edition), 139p.